P9-BJT-400

INTRODUCTION TO INSECT PEST MANAGEMENT

ENVIRONMENTAL SCIENCE AND TECHNOLOGY

A Wiley-Interscience Series of Texts and Monographs

PRINCIPLES AND PRACTICES OF INCINERATION
Richard C. Corey

AN INTRODUCTION TO EXPERIMENTAL AEROBIOLOGY
Robert L. Dimmick

AIR POLLUTION CONTROL, Part I
Werner Strauss

AIR POLLUTION CONTROL, Part II
Werner Strauss

APPLIED STREAM SANITATION
Clarence J. Velz

PHYSICOCHEMICAL PROCESSES FOR WATER QUALITY CONTROL
Walter J. Weber, Jr.

ENVIRONMENTAL ENGINEERING AND SANITATION
Joseph A. Salvato, Jr.

NUTRIENTS IN NATURAL WATERS
Herbert E. Allen and James R. Kramer, Editors

pH AND pION CONTROL IN PROCESS AND WASTE STREAMS
F. G. Shinskey

INTRODUCTION TO INSECT PEST MANAGEMENT
Robert L. Metcalf and William H. Luckmann, Editors

INTRODUCTION TO INSECT PEST MANAGEMENT

EDITED BY

ROBERT L. METCALF

WILLIAM H. LUCKMANN

A WILEY-INTERSCIENCE PUBLICATION

JOHN WILEY & SONS

NEW YORK · LONDON · SYDNEY · TORONTO

Library of Congress Cataloging in Publication Data:

Metcalf, Robert Lee, 1916-
Introduction to insect pest management.

(Environmental science and technology)
"A Wiley-Interscience publication."
Includes bibliographies.
1. Insect control. I. Luckmann, William Henry, 1926- joint author. II. Title. III. Title: Insect pest management.
SB931.M48 632'.7 74-34133
ISBN 0-471-59855-0

Printed in the United States of America

10 9 8 7 6 5 4 3 2 1

CONTRIBUTORS

P. L. Adkisson, Professor and Head
Department of Entomology
Texas A. & M. University
College Station, Texas

E. J. Armbrust
Associate Entomologist
Illinois Natural History Survey, and Associate Professor
Illinois Agricultural Experiment Station
Urbana, Illinois

B. A. Croft, Assistant Professor
Department of Entomology
Michigan State University
East Lansing, Michigan

G. G. Gyrisco, Professor
Department of Economic Entomology
Cornell University
Ithaca, New York

J. C. Headley, Associate Professor
Agricultural Economics
University of Missouri
Columbia, Missouri

Marcos Kogan
Associate Entomologist
Illinois Natural History Survey, and
Associate Professor
Illinois Agricultural Experiment Station
Urbana, Illinois

W. H. Luckmann, Professor and Head
Section of Economic Entomology
Illinois Natural History Survey
and Agricultural Entomology
Illinois Agricultural Experiment Station
Urbana, Illinois

J. V. Maddox
Associate Entomologist
Illinois Natural History Survey, and Associate Professor
Illinois Agricultural Experiment Station
Urbana, Illinois

R. A. Metcalf
Research Fellow
Harvard University
Cambridge, Massachusetts

R. L. Metcalf
Professor of Entomology and Agricultural Entomology
University of Illinois, and
Principal Scientist
Illinois Natural History Survey
Urbana, Illinois

P. W. Price
Assistant Professor of Entomology and Agricultural Entomology
University of Illinois
Urbana, Illinois

H. T. Reynolds, Professor
Department of Entomology
University of California
Riverside, California

W. G. Ruesink
Assistant Entomologist
Illinois Natural History Survey,
and Assistant Professor
Illinois Agricultural Experiment Station
Urbana, Illinois

R. F. Smith
Professor of Insect Ecology
University of California
Berkeley, California

R. W. Stark, Dean of Graduate School
Coordinator of Research
University of Idaho
Moscow, Idaho

F. W. Stehr, Associate Professor
Department of Entomology
Michigan State University
East Lansing, Michigan

G. P. Waldbauer, Professor of Entomology and Agricultural Entomology
University of Illinois
Urbana, Illinois

To Mrs. Sue E. Watkins

A wonderful and talented woman who for 32 years served the Section of Economic Entomology, Illinois Natural History Survey, with competence, devotion, and loyalty.

SERIES PREFACE

Environmental Sciences and Technology

The Environmental Sciences and Technology Series of Monographs, Textbooks, and Advances is devoted to the study of the quality of the environment and to the technology of its conservation. Environmental science therefore relates to the chemical, physical, and biological changes in the environment through contamination or modification, to the physical nature and biological behavior of air, water, soil, food, and waste as they are affected by man's agricultural, industrial, and social activities, and to the application of science and technology to the control and improvement of environmental quality.

The deterioration of environmental quality, which began when man first collected into villages and utilized fire, has existed as a serious problem since the industrial revolution. In the last half of the twentieth century, under the ever-increasing impacts of exponentially increasing population and of industrializing society, environmental contamination of air, water, soil, and food has become a threat to the continued existence of many plant and animal communities of the ecosystem and may ultimately threaten the very survival of the human race.

It seems clear that if we are to preserve for future generations some semblance of the biological order of the world of the past and hope to improve on the deteriorating standards of urban public health, environmental science and technology must quickly come to play a dominant role in designing our social and industrial structure for tomorrow. Scientifically rigorous criteria of environmental quality must be developed. Based in part on these criteria, realistic standards must be established and our technological progress must be tailored to meet them. It is obvious that civilization will continue to require increasing amounts of fuel, transportation, industrial chemicals, fertilizers, pesticides, and countless other products and that it will continue to produce waste products of all descriptions. What is urgently needed is a total systems approach to modern civilization through which the pooled talents of scientists and engineers, in cooperation with social scientists and the medical profession, can be focused on the development of order and equilibrium to the presently disparate segments of the human environment. Most of the skills and tools that are needed are already in existence. Surely a technology that has created such manifold environmental problems is also capable of solving them. It is our hope that this Series in Environmental Sciences and Technology will not only serve to make this challenge more explicit to the established professional but that it also will help to stimulate the student toward the career opportunities in this vital area.

Robert L. Metcalf
James N. Pitts, Jr.

PREFACE

The term "pest management" has become an important part of the vocabulary of entomologists, weed scientists, and plant pathologists during the past decade. We believe the practice of pest management is absolutely essential to the future, indeed, perhaps even the survival, of modern plant protection and pest control. The purpose of this book is to bring together some of the thinking in insect pest management to serve as a guideline to students and others. The book for the most part is purposely oriented to insect problems in agriculture, but it should provide a foundation on which programs can be developed in other specialty areas in agriculture and in public-health entomology.

Modern insect pest management is pest control based on sound biological knowledge and principles. Early twentieth-century entomologists were pioneer workers in insect pest management, but their programs were often unsuccessful. Today, with rapid communication, transportation, extremely effective and selective pesticides, computers, and other scientific aids, we can overcome pest-control obstacles that were insurmountable only a few decades ago.

We believe this book will serve a purpose. We and the other authors were motivated by a strong desire to prepare a book that would introduce the subject and provide examples of developing programs and techniques in pest management for use by students and field workers in pest control. We hope that this book will further stimulate the adoption and use of pest-management systems. Further, the references and selected readings at the end of each chapter have been especially chosen to provide the reader with an opportunity to become acquainted with scientific articles and books written by world leaders and experts in pest control.

We want to express our thanks to the many scientists and editors of scientific journals who have freely granted permission for the use of published material, to Mr. Lloyd LeMere and Mr. Larry Farlow of the Illinois Natural History Survey for illustrations and photographs, and to

Mrs. Alice Prickett, University of Illinois, for illustrations. We are very grateful to the coauthors who willingly agreed to contribute to the book.

To Mrs. Sue E. Watkins, Administrative Assistant, Illinois Natural History Survey, we are especially thankful for advice and guidance and the many hours she devoted to the preparation and editing of this book. We dedicate this book to her.

Robert L. Metcalf
William H. Luckmann

Urbana, Illinois
November 1974

CONTENTS

PRINCIPLES

1. The Pest-Management Concept 3
W. H. Luckmann and R. L. Metcalf

2. Ecological Aspects of Pest Management 37
P. W. Price and G. P. Waldbauer

3. The Economics of Pest Management 75
J. C. Headley

TACTICS

4. Plant Resistance in Pest Management 103
Marcos Kogan

5. Parasitoids and Predators in Pest Management 147
F. W. Stehr

6. Use of Diseases in Pest Management 189
J. V. Maddox

7. Insecticides in Pest Management 235
R. L. Metcalf

8. Attractants, Repellents, and Genetic Control in Pest Management 275
R. L. Metcalf and R. A. Metcalf

STRATEGY

9. The Quantitative Basis of Pest Management: Sampling and Measuring 309
W. G. Ruesink and Marcos Kogan

10. Analysis and Modeling in Pest Management 353
W. G. Ruesink

EXAMPLES

11. Cotton Insect Pest Management 379
H. T. Reynolds, P. L. Adkisson, and R. F. Smith

12. Forage Crops Insect Pest Management 445
E. J. Ambrust and G. G. Gyrisco

13. Tree Fruit Pest Management 471
B. A. Croft

14. Forest Insect Pest Management 509
R. W. Stark

15. Pest-Management Strategies for the Control of Insects Affecting Man and Domestic Animals 529
R. L. Metcalf

EPILOGUE

16. Pest Management and the Future 567
W. H. Luckmann

INDEX 571

INTRODUCTION TO INSECT PEST MANAGEMENT

PRINCIPLES

1

THE PEST-MANAGEMENT CONCEPT

William H. Luckmann and Robert L. Metcalf

I. WHAT IS PEST MANAGEMENT?

A nation, a state, or even a county cannot leap from a quarter of a century of overcommitment to preventive pest control practices into as complex a professional endeavor as pest management. Clearly, there must be a period of transition, perhaps from 5 to as much as 25 years or more, during which professional, educational, and extension philosophies are applied, new tools are readied, economic and social benefits/costs are tallied, and management strategies are refined. However, given the present awareness of the crises in some major pest-control programs and the concern about environmental quality a start must be made, and the present is none too soon. This, then, is the major purpose of this book.

Insect pest management is an idea whose time has come (Smith 1972). The concepts are not new to applied entomology, but growing awareness of overreliance on insecticides has rekindled enthusiasm for sound fundamental principles of control, thus greatly enhancing the chances for public acceptance and success of pest-management programs. This approach to pest control, which seeks compatibility of control interventions, has acquired various names. *Integrated control,* originally coined to define the blending of biological control agents with chemical control interventions (Bartlett 1956), has now assumed wider meaning. Geier and Clark (1961) have called this conception of pest control *protective management of noxious species* or *pest management* for short, in which all available techniques are evaluated and consolidated into a unified program to manage pest populations so that economic damage is avoided and adverse side effects on the environment are minimized (NAS 1969).

The term *pest* is a label applied by man and has no ecological validity. Man places an insect in the pest category, and some species can even be considered pests at certain times and beneficial insects at other times. Therefore an insect is a pest because we call it one, and a pest problem exists because an insect is competing with man. Insect pests are generally regarded as destructive or noxious in proportion to the number present and in competition with man.

Generally, because of the complexities of human society, it is impossible to eliminate pest problems by ceasing the activities that encourage them, but clearly we have often been too hasty and inclusive in our definitions and too impetuous in our efforts to exterminate and eradicate. Pest-management concepts dictate a tolerant approach to pest status. Indeed, it may be that not all pests are bad, and that not all pest damage is intolerable.

Further, we can readily use an old practice such as crop rotation and call it pest management, so long as the manager accepts and understands the philosophy of pest management. This more than anything else will determine the fate of insect pest-management programs. Insects can be managed, but management is people-oriented, and successful pest management depends largely on influencing the people who control the pest. This fact is explicit, although frequently not identified in examples and principles outlined by those writing about pest management during the past decade. The pest-management philosophy is relevant in all pest-control actions.

Pest management is the intelligent selection and use of pest-control actions that will ensure favorable economic, ecological, and sociological consequences (see Rabb 1972). Pest-control actions include the monitoring of pest increase, the judicious use of a pesticide, or the effective communication that no action is necessary. In agriculture pest management should ensure a strong agriculture and a viable environment. In public health it should ensure the protection of man and his domestic animals, and the maintenance of a suitable environment in which they may live. The practice of pest management has been described by Geier (1966) as: (1) determining how the life system of a pest needs to be modified to reduce its numbers to tolerable levels, that is, *below the economic thresh-*

old; (2) applying biological knowledge and current technology to achieve the desired modification, that is, *applied ecology;* and (3) devising procedures for pest control suited to current technology and compatible with economic and environmental quality aspects, that is, *economic and social acceptance.*

This book is about insect pest management, but the philosophy, many of the concepts, and the practice of pest management can also apply to many other kinds of pests.

II. WHY PEST MANAGEMENT?

A. Collapse of Control Systems

The enormous success of synthetic organic insecticides such as DDT and BHC following the conclusion of World War II began a new era of pest control. These two products were followed by hundreds of effective synthetic pesticides: acaricides, fungicides, herbicides, insecticides, nematocides, and rodenticides. The number of registered pesticides increased from about 30 in 1936 to more than 900 in 1971, and the annual United States production from less than 100 million pounds to more than 1.1 billion pounds in 1971 (*Pesticide Review* 1972).

This growth was to be expected, since the new chemicals are effective and easy to use. In the first flush of enthusiasm, it seemed that exclusive reliance on broad-spectrum insecticides could eliminate pest problems as far-ranging as those involving the housefly, *Musca domestica* L., in cities, the gypsy moth, *Porthetria dispar* (L.), in eastern forests of the United States, and malaria on a worldwide basis. As a result, regular spray programs were developed on a routine preventive basis, which provided a shield of pesticide protection whether the pest was present in damaging numbers or not. The onset of insecticide resistance, first experienced worldwide with DDT in the housefly within 2 years after its widespread use (Brown and Pal 1971), demonstrated the first flaw in exclusive reliance on insecticides. This was followed by a 20-year struggle in California to control floodwater mosquitoes, *Aedes* spp., by successive use of DDT, lindane, aldrin, dieldrin, toxaphene, EPN, methyl parathion, fenthion, temephos, chlorpyrifos, carbamates, and finally insect-growth regulators such as juvenile hormone analogs.

A parallel struggle was taking place in California citrus orchards against the citrus red mite, *Panonychus citri* (McGregor), and in apple orchards against the European red mite, *Panonychus ulmi* (Koch). Mite predators were eliminated, and resistance to various acaricides developed almost seasonally; orchardists exhausted in succession Neotran, ovex, tedion, sulphenone, chlorbenside, DMC, dicofol, schradan, demeton, and a variety of other organophosphorus compounds.

Perhaps the most alarming example of the endless spiral of more and more frequent treatments has taken place in cottonfields in Peru, Egypt, Central America, and Texas (see Chapter 7). The cotton bollworm, *Heliothis zea* (Boddie), and the tobacco budworm, *Heliothis virescens* (Fabricius), for example, have developed resistance (Adkisson 1969) and today are practically immune to all available insecticides. Some insects have changed from secondary pests, usually kept below damaging numbers by beneficial insects, into primary pests which have virtually destroyed cotton production in some areas. In efforts to control the resurgence of pests, growers have increased applications of such highly toxic materials as methyl parathion and parathion to 10, 20, and in extreme cases up to 60 applications during the growing season, with total applications of 30 to 40 lb or more per acre. Under these conditions the cost of pest control has made the production of cotton profitless, and the industry has collapsed in certain areas. In addition, such prodigious use of pesticides has had highly deleterious effects on environmental quality and has posed serious hazards to the health of agricultural workers.

B. Patterns of Crop Protection

Smith (1969) has classified worldwide patterns of crop protection in the cotton agricultural ecosystem into the following five phases, which are also applicable to many other crops.

1. *Subsistence Phase*

The crop, usually grown under nonirrigated conditions, is part of a subsistence agriculture. Normally, the crop does not enter the world market and is consumed in the village or in barter in the marketplace. Yields are low. There is no organized program of crop protection. Whatever crop protection is available results from natural control, inherent resistance of the cotton plant, hand picking, cultural practices, rare insecticide treatments, and luck.

2. *Exploitation Phase*

Crop protection programs are developed to protect expanded new acreage, new varieties, or new markets. Growers have observed the spectacular kill of insects with the new synthetic insecticides, and in most instances the pest-control program is dependent solely on chemical pesticides. They are used intensively, often on fixed schedules, and often as prophylactic treatments whether or not the pest is present. At first these programs are successful, resulting in high yields of food and fiber, and chemical pesticides are exploited to the maximum.

3. *Crisis Phase*

After a variable number of years in the exploitation phase and heavy use of insecticides, a series of events occurs. More frequent applications of pesticides and higher dosages are needed to obtain effective control. Insect populations often resurge rapidly after treatments, and the pest population gradually becomes tolerant to the pesticide. Another pesticide is substituted, and the pest population becomes tolerant to it, too. At the same time, insects that never cause damage or are only occasional feeders become serious primary pests. This combination of pesticide resistance, pest resurgence, and unleashed secondary pests causes greatly increased production costs.

4. *Disaster Phase*

The pesticide usage increases production costs to the point where the crop can no longer be grown and marketed profitably. Pesticide residues in the soil may be at such high levels that other crops cannot be successfully grown and meet legal residue tolerances. Repeated applications of insecticides and often mixtures of two insecticides no longer produce a crop acceptable to processors or the fresh market. There is a collapse of the existing pest-control program.

5. *Integrated Control Phase*

Insect-control programs are implemented that accept and utilize ecological factors and compatibility in control measures. The concept is one of optimizing control rather than maximizing it. It is pest management.

Not all pest-control programs fit neatly into the above phases, and some may exist side by side or circumvent some phases altogether. Currently, most pest control is in the exploitation phase, and pest-management concepts should be quickly adopted to avoid the crisis and disaster phases. Further, developing countries that are implementing or revising crop protection schemes can profit from the mistakes of others and adopt sound pest-management concepts to avoid control problems that will almost certainly arise.

C. Environmental Contamination

The ubiquitous presence of pesticide residues in foods, feeds, and organisms occupying every part of the ecosystem has caused widespread concern among scientists and thoughtful citizens alike about contamination of the environment. The effects of DDT transfer and magnification in the environment are well known, and the Clear Lake, California, incident, in which DDD applied at 20 ppb to control the larvae of the Clear Lake gnat, *Chaoborus astictopus* Dyar and Shannon, accumulated to more than 2000 ppm in carnivorous fish and western grebes, has become a classic (Hunt 1966). From such examples we have come to realize that the single-factor approach to insect control, involving sole reliance on insecticides, has the following limitations: (1) selection of resistance in pest populations, (2) destruction of beneficial species, (3) resurgence of treated populations, (4) outbreaks of secondary pests, (5) residues in feeds, foods, and the environment, and (6) hazards to humans and the environment.

It is unlikely that the many adverse events of the past 2 decades could have been prevented, as there was and still is sincere effort on the part of many people and governments to use the miracle insecticides to the benefit of people and there are still many pest problems for which the use of chemicals provides the only acceptable solution. "Contrary to the thinking of some people, the use of pesticides for pest control is not an ecological sin. When their use is approached from the sound base of ecological principles, chemical pesticides provide dependable and valuable tools and such use is indispensable in modern society" (NAS 1969). However, sole reliance on insecticides as the only control agent has created problems in insect control and the environment, and these in turn have strengthened the need for pest management. It is likely that most insect pest-management programs will utilize insecticides, but this use must be compatible with other controls and consistent with pest-management concepts.

III. CONCEPTS OF PEST MANAGEMENT

A. Understanding the Agricultural Ecosystem

Ecosystems are self-sufficient habitats where living organisms and the nonliving environment interact to exchange energy and matter in a continuing cycle (NAS 1969). Ecosystems are entities, such as forests, ponds, and fields, and in general they are self-regulating. Ecosystems and the ecological aspects of insect pest management are discussed in more detail in Chapter 2.

Agricultural ecosystems (agroecosystems) contain less diversity of animal and plant species than natural ecosystems such as forests and prairies. Usually, there are a few major species and numerous minor species and, in a pest outbreak, usually only 1 pest species at a time (often a major species) is present in large numbers. A typical agricultural unit may contain only 1 to 4 major crop species and 6 to 10 major pest species. Yet one need only walk into a crop field to recognize that the diversity of plants and insects is not as limited as conditions suggest.

The agroecosystem is intensively manipulated by man and subjected to sudden alterations such as plowing, mowing, and treatments with pesticides. Agronomic practices are critical in pest managment, since the need for pest control or the intensity of a pest problem is often directly related to agronomic practices. The magnitude of the agroecosystem is illustrated by the fact that today about 10% of the land (about 3 billion acres) supports 3.5 billion people. Obviously, these acres must be intensively managed.

Agroecosystems can be more susceptible to pest damage and catastrophic outbreaks, because of the lack of diversity in species of plants, species of insects, and the sudden alterations imposed by weather and man. However, the agroecosystem is a complex of food chains and food webs which interact together to produce a surprisingly stable unit. Diversity of species is frequently offset by homogeneity of plant species and uniformity of agronomic practices. Often an insect can attack, establish, and survive only during a short period of time, and uniformity in planting, plant development, and maturation can restrict rapid increase of a pest. Further, lack of diversity of plant species in agroecosystems is often offset by density; increased density of plants per acre can dilute pest attack or provide conditions unfavorable to pest increase, and plant species that are tolerant or resistant to insects are better able to withstand pest damage or suppress pest establishment and increase. Few of the multiple interactions that exist have ever been examined or explained, but it is important in pest management to recognize the existence of complex biological systems in the agroecosystem.

Case History

Weires and Chiang (1973) provide an excellent example of the food web associated with cabbage plants in Minnesota. The web illustrated in Fig. 1.1 is composed of food meshes. A food mesh is

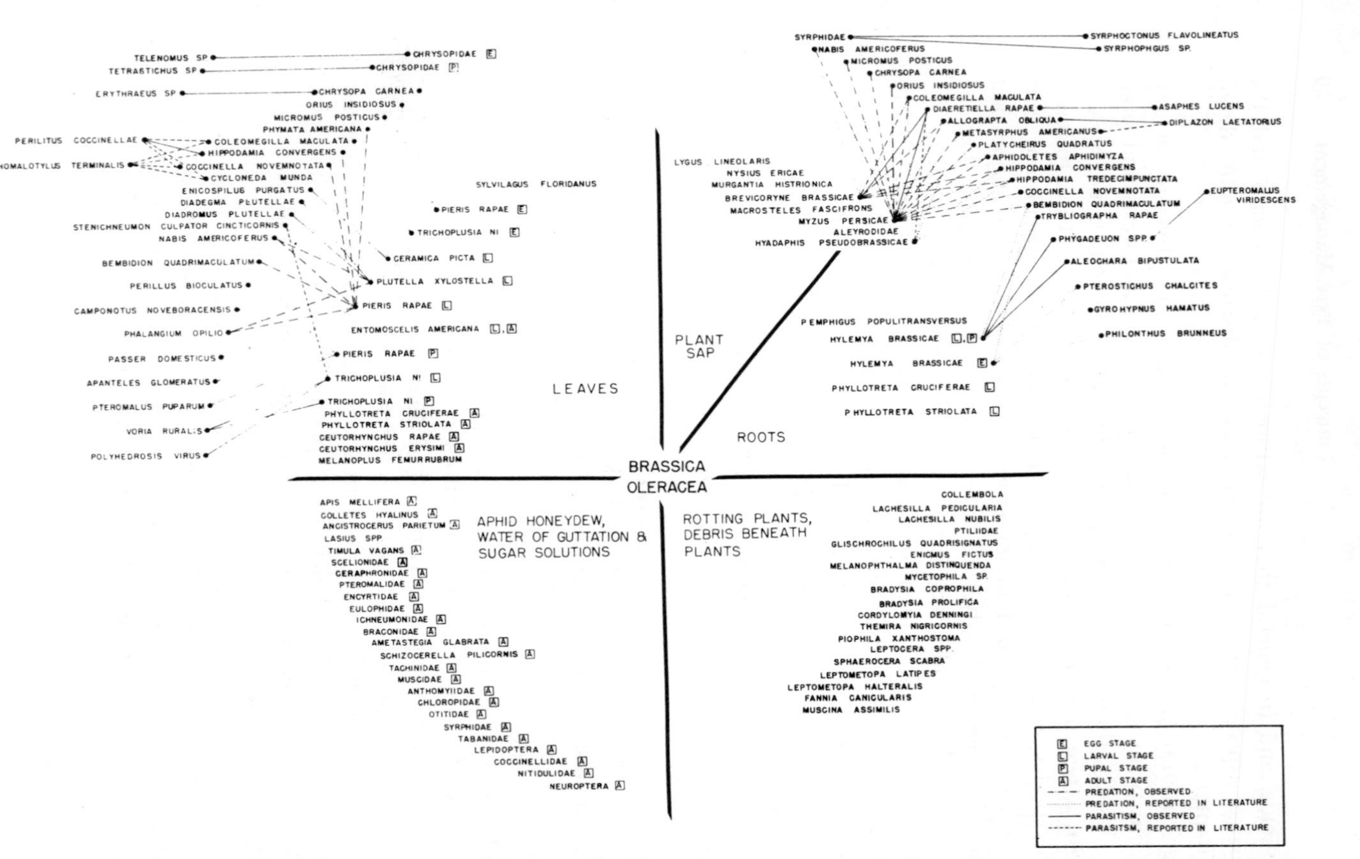

Fig. 1-1 Food web associated with cabbage plants in Minnesota. (Weires and Chiang 1973. Courtesy of Prof. H. C. Chiang and the University of Minnesota Agricultural Experiment Stations.)

defined by Allee et al. (1949) as "a taxonomic entity in a food web; for example a species or subspecies at a particular stage in its life cycle." The larval stage of the cabbage looper, *Trichoplusia ni* (Hübner), is one mesh feeding on cabbage leaves; the adult feeding on nectar constitutes another mesh. Quiescent stages such as eggs and pupae are not feeding meshes, but they constitute a part of the total food web.

In Fig. 1.1 the herbivorous, saprophagous, and saccharophilous meshes occupy the web's inner circle. First-order carnivore, predator, and parasite meshes occupy the second circle. Second-order carnivores occupy the outer circle. The cabbage food web contains 1 plant species, 11 leaf feeders, 10 sap feeders, 4 root feeders, 21 saprobes, 79 saccharophiles, and 85 carnivores interacting in the community.

B. Planning the Agricultural Ecosystem

It is unfortunate that one would try to grow a crop where that crop cannot be successfully grown. In insect pest management applied agroecosystem planning should anticipate pest problems and ways to avoid them. For example, a crop variety should not be grown if it is known to be unusually susceptible or potentially vulnerable to pest attack, thereby intensifying the need for control activity. Conversely, crops should be grown in a manner to avoid or reduce difficult pest problems. It behooves the plant protectionist working with plant and soil scientists to insist on agroecosystem planning that satisfies the world need for food and at the same time minimizes pest problems and avoids catastrophic events. This concept does not condemn monoculture, nor does it reduce the efficiency of specialized technical agriculture.

Case Histories

The soybean, *Glycine max* (L.) Merr., is well adapted to the midwestern United States, despite the presence of the destructive potato leafhopper, *Empoasca fabae* (Harris). Pubescent-type 'Harosoy' soybean plants grow waist high without chemical protection from this pest, whereas glabrous-type 'Harosoy' soybean plants are so severely attacked that they attain a height of only 8 to 10 in. Typical stunting of glabrous (smooth, nonpubescent) varieties caused by the potato leafhopper is illustrated in Fig. 1.2.

Fig. 1-2 Stunting of glabrous soybean varieties caused by feeding of the potato leafhopper; undamaged pubescent varieties in the background.

Planting pubescent soybean varieties adapted to the area is good agroecosystem planning.

Less obvious than the above is crop planting to avoid certain pests, while anticipating and planning for others. Planting corn in rotation with other crops provides complete control of the western and northern corn rootworms, but it aggravates other pest problems (Table 1.1). Petty (1972) shows that only the western corn rootworm, *Diabrotica virgifera* L., and the northern corn rootworm, *D. longicornis* (Say), increase where corn is grown continuously on the same acres, whereas white grubs, *Phyllophaga* spp., and the black cutworm, *Agrotis ipsilon* (Hufnagel) increase

Table 1.1 Effect of Crop Rotation of Corn on Insect Populations or Potential Damage

	Corn Rotation		
	None	Soybeans	Pasture and Hay Crops
Seed corn beetles	0	0	+[a]
Seed corn maggot	0	0	+
Wireworms	–	–	+
White grubs	–	+	+
Corn root aphid	–	–	+
Grape colaspis	–	–	+
Northern corn rootworm	+	–	–
Western corn rootworm	+	–	–
Southern corn rootworm	0	0	0
Black cutworm	0	+	0
Billbug	–	–	+
Slugs	–	–	0
Thrips	0	?	+
Mites	0	0	0
European corn borer	0	0	0
Southwestern corn borer	0	0	0
Corn earworm	0	0	0
Fall armyworm	0	0	0
True armyworm	0	0	+
Chinch bug	0	0	+
Corn leaf aphid	0	0	0
Totals +	2	2	10
–	6	7	2
0	13	11	9
?	0	1	0

[a] + means the practice will increase the population or damage from that insect; – means it will reduce the population or damage; 0 means no effect; ? means effect unknown.

when corn is grown in rotation with soybeans. Ten insects increase when corn is grown in rotation with pasture and hay crops. Of all these corn insects, rootworms are the easiest to control, and predictive guidelines are available for the farmer to measure the potential for high or low infestation in fields. Thus, in the United States corn belt, continuous corn appears to be a good insect pest-management practice. The grower would anticipate a possible need to treat some acres annually for corn rootworms, but probably would not need to treat all acres.

C. Cost/Benefit and Benefit/Risk

The economics of insect pest management is discussed in Chapter 3. An entire chapter in this book is devoted to this subject, as the agriculturalist can no longer think solely in terms of costs and benefits but must consider environmental effects as well—the benefits/risks of pest control. Many pest-control activities have social and environmental impacts.

1. Cost/Benefit

For the most part, pest-control recommendations are directed to the private sector, since control is most often the result of activities by the individual or custom applicator contracted to apply pesticides. Faced with the possibility of pest damage, the individual is interested in actions that reduce that uncertainty, as long as the amount of expenditure is commensurate with the amount of the probable damage. By using pesticides the possibility of pest problems can be greatly diminished, and catastrophic damage can be prevented. In agriculture, the implication of yield increase, often used to show benefit from treatment, is usually erroneous. The use of pesticides rarely increases yield, rather use prevents loss of yield.

In most agricultural pest-control activities, the benefits usually are not known, as they are usually not measured, and the costs of prevention become costs of production. Improving capabilities for predicting pest problems and defining economic thresholds will place increased emphasis on costs and benefits. Crop life tables provide a solid foundation for analysis of pest damage and cost/benefit in pest management.

Case History

Crop life tables provide excellent guidelines in the planning of pest-management strategies, particularly when coupled with meaningful cost/benefit analysis. Harcourt (1970) presents a typical life table (Table 1.2) and loss statistics (Table 1.3) for a planting of early-market cabbage. In Table 1.2 the first column gives the sampling period, namely, the stage of growth attained. The *lx* column represents the number alive (or potentially marketable) at the beginning of the period, and the *dx* column, the number dying

Table 1.2 Life Table for a Planting of Early Market Cabbage, Ottawa, 1968[a]

Growth Period x	Mean Number Living per Plot, lx	Mortality Factor dxF	Mean Number Dying per Plot, dx	Percent Mortality, $100rx$
Establishment	319.2 ± 4.2	Drought	7.2 ± 1.0	2.2
		Cutworms	56.1 ± 7.1	17.6
		Root maggot	1.5 ± 0.4	0.5
		Other[b]	0.8 ± 0.3	0.3
		Total	65.6 ± 7.1	20.6
Preheading	253.6 ± 6.0	Cutworms	8.6 ± 1.1	2.7
		Root maggot	9.6 ± 1.7	3.0
		Flea beetles	0.3 ± 0.1	0.1
		Rodents	1.3 ± 1.0	0.4
		Clubroot	1.3 ± 0.8	0.4
		Other[b]	0.9 ± 0.2	0.3
		Total	22.0 ± 4.5	6.9
Heading	231.6 ± 7.1	Cabbage caterpillars	29.4 ± 1.0	9.2
		Root maggot	0.6 ±0.1	0.2
		Clubroot	3.2 ± 1.6	1.0
		Soft rot	0.4 ± 0.1	0.1
		Total	33.6 ± 1.7	10.5
Harvest	198.0 ± 6.7	Cabbage caterpillars	18.4 ± 2.2	5.8
		Clubroot	5.5 ± 1.8	1.7
		Soft rot	3.0 ± 0.5	0.9
		Total	26.9 ± 2.1	8.4
Yield	171.1 ± 7.1		148.1 ± 4.5	46.4

[a] Harcourt 1970. Courtesy of D. G. Harcourt and the Entomological Society of Canada.

[b] Miscellaneous factors such as frost, hail, and mechanical damage.

(or "written off") within the period. The *dxF* column shows the agent or factor responsible for *dx*, and 100*rx* is percentage mortality based on the initial population. It is obvious that the young plants have highest mortality, and that cutworms, cabbage caterpillars, and root maggots were major mortality factors. Insects caused losses of $317.47, diseases $34.18, and miscellaneous factors (mechanical damage, rodents, weather, etc.) $26.01. Operating profit at $436.30 was just over 50% of potential revenues at the time of planting.

Table 1.3 Operating Statistics for a Planting of Early Market Cabbage, Ottawa, 1968[a]

Growth Period	Potential Revenue ($/acre)	Hazard	Loss of Revenue ($/acre)
Establishment	$813.96	Drought	18.36
		Cutworms	143.05
		Root maggots	3.83
		Other	2.04
		Total	167.28
Preheading	$646.68	Cutworms	21.93
		Root maggots	24.48
		Flea beetles	0.76
		Rodents	3.32
		Clubroot	3.32
		Other	2.29
		Total	56.10
Heading	$590.58	Cabbage caterpillars	74.97
		Root maggot	1.53
		Clubroot	8.16
		Soft rot	1.02
		Total	85.68
Harvest	$504.90	Cabbage caterpillars	46.92
		Clubroot	14.03
		Soft rot	7.65
		Total	68.60
Operating profit	$436.30		

[a] From Harcourt 1970. Courtesy of D. G. Harcourt and the Entomological Society of Canada.

2. *Benefit/Risk*

The social economics of pest control are necessary considerations in developing pest-control strategy, particularly when pesticides are used. Benefit/risk analysis provides a means for assessing the relevant economic benefits versus the risks in pest control. The consideration and assessment of benefit/risk is fundamental to pest management. A grower carefully considers the hazard of a highly toxic pesticide and takes action to ensure safety for himself and his workers in handling and in application. Similarly, a grower must consider the effects on society and on the environment of a pesticide that is applied.

The use of insecticides when they are not needed is contrary to the pest-management philosophy. The treatment of 1 million acres when only 100,000 acres needs protection imposes risks that exceed the benefits. Further, chemical treatment is seldom a process so efficient that all the input is fully utilized. As a rule, more than 90% of the insecticide applied to control insects does not hit the target pest, and it becomes incorporated into the environment in various ways. Parasitoids and predators are reduced. Persistent pesticide residues concentrate in foods and in the environment. These additional unwanted nonmarket effects are called externalities, and someone must pay for them. Externalities are adverse side effects which have economic, ecological, and sociological consequences. A case history is the magnification of DDT in fish in Lake Michigan, from about 0.000002 ppm in the water to as much as 10 ppm or more in game fish, so that the U.S. Food and Drug Administration (FDA) has declared these fish illegal for commercial sale for human consumption (see Chapter 7). Another example is the presence of persistent residues in soil at levels high enough that subsequent crops grown in these soils may be legally unacceptable in the marketplace. The oil in soybean grain and the waxy rind and seeds of pumpkins can contain objectionable residues of lipid-soluble pesticides when these crops are grown on soils that were treated in previous years and still contain residues of organochlorine insecticides (Bruce et al. 1966, 1967).

D. Tolerance of Pest Damage

Complete freedom from insect attack is neither necessary in most cases for high yields nor appropriate for insect pest management. Nearly all plants can tolerate a substantial degree of leaf destruction without appreciable effects on plant vigor. One needs only to examine the holes, blotches, and mines in the leaves of healthy forest trees to appreciate this statement. Quantitative studies of the degree of damage versus reduction in crop yield are urgently needed, so that thresholds can be established for allowable damage. In assessing this factor it should also be remembered that an important exception occurs in the case of plant diseases transmitted during a brief feeding period, sometimes as short as 15 seconds, by an insect vector.

1. *Economic Injury Level*

This quantitative measure of insect pest density determines if an insect component of an agroecosystem is to be classified as a pest. Without an estimate of the pest density that can be tolerated without significant crop loss, there can be no reasonable safeguard against either overtreatment with insecticides or unacceptable crop damage. Thus determination of the economic injury level is critical in defining the ultimate objective of any pest-management program and in delineating the pest population level below which damage is tolerable and above which specific interventions are needed to prevent a pest outbreak and to avert significant crop injury.

Various definitions have been proposed for the economic injury level, including "the lowest pest population density that will cause economic damage" (Stern et al. 1959), "the level at which damage can no longer be tolerated and therefore the level at or before which it is desirable to initiate deliberate control activities," or a "more critical density . . . where the loss caused by the pest equals in value the cost of available control measures" (NAS 1969). Headley (1972) has defined the economic injury level as the pest "population that produces incremental damage equal to the cost of preventing the damage."

2. *Economic Threshold*

This is another important parameter, defined as "the density at which control measures should be applied to prevent an increasing pest population from reaching the economic injury level" (Stern et al. 1959). An increasingly damaging population can occur as the result of an increase in density (number of individuals), or an increase in biomass (size of individuals). An increase in biomass can occur with a decreasing population density. The economic threshold always represents a pest density lower than that of the economic injury level, to allow the initiation of control measures so that they can take effect before the pest density exceeds the economic injury level. These relationships are illustrated in Fig. 1.3.

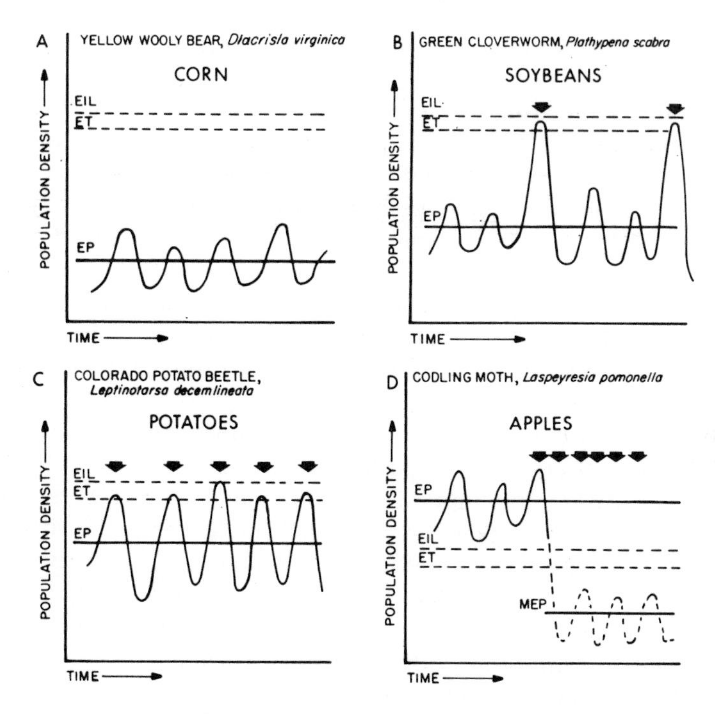

Fig. 1-3 Economic injury levels and economic thresholds for typical insect pest situations. EIL, economic injury level; ET, economic threshold; EP, equilibrium position; MEP, modified equilibrium position; arrowheads, pest-control intervention. (Modified after Stern 1965.)

3. *General Equilibrium Position*

This is the average population density of an insect population over a long period of time, unaffected by the temporary interventions of pest control. The population density fluctuates about this mean level as a result of the influence of density-dependent factors such as parasitoids, predators, and diseases. The economic injury level may be at any level from well below to well above the general equilibrium position. Insects can be grouped in four general categories in this regard, as shown in Fig. 1.3.

1. Many insect species feed on cultivated crops without ever reaching densities high enough to cause economic injury (Fig. 1.3*A*), and consequently are rarely if ever noticed. Familiar examples include the cowpea aphid, *Aphis craccivora* Koch, on alfalfa; the yellow woollybear,

Diacrisia virginica (Fabricius), on corn; and the painted lady, *Vanessa cardui* (L.), on soybeans.

2. Another large group of insects are occasional pests (Fig. 1.3*B*), and exceed economic injury levels only when their population densities are affected by unusual weather conditions or injudicious use of insecticides. Examples include forest insect pests such as the fall webworm, *Hyphantria cunea* (Drury), which becomes epidemic in 5- to 10-year cycles; the green cloverworm, *Plathypena scabra* (Fabricius), on alfalfa or soybeans; and the white-lined sphinx, *Hyles lineata* (Fabricius), of California deserts. At their peaks of population density, some sort of intervention, usually insecticides, is required to reduce their numbers to tolerable levels.

3. A third group of insects has economic injury levels only slightly above the general equilibrium position (Fig. 1.3*C*), and intervention is necessary at nearly every upward population fluctuation. These insects are perennial pests, for example, the gypsy moth, *Porthetria dispar,* in hardwood forests; the cotton boll weevil, *Anthonomus grandis* Boheman; the Colorado potato beetle, *Leptinotarsa decemlineata* (Say); and the Mexican bean beetle, *Epilachna varivestis* Mulsant, on beans. The general practice is to intervene with insecticides whenever necessary to produce a modified average population density well below the economic injury level (Fig. 1.3*C*).

4. Severe pests are found in a group of insects having economic injury levels below the general equilibrium position (Fig. 1.3*D*). Classic examples include the codling moth, *Laspeyresia pomonella* (L.), on apples; the corn earworm, *Heliothis zea,* on sweet corn; the asparagus beetle, *Crioceris asparagi* (L.), on asparagus; and the artichoke plume moth, *Platyptilia carduidactyla* (Riley), on artichokes. Regular and constant interventions, usually with insecticides, are required to produce marketable crops.

The same insect pest attacking several crops may have greatly differing economic injury levels. *Heliothis zea* feeding on alfalfa never reaches a population density sufficient to cause economic injury and belongs in category 1. As the cotton bollworm, *Heliothis zea* is a major pest of cotton, has an economic threshold of four larvae per plant (Stern 1965), and generally requires insecticidal treatments several times yearly; thus it belongs in category 3. *Heliothis zea,* the corn earworm attacking sweet corn, is a severe pest; it has an economic threshold approaching zero population density and falls in category 4.

Determining the economic injury level and economic threshold is generally a complex matter based on detailed operations of pest ecology as it relates to bioclimatology, predatorism, diseases, the effects of host-plant resistance, and the environmental consequences of applied control interventions. The economic injury level concept is flexible and may vary from area to area, crop variety to crop variety, and even between two adjoining fields, depending on specific agronomic practices (Reynolds 1972). The economic injury level decreases as the value of the crop increases, and is also a function of consumer standards. Thus for tree fruits, sweet corn, asparagus, potatoes, cut flowers, and the like, the threshold may be very low, as a single codling moth, scale insect, or earworm attack drastically affects consumer acceptance of produce. Changes in marketing developments, such as the rapid growth of the frozen-food industry, and FDA laws regulating the presence of insect fragments in canned food products, can produce decisive changes in economic injury levels for vegetable and fruit crops. The economic injury level is inversely related to the product price, and directly related to the cost of control (Headley 1972).

For the individual grower, assuming no external costs, Rabb (1972) has suggested the following factors as essential for determination of the economic injury level:

1. Amount of physical damage related to various pest densities
2. Monetary value and production costs of the crop at various levels of physical damage
3. Monetary loss associated with various levels of physical damage
4. Amount of physical damage that can be prevented by the control measure
5. Monetary value of the portion of the crop that can be saved by the control measure
6. Monetary cost of the control measure.

From this information it is possible to determine the level of pest density at which control measures can be applied to save crop value equal to or exceeding costs of control. This simplified approach, however, does not consider important externalities, for example, adverse environmental effects such as increasing soil pesticide residues, which may make subsequent crop production less profitable, or ecological effects on natural enemies, which may result in increased frequency of pesticide intervention or in outbreaks of secondary pests. Therefore economic injury levels for individual insect pests are almost always higher than superficial evidence suggests. In pest management great care must be taken not to equate the visual threshold with the action economic threshold where the level of pest population is such that intervention must occur to prevent the population from rising above the economic injury level.

A special category of economic injury level must be applied to insects serving as vectors of plant and animal diseases. Here a single insect attack may cause the death of a valuable tree, a domestic animal, or a human. Examples include the smaller European elm bark beetle, *Scolytus multistriatus* (Marsham), and Dutch elm disease; the tsetse flies, *Glossina* spp., and African trypanosomiasis; and *Aedes aegypti* (L.) and yellow fever. Economic values can scarcely be set for the costs of such depredations, and the economic injury level approaches zero population density.

Case History

Alfalfa, *Medicago sativa* L., is the principal hay crop of the United States; about 125 million tons are produced annually, or 43% of world production. Most alfalfa production is for hay, which is the staple feed of the dairy industry, and lesser amounts are processed as dehydrated alfalfa meal and as alfalfa seed. As many as 13 crops of hay are produced during a single growing season. Alfalfa fields produce and harbor very large populations of insects, because of their lush growth, dense foliage, and protected microclimate. Pest-control programs are complex, because persistent insecticides contaminate hay and consequently milk, which has a "zero tolerance" (established by the FDA) for pesticide residues. In addition, alfalfa seed production depends on pollination by bees which are destroyed by most insecticide applications. Therefore Stern (1965) has pioneered in the development of successful pest-management programs for major alfalfa pests, which minimize insecticide use and conserve biological control agents. The programs are based on accurate determinations of economic thresholds by direct counting of the number of pest individuals in sample units, for example, a 180° sweep by an insect net, and relating pest densities to crop damage and to costs of control intervention. These economic thresholds, natural enemies, and compatible insecticide treatments form the framework of successful pest-management programs (Stern, 1965):

Pest	Economic Threshold	Natural Enemies	Compatible Insecticides
Alfalfa caterpillar *Colias eurytheme* Boisduval	10 nonparasitized larvae per sweep	*Apanteles* *Trichogramma*	*Bacillus thuringiensis* toxin, trichlorfon
Beet armyworm *Spodoptera exigua*	15 larvae (0.5 in.) per sweep	Ichneumonids, braconids, tachinids	*Bacillus thuringiensis* toxin, trichlorfon
Alfalfa weevil *Hypera postica* (Gyllenhal)	25 larvae per sweep	*Bathyplectes*	Parathion
Egyptian alfalfa weevil *Hypera brunneipennis* Boheman	25 larvae per sweep	?	Methoxychlor
Spotted alfalfa aphid *Therioaphis maculata* (Buckton)	Spring, 40 aphis per stem Summer, 20 aphis per stem	*Trioxys*, *Praon*, *Aphelinus*, coccinellids	Demeton
Pea aphid *Acyrthosiphon pisum* (Harris)	<10 in., 40–50 per stem 15 in., 70–80 per stem 20 in., 100 per stem	*Aphidius*	Demeton

E. Leave a Pest Residue

The ecological balances sought in pest-management programs necessitate the widespread encouragement of beneficial insects that are effective natural enemies of the pest species. These are often effectively removed by direct contact with broad-spectrum insecticides regularly applied to fields and orchards, and are also destroyed by starvation when their prey is totally eliminated by chemical control. Therefore an important concept of pest management is the necessity for leaving a permanent pest residue, below the economic threshold, in an area where control measures are conducted. The concept is to suppress a pest but not annihilate it. Needless to say, this idea is at variance with the general grower attitude and consumer insistence on unblemished fresh produce and canned products free of insect fragments.

The exceptions to this concept are (1) the complete absence of parasitoids, predators, or diseases of a pest, and (2) the practice of eradication where eradication is truly feasible within acceptable ecological parameters. Elimination of the screwworm from the southeastern United States by the sterile-male technique is a classic example of exception 2, and other equally sophisticated programs should be developed and implemented (Chapter 8). However, at best, most eradication programs have only retarded the spread of the pest and led to the generalization that a well-entrenched insect pest cannot be eradicated by massive applications of insecticides alone. Regrettably, insect pest control over the past 25 years has all too frequently diverged from the idea of economic thresholds and tolerable levels toward pest eradication in a field, orchard, state, or country.

Case History

The European red mite, *Panonychus ulmi* (Koch), was only an occasional pest of apples and other deciduous fruits before the introduction of DDT into fruit orchards in 1946 for the control of the codling moth. DDT, however, adversely affects predators that largely controlled the European red mite population, and this mite has become a limiting factor in deciduous fruit production, feeding on foliage and causing loss of chlorophyll and leaf drop, weakening buds, and producing undersized fruit. Because of a very rapid life cycle, as little as 4 days from egg to adult at 77°F, the mite has a great capacity for rapid development of races resistant to pesticides and, during the past 25 years, orchardists have used in succession a large range of special acaricides and organophosphorus pesticides. Further, mite control on apple is complicated by the large numbers of other pests for which regular spray applications are made during the apple-growing season (Chapter 13). As a result of this complex of enemies and complications of pest control, as many as 20 sprays are applied during a season in some apple orchards. Pest control in apple orchards has been largely a "preventive" chemical program, sprays being applied at regular intervals as insurance against possible attack (Glass and Hoyt, 1972).

Luckmann et al. (1971) and Meyer (1974) have demonstrated a unique management program for the control of phytophagous mites in apple orchards in Illinois (Fig. 1.4). Mite control was previously a difficult problem, requiring the application of six to seven sprays each season; the use of carbaryl, moreover, had completely eliminated predaceous mites from some orchards. In the mite program a single spray applied to the periphery of the tree suppresses the European red mite, *Panonychus ulmi,* which overwinters and begins early-season increase on peripheral twigs and branches. The spray is applied 3 to 5 weeks after bloom. The predatory mite *Amblyseius fallacis* (Garman), which overwinters on the trunk and in the grass and debris at the base of the tree, is allowed to increase by preying on spotted mites, *Tetranychus* spp., which also overwinter at the base of the tree. The increasing population of the predator mite then spreads upward throughout the tree to control all phytophagous mites for the remainder of the season. The predatory mite is resistant to some insecticides, and these are used to control insect pests of apple such as the codling moth.

Amblyseius fallacis has been reared in the laboratory and successfully released and established in orchards devoid of the predator.[1] Orchardists are advised to avoid clean cultivation around the base of the tree, to leave some debris and groundcover (e.g., grass) to aid the predators in overwintering and increasing the following spring.

F. Timing of Treatments

A crucial problem in successful pest management is the proper timing of insecticide treatments. Virtually every group that has critically evaluated the need for pest management has commented forcefully on this. The President's Science Advisory Committee (1965) recommended replacement of wasteful routine-treatment schedules by treat-when-necessary schedules and commented that "substantial reduction in pesticide use, in specific cases as much as 50%, can be made by applying our present knowledge of pests and their control." The American Chemical Society (1969) recommendations on minimizing contamination of the environment with pesticides state, "Optimum methods of pest control will involve careful integration of chemical, biological, and cultural techniques—only in this way can the objective of economic control of pests in crops and animals be obtained with minimal environmental and ecological impact."

[1] R. H. Meyer, personal communication.

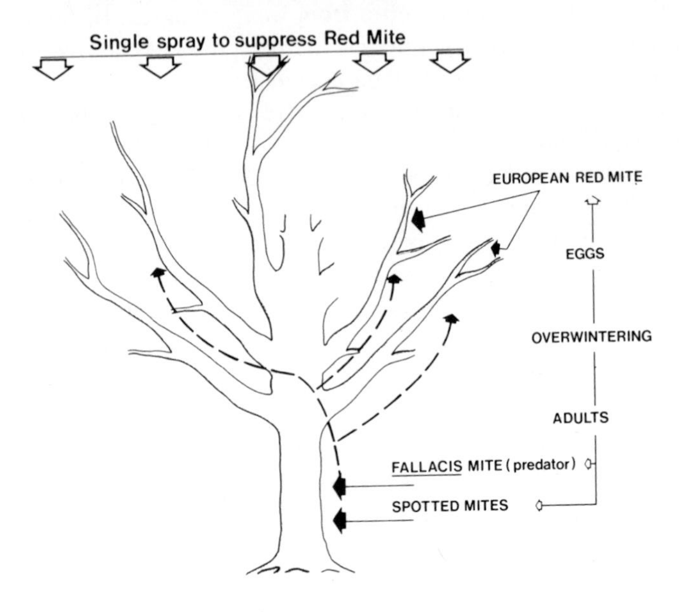

Fig. 1-4 Diagram showing salient ecological facts concerning life histories of mites on orchard tree.

Treatment should be based on need, and a single spray properly timed can often prevent excessive spraying. More efficient use results from careful timing of treatments based on improved techniques of monitoring pest populations and crop development. An example of this is the use of pheromone traps for monitoring the appearance and intensity of the codling moth on apple at various stages of plant growth (Chapters 7 and 8).

Case History

Classifying crop susceptibility to establishment, survival, and control of European corn borer, *Ostrinia nubilalis* (Hübner), is a standard practice. Corn borer larvae do not survive on small corn because of the presence of the chemical DIMBOA (see Chapter 4), so treatment is justified only on taller corn (36 in. extended height) when 50% of the plants show leaf feeding by the borers. But corn varieties differ in growth habits, and some mature in 70 days and

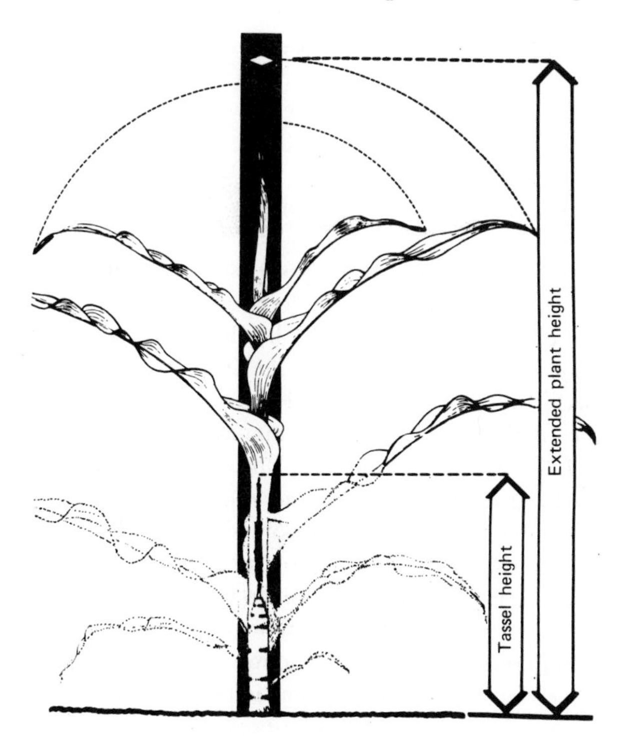

Fig. 1-5 Illustration of measurements needed to obtain tassel ratio. (TR).

$$\frac{\text{Tassel height}}{\text{Extended plant height}} \times 100 = \text{tassel ratio}$$

some in 120 days; some are 4 ft tall at maturity, while others are 8 ft. A 36-in. plant of one variety might be highly susceptible to attack by the corn borer, while a 36-in. plant of another is still immune. The tassel ratio technique (Fig. 1.5) of Luckmann and Decker (1952) is an aid to insect scouts and growers in monitoring corn development, predicting potential for damage, and selective timing of treatment for first-generation corn borers. The ratio compares the height of the developing tassel inside the plant to the extended height of the plant. Suppose the extended height of the plant is 50 in. and the height of the tassel is 15 in. The tassel ratio is 30.

$$\frac{\text{Tassel height}}{\text{Extended plant height}} \times 100 = \text{TR} \qquad \frac{15\text{ in.}}{50\text{ in.}} = 0.3 \times 100 = 30$$

Corn should receive a single treatment of insecticide between tassel ratio 40 and 60 if 50% or more of the plants show evidence of fresh borer feeding on the leaves.

G. Public Understanding and Acceptance

Bringing people to an understanding of pest management is the best way to deal with insect pest problems. No program is any more successful than the degree of commitment made by the people involved. This concept is of critical importance in the transition era, when special efforts are being made by extension entomologists to educate growers and the public in the methodologies of pest management and the reasons for using them. Effective communication and salesmanship are the key to successful public understanding and acceptance.

Pest management, as Rabb (1972) points out, involves *first* a scientific judgment, "How can insect control be achieved?," and *second* a social judgment, "How should insect control be achieved?" The answer to the second question involves both economic and social acceptance (Geier 1966), but acceptance and even conviction that a pest can best be controlled via a pest-management program is useless unless put into action. Acceptance without action will undoubtedly be the greatest single impediment to pest management. Leadership in pest management is a critical factor.

It is safe to say that pest-management concepts and philosophy are not fully understood and appreciated by all key people who are in a position to make recommendations. One of the barriers is the erroneous assumption that a pest-management program must be complete, with all parameters known, before implementation can begin. This sort of thinking constitutes a serious waste of available knowledge and experience. A satisfactory program can be started on a single pest of a complex of 10 to 20 that attack a crop, for example, the European red mite on apples. It will be far easier to add components directed at other pests in the complex if some sort of program is already in operation and if growers already accept the pest-management philosophy.

Furthermore, it is not necessary to know everything about pest biology and the economic injury level of pest populations before beginning a pest-management program. From accumulated experience we can be confident that, if the economic injury level lies somewhere between 10 and 20 insects per plant or tree, a grower having a population of 25 to 30 or more insects per unit has a potential for high damage, while a grower with 0 to 5 insects per unit has a potential for little or no damage. Thus the initial phase of a pest-management program could be directed only to situations with potentials for very high or very low damage, while researchers and students are in the process of defining more precise economic injury levels and are refining methods of sampling and procedures of survey. It is certain that pest damage will not occur if the pest is not present. Limited pest-management guidelines, provided they are accurate, can be the vehicle for education and implementation of more inclusive guidelines in the future.

A very substantial step in pest management was taken in 1972 with *Implementing Practical Pest-Management Strategies,* the Proceedings of a National Extension Pest-Management Workshop (1972). There is presently a wide divergence in philosophy between the specialist who seeks to promote pest-management practices and the specialist who in general seeks to promote the maximum application of pesticides. The bridging of these philosophies and the education of a vital third party, the public as producer of commodities and consumer of pesticides, is a facet of a pest-management program. A period of transition, hopefully leading to mutual understanding of the objectives of pest management, is predicted.

One strategy in pest management is the formation of a legal, quasi-legal, or voluntary cooperative district to implement pest-management strategies on an areawide basis. These districts offer many opportunities for public understanding, acceptance, and support. They may be empowered to levy modest taxes on land or on agricultural produce to provide support for hiring of professional insect scouts and control personnel, and a districtwide program is valuable in enforcing public cooperation in desirable pest-management practices, for example, removal of alternative hosts, drainage of standing water, or disposal of crop refuse. The district provides a visible symbol and focus for public educational programs and forums, ranging from elementary school demonstrations and field days to short courses at the college level; and for cooperative work with land grant college and federal extension programs for better pest control.

Case History

The green peach aphid, *Myzus persicae* (Sulzer), is the most important insect vector of plant diseases, and is known to transmit more than 50 plant viruses including aster yellows, cranberry false blossom, curly top of sugar beets, peach yellows, and potato leaf roll. It is chiefly responsible for the transmission of a complex

Fig. 1-6 Advertising Idaho's green peach aphid pest-management program. The bumper sticker slogan "*A*pricots & *P*eaches *H*arbor *I*nsects *D*amaging to *S*puds" is a unique way to enlist public support.

group of potato viruses which may cause 60 to 95% losses in potato production. Virus transmission may occur with aphid feeding periods of only 10 to 15 seconds. The aphid overwinters as black, shiny eggs on the bark of peach, apricot, plum, and cherry trees, and the young aphids begin to hatch about bloom time. When full-grown, the wingless, parthenogenic females give birth to living young. These aphids remain on the peach for two to three generations, when most of the individuals acquire wings and migrate to potato, beet, or other summer host plants in the late spring. At the onset of cold weather in the fall, the female aphids produce true sexual females. These mate with males flying from summer host plants, and the fertilized females deposit the overwintering eggs on peach and apricot trees.

Idaho, the Potato State, is presently involved in a multicounty campaign to minimize transmission of potato virus diseases by *Myzus persicae*. In part this program involves all residents in Idaho potato seed and potato production districts through a public information campaign highlighted by components such as the bumper sticker shown in Fig. 1.6. The plea "*A*pricots and *P*eaches *H*arbor *I*nsects *D*amaging to *S*puds" seeks to enlist public support in destroying or in spraying all domestic and wild peach and apricot trees which provide essential overwintering hosts for the green peach aphid. It is sure to elicit public pride and acceptance of the program.

IV. TOOLS OF PEST MANAGEMENT

The nearly 100 years of the formal existence of a federal Bureau of Entomology, of the Agricultural Experiment Station system, and of the Cooperative State Extension Service have given the United States an unexcelled background of knowledge and experience in pest control and a framework within which pest-management strategies can be applied. As emphasized repeatedly in this book, our present knowledge is adequate to support major beneficial changes in pest-control practices. The major innovations necessary for designing pest-management programs will relate to the concepts of pest management (Section III) and to their economic and social acceptance.

The available techniques for controlling individual insect pests are almost inexhaustible and involve a very wide range of applied science and technology. They are conveniently categorized, in increasing order of complexity, as cultural, mechanical, physical, biological, chemical, genetic, and regulatory methods, as shown in the following outline, modified from Metcalf et al. (1951).

Inventory of Methods of Applied Insect Control
(arranged in approximate order of complexity)

I. Cultural methods or the use of agronomic practices
    A. Use of resistant varieties of domestic plants and animals
    B. Crop rotation
    C. Crop refuse destruction

   D. Tillage of soil
   E. Variation in time of planting or harvesting
   F. Pruning or thinning
   G. Fertilization
   H. Sanitation
   I. Water management
   J. Planting of trap crops

II. Mechanical methods
   A. Hand destruction
   B. Exclusion by screens, barriers
   C. Trapping, suction devices, collecting machines
   D. Crushing and grinding

III. Physical methods
   A. Heat
   B. Cold
   C. Humidity
   D. Energy—light traps, light regulation
   E. Sound

IV. Biological methods
   A. Protection and encouragement of natural enemies
   B. Introduction, artificial increase, and colonization of specific parasitoids and predators
   C. Propagation and dissemination of specific bacterial, virus, fungus, and protozoan diseases

V. Chemical methods
   A. Attractants
   B. Repellents
   C. Insecticides
   D. Sterilants
   E. Growth inhibitors

VI. Genetic methods
   A. Propagation and release of sterile or genetically incompatible pests

VII. Regulatory methods
   A. Plant and animal quarantines
   B. Eradication and suppression programs.

The incorporation of some of these techniques into pest-management programs is discussed in subsequent chapters of this book.

REFERENCES

Adkisson, P. L. 1969. How insects damage crops. *In* How crops grow—A century later. *Conn. Agr. Exp. Sta. Bull.* 708, pp. 155–164.

Allee, W. C., O. Park, T. Park, A. E. Emerson, and K. P. Schmidt. 1949. Principles of animal ecology. W. B. Saunders, Philadelphia. 837 pp.

American Chemical Society. 1969. Cleansing our environment. The chemical basis for action. American Chemical Society, Washington, D.C. 250 pp.

Arundel, G. E. 1948. Entire towns abolish flies. *Reader's Dig.* **52**:22.

Bartlett, B. R. 1956. Natural predators. Can selective insecticides help to preserve biotic control? *Agr. Chem.* **11**:42–44, 107.

Brown, A. W. A., and R. Pal. 1971. Insecticide resistance in arthropods. World Health Organization, Geneva. 491 pp.

Bruce, W. N., G. C. Decker, and W. H. Luckmann. 1967. Residues of dieldrin and heptachlor epoxide found in pumpkins growing on soil treated with aldrin and heptachlor. *J. Econ. Entomol.* **60**(3):707–709.

Bruce, W. N., G. C. Decker, and Jean G. Wilson. 1966. The relationship of the levels of insecticide contamination of crop seeds to their fat content and soil concentration of aldrin, heptachlor, and their epoxides. *J. Econ. Entomol.* **59**(1):179–181.

Geier, P. W. 1966. Management of insect pests. *Ann. Rev. Entomol.* **11**:471–490.

Geier, P. W., and L. R. Clark. 1961. An ecological approach to pest control. *Proc. Tech.* Meeting *Inter. Union Conserv. Nature Nat. Res. 8th, 1960, Warsaw,* pp. 10–18.

Glass, E. H., and S. C. Hoyt. 1972. Insect and mite pest management on apples. Pages 98–106 *in* Implementing practical pest management strategies. Proceedings of a national extension pest-management workshop. Purdue University, Lafayette, Indiana.

Harcourt, D. G. 1970. Crop life tables as a pest management tool. *Can. Entomol.* **102**(8):950–955.

Headley, J. C. 1972. Economics of pest control. Pages 180–187. *in* Implementing practical pest management strategies. Proceedings of a national extension workshop. Purdue University, Lafayette, Indiana.

Hunt, E. R. 1966. Biological magnification of pesticides. Pages 251–262 *in* Scientific aspects of pest control. *Nat. Acad. Sci. Publ.* 1402.

Luckmann, W. H., and G. C. Decker. 1952. A corn plant maturity index for use in European corn borer ecological and control investigation. *J. Econ. Entomol.* **45**(2):226–232.

Luckmann, W. H., W. M. Bever, B. J. Butler, H. J. Hopen, R. L. Metcalf, H. B. Petty, and F. W. Slife. 1971. Pesticides and pest control systems.

Pages 13–16 *in* Agriculture's role in environmental quality. Proceedings of the 1st Allerton conference. *Uni. Ill. Coll. Agr. Spec. Publ.* 21.

Metcalf, C. L., W. P. Flint, and R. L. Metcalf. 1951. Destructive and useful insects. 3rd ed. McGraw-Hill, New York. 1071 pp.

Meyer, R. H. 1974. Management of phytophagous and predatory mites in Illinois orchards, *Environ. Entomol.* **3**(2):333–340.

National Academy of Sciences. 1969. Insect pest management and control. Publ. 1695. Washington D.C. 508 pp.

National Extension Insect Pest-Management Workshop. 1972. Implementing practical pest management strategies. Purdue University, Lafayette, Indiana. 206 pp.

President's Science Advisory Committee. 1965. Restoring the quality of our environment. The White House, Washington, D.C. 317 pp.

Petty, H. B. 1972. Corn insect pest management, pp. 107–115 *in* Implementing practical pest management strategies. Proceedings of a national extension pest-management workshop. Purdue University, Lafayette, Indiana.

Rabb, R. L. 1972. Principles and concepts of pest management, pages 6–29 *in* Implementing practical pest management strategies. Proceedings of a national extension pest-management workshop. Purdue University, Lafayette, Indiana.

Reynolds, H. T. 1972. Practical application of suppression methods in pest management. Pages 30–6 *in* Implementing practical pest management strategies. Proceedings of a national pest-management workshop. Purdue University, Lafayette, Indiana.

Smith, E. H. 1972. *in* Implementing practical pest management strategies. Proceedings of a national extension pest-management workshop. Purdue University, Lafayette, Indiana.

Smith, R. F. 1969. The new and the old in pest control. *Proc. Accad. Nazion. Lincei, Rome* (1968) **366**(128):21–30.

Soper, F. L., and W. D. Bruce. 1943. *Anopheles gambiae* in Brazil 1939 to 1940. Rockefeller Foundation, New York. 262 pp.

Stern, V. M. 1965. Significance of the economic threshold in integrated pest control. *Proc. FAO Symp. Integrated Control* **2**:41–56.

Stern, V. M., R. F. Smith, R. van den Bosch, and K. S. Hagen. 1959. The integrated control concept. *Hilgardia* **29**(2):81.

U. S. Department of Agriculture. 1972. The pesticide review, Washington, D. C. 56 pp.

Weires, R. W., and H. C. Chiang. 1973. Integrated control prospects of major cabbage insect pests in Minnesota—Based on the faunistic, host varietal, and trophic relationships. *Univ. Minn. Agr. Exp. Sta. Tech. Bull.* 291. 42 pp.

SELECTED READINGS

Adkisson, P. L. 1972. Use of cultural practices in insect pest management. Pages 37–50 *in* Implementing practical pest management strategies, Proceedings of a national extension pest-management workshop. Purdue University, Lafayette, Ind.

Begum, A., and W. G. Eden. 1965. Influence of defoliation on yield and quality of soybeans. *J. Econ. Entomol.* **58**:591–592.

Loomis, E. C. 1972. Mosquitoes. Pages 166–174 *in* Implementing practical pest management strategies, Proceedings of a national extension pest-management workshop. Purdue University. Lafayette, Ind.

McAlister, D. F., and O. A. Krober. 1958. Response of soybeans to leaf and pod removal. *Agron. J.* **50**:674–677.

National Academy of Sciences. 1972. Pest control strategies for the future. Washington, D.C. 376 pp.

National Academy of Sciences. 1966. Scientific aspects of pest control, Pub. 1402. Washington, D.C. 470 pp.

Oatman, E. R., and G. R. Platner. 1971. Biological control of tomato fruitworm, cabbage looper, and hornworms on processing tomatoes in southern California using mass releases of *Trichogramma pretiosum*. *J. Econ. Entomol.* **64**:501–506.

Quayle, H. J. 1938. Insects of citrus and other subtropical fruits. Comstock, Ithaca, N. Y. 583 pp.

Ridgway, R. L., and S. L. Jones. 1969. Inundative releases of *Chrysopa carnea* for control of *Heliothis* on cotton. *J. Econ. Entomol.* **62**:177–180.

Science. 1971. Mirex and the fire ant: Decline in fortunes of "perfect" insecticide. *Science* **172**:357–360.

Steiner, L. F., W. C. Mitchell, E. J. Harris, T. T. Kozama, and M. S. Fujimoto. 1965. Oriental fruit fly eradication by male annihilation. *J. Econ. Entomol.* **58**:961–964.

U.S. Department of Agriculture. 1960. Insecticide recommendations. *Agr. Hndbk.* 120, Washington, D.C.

U.S. Department of Agriculture. 1971. Cooperative economic insect reports **21**: 437. Plant Pest Control Division, Washington, D.C.

U.S. Tennessee Valley Authority. 1947. Malaria control on impounded water. U.S. Government Printing Office, Washington, D.C. 422 pp.

Weber, C. R. 1955. Effect of defoliation and topping simulating hail injury to soybeans. *Agron. J.* **47**:262–266.

2

ECOLOGICAL ASPECTS OF PEST MANAGEMENT

Peter W. Price and Gilbert P. Waldbauer

The purpose of this chapter is to form a conceptual basis with which to examine agricultural systems, and to introduce the reader to some ecological concepts that seem to be particularly useful in suggesting ways in which we might analyze these systems with a view to developing efficient pest-management procedures. Further insight into ecological aspects of pest management may be obtained from Huffaker and Messenger (1964a,b), Doutt and DeBach (1964), Southwood and Way (1970), Huffaker et al. (1971), and Varley and Gradwell (1971). For aspects of general ecology, E. P. Odum (1971), Krebs (1972), and Collier et al. (1973) may be consulted. Insect population ecology has been treated by Clark et al. (1967).

I. THE ECOSYSTEM CONCEPT

Individual organisms of the same species live together as a population, populations of different species live together and form a community, and the community is influenced by its physical environment. We call such a complex system of biotic and abiotic factors an ecosystem—for example, an experimental ecosystem in a laboratory, an urban area, an agricultural system, or a natural ecosystem such as a woodland, a watershed, or a hot spring. An expanse of any convenient size may be chosen for study.

The value of the ecosystem concept is that it emphasizes the interaction of all factors in a given area, and that it forces us to look further than studies of isolated pockets or aspects of biological activity (see also Van Dyne 1969). The danger is that we become blinded by the complexity of the interactions, and that we are led to superficiality in our studies. However, it must be emphasized that the individual, the population, the community, and the physical factors are the building blocks of the ecosystem, and that understanding each level of organization is essential to understanding the whole system (see Price 1971).

The fundamental components of an ecosystem are the individual organisms. We must remember that natural selection works principally on reproducing individuals; through this selection populations become adapted, and the results of adaptation lead to evolutionary change. Thus the study of individuals is of prime importance—their biology, behavior, physiology and morphology, and response to other members of the same species, to other organisms, and to abiotic factors in the environment. The study of individuals offers a potent method for the analysis of population change and, until we understand the basic needs and responses of individuals, we are ignoring the major forces of selection that influence the whole population and community.

Just as the individual is a natural and basic unit for study, so is the population. While the recognition of a population is no doubt more difficult, and its limits harder to discern, such recognition is none the less critical to any study of population dynamics. Study methods should be designed to permit the detection of populations—that is, groups of interbreeding individuals—the extent of the populations as well as their degree of isolation, sources from which emigration can proceed, and areas for colonization in the event that dispersal should occur. Even so, each individual organism must be considered unique, carrying its own genetic contribution to the gene pool of the population. Thus, when we consider populations, the genetic contributions of individuals and the consequent individual qualities (behavioral, physiological) that contribute to variation within the population are no less important than numbers of individuals. It is on this variation that natural selection acts, and without it evolution stagnates. Since populations must constantly adapt to changing conditions, either biotic or abiotic, and since population size depends to a larget extent on the success of this adaptation, we must place equal emphasis on the study of qualitative and quantitative change in populations.

Populations of different species coexist and interact with each other, thus demanding study at the community level. As we see later, any given habitat seems to have a finite saturation level for species, and this limitation forces an organization within the community which could not be recognized other than by the study of the whole system. Here intra- and interspecific competition play an important role, and comparative studies between species that are closely related either taxonomically or ecologically contribute greatly to data interpretation. Environmental effects on different organisms can be compared, and then interaction through competition for a limiting resource observed. Changes in species numbers, relative abundance, and diversity in space and time can be compared. Distribution of individual species may reveal differences in niche occupation or niche breadth (see definition later) and the extent to which species overlap and interact. These considerations bear directly on the understanding of population fluctuations.

Naturally, each individual, as well as all interactions between individuals and populations at the community level, are influenced by physical factors. Temperature and humidity, soil or water conditions, topography, drainage, aspect, and shading may all influence community members, and certain members may modify these factors to the benefit or detriment of other members. Since physical factors act directly on the physiology of the organism, much information can be obtained from physiology texts (e.g., Chapman 1969; Prosser and Brown 1961). Therefore we consider mainly biotic factors which we feel have received too little attention in relation to pest management.

These considerations bring us, by logical steps, to the only conclusion possible for understanding insect populations, the need for studying them at the ecosystem level, not forgetting that all components of the ecosystem provide insight into the basic mechanisms involved in insects and agricultural ecosystems. Since insects are such mobile creatures, and man has increased their mobility and that of their food plants, interactions are extensive; thus we must define our ecosystem for study as a similarly extensive area. Therefore the minimum size we should consider probably involves several crop fields plus the neighboring uncultivated areas. But we will see that interaction sometimes comes from far beyond this local area in the form of migrating insects on the same land mass, and even invaders from other continents.

Thinking in terms of the ecosystem concept and its component parts dictates certain approaches to the sampling and general study of insects. If variation within a population is to be identified, many sample stations must be employed. If movement of organisms must be studied, extensive sampling is necessary. In view of these needs, the use of single-plot studies is not satisfactory where mean abundances are derived from random samples. Such randomization suppresses information on population variation and is an inefficient way of testing hypotheses (see also Price 1971). Samples sited on transects through areas occupied by populations or across extensive tracts of land covering at least a field and its adjoining areas are more useful. Stratified random sampling may be used with strata carefully selected in order to test a hypothesis. With the adoption of extensive sampling methods, we are more likely to reach an understanding of pest insects at the ecosystem level.

II. THE ECOLOGICAL NICHE CONCEPT

Each individual—the smallest natural unit in the ecosystem—requires a variety of resources in certain quantities, neither too much nor too little, in order to survive and reproduce. By looking at many individuals of a species in a community, we can determine what these resources are for the species. Major resources are food, breeding site, space, time, temperature, humidity, and a habitat that provides protection from predators, but species of course differ in their quantitative and qualitative utilization of these basic resources. The ecological niche of a species can be defined as the set of resources that provides a species with all its requirements for existence and reproduction. Species that can reproduce over a broad range of conditions are said to have a broad niche, and those that tend to be specific in any requirement are said to have narrower niches.

Once we realize that each species has a definite set of resources essential to its perpetuation, it becomes clear that there must be a limit to the number of species that can occupy any given area. In other words, there must be a saturation level for numbers of species, set by the abiotic and biotic components of the area. It appears that the more heterogeneous the physical environment, particularly in terms of soil quality and topography, the more plant species occupy an area.

Of course, when we deal with insects—which tend to be host-specific—the more plant species there are in the community the more kinds of insect herbivores there are likely to be, and these in turn support a greater variety of parasitoids and predators. Thus the complexity of the community increases rapidly as more plant species coexist.

But the variety of plant species is not the only component of the community that sets limits on its carrying capacity for insect species. As

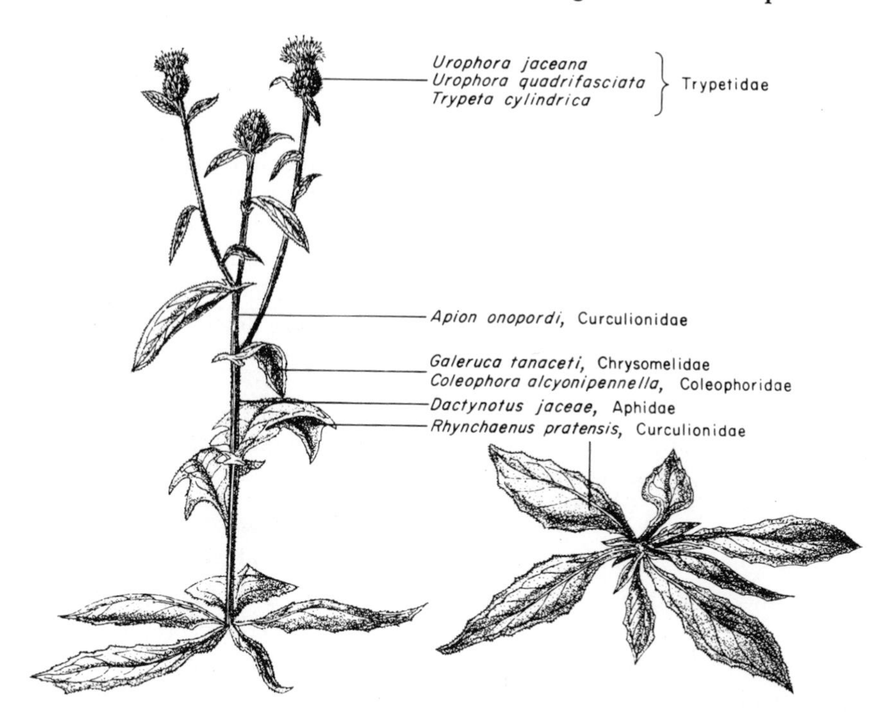

Fig. 2-1 Lesser knapweed, *Centaurea nigra* L., before and during flowering, showing the change in structural diversity due to flowering, and the increase in insect species that this permits. The insect herbivores, the family to which they belong, and the positions in which they feed are indicated. Grazing maintains the plant in the nonflowering condition. (After Morris 1971a. Courtesy of E. Duffey and A. S. Watt (eds.) The scientific management of animal and plant communities for conservation. Blackwell, Oxford.)

plants pass through their various growth stages, they may produce completely new resources which can be exploited by insects. For example, in the rosette stage the lesser knapweed, *Centaurea nigra* L., is exploited by only one herbivorous insect (Morris 1971a) but, when the plants send up stems and flowers, another seven species are able to exploit them (Fig. 2.1). Here complexity of the community is increased not by adding more species but by increasing *structural complexity*. Thus we should expect a plant community's carrying capacity for insect species to be determined by two components–the number of plant species and the structural complexity the plants provide. As these two components increase in a plant community, so the number of ecological niches for insects increases.

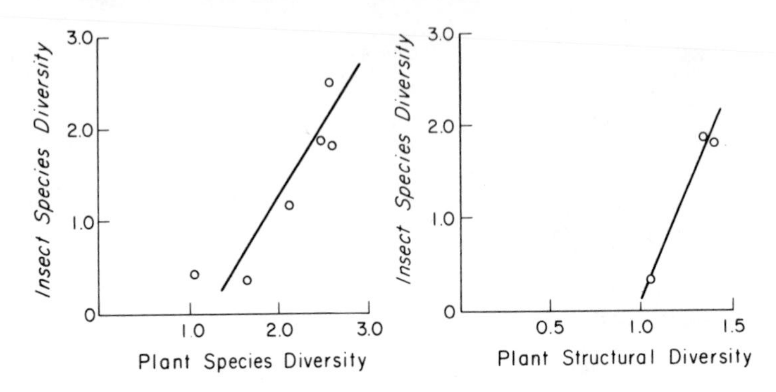

Fig. 2-2 The relationships between plant species diversity and insect species diversity (left), and plant structural diversity and insect species diversity (right). Plantsucking bugs (Homoptera) were studied in old-field communities by Murdoch et al. (1972) from whom the data are taken.

The only field data available so far to support these conclusions for insects were collected in old fields in Michigan where the relationship between the community structure of plants and the community structure of plant bugs (Homoptera) was examined. The species diversity of the plants (see Appendix for a discussion of diversity) and their structural diversity were equally well correlated with insect species diversity; together they accounted for 79% of the variance in insect species diversity (Fig. 2.2) (Murdoch et al. 1972). Thus we can conclude that the plant community has a profound influence on the insect community that exploits it.

Perhaps a more dramatic method of demonstrating that each plant community has a natural limit for species is to remove all insects from an area and to observe how many species eventually recolonize the area. This has been done with small, red mangrove, *Rhizophora mangle* L., islands of similar size in Florida Bay, (Wilson and Simberloff 1969; Simberloff and Wilson 1969). After a detailed census had revealed the number of arthropod species originally present, each island was covered with a tent and fumigated with methyl bromide. A subsequent census revealed that the islands had been almost completely defaunated; only a few individuals of two species of wood-boring beetles survived, but the mangrove trees had received only minor damage. Frequent censuses monitored the recolonization of these islands. Immigration was rapid (Fig. 2.3), and it was soon evident that the rate at which additional species became established declined rapidly and reached

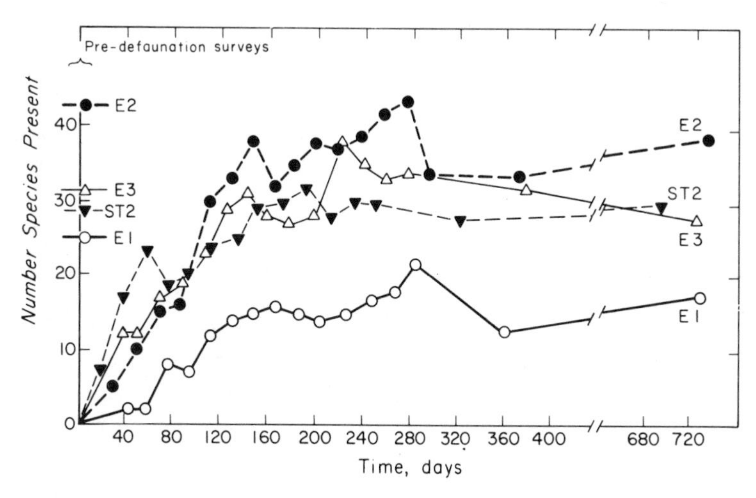

Fig. 2-3 The progress of colonization by arthropods on four small mangrove islands with time, after defaunation. Species numbers before defaunation are indicated on the ordinate by the position of the island's symbol. Note that, apart from the most distant island from the mainland (*E*1), the species numbers were similar before defaunation and after 720 days. *E*2 was the nearest island and supported the most species, while *E*3 and *ST*2 were intermediate in distance and species-supported. (After Simberloff and Wilson 1970. Copyright 1970 by the Ecological Society of America. Reprinted by permission of Duke University Press.)

almost zero when the number of species established was about equal to the number of species present before defaunation (cf. islands *E*2, *E*3, *ST*2) (Simberloff and Wilson 1970). Only on the island most distant from the mainland—the major source of colonists—had the predefaunation numbers not been reached after 720 days.

We look at these colonization curves in more detail later, but the important thing to notice now is that each island had its own limit for species, and that this limit did not change even after nearly complete defaunation. Since these islands had only one species of plant, the plant communities of each island must have been essentially the same, both in species and structure, and therefore in saturation level also. Therefore, we must look for another explanation for the differences in species numbers among the islands seen in Fig. 2.3. Actually, island *E*2 was only 2 m from the Florida mainland, island *E*1 was 500 m from the mainland, and *E*3 and *S*T2 were at intermediate distances. We see that *E*2 supported the most species, and *E*1 the fewest. Clearly, distance from the source of colonists is a third important factor in determining the number of species that occur in a plant community. The theoretical explanation for this is given in Section III.

Before leaving the niche concept, we must examine another organizing influence within the community—energy flow. In green plants the sun provides energy which drives the photosynthetic process. Herbivores derive their energy by feeding on plants, and carnivores feed on herbivores or other carnivores. Thus the community can be divided into levels composed of organisms of a certain feeding category, called trophic levels. The primary producers are the plants which actually produce food by photosynthesis; next come the herbivores; then the primary carnivores, secondary carnivores, and so on. Energy passes from one trophic level to another along a series of feeding links—the food chain. One example follows.

Food Chain	Sun →	Corn →	Corn borer →	Nabid bug →	Wren →	Hawk
Trophic level		Primary producer	Herbivore	Primary carnivore	Secondary carnivore	Tertiary carnivore

The biomass of the hawk population is much smaller than the biomass of the corn population and therefore contains much less energy. As energy passes from trophic level to trophic level, there is an inevitable loss with each transfer; the organisms in the food chain use some of the energy to support their own activity and physiological functions, and much energy remains unutilized and is lost to the food chain by decay. Thus energy flows through the community, being rapidly depleted until there is not enough to support an additional feeding link.

This food chain is, however, an oversimplified example, for there are many organisms that feed on corn, and many predators that in turn feed on them. The interactions are complex, and it is more realistic to consider the passage of energy through a community as traveling along a food web rather than along a unidirectional food chain. We return to these concepts later.

III. COLONIZATION OF ISLANDS

The application of the studies on island defaunation to agricultural ecosystems is obvious. We can regard annual crops, for example, as islands which may have a life-span of as little as one growing season and which are colonized by arthropods from the surrounding area. We must regard crops as islands both on a short time scale and a long, evolutionary, time scale (Janzen 1968). Each growing season an annual crop becomes available for colonization, and we are likely to see colonization curves similar to those on mangrove islands. But in addition, in evolutionary time new plant species evolve and become available to insects which can adapt to exploiting them. As evolutionary time progresses, more and more species adapt, and we are again likely to see a rate—in this case of adaptation to the new plant species—that is rapid at first and falls off as the new ecological niches provided by the plant are filled. As man changes the genetic makeup of his crop plants, the spectrum of insects adapted to exploit these plants also changes. In this way we can exercise some control over the colonization rates and colonizing species that affect the productivity of our crops. Therefore it is worth studying some theoretical considerations of colonization of islands, so that we may understand the colonization process better. Thus we turn to the theory of island biogeography (MacArthur and Wilson 1967; see also MacArthur 1972).

As we have seen in the defaunation studies, each island or each community has its own limit for species of arthropods and, once this has been reached, an equilibrium exists. This equilibrium is not static, but dynamic in the sense that new species may arrive and resident species may become extinct—however, the number of species remains at equilibrium. Thus the equilibrium number clearly depends on the immigration rate of arthropods into the community and the rate of loss from the community by emigration and extinction. In this context we consider immigrants to be those species represented by individuals capable of reproducing, either because an inseminated female arrives, or because both sexes colonize, that is, they are capable of founding a colony.

Naturally, the rate at which new species become established is rapid at first when many niches are vacant, but declines as the niches are filled (Fig. 2.4). In the early stages of colonization, most of the species that arrive are new to the island. At this stage the rate of colonization also tends to be rapid, because the species capable of rapid dispersal are arriving, while the slow dispersers arrive later. These factors combine to produce a concave immigration curve. The extinction curve is also concave, because the more species there are present, the more chance there is of species becoming extinct—simply because there are more species and also because there is more interactive pressure (principally competition and predation). Thus we see that the equilibrium number of species $\hat{S}$ is defined by the point at which immigration and extinction rates balance each other (Fig. 2.4). For the purposes of this discussion, extinction refers to both mortality and emigration.

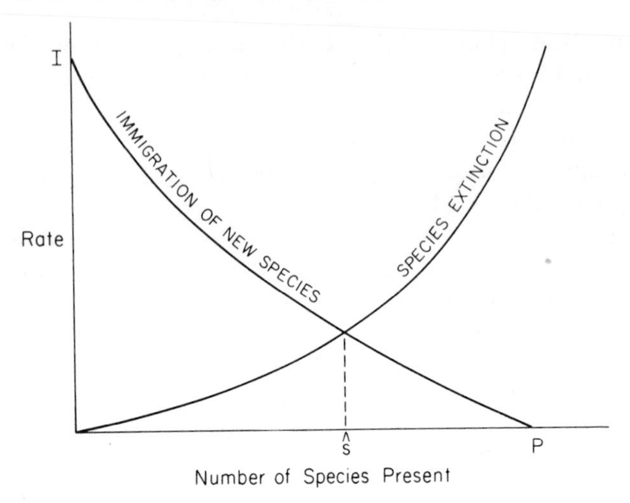

Fig. 2-4 The equilibrium model of species on an island. The equilibrium number of species $\hat{s}$ is at the intersection of the curves of the rates of immigration and the rates of extinction. I is the initial rate of immigration and P is the total number in the species pool on the mainland (Robert H. MacArthur and Edward O. Wilson. 1967. *The Theory of Island Biogeography. Copyright* © 1967 by Princeton University Press. Reprinted by permission of Princeton University Press.)

The limits of the immigration curve are set by the total number of species on the mainland that are available as colonists, that is, the species pool P, the initial rate of immigration I which depends on P, and the distance of the island from the mainland. Immigration to a near island is rapid, because the chances of an individual leaving the mainland and reaching the island are high; immigration to a distant island is slow, as chances of arrival are greatly reduced. Thus we can develop a set of immigration curves which differ according to the distance of the island from the mainland (Fig. 2.5).

Extinction rates on islands are also likely to differ, not with the distance of the island from the mainland, but with the size of the island. Naturally, as island size increases, the resource base for species increases, as both species diversity and structural diversity are likely to increase. Thus extinction rates are likely to be lower on larger islands, and can be expressed by a series of curves such as those in Fig. 2.5.

Now we can appreciate the full impact of the plant community on the arthropods that are able to exploit it. As the area of the community increases, so its species and physical structure become more complex and,

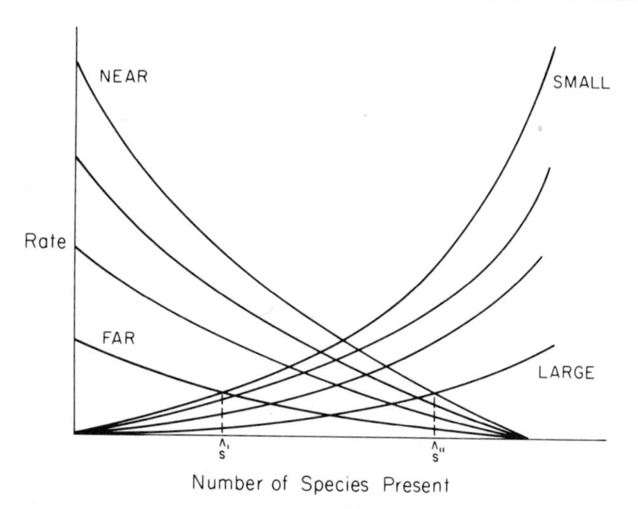

Fig. 2-5 A comparison between equilibrium models for near and distant islands and small and large islands using the basic model shown in Fig. 2-4. Note that the equilibrium number of species on near, large islands $\hat{S}''$ is much greater than that for distant small islands $\hat{S}'$. (From Robert H. MacArthur and Edward O. Wilson. 1967. *The Theory of Island Biogeography*. Copyright © 1967 by Princeton University Press. Reprinted by permission of Princeton University Press.)

as it is located closer and closer to the source of immigrants, so the equilibrium number of species increases (cf. $\hat{S}''$, Fig. 2.5). The opposite reasoning produces an equilibrium number of species for distant small islands ($\hat{S}'$) far below $\hat{S}''$.

Now that we have looked at some characteristics of the island habitats, we must look at the qualities of the colonizing species. Of course, efficient colonizers must have two vital characteristics: large numbers of progeny because colonization is hazardous, and a high dispersability of progeny so that vacant sites for colonization can be reached. We are already familiar with good colonizers among plants, since they are the weeds that grow on disturbed land, that is, vacant sites—be they produced by gardening, farming, or construction work. One need only think of the common dandelion, *Taraxacum officinale* W., and its large production of small seeds, each seed with a pappus or parachute which provides an efficient means of dispersal. The dandelion is always an early colonizer, because the chances of its small airborne seeds reaching a vacant site are very high. In contrast, we can think of oak trees which produce large seeds too heavy to be dispersed by wind, seeds that drop to the ground and may be dispersed over short distances by small mammals. Thus it may be several decades before an acorn reaches a vacant site. We see that the strategy of the oak is not to avoid competition by reaching a vacant spot before any other plant, but to arrive much later and to succeed in an already established community by virtue of its competitive ability; hence the large energy store that the parent plant invests in each acorn.

Thus we see two extremes on a continuum of reproduction strategy in plants: the production of many small seeds of high dispersability and low competitive ability, or the production of relatively small numbers of large seeds with low dispersability and high competitive ability. In the parlance of evolutionary ecologists, these are spoken of as *r* strategies and *K* strategies, respectively; it is believed that they evolved in response to *r* and *K* selection (see MacArthur and Wilson 1967; Pianka 1970, 1972). It is clear that each type is adapted for exploiting very different ecological conditions, as we shall see when we consider community succession.

Just as plants can be divided into good colonizers (*r* strategists) and good competitors (*K* strategists), so can insects. As we should expect, many pests of crops are *r* strategists. They are small and can be dispersed great distances by wind, and are often present in crops very soon after germination of the host plant. Aphids, thrips, cicadellids, and flies are a few examples of some of the most effective early colonizers, and numbered among them are some of our worst pests.

In contrast to the herbivores, their predators and parasitoids are slower in colonizing vacant sites. There is no selective pressure for them to colonize early, as they would encounter a shortage of food. This difference in colonizing ability, coupled with differences in reproduction rates, frequently leads to pest outbreaks early in the season or after the application of insecticides.

We should also remember that early colonists must be adapted to harsh, unsheltered sites where physical conditions may be extreme. Thus the mortality of colonizers usually results from factors associated with severe weather conditions. Such factors are termed density-independent, because their operation does not depend on the population density of the organism involved. The best way to adapt to such factors is to produce many offspring, that is, to be an *r* strategist. In contrast, after establishment of a community, the microclimate is ameliorated, and the severest threat to survival becomes predation and competition from other species, often density-dependent factors which increase in severity with increasing population density of the affected organism. The best way to adapt to this situation is to invest a lot of energy in improving the competitive ability of progeny, and in defending them against predation, that is, to be a K strategist.

We can see the same trend if we look at closely related insects in a relatively unpredictable, harsh temperate climate and in a predictable, equable, tropical climate. For example, the large milkweed bug, *Oncopeltus fasciatus* (Dallas), occurs in temperate North America, and its close relative, *Oncopeltus unifasciatellus* Slater, is found in tropical South America. When egg production is compared, we see that *O. fasciatus* females start to oviposit much sooner after reaching maturity than do their relatives; egg production per day is much higher in their early life, and they produce more eggs, 556 versus 324 eggs per female (Fig. 2.6) (Landahl and Root 1969). The result is that the population growth rate of the temperate species is much more rapid than that of the tropical species, supporting the assertion that species in harsh environments are likely to be relatively more r-selected than those in equable environments. Even within one species individuals in northern populations may lay more eggs than those in southern populations (e.g., the Klamath beetle, *Chrysolina quadrigemina* (Suffr.), Peschken 1972).

During the colonization process, once the species equilibrium has been reached, we are likely to see a rapid shift from the dominance of r strategists in the community to the dominance of K strategists. Thus two

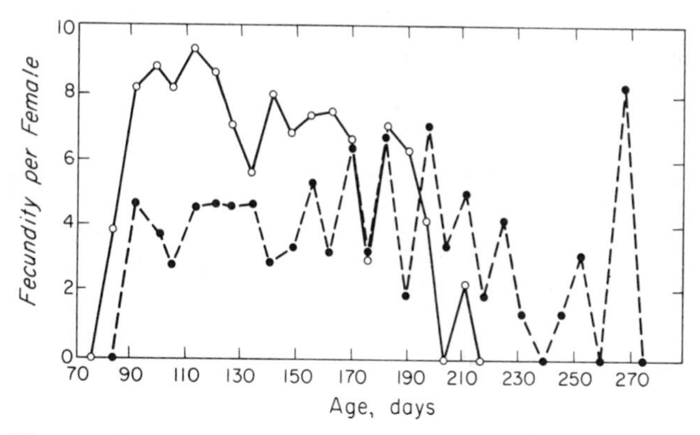

Fig. 2-6 The number of eggs laid per week by milkweed bugs, the temperate *Oncopeltus fasciatus* (Dallas) (solid line) and the tropical *Oncopeltus unifasciatellus* (dashed line). Note the earlier start and higher rate of egg laying in the temperate species. (After Landahl and Root 1969. Copyright 1969 by the Ecological Society of America. Reprinted with permission of Duke University Press.)

phases in colonization can be recognized (Wilson 1969). The *noninteractive phase* occurs during colonization before equilibrium is established, and may often lead to overshooting the equilibrium number. This is seen in the defaunation studies on the nearer islands, *E*2, *E*3, and *ST*2 (Fig. 2.3). Later, as competition increases because of increasing population densities, and as more species of predators and parasitoids arrive, biotic interactions in the community assume a more important role. This *interactive phase* brings the number of species to equilibrium. When wheat was first grown on the Siberian steppes, recolonization of the cultivated land by predators played an important role in the establishment of stability in pest populations (Grigoryeva 1970).

It has been postulated that, given longer periods of time, there is an *assortative phase* in the community, essentially a reshuffling of *K* strategists, resulting in the selection of those species that can coexist most efficiently. This may allow the species equilibrium number to increase slowly (Wilson 1969).

Finally, given even longer time spans, we can conceive of an *evolutionary phase* in community development in which species are not only sorted but undergo genetic change as a result of selection during coexistence, so that species become mutually adapted for living together (Wilson 1969). We have already mentioned earlier in this section the increase in number of herbivores per plant species with evolutionary time. Evidence for this phenomenon is seen clearly in the comparison of insect faunas on different species of trees in Great Britain (Southwood 1961). The longer the tree has been resident on that island, and the more abundant—estimated by an analysis of all Quaternary records of plant remains—the more insect species it has associated with it (Fig. 2.7).

Thus a crop grown for the first time in an area away from closely related species may be occupied by a relatively depauperate herbivore fauna. Soybeans in Illinois are an example. With time, however, we may expect more species of herbivores to adapt to soybeans, thereby increasing the potential for pest problems. However, when a crop is grown close to wild relatives, there is the probability that many herbivores will transfer, and that pest problems will immediately be critical. In other words, in the latter situation there is a nearby source of many herbivores preadapted for exploiting the crop. The most dramatic example is the transfer of the Colorado potato beetle, *Leptinotarsa decemlineata* (Say), from its native hosts, wild species of *Solanum*, to the potato, *Solanum tuberosum* (L.), when this crop was first cultivated in the foothills of the Rocky Mountains.

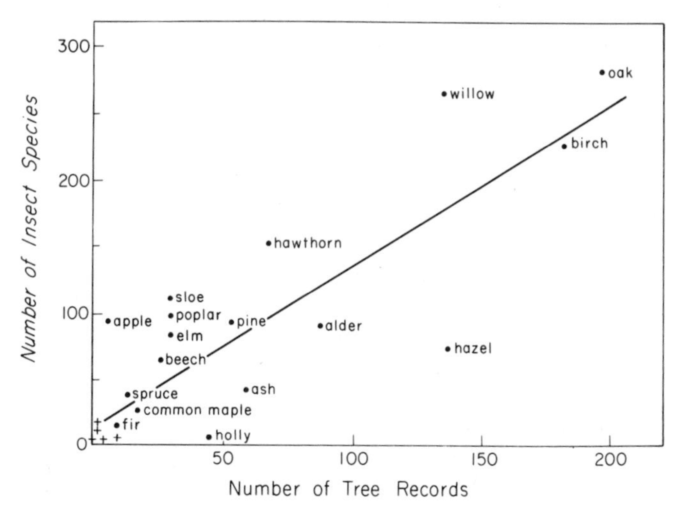

Fig. 2-7 Relationship between the number of records of trees, based on their relative abundance and geological time over which they were resident in Great Britain, and the number of insect species that have adapted to exploiting these tree species. The trees that have been establishd longest; for example, oak, birch, and willow, have the largest insect faunas. Note that naturally colonized trees (•) support many more species than recently introduced trees (+). (After Southwood 1961. *J. Anim. Ecol.* **30.** Blackwood Scientific Publications Ltd.)

Of course, the plant species will respond in evolutionary time to colonization by insects. Through natural selection induced by heavy herbivore pressures, a plant population may show an increase in the production of toxic substances (secondary metabolic products). This may limit to a few species those insects that can exploit the plant, but with time more herbivorous species will adapt as they evolve mechanisms that enable them to tolerate or detoxify the chemical defenses of the plant. Thus in plant–herbivore relationships we are dealing with a coevolutionary process (see Ehrlich and Raven 1964, 1967). The plant evolves in response to the herbivore, and the herbivore evolves in response to the plant. In this way many herbivorous insects have become highly specific to groups of plants that contain similar defensive chemicals and that are toxic to the majority of other organisms. For example, the flea beetles *Phyllotreta cruciferae* (Goeze) and *Phyllotreta striolata* (Fabricius) are restricted in their food selection to plants in the families Caparidaceae, Cruciferae, Tropaeolaceae, and Limnanthaceae (Feeny et al. 1970). The common quality of these families is that all species contain mustard oils, potent antibiotics which are toxic to most animals but not to the flea beetles. Thorsteinson (1953) demonstrated a feeding response to mustard oil glucosides by the imported cabbage worm, *Pieris rapae* (L.) and the diamondback moth, *Plutella xylostella* (L.). Again we see that, if we plant crops closely related to indigenous plants, we are likely to witness the rapid adoption of the crop by the coevolved fauna because of a lack of chemical barriers and the presence of the correct chemical cues to initiate feeding.

In summarizing this section, we can say that the number of arthropod species in a community is closely correlated with five major factors:

1. Plant species diversity
2. Plant structural diversity
3. Distance of community from source of colonists
4. Length of contemporary time available for colonization
5. Evolutionary time available for coevolution between herbivores and their host plants.

IV. CROP ISLANDS IN AGRICULTURAL ECOSYSTEMS

During the early settlement of North America by European colonists, agricultural crops were grown in small patches surrounded by large tracts of uncultivated land. Crops were like small islands in a matrix of natural vegetation. As agriculture developed, this matrix dwindled until, at first sight, it may now seem to be an insignificant component of agricultural ecosystems. It is obvious that this is a misconception if we remember two things about insects, their ability to fly long distances, and their need for protected sites in which to pass inclement seasons.

Protected sites are found in uncultivated areas, and these may provide cover and insulation from the cold in winter, moisture and milder temperatures during very hot dry periods of the summer, and food when crops are not available. Adult insects must frequently live through such periods, and the majority do so in uncultivated areas around crops, necessitating emigration from the crop and recolonization. These areas therefore constitute a vital part of the ecological niche of many pest and beneficial insects.

Overwintering sites include under the bark of trees, under fallen leaves, logs, and other litter, under stones, and at the bases of various plants. For example, boll weevils, *Anthonomus grandis* Boheman, and bean leaf beetles *Cerotoma trifurcata* (Forster), winter under ground litter in the woods; chinch bugs, *Blissus leucopterus* (Say), in bunch grasses; and squash bugs, *Anasa tristis* (De Geer), under rocks and other large objects. During the hot periods of summer, many species leave lowland crops to aestivate, and later to hibernate, in nearby mountain areas, returning in spring to the lowland breeding areas. Many cutworms, other noctuid moths, and coccinellids make such migrations (Johnson 1969). One well-known example is the massive aggregations of the ladybird beetle *Hippodamia convergens* G.-M. in mountainous areas of the western United States. The beet leafhopper, *Circulifer tenellus* (Baker), overwinters on wild annual plants; as these die in April and May, the insects migrate to sugar beets in the San Joaquin Valley, California. Hedges and windbreaks harbor a much more diverse fauna than neighboring crops, even during summer weather, because of the shelter and food provided by the wild plants (Lewis 1965, 1969a,b).

At this point we must note that by no means do all insects associated with crops retreat to uncultivated areas during unfavorable seasons. For example, the European corn borer, *Ostrinia nubilalis* (Hübner), winters in corn stalks; the Hessian fly, *Mayetiola destructor* (Say), behind the leaf sheaths of wheat; the northern and western corn rootworms, *Diabrotica longicornis* (Say) and *D. virgifera* LeC., as eggs in the soil in the crop field; and the codling moth, *Laspeyresia pomonella* (L.), under the bark of apple trees.

Of course, uncultivated areas also offer resources other than just protection. For example, parasitoids have been found to be much more effective in areas where there are abundant wildflowers which provide nectar and pollen (Leius 1967); northern corn rootworms, although highly host-specific as larvae, as adults feed on the flowers of a wide array of plants, including many Compositae.

In many cases it is clear that the theory of island biogeography is applicable to the recolonization of crops after hibernation or aestivation by insect pests. For example, the incidence of sugar beet mild yellows virus declined rapidly with increasing distance from storage pits for beets in which the aphid vectors *Myzus persicae* (Sulzer) and *Rhopalosiphoninus staphyleae* (Koch) overwintered and contracted the virus (Fig. 2.8) (Heathcote and Cockbain 1966). Other examples are given by Wolfenbarger (1940, 1946, 1959). This indicates the important role that distance from source plays in colonization.

Clearly, we can profitably consider a crop field an island in a matrix composed of a mosaic of land types—uncultivated land, fallow fields, and land planted in other crops. It is time we fully realized the importance of this matrix—particularly uncultivated land or semipermanent crops

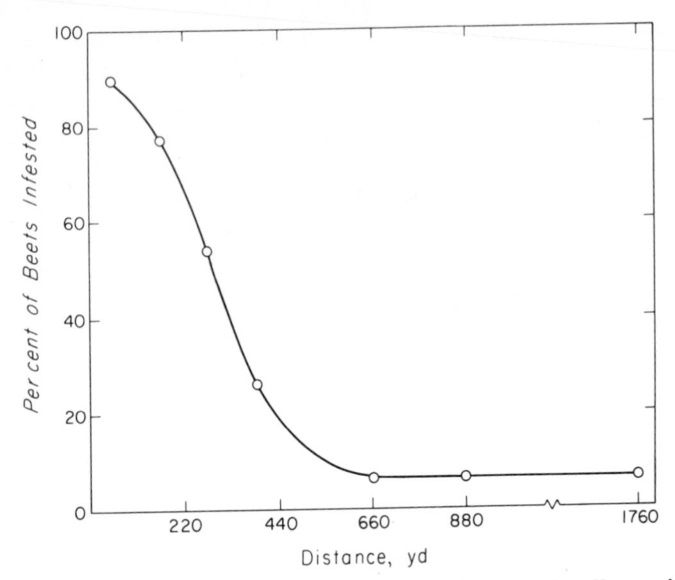

Fig. 2-8 The percent of beets infested by sugar beet mild yellows virus with distance from storage sites for beets in uncultivated areas, indicating the distance traveled by the disease vectors *Myzus persicae* (Sulzer) and *Rhopalosiphoninus staphyleae* (Koch). (Data from Heathcote and Cockbain 1966.)

like alfalfa—in the dynamics of both the destructive and the beneficial insects associated with crops (see Lewis 1965; Van Emden 1965). Its impact on the development of the whole insect community cannot be overrated, and we should keep in mind the possibilities available to us for manipulation of this matrix for the improvement of insect pest-management practices. One enlightened use of wild plants has been advised by Doutt and Nakata (1973), who found that wild blackberries, *Rubus* spp., support the rubus leafhopper and its egg parasitoid. When grape leafhoppers become abundant, the egg parasitoid invades vineyards and plays a major role in limiting pest numbers.

Next we must consider the nature of the crop islands in the matrix. In view of the determinants of the number of species in a community listed in Section III, it is clear that there are major differences among crops of short duration (e.g., 1-year fields of corn, wheat, soybeans, or cotton), moderate duration (e.g., alfalfa or pasture), and long duration (e.g., orchards or forests). The most influential differences among these systems result largely from differences in their longevity, and these are: (1) plant species diversity, (2) plant structural diversity, and (3) length of time available for colonization.

Plant species diversity is likely to increase as crop duration increases. Only annual crops receive such intensive care through their duration that weedy species are kept at a minimum—thus keeping the plant community in a very simple state. In crops that last for 2 or 3 years, little can be done to exclude the invasion of other species with time, thus the plant community is likely to become progressively more diverse. With orchards and forests often no attempt is made to exclude all other plant species, so initial diversity is high and may continue to increase with time.

There is likely to be a concomitant increase in structural diversity with age, because each new plant species added to the community is likely to differ structurally from those already present. This increase will be particularly striking if trees are added. The result is of course that the resources of the plant community can be divided so as to yield a greater number of niches for insects.

Finally, the length of time available for colonization is very different among the three crop types. We saw in the defaunation studies that the arthropod community on an island only 2 m from the mainland took almost 300 days to reach dynamic equilibrium. This is a period of time far longer than the growing season for many crops, so we must wonder whether equilibrium is ever reached in annual crops. The failure of the community to develop an equilibrium certainly reduces the predictability of the community in terms of number of species present and stability of populations, since in the colonization phase competition and predation have little regulating influence. In longer-lived crop communities such as alfalfa and pasture, equilibrium is probably reached, since many species are capable of overwintering *in situ,* so that each spring the insect community becomes reestablished more rapidly than in the first year. In orchard and forest communities, most arthropods are likely to be resident. Thus equilibrium should be established, and some predictability should be evident both in the number of species present, dictated by the equilibrium, and the kinds of species as determined during the interactive phase.

Colonizing arthropods may be conveniently classified according to their origin immediately before reaching the crop. They may be present in the crop field at the time of planting, may invade from nearby communities in the matrix, or may migrate from distant areas. We refer principally to annual crops, but the parallels in perennial crops of either short or long duration can readily be seen.

Resident species, in other words, species present in the field when the crop is planted, may have a headstart on those that must invade the crop, and may preempt resources so that another species cannot become established. Colonization of the crop is almost immediate, and the population density in any part of the field is probably independent of its distance from the crop perimeter. However, population density is profoundly influenced by the success of the insect in a particular field the previous year. Examples include species that originally become established when the land is in another crop, such as white grubs and wireworms which may attack corn planted on land previously in sod. There are also species that are particularly successful when land is continuously planted to the same crop, such as northern and western corn rootworms. In rare instances insects may be introduced by man when the crop is planted. Gladiolus thrips, *Taeniothrips simplex* (Morison), for example, may winter on corms stored indoors and become active after the corms are planted.

In contrast, insects that must invade from the surrounding matrix probably colonize gradually, and there is likely to be a gradient in population density from perimeter to center, particularly leeward of overwintering or aestivation sites, and when small insects which are likely to be windblown are involved—such as aphids, cicadellids, or flea beetles. Population densities are much less likely to depend on breeding success at the same site the previous year. Insects may move to the crop from wintering quarters, as do the Mexican bean beetle, *Epilachna varivestis* Mulsant; the bean leaf beetle; or the European corn borer. These movements may be synchronous, but are usually relatively gradual and may extend over a fairly long period of time. Movement may be from other crops, for example, chinch bugs from small grains to corn, or grasshoppers and armyworms, *Pseudaletia unipuncta* (Haworth), to corn from small grains or weedy patches. These are likely to be synchronized mass movements which may have a dramatic impact on the colonized crop. They may be triggered by a natural event, as when chinch bugs respond to the ripening of small grains, or by cropping procedures, as when potato leafhoppers, *Empoasca fabae* (Harris), are driven to nearby soybean fields when alfalfa is cut. Such movements are most likely to have an important impact when related crops or complementary crops are grown in close proximity.

The above discussion emphasizes the point that planning of the spatial relationships of crops and uncultivated land should be done at the ecosystem level, in anticipation of insect movements over a period of years in relation to their densities in crops, prevailing winds, and cultural practices. In this way large colonizing populations may be avoided, or at least predicted, as explained in Fig. 2.9.

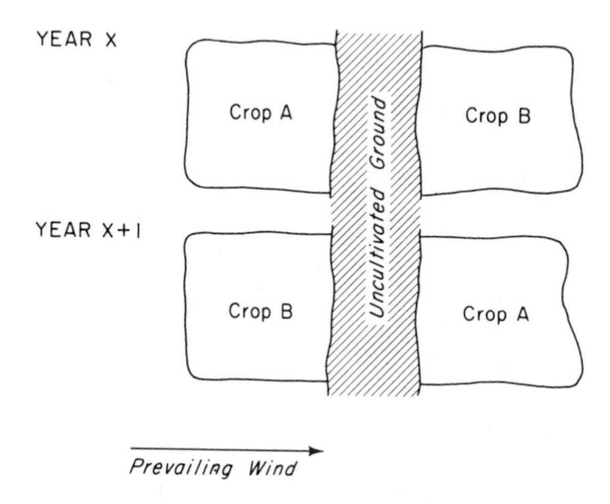

Fig. 2-9 In year x uncultivated ground provides overwintering sites for adults of pest species on crop A. Thus in year $x + 1$, with crops arranged as shown, conditions are ideal for the pest to colonize crop A from its overwintering quarter if the prevailing wind is as indicated. This provides a very simple example of the type of planning that should be conducted on a very extensive basis.

Next we must consider insects that colonize from a considerable distance and may be loosely termed migrants. In the United States many pest species overwinter in the Gulf states and migrate with the prevailing winds into northern states. Some of these migrants are small and can be dispersed rapidly over long distances by wind, for example, the leafhoppers *Empoasca fabae* and *Macrosteles fascifrons* (Stål), grain aphids (Johnson 1969), and some species of thrips. Other species are larger and migrate more slowly, for example, the large milkweed bug, *Oncopeltus fasciatus* (Dallas) (Dingle 1972). The adaptive features of this migration make some species exceedingly important pests. Overwintering may be very successful in a mild climate, and reproduction may continue during the winter. Mass migration may occur under favorable conditions, arrival times and places may be unpredictable, and relatively high densities may be established suddenly. All these characteristics mitigate against the early establishment of predator and parasitoid populations large enough to contribute effectively to population regulation. In fact, a sweeping generalization which may nevertheless be worth considering is that, the greater the distance from which an insect colonizes, the more likely it is to become a serious pest if it can become established at all.

From time to time species invade areas in which they did not previously occur, usually because they are accidentally transported by man. A quick look at a few such invaders should convince us of their potency as pests: alfalfa weevil, *Hypera postica* (Gyllenhal); Japanese beetle, *Popillia japonica* Newman; European spruce sawfly, *Diprion hercyniae* (Hartig); European corn borer, *Ostrinia nubilalis* (Hübner); cottony cushion scale, *Icerya purchasi* Maskell; and gypsy moth, *Porthetria dispar* (L.). Of course, it is most unlikely that the appropriate parasites, parasitoids, and predators will be accidentally introduced at the same time, or that endemic species are well adapted to prey on or parasitize the introduced species. Furthermore, the colonists may be preadapted for exploiting a vacant niche, and may thus avoid severe competition from indigenous species.

The extent of man's inadvertent movement of insects is enormous. One example is particularly informative. If the distribution of carabid beetles in North America is examined, we see that many species also exist in Europe (Fig. 2.10) (Lindroth 1957). The most notable thing about their distribution in North America is that there seems to have been diffusion from small centers of origin on the coasts of Nova Scotia and Quebec in the east, and British Columbia and Washington in the west, toward the center of the continent. We can only conclude that introductions were very local, on a massive scale, and fairly recent.

The basis for this distribution is to be found in the practices of sailing vessels which visited ports in the locations mentioned above from the early 1600s to about 1850. Ships visited these ports to load various products, but since the human populations at these sites were too low to constitute a lucrative market, there was essentially a one-way traffic of goods from North America to Europe. On the outward-bound trip, the ships carried a load of ballast to maintain stability. The ballast, soil, and rocks dug up at ports in Europe, containing many insects and plant seeds, was dumped in the New World, and no doubt accounts for the many ground beetles common to both continents.

The sort of pattern seen in Fig. 2.10 is typical of the dispersal of pest species with time. We frequently find that a pest radiated out from a port of entry, or some other point where the initial insects were released; for example, the Japanese beetle; Argentine ant, *Iridomyrmex humilis* (Mayr); the felted beech scale, *Cryptococcus fagi* Baerensprung; the European elm bark beetle, *Scolytus multistriatus* (Marsham); and the gypsy moth in North America (Elton 1958); and the Colorado potato beetle in Europe (Johnson 1969). It is unfortunate for the scientist who

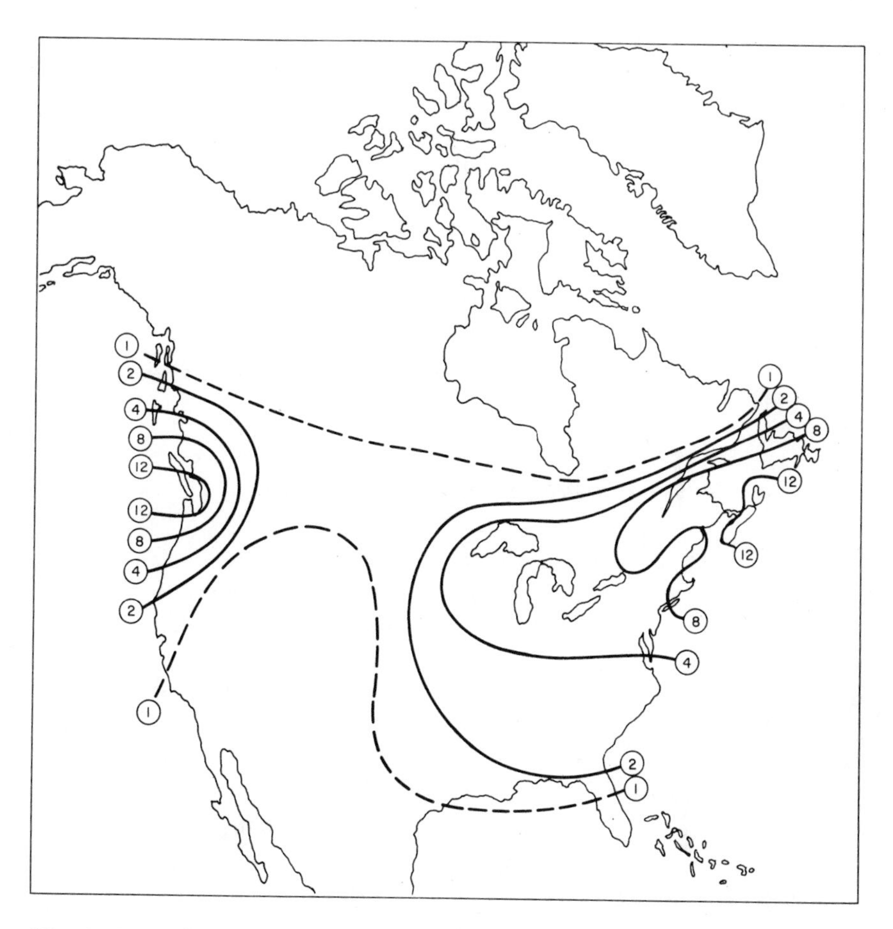

Fig. 2-10 The distribution of European carabid beetles in North America. Each line shows the limit to which the number of species indicated at its ends have extended. Note the centers of origin in Nova Scotia and Quebec in the East and Washington and British Columbia in the West. Newfoundland has the highest number of European species (not shown), with 16 species on the Avalon Peninsula. (From Carl H. Lindroth 1957. The faunal connections between Europe and North America. Almqvist & Wiksell, Stockholm and John Wiley & Sons, New York.)

wishes to learn more about colonizing species, but not so unfortunate for those concerned with the protection of crops and trees, that these introductions do not occur more frequently. We have not yet made sufficient use of these unique events to learn more of the community interactions involved in establishment.

Man has an impact on the distribution of insects not only by transporting exotic species, but also by introducing new food plants which insects may exploit. Both the Colorado potato beetle and the cotton boll weevil were able to expand their ranges enormously when vast areas in which potato and cotton were planted became contiguous with the original ranges of these insects.

V. COMMUNITY SUCCESSION IN COLONIZATION

We have already seen that as the age of the community increases conditions tend to change from severe to equable, and that we are likely to see a trend for the establishment of *K*-selected species rather than *r*-selected species. This trend constitutes only one of many we see in the succession of communities through time, some of which are summarized in Table 2.1 and Fig. 2.11 (see Odum 1969).

The main driving forces for this succession are:

1. Plant species that colonize an area actually change that area and thus make it more suitable for other colonists.
2. Different species of colonists arrive at different times.

The first factor can be broken down into four main influences plants have on the development of the succeeding community. (1) They create more shade and thus ameliorate the microclimate, making it possible for shade-tolerant species to colonize. (2) They contribute organic matter which changes soil texture and nutrient status. (3) They produce chemicals (secondary metabolic compounds) which may be toxic to themselves or to other plants. (4) They attract animals, including insects, which change factors in the environment by burrowing in soil, leaving excrement, trampling ground, selectively eating plants, dispersing seeds, pollinating flowers, and attracting their own predators and parasites.

With regard to the second factor, it is obvious that plant species with tiny, wind-dispersed seeds arrive very early, for example, dandelions. Eventually, enough large seeds arrive to outcompete the resident species, and succession results. Similar trends may be seen in the insect species

Table 2.1 Comparison between Early Stages of Succession (Agricultural Situation) and Mature (Climax) Stages of Succession[a]

Characteristics compared	Early (Agriculture)	Late (Climax)
Community energetics		
Gross production, standing crop biomass	High	Low
Biomass supported, unit energy flow	Low	High
Net community production (yield)	High	Low
Food chains	Linear (simple)	Weblike (complex)
Community structure		
Total organic matter	Small	Large
Species diversity	Low	High
Structural diversity	Low	High (stratification)
Biochemical diversity	Low	High
Life histories		
Niche specialization	Broad	Narrow
Size of organism	Small	Large
Life cycles	Short, simple	Long, complex
Nutrient cycling		
Mineral cycles	Open (wasteful)	Closed (conservative)
Nutrient exchange rate between organisms and environment	Rapid (annual plants)	Slow (long-lived plants)
Overall homeostasis		
Internal symbiosis	Undeveloped	Developed
Nutrient conservation	Poor	Good
Stability (resistance to change)	Poor	Good
Complexity of organization	Low	High
Selection pressure		
Strategy for reproduction	*r*-selected	*K*-selected
Numbers and size of progeny	Many small progeny	Few large competitive progeny
Progeny production	Quantity	Quality

[a] Adapted from Odum 1969. See Fig. 2.11 for explanations. Reproduced by permission from E. P. Odum (1969) *Science* 164(3877):262–270. Copyright 1969 by the American Association for the Advancement of Science.

that colonize an area, although insect size does not correlate so well with competitive ability.

Thus in time we see changes in the species composition and the complexity of the community. As the microclimate is ameliorated, conditions become more favorable, and more and more species are able to tolerate them. *r*-Selected species are replaced by *K*-selected species, good competitors which tend to become permanent members of the community. As succession progresses, the community acquires an increasing proportion of permanent members. The latter species tend to be more long-lived than those present earlier in the succession. Eventually, changes in species composition occur so seldom that there is little noticeable change. We consider this a mature community—also called a climax community. Other trends are summarized in Table 2.1 and Fig. 2.11.

Thus we see that community succession leads to an increase in the diversity of plants and animals with time. The concept of diversity and its measurement are important in ecology, and are explained in the Appendix. When there are many kinds of plants and animals in a community, it is clear that there are many sources of food for animals to exploit; in simple communities there are few. Thus a diverse community contains more buffers against environmental change. Should one food

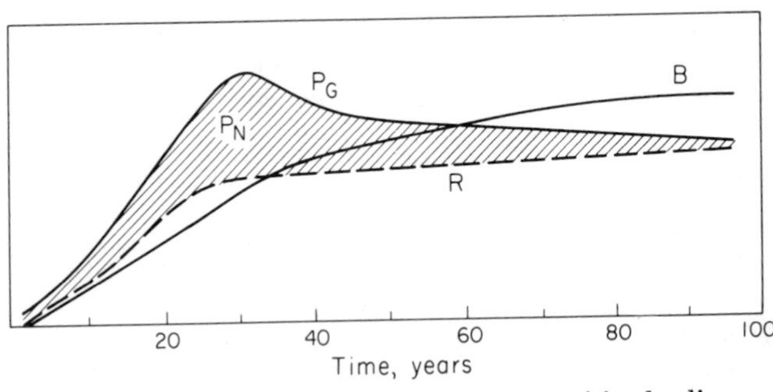

Fig. 2.11. A comparison of the energetics of communities leading to a climax forest community. Total biomass *B* increases rapidly and then reaches a steady state. Total respiration *R* follows a similar trend, although it reaches the plateau stage earlier. Gross production P_G is the amount of energy fixed by the green plants, and this reaches a peak in early stages of succession and then declines. Since net production P_N is what remains of gross production after some has been used in respiration, this is shown by the shaded area between lines for P_G and *R*. (Reproduced by permission from E. P. Odum. 1969. *Science* **164** (3877): 262–270. Copyright 1969 by the American Association for the Advancement of Science.)

source become scarce, another may be exploited. A change in the status of one species is less likely to have a severe impact on another species; thus the system is likely to be more stable. Simple systems, with few or only one food source for each animal, are correspondingly less stable.

We can estimate the stability of a community by counting the number of possible routes along which energy can travel through its food web. In Fig. 2.12*a* and *b* we see two possible arrangements of feeding relationships, each with four routes along which energy can pass. Each route is assumed to carry one-quarter of the total energy. Applying the formula for diversity, MacArthur (1955) calculated the stability of these systems, 1.38 in each instance. Adding another predator to the system doubles the number of routes along which energy may pass, and increases the stability to 2.08 (Fig. 2.12*c* and *d*). A notable feature of this system of estimating stability is that it generates the same index of stability if four highly specialized predators each feed on a single species of prey (Fig. 2.12*e*) or when one general predator feeds on four species of prey (Fig. 2.12*a*). In this context plants may be considered prey species.

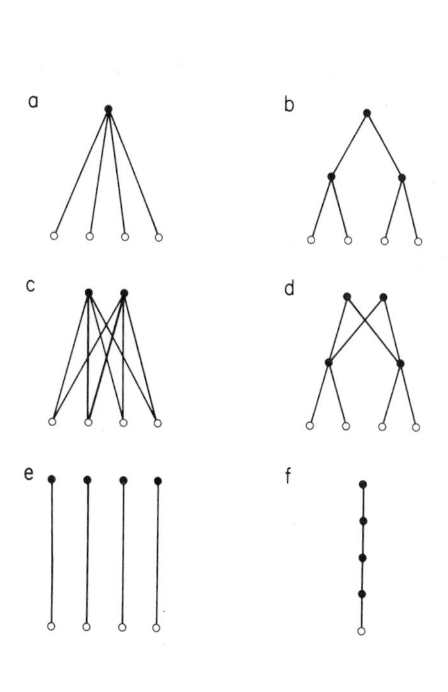

Fig. 2-12 The paths along which energy travels determine the stability of the system (*a* and *b*). Four possible routes along which energy travels. In (*a*) the lower level represents four herbivores (o) and the upper level 1 predator (•). In (*b*) an additional predator level is added. By adding another predator to the system, the number of feeding links is doubled (*c* and *d*). The same stability may be achieved by a generalized predator feeding on four prey species (*a*) or four specialized predators each feeding on a single prey species (*e*). Maximum stability can be achieved by having one consumer on each trophic level, each of which feeds on all lower trophic levels (*f*). Minimum stability for the same number of species exists when one predator feeds on four prey species (*a*). In this context plants may also be considered prey species.

Several properties in relation to stability are worth remembering (from MacArthur 1955):

Stability increases as the number of links increases, thus:

1. If the number of prey species for each predator remains constant, an increase in numbers of species in the community will increase the stability.
2. A given stability can be achieved either by a large number of species each with a fairly restricted diet, or by a smaller number of species each eating a wide variety of other species.
3. Stability is at the maximum possible for N species if there are N trophic levels with one species on each, each species eating all species on lower levels. Thus, if $N = 5$, there are eight possible paths up which energy can travel (Fig. 2.12*f*). *Similarly,* stability is at the minimum for N species if one carnivore feeds on $N - 1$ herbivores. Thus, if $N = 5$, there are four possible paths up which energy can travel (Fig. 2.12*a*).

It is clear that restricted diets lower stability, but dietary specialization makes for the efficient exploitation of food. Efficiency and stability are essential for survival. Efficiency enables species to outcompete others, but stability allows individual communities to outsurvive less stable ones. So there must be one of two evolutionary compromises between efficiency and stability:

1. When there are few species, stability can be achieved only if these species eat a wide variety of foods on many trophic levels. We see this pattern in northern latitudes.
2. When there is a large number of species, for example, in the tropics, stability can be achieved even if diets are fairly restricted. Thus species can specialize in their feeding habits, and may feed on only one or at most two trophic levels. These generalizations are of course relevant to the strategy of biological control and to an understanding of the impact of insecticide application at different latitudes. Southwood and Way (1970) have considered other aspects of diversity and stability.

Experimental support for the relationship between diversity and stability was provided by Pimentel (1961). He planted some collards in a pure stand, representing a very simple plant community, and others mixed in with natural old-field vegetation, representing a much more diverse environment. At frequent intervals he sampled the arthropods on the collards. In the pure stand several species reached outbreak proportions and did a great deal of damage to the crop. They included aphids, flea beetles, and caterpillars [imported cabbage worm, *Pieris rapae* (L.), and cabbage looper, *Trichoplusia ni* (Hübner)]. The collards in the diverse community suffered no such pest outbreaks.

We must also consider diversity in terms of the environment insects must exploit. This involves understanding the homogeneity or heterogeneity of sites, particularly in relation to the feeding habits (and perhaps oviposition habits), and the size of individual species. One species, for example, a carabid beetle, may feed on many food items and select them more or less in the proportion in which they occur in the habitat. Each food item may be viewed as a small grain of food in a multitude of grains, and we can call this a fine-grained food resource for a carabid. However, a monophagous herbivore in the same habitat can utilize only one plant species of the many that are present, and we may visualize individual plants as relatively large grains of food, and the food resource as coarse-grained (see MacArthur and Levins 1964; MacArthur and Wilson 1967). With this approach it becomes obvious that increasing diversity has little impact on a species utilizing fine-grained resources, but makes food-finding much more difficult for a species exploiting a coarse-grained resource because its food becomes relatively less frequent in the community. Conversely, simplifying the community is more favorable to a species requiring a coarse-grained resource, as the grains become relatively dense. The extreme case is in agricultural crops with one plant species predominating. Species requiring a coarse-grained resource, often pests, are favored far more than those needing a fine-grained resource, often their predators.

In a similar vein we must consider diversity in terms of the distance insects are capable of traveling. For highly mobile insects diversity must be viewed on an extensive areal basis. Other insects are much more confined in their movements. If cultural methods are used for increasing diversity with a view to reducing pest problems, the distinction becomes important. For example, during the spring colonization of crops, the mobility of insects depends very much on the stage in development they have reached. The European corn borer emerges as an adult in the spring and flies considerable distances in order to lay eggs on new corn. Thus it is difficult to create an agricultural ecosystem diverse enough to lessen seriously the chance that this insect will discover a host plant. In contrast, the northern corn rootworm overwinters in the egg stage, and it is the first-instar larva that must discover food; in this instance we can measure diversity in terms of the few centimeters the larva is able to travel in the soil before starvation ends its search for food. For the farmer of course it is easiest to increase diversity in time, by crop rotation, which works admirably for the northern corn rootworm, but not for the European corn borer.

Diversity is also influenced by grazing animals, but in different ways according to the intensity of grazing. Lightly grazed pastures frequently increase in diversity, particularly if the preferred food of the grazer is the dominant plant in the pasture community (Harper 1969). This is because grazing pressure prevents the dominant plant from outcompeting other species. As stated earlier, herbivores also increase the number of sites available to colonizers by leaving excrement and by changing the compaction of the soil and microtopography. Heavy grazing tends to reduce plant species diversity by strong selection for a few species with resistance to grazing pressure. Not only is species diversity reduced, but so is structural diversity, because many species never produce flowering stems under grazing pressure (see Fig. 2.1). For further discussion of influences on plant species diversity in managed systems, see Duffey and Watt (1971). Naturally, as plant species diversity and structural diversity decline, there is a similar decline in the insect species that can be supported (e.g., Morris 1967, 1971a,b). Dempster (1971) concluded that either overgrazing or undergrazing may cause local extinction of the cinnabar moth, *Tyria jacobaeae* L.

Finally, as we have seen earlier, the diversity of the matrix in which a crop is grown has an enormous effect on the diversity of plant and animal life in the crop. The reader should refer to Section IV where this is discussed in detail.

VI. STRATEGIES OF THE FARMER IN RELATION TO ECOLOGICAL CONCEPTS

The farmer's major activities, expressed in ecological terms, are to start the successional process by planting seed, to keep succession in a very early state by cultural activities, and to truncate succession by harvesting and plowing. Thus the farmer maintains an agricultural ecosystem typified by the set of characteristics given in Table 2.1 for early successional stages. Nevertheless, the more desirable characteristics listed in Table 2.1 are usually properties of mature communities. Thus any pest-management approach should try to develop an ecosystem that emulates later stages of succession as much as possible, for this is how stability can be achieved.

The farmer actually uses a great deal of energy performing jobs that would normally be done naturally, and thus requires large subsidies in terms of fossil fuel and services from the city, also supported by a fossil fuel economy. The work that supports the farmer and that the farmer does may be divided into six main activities (H. T. Odum 1971).

1. Mechanized and commercial preparation of seeds and planting to replace the natural dispersal system
2. Fertilizer application which augments and largely replaces the natural system of mineral cycling
3. Chemical and power weeding which largely replace the natural system of competition and extinction
4. Soil preparation and treatment to augment natural soil-building processes
5. Application of insecticides which replaces the system of chemical diversity, and carnivores in preventing epidemic grazing and disease
6. Development of varieties capable of passing on the savings in work to net food storages. New varieties are developed as disease and insect pests appear, thus providing the genetic selection formerly provided by natural selection.

The farmer creates a relatively simple ecosystem by tile-draining to provide uniform soil moisture, by removing rocks, stones, trees, and shrubs, by channeling streams and rivers, and by growing only a few crops. And yet stability can be most easily achieved in a complex ecosystem. As Odum (1963) states, "The only way man can have both a productive and a stable environment is to insure that a good mixture of early and mature successional stages are maintained, with interchanges of energy and materials. Excess food produced in young communities helps feed older stages that in return supply regenerated nutrients and help buffer the extremes of weather."[1]

Even though the farmer operates in a simple ecosystem, he still has a good deal of control. He can decide when to start succession and when to terminate it, and this must be decided in terms of damage by weather and pests. Late planting of winter wheat to avoid attack by the Hessian fly, *Mayetiola destructor* (Say), is a good example. Another is the delay of plowing until spring to avoid soil erosion over winter. Timing of harvesting may be influenced by the possible migration of insects from one crop to another, or in order to prevent the further buildup of a pest population, in alfalfa for example. Thus the farmer has already exercised considerable control over the start and termination of the successional process, although the potential for its further development is great.

[1] E. P. Odum (1963) Ecology. Holt, Rinehart and Winston, New York, p. 88.

The profound effect agricultural practices may have on insect populations is well illustrated by the European corn borer. This species overwinters only as a larva in the stalk or, occasionally, in the cob or shank of the ear. Although European corn borers are not host-specific, in the Midwest they are practically confined to corn. It should therefore be possible to control them by clean-plowing cornfields before the adults emerge in spring. However, clean-plowing will probably fail to give good control as long as corn is harvested on the ear and stored in cribs. Many larvae survive the winter in cribs, and in southern Minnesota at least 26% of the moths in the field in spring came from this source (Chiang 1964). The more recent use of the picker-sheller harvester should avoid this problem, since with this method of harvest only the kernels are removed and the rest of the ear is discarded in the field, where it can be plowed down.

Another major influence the farmer may exert on the agricultural ecosystem is the control of diversity. We should remember that small crop islands are likely to have higher extinction rates, and distant stands to have lower colonization rates. Small crop islands distant from a source of colonizing insects are possible only in a diverse ecosystem. Of course, a major trend in agriculture has been to reduce diversity to increase the efficiency of mechanization. But when pest problems need control, careful cost accounting is necessary to establish whether simplicity or diversity will yield the greatest long-term benefit. We have already discussed the influence of diversity on pests and beneficial insects in Section V.

As a very general conclusion, agricultural ecosystems can be viewed in terms of two central concepts of ecology—island biogeographical theory and succession of communities. Current thinking and developments in these fields should have an heuristic influence on research in the improvement of pest-management strategies.

APPENDIX: THE CONCEPT OF DIVERSITY AND ITS MEASUREMENT

Two components contribute to diversity. As the number of species of plants or insects in a community or the number of communities in an ecosystem increases, so the diversity increases. However, it is also essential to consider the evenness in distribution of the number of individuals per species in the system under study. It is intuitively obvious that as the chances increase of finding an individual of any particular species among a given number of species, that is, as evenness increases, so the diversity increases. Thus we need to quantify both number of units and evenness of distribution of these units.

A commonly used index of diversity is H', known as the Shannon-Wiener index. It and other indexes are discussed in the general ecology texts cited at the beginning of this chapter. The formula for H' is

$$H' = -\Sigma p_i \log_e p_i$$

where p_i is the proportion of the ith species in the total sample. In practice any log base may be used, and any unit other than species may be used in this formula. But a few examples using species distribution are given below to help establish that both number and evenness are essential components of diversity.

Example	Numbers of Individuals			H'
	Species 1	Species 2	Species 3	
1: Two-species system	90	10	—	0.33
2: Two-species system	50	50	—	0.69
3: Three-species system	80	10	10	0.70
4: Three-species system	33.3	33.3	33.3	1.10

It can be seen that example 2 has an H' much larger than example 1, although the number of species has not changed. Also, keeping evenness the same but adding one species, we see that example 4 yields a much higher H' than example 2.

The ability to express diversity in this and other ways has had a major impact on the development of ecology. MacArthur (1955) also used H' for calculating the stability of a system by considering the p_i's the proportions of energy passing up each feeding link. Examination of any textbook on ecology should convince the reader how deeply entrenched the concept of diversity is in ecological thinking.

ACKNOWLEDGMENT

The assistance of Ms. Alice Prickett, Scientific Artist, is gratefully acknowledged.

REFERENCES

Chapman, R. F. 1969. The insects. Structure and function. American Elsevier, New York. 819 pp.

Chiang, H. C. 1964. Overwintering corn borer, *Ostrinia nubilalis*, larvae in storage cribs. *J. Econ. Entomol.* **57**:666–669.

Clark, L. R., P. W. Geier, R. D. Hughes, and R. F. Morris. 1967. The ecology of insect populations in theory and practice. Methuen, London. 232 pp.

Collier, B. D., G. W. Cox, A. W. Johnson, and P. C. Miller. 1973. Dynamic ecology. Prentice-Hall, Englewood Cliffs, N.J. 563 pp.

Dempster, J. P. 1971. Some effects of grazing on the population ecology of the Cinnabar moth (*Tyria jacobaeae* L.). Pages 517–526 *in* E. Duffey and A. S. Watt, eds., The scientific management of animal and plant communities for conservation. Blackwell, Oxford.

Dingle, H. 1972. Migration strategies of insects. *Science* **175**:1327–1335.

Doutt, R. L., and P. DeBach. 1964. Some biological control concepts and questions. Pages 118–142 *in* P. DeBach, ed., Biological control of insect pests and weeds. Reinhold, New York.

Doutt, R. L., and J. Nakata. 1973. The *Rubus* leafhopper and its egg parasitoid: An endemic biotic system useful in grape-pest management. *Environ. Entomol.* **2**:381–386.

Duffey, E., and A. S. Watt, eds. 1971. The scientific management of animal and plant communities for conservation. Blackwell, Oxford. 652 pp.

Ehrlich, P. R., and P. H. Raven. 1964. Butterflies and plants: A study in coevolution. *Evolution* **18**:586–608.

Ehrlich, P. R., and P. H. Raven. 1967. Butterflies and plants. *Sci. Amer.* **216** (6):104–113.

Elton, C. S. 1958. The ecology of invasions by animals and plants. Methuen, London. 181 pp.

Feeny, P., K. L. Paauwe, and N. J. Demong. 1970. Flea beetles and mustard oils: Host plant specificity of *Phyllotreta cruciferae* and *P. striolata* adults (Coleoptera: Chrysomelidae). *Ann. Entomol. Soc. Amer.* **63**:832–841.

Grigoryeva, T. G. 1970. The development of self-regulation in an agrobiocoenosis following prolonged monoculture. *Entomol. Rev.* **49**:1–7.

Harper, J. L. 1969. The role of predation in vegetational diversity. *Brookhaven Symp. Biol.* **22**:48–62.

Heathcote, G. D., and A. J. Cockbain. 1966. Aphids from mangold clamps and their importance as vectors of beet viruses. *Ann. Appl. Biol.* **57**:321–336.

Huffaker, C. B., and P. S. Messenger. 1964a. Population ecology—Historical development. Pages 45–73 *in* P. DeBach, ed., Biological control of insect pests and weeds. Reinhold, New York.

Huffaker, C. B., and P. S. Messenger. 1964b. The concept and significance of natural control. Pages 74–117 *in* P. DeBach, ed., Biological control of insect pests and weeds. Reinhold, New York.

Huffaker, C. B., P. S. Messenger, and P. DeBach. 1971. The natural enemy component in natural control and the theory of biological control. Pages 16–67 *in* C. B. Huffaker, ed., Biological control. Plenum Press, New York.

Janzen, D. H. 1968. Host plants as islands in evolutionary and contemporary time. *Amer. Nat.* **102**:592–595.

Johnson, C. G. 1969. Migration and dispersal of insects by flight. Methuen, London. 763 pp.

Krebs, C. J. 1972. Ecology: The experimental analysis of distribution and abundance. Harper and Row, New York. 694 pp.

Landahl, J. T., and R. B. Root. 1969. Differences in the life tables of tropical and temperate milkweed bugs, genus *Oncopeltus* (Hemiptera: Lygaeidae). *Ecology* **50**:734–737.

Leius, K. 1967. Influence of wild flowers on parasitism of tent caterpillar and codling moth. *Can. Entomol.* **99**:444–446.

Lewis, T. 1965. The effects of shelter on the distribution of insect pests. *Sci. Hort.* **17**:74–84.

Lewis, T. 1969a. The distribution of flying insects near a low hedgerow. *J. Appl. Ecol.* **6**:443–452.

Lewis, T. 1969b. The diversity of the insect fauna in a hedgerow and neighbouring fields. *J. Appl. Ecol.* **6**:453–458.

Lindroth, C. H. 1957. The faunal connections between Europe and North America. John Wiley, New York. 344 pp.

MacArthur, R. H. 1955. Fluctuations of animal populations, and a measure of community stability. *Ecology* **36**:533–536.

MacArthur, R. H. 1972. Geographical ecology: Patterns in the distribution of species. Harper and Row, New York. 269 pp.

MacArthur, R., and R. Levins. 1964. Competition, habitat selection, and character displacement in a patchy environment. *Proc. Nat. Acad. Sci. U.S.* **51**: 1207–1210.

MacArthur, R. H., and E. O. Wilson. 1967. The theory of island biogeography. Princeton University Press, Princeton, N.J. 203 pp.

Morris, M. G. 1967. Differences between the invertebrate faunas of grazed and ungrazed chalk grassland. I. Responses of some phytophagous insects to cessation of grazing. *J. App. Ecol.* **4**:459–474.

Morris, M. G. 1971a. The management of grassland for the conservation of invertebrate animals. Pages 527–552 *in* E. Duffey and A. S. Watt, eds., The scientific management of animal and plant communities for conservation. Blackwell, Oxford.

Morris, M. G. 1971b. Differences between the invertebrate faunas of grazed and ungrazed chalk grassland. IV. Abundance and diversity of Homoptera—Auchenorhyncha. *J. Appl. Ecol.* **8**:37–52.

Murdoch, W. W., F. C. Evans, and C. H. Peterson. 1972. Diversity and pattern in plants and insects. *Ecology* **53**:819–829.

Odum, E. P. 1963. Ecology. Holt, Rinehart and Winston, New York. 152 pp.

Odum, E. P. 1969. The strategy of ecosystem development. *Science* **164**:262–270.

Odum, E. P. 1971. Fundamentals of ecology. 3rd ed. W. B. Saunders, Philadelphia. 574 pp.

Odum, H. T. 1971. Environment, power and society. Wiley-Interscience, New York. 331 pp.

Peschken, D. P. 1972. *Chrysolina quadrigemina* (Coleoptera: Chrysomelidae) introduced from California to British Columbia against the weed *Hypericum perforatum:* Comparison of behaviour, physiology, and colour in association with post-colonization adaptation. *Can. Entomol.* **104**:1689–1698.

Pianka, E. R. 1970. On *r*- and *K*-selection. *Amer. Nat.* **100**:592–597.

Pianka, E. R. 1972. *r* and *K* selection or *b* and *d* selection? *Amer. Nat.* **106**: 581–588.

Pimentel, D. 1961. Species diversity and insect population outbreaks. *Ann. Entomol. Soc. Amer.* **54**:76–86.

Price, P. W. 1971. Toward a holistic approach to insect population studies. *Ann. Entomol. Soc. Amer.* **64**:1399–1406.

Prosser, C. L., and F. A. Brown. 1961. Comparative animal physiology. W. B. Saunders, Philadelphia. 688 pp.

Simberloff, D. S., and E. O. Wilson. 1969. Experimental zoogeography of islands: The colonization of empty islands. *Ecology* **50**:278–296.

Simberloff, D. S., and E. O. Wilson. 1970. Experimental zoogeography of islands. A two-year record of colonization. *Ecology* **51**:934–937.

Southwood, T. R. E. 1961. The number of species of insect associated with various trees. *J. Anim. Ecol.* **30**:1–8.

Southwood, T. R. E., and M. J. Way. 1970. Ecological background to pest management. Pages 6–28 *in* R. L. Rabb, and F. E. Guthrie, eds., Concepts of pest management. North Carolina State University, Raleigh.

Thorsteinson. A. J. 1953. The chemotatic responses that determine host specificity in an oligophogous insect (*Plutella maculipennis* (Curt.) Lepidoptera). *Can. J. Zool.* **31**:52–72.

Van Dyne, G. M., ed. 1969. The ecosystem concept in natural resource management. Academic Press, New York. 383 pp.

Van Emden, H. F. 1965. The role of uncultivated land in the biology of crop pests and beneficial insects. *Sci. Hort.* **17**:121–136.

Varley, G. C., and G. R. Gradwell. 1971. The use of models and life tables in assessing the role of natural enemies. Pages 93–112 *in* C. B. Huffaker, ed., Biological control. Plenum Press, New York.

Wilson, E. O. 1969. The species equilibrium. *Brookhaven Symp. Biol.* **22**:38–47.

Wilson, E. O., and D S. Simberloff. 1969. Experimental zoogeography of islands: Defaunation and monitoring techniques. *Ecology* **50**:267–278.

Wolfenbarger, D. O. 1940. Relative prevalance of potato flea beetle injuries in fields adjoining uncultivated areas. *Ann. Entomol. Amer.* **33**:391–394.

Wolfenbarger, D. O. 1946. Dispersion of small organisms. Distance dispersion rates of bacteria, spores, seeds, pollen and insects; incidence rates of diseases and injuries. *Amer. Midl. Nat.* **35**:1–152.

Wolfenbarger, D. O. 1959. Dispersion of small organisms. Incidence of viruses and pollen; dispersion of fungus spores and insects. *Lloydia* **22**:1–106.

3

THE ECONOMICS OF PEST MANAGEMENT

J. C. Headley

As people try to harvest from nature materials that are necessary to society, such as food, fiber, and lumber, and to protect values such as a pleasant environment, one fact of life emerges. People find themselves in competition with other species of life for these things. As the number of people has increased and more pressure has been placed on the available resources for producing the things they need and want, this competition has intensified. Just as intraspecific human competition has required the allocation of resources for national defense, so interspecific competition has also required the allocation of resources to deal with the relations between people and other organisms such as insects, weeds, fungi, rodents, and large predators.

This situation exists in part because the actions of people to increase production of food, fiber, and timber have also created more food and a more desirable habitat for the competing species. The increase in the acreage of corn has made it possible to support a larger population of the western corn rootworm, *Diabrotica virgifera* LeConte. Better seed beds and fertilization for growing food crops encourage larger weed populations, and so on. The reasons for pest competition should be clear. The question is, "How does one deal with this competition in the most efficient way, given the needs and desires of people?"

In this chapter the economic problems associated with man's competition with other species are outlined to provide a basis for developing strategies of pest management that satisfy the objectives of people both as individuals and as a collective society. Pest control through the various management techniques currently available or to be developed can only be justified in terms of its net contribution to human values, that is, the difference between positive and negative values.

I. PEST CONTROL AT THE MICROLEVEL

Individual producers of food and fiber are constantly faced with decisions concerning pest control. While the importance and the number of these decisions vary with the crops produced, the methods of production, and the geographical location of the production unit, almost every producer is required to do something about pests.

A. Simple Benefit-Cost Analysis

In Chapter 1 it was stated that pest-control activities almost never increase yields when viewed from a noncompetitive standpoint. Rather, pest control serves to defend or protect what would be produced in the absence of pest competition. The benefits for the producer then come from the value of damage prevented, much like the benefits from a dam on a flooding river arise from the value of flood damage prevented.

The individual farmer, then, in the simplest situation, is not willing to allocate any more resources to pest control than can be justified by expected[1] damages to be prevented by the use of such resources. If the expected value of the crop is Py, where P is the unit price and y is the physical yield of a given quantity, and if pest competitors are expected to reduce this yield by 10%, the farmer will not be willing to spend more for pest control than an amount given by $0.10Py$. Actually, he is only willing to spend less than this amount, because if he spends $0.10Py$ the value of damages prevented just equals the cost of prevention and he is indifferent as to whether he gives this amount to the pests or to the suppliers of pest-control inputs. In the case in which the damages expected are $0.10Py$ and the cost of preventing such damages is $0.10Py$,

[1] Note the use of the word *expected.* The analysis that follows assumes that the expectation is perfect or single-valued.

one can say that the benefit/cost ratio of these pest control activities is equal to 1, that is, $0.10Py/0.10Py = 1$.

Actually, the farmer faced with this situation and with an alternative of spending less for insect control but allowing a higher level of damage might decide to seek a more satisfactory solution. If he can use less insecticide, either in terms of dosage per application or fewer applications of a given dosage, or substitute predators, parasitoids, or cultural practices for insecticides, he could justify using pest-control inputs up to the point where the incremental value of damages prevented just equals the incremental cost of the measures required to control pests. This is shown graphically in Fig. 3.1.

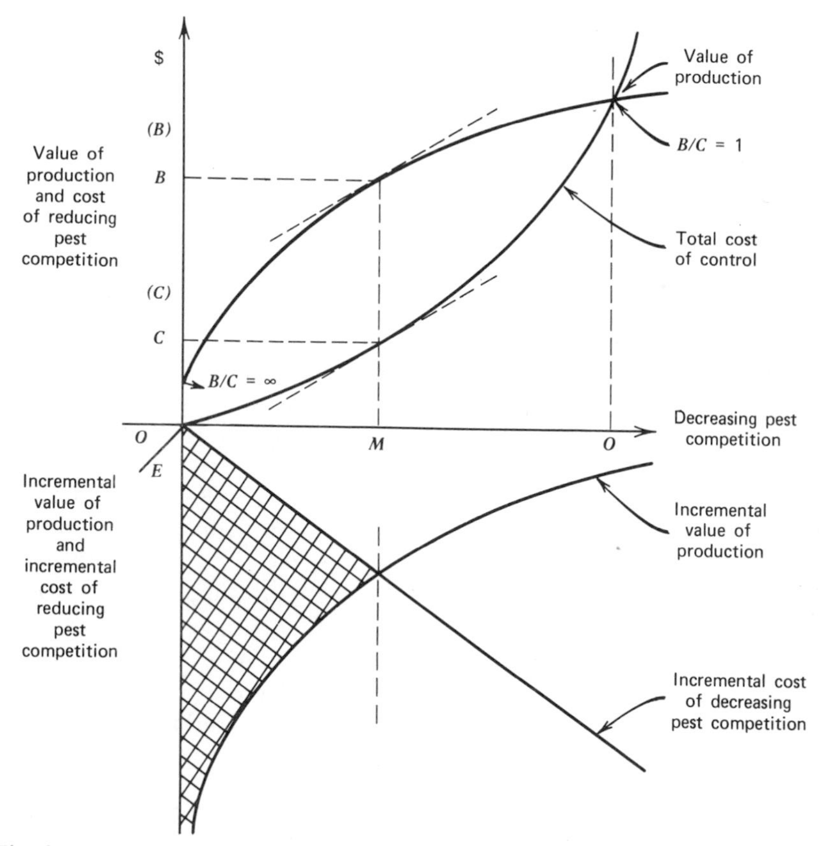

Fig. 3-1 Relation of value of production, cost of reducing pest competition, and pest competition.

In Fig. 3.1 the origin represents zero output and the equilibrium level of pest competition so that, as pest competition is reduced by management, the output increases as a result of damages prevented. Conceptually, at some point the competition can be reduced to zero (no economic damage), but the total costs of pest control would exceed the total value of the crop. If the farmer reduces the competition to the level M, he maximizes the difference between the value of production and the total cost of pest control. That is, the difference $B - C$ is a maximum. Note that this occurs when the incremental value of production with respect to pest competition equals the incremental cost of reducing competition shown in the lower part of the figure, where the two incremental curves intersect. If one were to find the area of the cross-hatched section between the two incremental curves, this area would equal $B - C$ in the upper part of the figure.[2]

The main thrust of the above simplified analysis is that if the individual farmer wishes to maximize returns above the cost of pest control, he will not be satisfied with the simple result that $B/C > 1$. Rather he will want to maximize $B - C$. However, where $B - C$ is a maximum, the ratio of *incremental* benefits and *incremental* costs equals 1.

B. The Farmer Faces Reality

While the above analysis treats the major parts of the resource allocation decision in pest management, it has abstracted from reality in several ways. First, it has assumed that the farmer knew with certainty the pest–damage relationship, the pest response to the various alternative control methods, the output relationship that would prevail, that is, weather conditions and fertility, and the price to be received for the product at harvest. Second, it assumes that the damage only reduces yield and has no effect on quality—not at bad assumption for grain crops, but not a good assumption for fruit or vegetable crops. Third, it assumes that each pest-control decision made by the farmer is independent of every other pest-control decision, as well as other input decisions such as fertility level, tillage practices, plant population, and choice of variety.

In reality the individual farmer is faced with uncertainty concerning weather and its effect on crop development, as well as on the pest. In

[2] A good treatment of profit maximization principles is found in Richard Leftwich, *The Price System and Resource Allocation,* Holt, Rinehart and Winston, New York, 1966, pp. 156–182.

addition, farmers may not have sufficient knowledge about how other factors, such as cultural practices and plant varieties, influence the potential for pest damage.

Another area of reality not represented in the above analysis is the relationship of decisions in one time period to the problem in the future. For example, control methods used against one species early in the season may produce favorable conditions for the development of another pest species later in the season or in the next season. In other words, the above analysis is timeless or static, while the real problem is dynamic.

The questions of uncertainty and dynamics are dealt with in the two following subsections.

C. The Farmer and Uncertainty

Since competitive species do not just materialize full grown in an instant, each producer must deal with the uncertainty surrounding whether or not the competition will come, and to what extent. Entomological knowledge, for example, can suggest that, given a certain number of larvae in a certain stage, the potential for a damaging infestation exists. But these are not single-valued expectations. An unforeseen turn of events in weather conditions or an unpredicted buildup of parasites and predators could cause the infestation to fall short of its *expected* potential. How does a farmer proceed in the face of such a situation?

Risk and uncertainty are not new phenomena in economic activity. They are the rule rather than the exception. Therefore it is useful to examine an example pest-management decision under uncertainty.

First, more often than not farmers select varieties, cultural practices, and fertility levels more or less independent of the threat of pest competition. This being the case, the potential yield as a result of these decisions represents an asset to be protected by pest management. Second, faced with the uncertainty of the threat of pest damage, the farmer is interested in expenditures that reduce that uncertainty, as long as the amount of the expenditures is commensurate with the amount of probable damage. That is, he is interested in turning the expected value of the damage (the size of the damage times the probability of the damage occurring) into a part of his cost of production. This is the function of insurance, to turn a probabilistic loss into a certainty, hopefully at a lower cost. It should be emphasized now that the use of the insurance principle is valid only over a series of events, such as a large number of people whose lives are insured, or in the case of pest management over several seasons. For a given farmer in a given year, either a damaging infestation materializes or does not. The posterior probability is either 1 or 0, but over several seasons other probabilities are possible.

If a farmer has a policy saying that, if he finds four larvae of *Heliothis zea* (Boddie) per plant in his cotton crop, he will treat the field to keep a damaging infestation from developing, then this cost of treatment becomes part of his cost of doing business. However, if the farmer takes no action, any one of a variety of possibilities may follow. These are summarized in Table 3.1.

Table 3.1 Schedule of Hypothetical Damages to Cotton per Acre and Their Probabilities, Given Four Larvae Per Plant

Level of Economic Damage ($/acre)	Probability
0	0.05
10	0.60
20	0.20
30	0.10
40	0.03
50	0.02
Sum of probabilities	1.00

$$\text{Average expected loss} = 0(0.05) + 10(0.60) + 20(0.20) + 30(0.10) + 40(0.03) + 50(0.02) = \$15.20$$

Faced with this sort of situation and the possibility for repeating it several times, the first piece of information needed is the mathematical expectation of the losses.[3] This is found by finding the algebraic sum of the product of damage levels and their respective probabilities. What this computation says is that, given the set of probabilities in Table 3.1, the farmer's decision to take no action given four larvae per plant will result on the average in a loss of $15.20. Therefore if the control costs necessary

[3] The term *mathematical expectation* refers to the average outcome of a series of probable events. If in a large number of tosses of an unbiased coin, one wins 5¢ for heads and loses 10¢ for tails, the mathematical expectation of the game is $\frac{1}{2}(5) - \frac{1}{2}(10) = -2.5$¢. That is, one will lose 2.5¢ per toss on the average.

to ensure that the loss is zero are less than $15.20 per year, then there will be a positive return from control activities. If, then, the farmer can apply control measures for $10 per acre, he can prevent losses of $15.20 per acre on the average, with the assumption that the $10 expenditure will guarantee maintaining the population below the economic injury level.

According to Table 3.1, the $10 control costs will be wasted 5% of the time, but 95% of the time will prevent losses having a value larger than or equal to $10 per acre. Faced with this sort of situation, the farmer accepts a small probability of a small loss if he controls the pest whereas, if he does not apply control measures, the probability of losses greater than control costs is large. The only rational reason a farmer would not attempt to control the pests by choice is that he obtains increasing satisfaction from the risk. This describes the compulsive gambler. Of course, if the farmer were so limited in capital that he could not afford the $10 per acre, he would be forced to gamble and take the consequences.

It should be emphasized that the above uncertainty analysis is carried out with only one control method, and may not in fact be the optimum even though the average return above costs is sizable. There may be other methods available which cost less for the same level of control, or which cost less but provide less control. The following hypothetical data serve to illustrate.

The analysis shown in Table 3.2 illustrates both dimensions of uncertainty involved in pest management in a static situation. Namely, insofar as a management technique does not absolutely hold pest competition below the economic injury level, there may be some losses in yield from pest competition. If the management method allows for the possibility of damages, there exists a series of probabilities of various levels of losses for each alternative method. In this case the expected cost of the management policy is the cost of the methods used plus the mathematical expectation of the losses that may be experienced from less than perfect control.

In Table 3.2 the hypothetical discrete probabilities are shown for the various management alternatives including no action ("let nature take its course"). Alternative 1 completely suppresses the pest below the economic injury level. Alternatives 2 and 3 suppress the pest, but some possibility for damage exists if conditions are right. Therefore 2 and 3 have possibilities for losses in yield, but with probabilities reduced compared to the no-action alternative. Over a series of seasons, then, one could expect losses to average $5.60 per acre with alternative 2 and $7.80 per acre with

Table 3.2 Total Costs of Pest Management for Alternative Methods of Management of *Heliothis zea* in Cotton Given Four Larvae per Plant (Hypothetical)

	Management Alternatives			
	1	2	3	No Action
Degree of control	100%	95%	90%	0%
Losses $/acre	Probabilities of Losses			
0	1.00	0.60	0.40	0.05
10	0	0.30	0.40	0.60
20	0	0.05	0.10	0.20
30	0	0.04	0.03	0.10
40	0	0.01	0.01	0.03
50	0	0.00	0.01	0.02
Sum of loss probabilities	1.00	1.00	1.00	1.00
Expected value of losses ($/acre)	0	5.60	7.80	15.20
Management costs ($/acre)	10.00	4.00	2.00	0
Total cost ($/acre)	10.00	9.60	9.80	15.20

alternative 3, while with alternative 1 the expected value of the losses is zero.

The management alternatives have costs shown in Table 3.2, which decline with effectiveness. Therefore, given these data as an accurate description of the situation, the farmer interested in minimizing the total cost over time of managing the pest would choose alternative 2, because the total cost is less. This alternative does not prevent as much loss as alternative 1, but its cost is only 40% of the cost of alternative 1.

One should remember that the conditions surrounding this decision are that (1) the probabilities are known; (2) the effectiveness of the alternatives is known; (3) the decision can be replicated several times; (4) the decision for each season is independent of the decision for every other season; and (5) the farmer can financially withstand the loss in the event that the unlikely, but possible, $40-loss event occurs, not just once but perhaps several times and perhaps consecutively. If the last condition is not met, it would be rational to choose alternative 1 where zero losses are assured in return for an insurance premium of $10 per acre per year—in other words, complete coverage.

The above analysis is an oversimplified version of the situation under uncertainty, but does include the major conceptual aspects of the problem. In reality the loss-probability function is not a discrete one but perhaps continuous, representing an infinite number of possibilities. This does not change the analysis in any major respect—only the mathematical methods of dealing with the problem.[4]

D. The Farmer and Dynamic Pest-Management Decisions

One of the basic decision parameters used by entomologists is the economic threshold. This concept involves the idea that when a pest population reaches a certain level it is profitable to reduce it by control. That is, the costs of the control method chosen will be less than the value of the damages prevented. This is, as shown earlier in this chapter, an application of standard economic costs and returns analysis.

The problem with the economic threshold idea, in addition to the fact that the parameters have been treated nonprobabilistically, that is, as single-valued expectations, is that historically it has ignored the interdependence of decisions at different time periods. For example, pests may evolve to combat extrinsic pressures, or pest populations which are currently suppressed may explode in later periods; therefore the population during some period subsequent to the one in which the control decision is made is a function of the initial population level. This suggests that the problem is a dynamic one (as opposed to a static one), and that more than one generation of the pest needs to be considered in developing a management strategy. The term *management strategy* is used here to mean a combination of single-period decisions over the relevant pest-management horizon. This planning horizon may be multiple generations during one season, or may encompass several seasons.

A simple hypothetical but illustrative example is used to clarify the

[4] A good example of the application of decision theory to pest control is found in Gerald A. Carlson, "A Decision Theoretic Approach to Crop Disease Prediction and Control," *Amer. J. Agr. Econ.*, 52(2):216–223 (1970).

assertions above.[5] Assume three alternative control methods with differing costs and mortality, as summarized in Table 3.3.

Table 3.3 Alternative Control Methods—Mortality and Costs

Method	Mortality (%)	Cost ($)
A	0	0
B	$33\frac{1}{3}$	20
C	75	100

Further assume that the gross benefits related to the pest population are given by

$$B_t = 200 - (1 - M_t)Y_t \tag{1}$$

where B_t = gross benefits in period t, M_t = mortality rate in period t resulting from control in t, and Y_t = level of pest population at beginning of period t. Thus the net benefits for a certain period are given by the difference between the gross benefits and the cost of each method.

Table 3.4 Net Benefits for Any Period t for Three Alternative Control Methods

Method	Net Benefits
A	$200 - Y_t$
B	$180 - \frac{2}{3}Y_t$
C	$100 - \frac{1}{4}Y_t$

The net benefits in Table 3.4 are found by applying the relevant mortality rates to the gross benefit function and subtracting the cost of the control method.

Finally, this model needs some function relating populations between periods. For illustrative purposes only, this is assumed to be an increase

[5] This example was taken from C. Robert Taylor, "Dynamic Economic Evaluation of Pest Control Strategies," unpublished Ph.D. thesis, University of Missouri, Columbia, 1972.

of three times the population after application of the control method. That is,

$$Y_{t+1} = 3(1 - M_t)Y_t \tag{2}$$

where Y_{t+1} = population in the next period after t, and M_t and Y_t are defined as before. The right-hand side of the above equation is three times the population remaining. While this model is purely hypothetical, it can be used to demonstrate dynamic economic evaluation.

There are essentially two ways to approach this problem. One way is to optimize the pest-control method one period at a time—that is, to select the method with the largest net benefits for each period separately. Another way is to optimize over both periods and choose the combinations of methods that maximize the sum of the net benefits from the two periods. It is shown that the two ways do not provide the same result, and that the latter way is the correct one.

It is useful to construct a tree diagram of the two-period process, and then evaluate the best strategy for one period at a time and compare this with the best two-period strategy. Assume that the initial population for the first period t_1 is 72. By making use of the net benefit equations in Table 3.4, the three methods can be evaluated. Each of these methods provides a new population for the beginning of the second period, each of which can again be evaluated according to the three alternative methods. Therefore, in the first period t_1 there are three possible outcomes, and in the second period t_2 there are nine possible outcomes. All these results are summarized in Fig. 3.2.

In Fig. 3.2 one can show the result of looking at each period and selecting the best alternative for each period. In this case, in period 1, method E would be chosen with net benefits of:

$$\begin{aligned} \$132 &= 180 - \tfrac{2}{3} Y_1 \\ &= 180 - \tfrac{2}{3}(72) \\ &= 180 - 48 \end{aligned}$$

The net benefit function comes from Table 3.3. However, having chosen E in period 1, one must accept the biological reality of a population of 144 at the beginning of period 2. The choices are then limited to the three methods, starting with Y_2 = 144. Again the selection is method E with net benefits of \$84 and a final population of 288 at the beginning of period 3. The population doubles each period, because the mortality is one-third and the recruitment to the population increases by a factor of 3. The total value of the net benefits from the two periods is

$$(B_1 - C_1) + (B_2 - C_2) = \text{two-period net benefits}$$
$$132 + 84 = \$216$$

Now consider looking at the problem as a two-period problem. That is, rather than obtaining

$$\max_{A,E,C} (B_1 - C_1) + \max_{A,E,C} (B_2 - C_2)$$

try

$$\max_{A,E,C} (B_1 - C_1) + (B_2 - C_2)$$

That is, look at all nine possible two-period outcomes at one time. These outcomes are summarized in Table 3.5. The entries in Table 3.5 result from following the lines of the decision tree in Fig. 3.2.

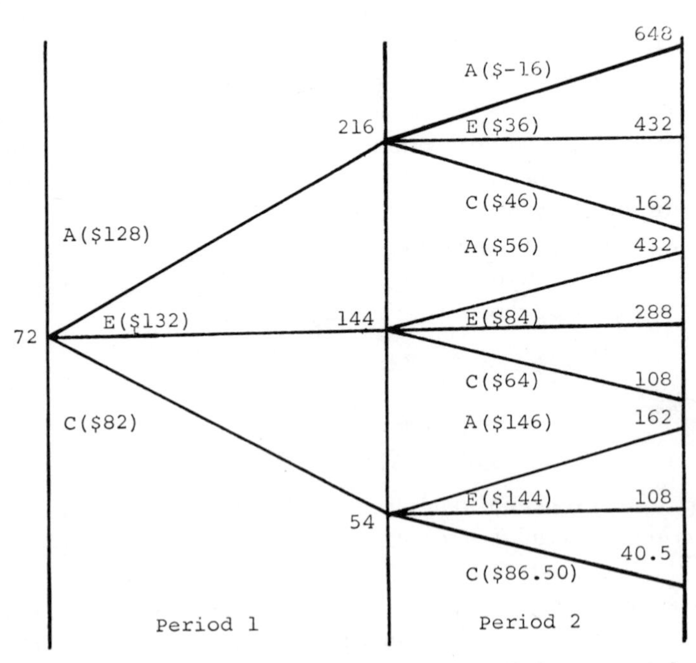

Fig. 3-2 Possible outcomes for a hypothetical two-period pest control problem with an initial population level of 72.

Table 3.5 Pest-Control Strategies and Net Benefits for a Two-Period Planning Horizon with an Initial Population of 72

Strategy	Net Benefits ($)
A,A[a]	112[b]
A,E	164
A,C	174
E,A	188
E,E	216
E,C	178
C,A	228
C,E	226
C,C	168.50

[a] The first letter in the pairs in this column gives the name of the method used in the first period, while the second letter gives the name of the method used in the second period.

[b] The net benefits are the sums of the single-period benefits shown in Fig. 3.2. For example, to evaluate *A,A* we have $128 + (-16) = \$112$.

A look at Table 3.5 shows that, over the two periods, use of method *C* in period 1 and method *A* in period 2 (*C,A*) provides maximum net benefits of $228. The single-period maximum approach yields a strategy using method *E* in both periods (*E,E*) and provides net benefits for the two periods of $216. This illustrates the difference that can occur between the one-period-at-a-time approach as compared to the complete planning horizon method.

If the time-dependent model used here is evaluated for a series of initial populations, one finds that static single-period decision making never provides higher net benefits than the two-period method.[6] This is characteristic in general of any problem with time dependencies, such as the pest-management problem. This conclusion leads one to raise serious questions about pest-management decisions made with the aid of eco-

[6] See C. Robert Taylor, "Dynamic Economic Evaluation of Pest Control Strategies" unpublished Ph.D. thesis, University of Missouri, Columbia, 1972, pp. 27 and 28.

nomic thresholds with the aim of maximizing the net benefits in one growing season. Recognition of the time dependencies in pest problems makes it more important to work toward the management of pests as a policy, than to follow the policy of "treat as needed."

Before concluding this discussion of dynamic aspects of pest-control decisions, a few cautions are in order. First, the knowledge necessary to implement multiple-period strategies is currently lacking or not appropriately organized. As shown in the earlier models, the notion of the probability of various outcomes is needed for essential reality. The model shown was deterministic. Second, when the number of alternatives becomes large and the periods many, the computational problem becomes immense. Witness how a two-period problem with three alternatives gave nine possible outcomes. A problem with three alternatives and five periods provides $3^5 = 243$ possible outcomes to be evaluated. There are limits on the size of manageable problems, even with large computers. This problem of dimensionality should not cause one to ignore the dynamic decision-making process, however, because often the alternatives can be reduced to a significant few and computer simulation used to indicate the direction of control strategies better than those used at present even if they are not optimal.

II. THE SOCIAL ECONOMICS OF PEST MANAGEMENT

The individual decisions of producers, which have been discussed in previous sections, deal with what is called the private economics of pest management. A significant part of the economics of pest management deals with what is called the social economics of pest management.

Pest management in whatever form is a technology that can have an impact on society as a whole. The benefits arising from pest management can take one of two forms. Either the benefits can arise from increases in output over what would normally occur, or the benefits can arise from a saving of the resources needed to produce a constant output. In general, the effects of pest control have been a mixture of the two forms. Output from agriculture has increased as farmers have substituted inputs produced off the farm, of which many pest-management inputs are a part, for labor and land. Certainly, there have been benefits to society from released labor and land which could be used otherwise.

Benefits and costs of pest management that accrue to individual producers are examples of private benefits and costs and are concerned only with the cost of pest-management resources to the producer and the revenue received by the producer. If, for example, the crop could be grown in another location where the total resources required to produce the output are less, the net private benefits (gross producer revenue minus producer costs) do not measure net social benefits (gross social benefits minus social costs). Or, if the market value of the resources used in pest management does not completely reflect the value of all the resources being used, then the accounting costs of producers measure private costs only and the net benefits are private benefits and do not represent net social benefits. This discrepancy in costs could be due to disruptions in the market due to government policy, monopolistic practices in the sale of pest-management resources, or the effects of pest management that influnce resource allocation but are not measured in the market, such as external effects on pollinators or other beneficial species.[7]

The importance of the distinction between private and social values in pest-management economics is the effect the divergence between social and private values can have on resource allocation, and therefore on economic welfare. If all costs and benefits, properly valued, are not considered by those making pest-management decisions, society does not receive the maximum output from these resources. That is, the system is not economically efficient. This is the question that has been at issue between environmentalists and those advocating pest control based principally on pesticides. Pesticide proponents allegedly impose external costs on others who are not able to bring their values to bear on pest-control decisions. In other words, the allegation is that the costs of pesticide chemicals and of application understate the full social cost, and therefore net benefits from pesticide use are overestimated.

A. Pest Management, Market Failure, and Public Policy

In the discussion above it was noted that there are results of economic activities in which there are important impacts which are not valued in existing markets. When this occurs, it constitutes market failure. In the particular case of pest-management technology, any action of a producer or producers conducting pest management that has an influence on other

[7] A good discussion of private versus social values can be found in Tibor Scitovsky, *Welfare and Competition*, Richard D. Irwin, Chicago, pp. 181–188; also, J. C. Headley and A. V. Kneese, "Economic Implications of Pesticide Use," *Ann. N.Y. Acad. Sci.*, **160**:30–39 (1969).

producers' technical production activities or on consumers who are not directly involved in the market constitutes a case of market failure if the affected producers or consumers are not able to reflect that influence through the market. This is true whether the effect is beneficial or detrimental. Such technical effects represent what economists call technical externalities and, to the extent that the externalities must be corrected by some sort of political decision, they constitute a problem for public policy. It is in this way that environmental issues surrounding pesticide use have become public policy questions.

To be more explicit, consider a hypothetical situation (hypothetical because data are not yet available to provide an empirical example) in which a group of producers, group A, in an industry are carrying out pest-management activities that result in technical effects on another group of producers, group B, such that the production costs for group B are increased because of destruction of beneficial insects. If group A produces cotton, for example, there is an industry supply relationship for cotton which shows how much cotton group A can produce at various prices. Let this relation be

$$P = 25Q$$

where P = price per 500-lb bale of lint cotton, and Q = millions of bales of cotton annually. In addition, society, in purchasing cotton, has a demand relation which shows the amounts of cotton that would be taken at various prices. Let the industry demand relation be

$$P = 500 - 25Q$$

where P = price per 500-lb bale of lint cotton, and, Q = millions of bales of cotton annually. These two relations are graphed in Fig. 3.3 which shows that at 10 million bales the price is \$250 per bale and supply equals demand. The market is then in equilibrium.

In the absence of the external effects (external costs) that cotton production imposes on group-B producers, this market solution is optimal. The supply relation reflects the incremental costs to group A of producing cotton, and the demand relation reflects the incremental satisfaction received by consuming cotton.

It is necessary now to examine the net social benefits from this solution. Since the demand relation represents the incremental satisfaction from consuming cotton at various prices, the area under the demand curve can be used as a monetary measure of the satisfaction derived by consumers. This is the area of the trapezoid $OP^{**}EQ_0$; it has a money value of \$3.75 $\times$ 10^9 and represents the gross social benefits of the pro-

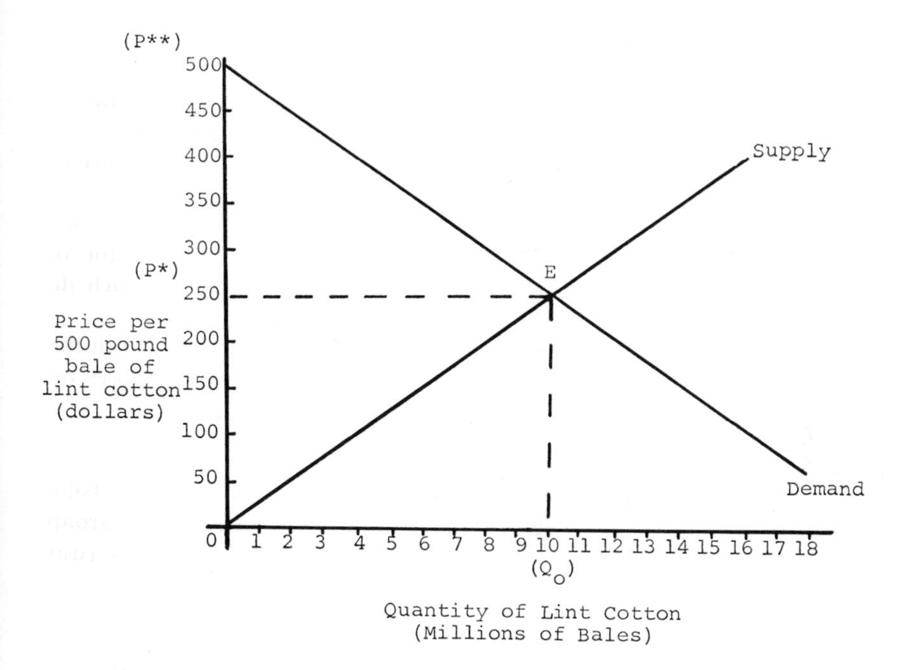

Fig. 3-3 Hypothetical demand and supply relations for cotton, annual.

duction. But consumers pay an amount equal to the area of the rectangle OP^*EQ_0 or \$250 × 10 million bales, which is equal to \$2.5 × 10^9, so there is a surplus here for consumers equal to the difference between the total monetary value of the satisfaction and the monetary value of what they paid for the 10 million bales. This difference is $(3.75 \times 10^9) - (2.5 \times 10^9) = \$1.25 \times 10^9$ and is the area of the triangle $P^*P^{**}E$. This is called the *consumer's surplus.*

It is necessary to compute the social costs of production by group A in the absence of the external effects. The logic is similar to that used for the social benefits. Therefore the area under the supply curve represents the costs of resources used to produce the cotton. This is the area of the triangle OEQ_0, and has a monetary value of \$1.25 × 10^9. But producers receive \$250 per bale for 10 million bales, or \$2.5 × 10^9, and therefore have a surplus of $(2.5 \times 10^9) - (1.25 \times 10^9) = \$1.25 \times 10^9$. This represents what is called the *producer's surplus.*

Finally, the net social benefits from the production of 10 million bales of cotton are the total social benefits (the area $OP^{**}EQ_0$) minus the social costs (the area OEQ_0) or $(3.75 \times 10^9) - (1.25 \times 10^9) = \$2.5 \times$

10^9. That is, the net social benefits are equal to the sum of the producer's and consumer's surplus.[8]

The net social benefits just computed are the benefits in the absence of any technical external effects, such as the one noted in which group A's action increased the costs of production for group B by destroying beneficial insects. When this externality exists, the total monetary value of the externality must be deducted from the net social benefits which do not include it.

One way of incorporating these external costs is to include them in the supply relation for group A. If the external costs group A imposes on group B are given by

$$P = 5.0Q$$

where P = price per 500-lb bale of lint cotton, and Q = millions of bales of lint cotton, these costs can be added to the supply relation for group A given previously. The new supply relation including external costs then becomes

$$P = 25Q + 5.0Q = 30Q$$

Figure 3.4 shows the situation when the external costs are included.

By including the external costs of cotton production in group A's supply relation, and therefore imposing these costs on group A, a new market equilibrium is found. The most obvious results are that the level of cotton output falls to slightly over 9 million bales and the price rises to $272.75 per bale.

The net social costs and the consumer and producer surpluses can be computed again after including external costs. Gross social benefits are the area under the demand curve $OP^{**}E_1Q_1$, equal to $\$3.5 \times 10^9$. Total revenue is the area of the rectangle $OP^{***}E_1Q_1$, equal to 272.75×9 million $= \$2.48 \times 10^9$. Therefore the consumer's surplus, the area of the triangle $P^{***}P^{**}E_1$, equals the difference between the gross social benefits and total revenue $[(3.5 \times 10^9) - (2.48 \times 10^9)] = \$1.02 \times 10^9$. The total social costs are given by the area of the triangle OE_1Q_1 and equal $\$1.24 \times 10^9$. Therefore the producer's surplus equals the difference between total revenue and total social cost, or $(2.48 \times 10^9) - (1.24 \times 10^9) = \$1.24 \times 10^9$. The net social benefits from cotton production after

[8] Mathematically, the net social benefits are given by the expression $[(\int_0^{10} (500 - 25Q)\, dQ)] - [(\int_0^{10} (25Q)\, dQ)]$ where the integrals represent the areas under the demand and supply curves, respectively. For a more complete discussion of net social benefits, see J. C. Headley and J. N. Lewis, *The Pesticide Problem: An Economic Approach to Public Policy*, Johns Hopkins Press, Baltimore, 1967.

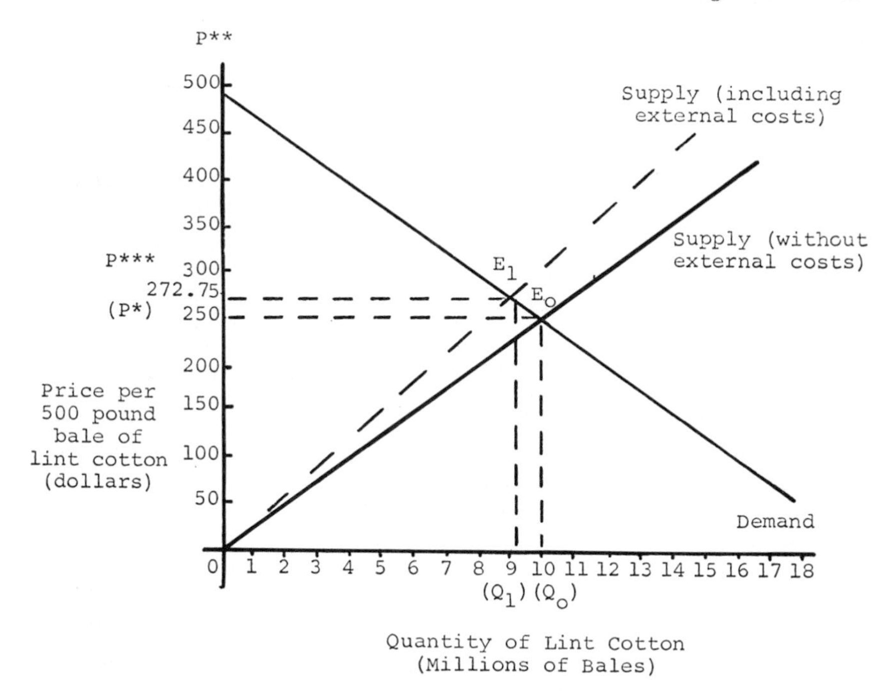

Fig. 3-4 Hypothetical demand and supply relations for cotton, including external costs from pest-management activities, annual.

including external costs, the sum of consumer's and producer's surplus, are $(1.02 \times 10^9) + (1.24 \times 10^9) = \$2.26 \times 10^9$. Table 3.6 summarizes the two situations for cotton production, as well as the level of external costs for group-B producers.

Recall that the external costs borne by group B were given by the relation

$$P = 5.0Q$$

where P = price per 500-lb bales of lint cotton, and Q = millions of bales. Then, before group A was required to consider the external effects, the costs imposed on group B were the area under the external cost relation. If $Q = 10$, then $P = \$50$ and the relation is a triangle with altitude 50 and base 10. Therefore the area of this triangle is $(50 \times \frac{10}{2} \times 10^6) = \$250 \times 10^6$. That is, at an output of 10 million bales of cotton, the external costs imposed on group B are \$250 million. After group A is forced to include or pay external costs, their output drops to 9.09 million bales,

Table 3.6 Comparison of Hypothetical Results of Including External Costs of Pest Management in the Market Solution for Cotton Production

Measures	No External Costs (1)	External Costs Included (2)	Difference (1 – 2)
Gross output	10 million bales	9.09 million bales	0.91 million bales
Price per bale	$250	$272.75	$–22.75
Gross social benefits	$3.75 billion	$3.5 billion	$0.25 billion
Total revenue	$2.5 billion	$2.48 billion	$0.02 billion
Consumer surplus	$1.25 billion	$1.02 billion	$0.23 billion
Producer surplus	$1.25 billion	$1.24 billion	$0.01 billion
Total social costs	$1.25 billion	$1.24 billion	$0.01 billion
Net social benefits	$2.5 billion	$2.26 billion	$0.24 billion
Total external costs generated	$250 million	$206.57 million	$43.43 million

and the external costs imposed on group B are the area under the external cost relation when $Q = 9.09$, providing group A continues to use the same method of pest management. This amounts to $(45.45 \times 9.09/2 \times 10^6) = \$206.57 \times 10^6$. So the external costs are reduced by $(250 \times 10^6) - (206.57 \times 10^6) = \$43.43 \times 10^6$.

The results of requiring group A to include external costs are (1) a reduction in cotton output, (2) an increase in the price of cotton, (3) a reduction in net social benefits from cotton production, and (4) a reduction in the external costs imposed on group B. It should be noted that this is only a partial list of the consequences of the action to impose the external costs created by group A on group A. As a result of the reduction in total external costs, group B can be expected to alter its output, since the cost reduction shifts the group-B supply relation to the right, resulting in an increase in the net social benefits from group B's production, an increase in group B's output, and a lowering of the price for group B's output.

It can be seen from Table 3.6 that the distribution of the losses in net social benefits resulting from imposing the external costs of cotton production on group A is shared by the producers and the consumers of cotton, with the consumers bearing most of the burden, that is, consumer surplus was reduced $0.23 billion as shown in Table 3.6, while the producer's surplus was reduced only $0.01 billion. One cannot generalize from these proportions, because they depend on the relative slopes of the supply and demand curves. What can be said is that, given a supply curve, the steeper the demand curve, the more the imposition of external costs will be borne by the consumers.

This analysis of the results of internalizing external costs shows how market failure can lead to inefficient resource allocation. In the case of the hypothetical external costs resulting from cotton production, the fact that the market fails to impose these costs on producers usually results in too many resources being allocated to cotton production. The opposite holds in general (although there are exceptions) when external benefits are involved—that is, too few resources are likely to be allocated to the externality-producing activity such as pest-management practices that increase populations of beneficial species.

Clearly, in a situation such as the hypothetical cotton production problem, which is industrywide, there is a need to correct for externalities to provide more efficient resource allocation. Such correction is a matter for public policy. Since the market has partially failed, it requires a political decision to provide an institution to substitute for the market. That is, the conditions necessary to justify public action are present.

There are alternative methods of dealing with externalities. Those who produce external costs can be taxed, or those who produce external benefits can be subsidized. The taxes or the subsidies can be applied to the output, or they can be applied to the inputs creating the external effects; for example, a tax might be levied on a broad-spectrum insecticide. Another method is by regulation of activities considered important generators of external effects. Regulation of pest-control chemicals is one method of controlling the level of external costs and benefits resulting from this form of pest control. All these are examples of public policy instruments for dealing with external effects related to pest management or any other activity.

While the discipline of economics can evaluate the consequences of various methods of dealing with external effects, usually the choice of method is a political choice. This is true because, as shown in Table 3.6, not only does a tax, for instance, change the resource allocation, it also alters the distribution of income, as between producers of group A and B and between producers and consumers. Economic efficiency is an objective goal. One can measure total benefits and decide if they have increased or decreased. But where income distribution is concerned, this is another matter and requires a subjective evaluation. While economists can show, as in Table 3.6, what the income distribution consequences are, they cannot say objectively that one distribution is superior to another. Therefore whether the income distribution changes that result from public policy are good or bad must be a political decision and not an expert decision.

B. Pest-Management Technology and the Fallacy of Composition

The adoption of technology based on the expectation of increased income by individual farmers can provide an aggregate result that is contrary to expectations. Individual producers adopting any technology do so because (1) they expect the technology to increase output with no increase in costs, or (2) they expect to decrease costs and produce the same output level, or (3) they expect to reduce the variation in output and therefore reduce variation in income. In reality a new technology may exhibit some of each of the three effects.

If the adoption of a technology results in increasing output, that is, raising the average yield for individual producers, there will be financial gains for the early adopters, since their increase in output will not be large enough to affect the price received by all farmers. But, as the technology becomes widely adopted, with the result that the output of the industry is increased enough to influence the price, the supply curve of the industry shifts downward and to the right, and lower prices result unless the demand has increased enough to offset the increase in supply.[9] For most agricultural commodities the lowering of price through an increase in supply with no change in the basic demand reduces total revenue to producers, because the demand is "inelastic." That is, a 1% increase in output reduces prices by more than 1%. The term *fallacy of composition* refers to this expectation that the action of individuals is independent of the result for the industry.

As an example, if producers expect that adoption of a new technology will increase output by 10%, from 100 to 110 units, and the expected price is $10, they will expect $100 increased total revenue and, if the addition to total costs were $50, there would be an expected $50 increase in net income. But if all producers adopt the technology, and it is effective, industry output will increase by 10% and the price may actually be $8 per unit. In this case the farmer producing 100 units, with a 10% in-

[9] A shift of the supply curve downward and to the right means that producers are willing to supply a larger quantity at the price that prevailed before the shift. The supply curves in Fig. 3.4 are examples of such a shift.

crease to 110 will receive a total revenue of \$880 (8 × 110), rather than \$1100 (10 × 110) expected. So, the plan backfires. The individual producer's situation resulting from the technology can be summarized as follows:

Change in output = +10 units
Change in total revenue = − \$120 (1000 − 880)
Change in total costs = +\$50
Change in net income = \$170 (120 + 50)

While this example is overstated, it is accurate in principle and illustrates the fallacy of composition. To the extent that pest-management technology tends to increase output, care should be taken so that expected income increases for individuals based on constant prices are not extrapolated to the industry. If it is assumed that the adoption of pest-management practices results in a reduction in technical external effects, increases in output will clearly benefit society as a whole, since consumers will have more products for less total cost. However, because of the inelastic nature of the demand for food, output increases without an offsetting increase in demand result in lower incomes for producers, and therefore it is the social benefit of more food at lower total cost that justifies the technology and not the effect on producers' incomes.[10]

The purpose of this discussion is not to condemn pest management as damaging to farm income. Much of pest management is aimed at reducing the cost of pest control, not increasing output. What the above argument does suggest is that, where pest-management practices can be expected to reduce pest competition compared to present methods, output can be expected to increase. When this is the case, practices should be analyzed to determine the distribution of benefits between producers and society as a whole as the basis for allocating the costs.

It should be emphasized again that, while economic thresholds are useful for individual decisions given an established technology of pest management, the aggregate result is not usually a linear sum of the expected individual benefits, and is therefore of no value for policy-making decisions affecting large groups of producers.

[10] A more detailed example of this problem is found in Zvi Griliches, "Research Costs and Social Returns: Hybrid Corn and Related Innovations," *J. Polit. Econ.* **66**(5):419–431 (1958).

III. SUMMARY AND CONCLUSIONS

The contents of this chapter are an attempt to survey the major elements of the economics of pest management. It covered the analysis of pest-management decisions at the microlevel under certainty and uncertainty, and in the static (timeless) and the dynamic (time-related) framework. The social economics of pest management were analyzed with respect to market failure and technical external effects as questions of public policy. Finally, pest-management technology was examined, considering the possible effects of a new technology on the aggregate outcome in terms of producer benefits and social benefits.

The major conclusions were that (1) the added cost of pest management must be less than or equal to the added benefits to justify its use; (2) the insurance principle is important given the uncertainty surrounding pest competition; (3) wherever possible, the time-related phenomenon of pest populations as a function of management activities should be considered to design optimal strategies over time; (4) external effects should be brought to bear on those persons responsible for generating such external effects in order to improve resource allocation; and (5) the use of individual results from pest management to generalize about societal and industry effects is likely to provide erroneous information for policy purposes.

The overriding conclusion is that the economic consequences of any technological development, and pest-management technology in particular, are the result of a complex interaction of technical and behavioral relations. Designing techniques to curb pest competition is one thing; designing pest-management techniques that advance the values of society is another. The latter requires the cooperation of biological, physical, and social science disciplines if the results are to be satisfactory.

REFERENCES

Carlson, G. A. 1970. A decision theoretic approach to crop disease prediction and control. *Amer. J. Agr. Econ.* **52**(2):216–223.

Griliches, Z. 1958. Research costs and social returns: Hybrid corn and related innovations. *J. Polit. Econ.* **66**(5):419–431.

Headley, J. C., and A. V. Kneese. 1969. Economic implications of pesticide use. *Ann. N.Y. Acad. Sci.* **160**:30–39.

Headley, J. C., and J. N. Lewis. 1967. The pesticide problem: An economic approach to public policy. Johns Hopkins, Baltimore. 141 pp.

Leftwich, R. 1966. The price system and resource allocation. Holt, Rinehart and Winston, New York. 402 pp.

Scitovsky, T. 1951. Welfare and competition. Richard D. Irwin, Chicago. 457 pp.

Taylor, C. R. 1972. Dynamic economic evaluation of pest control strategies. Unpublished Ph.D. thesis, University of Missouri, Columbia. 103 pp.

SELECTED READINGS

Carlson, G. A. 1973. Economic evaluation of producer pest management alternatives. *Proc. Nat. Ext. Pest Manage. Workshop, Baton Rouge, La., March 13–15.*

Shoemaker, C. 1971. The application of dynamic programming to agricultural ecology. *Dep. Math., Univ. S. Calif. Tech. Rep.* 71-29. Mimeo.

Smith, R. F. 1971. Economic aspects of pest control. *Proc. Tall Timbers Conf. Ecol. Anim. Control Habitat Manage.* 3:53–83.

TACTICS

4

PLANT RESISTANCE IN PEST MANAGEMENT

Marcos Kogan

INTRODUCTION

Definition

Resistance of plants to insects is the property that enables a plant to avoid, tolerate, or recover from injury by insect populations that would cause greater damage to other plants of the same species under similar environmental conditions. This property generally derives from certain biochemical and/or morphological characteristics of plants which so affect the behavior and/or the metabolism of insects as to influence the relative degree of damage caused by these insects.

From an evolutionary point of view, resistance traits are preadaptive characteristics of a plant. Plants with genes for these preadaptive features withstand the selective pressure of herbivore populations, thus increasing the chances for their survival and reproduction. Practical work in plant resistance is generally oriented toward revealing these preadaptive characteristics and using them in breeding programs.

Although work on resistance covers many insect–crop associations, in this chapter examples are drawn mostly from field and forage crops. The principles and basic techniques, however, apply to other crops as well.

There are numerous review articles that cover the various aspects of plant resistance (Snelling 1941; Auclair 1957; Painter 1958, 1968; Beck 1965; Van Emden 1966; Lupton 1967; NAS 1969; Pathak 1970; Luginbill 1969; Day 1972; Horber 1972; Sprague and Dahms 1972; Maxwell et al. 1973). The books by Painter (1951), and those edited by de Wilde and Schoonhoven (1969), Sondheimer and Simeone (1970), Rodriguez (1972), and Van Emden (1973) are recommended for a broad view of plant resistance and related fields.

Historical Development

The influence of man on plant evolution as a consequence of his agricultural practices is reflected in: (1) the domestication of certain plants, (2) the creation of a new class of plants called weeds, and (3) the interference with natural vegetation as a result of making room for crops to be grown or animals to be grazed (Baker 1972). Domesticated plants show both the intentional and accidental results of human actions. Unintentional selection for plant resistance against herbivore attacks probably occurred in the very early stages of agriculture.

The so-called rubbish-heap hypothesis of the origin of agriculture assumes that plants with weedy tendencies colonized kitchen middens and refuse heaps near man's dwellings. Those plants were gathered by man and gradually brought into cultivation (Hawkes 1970).

Of the thousands of weed species that colonized rubbish heaps, only a very few were domesticated. The ancestors of our domesticated cultivars were under heavy environmental pressures, including attacks by herbivores which certainly competed with man for the primitive crop. Under these conditions it is likely that only the least susceptible plants survived to produce viable seed for the next year's crop. This natural selection for resistant plants probably continued until man started to interfere actively with the process by favoring certain plants which were more tasty or produced more and bigger fruit or seed. At this point natural resistance was often inadvertently suppressed in favor of the other traits.

Improvement of the desired product frequently involved the intentional reduction of factors that coincidentally were involved in mechanisms of resistance. Starch reserves in the roots of primitive cassava plants are protected from herbivores by the presence of cyanogenetic glucosides. The sweet cassavas are low in glucosides and safer for man, but they also become more susceptible to herbivore attacks. In Africa bitter cassava may be the only crop plant that can be grown in regions where wild pigs, baboons, and porcupines are abundant. It is also resistant to grasshoppers (Baker 1972).

The selection of the bush habit in cultivated beans by American Indians involved major genetic changes in plant organization, which resulted in the loss of genes. According to Gentry (1969) ". . . as the bush bean diffused to other localities, the inadvertent selections resulted in many endemic varieties, narrowly tolerant, and genetically limited within themselves as breeding stocks. The genetically rich resource of the early ancestors was innocently subverted to dead ends. One obvious result is the weak, diseased and pest-ridden varieties common in many parts of Tropical America, and which can no longer supply a food need for the expanding human population." Modern breeding procedures have brought crops to extreme uniformity, rendering them genetically vulnerable. The genetic vulnerability of modern varieties is a major concern of plant geneticists (NAS 1971).

In modern times a wheat variety resistant to the Hessian fly, *Mayetiola destructor* (Say), was first reported in the late 1800s in the United States. In 1831, Lindley reported that in England the apple variety 'Winter Majetin' was resistant to the woolly aphid, *Eriosoma lanigerum* (Hausmann) (Painter 1951; NAS 1969).

The most dramatic early success in plant resistance, however, was the control of the grape phylloxera, *Phylloxera vitifoliae* (Fitch), in European grapevines. Balachowsky (1951) described the "brutal appearance" of this North American pest species in France in 1861 and its rapid spread to vineyards in other European and Mediterranean countries. The entire French wine industry was on the brink of collapse by 1880. Complete control of the pest was achieved by 1890, after French vineyards were reconstituted using grafts of the susceptible European grapevine scions on resistant North American rootstocks. The entire operation cost France 10 billion francs, but the French wine industry was saved.

Despite this spectacular beginning plant resistance attracted little attention during the first decades of this century. The systematic study of plant resistance to insects was initiated by Reginald H. Painter and coworkers at Kansas State University in the late 1920s. Awareness of the complexities and far-reaching implications of pest management (see Chapter 1) gave a new impetus to research and utilization of plant resistance as one of the soundest tactics in the repertoire of the agricultural entomologist.

I. INSECT–PLANT INTERACTIONS

The *host plant* of a phytophagous insect is the universe in which it finds nourishment and shelter. The set of plants with which an insect species is trophically associated is called its host–plant range. Species that are broadly adapted to fit successfully into a variety of habitats or eat a variety of foods are known as *eurytopic* (generalists) or *euryphagic* (food generalists) when referring to their food habits. Adaptive specialists are called *stenotopic* and, when referring to their food habits, *stenophagic* (food specialists) (Emlen 1973). If the taxonomic relationship of the host–plant range is considered, insects may be classified as *monophagous, oligophagous,* or *polyphagous* if their host ranges include plants of one or a few closely related species within a genus, several genera within a family, or several families in various orders of plants, respectively. These terms have also been used to define the nature of underlying chemical and physiological processes in insect–plant interactions (Thorsteinson 1953). Of course, these categories have no rigid boundaries and examples are found for a great variety of combinations in the constitution of host–plant ranges. One striking aspect of insect–plant relationships, however, is that, no matter how euryphagic an insect might appear, on closer examination it is frequently possible to uncover some underlying common traits among plants of its host range. These traits are in general of a chemical nature.

A. Evolution of Insect–Plant Interactions

It seems that phytophagous insects were originally polyphagous (Dethier 1954). They ate indiscriminately a wide variety of plants available in their pristine habitats. Some primitive plants evolved to produce and concentrate certain secondary metabolites which had an adverse effect on the insects feeding on them. Insects avoided feeding on these plants. However, certain biotypes evolved (mutations), which were able to bypass the barrier created by the odd compounds in the plants. The plants thus became acceptable food for these insects. The biotypes had exclusivity in the utilization of their new food plants, thus gaining an evident advantage over their competitors. In time these odd compounds became feeding excitants or stimulants (see Section I.B). What was in the beginning a chemical defense of plant against insect became the determinant of a more intimate association between insect and plant (Fraenkel 1959). One of the striking examples of such a development is offered by the Klamath weed and other plants of the genus *Hypericum*. These plants secrete a compound—hypericin—which causes photosensitivity and skin irritation, sometimes leading to blindness and starvation in animals that eat them. Klamath weed is avoided by most herbivores, but the beetle *Chrysolina brunsvicensis* Gravenhorts uses the hypericin secretions as an attractant to locate its food (Rees 1969).

This continuous process of development of new biochemical barriers by plants and adaptation by insects had a profound influence on the direction of the evolution of insects and plants. The mutually influenced evolutionary process exemplified by the relationships of butterflies and their host plants has been termed *coevolution* by Ehrlich and Raven (1964) (see Chapter 2). The chemicals involved in this process are called *allelochemics* (Whittaker and Feeney 1971), and they play a central role in host–plant resistance.

B. Behavioral and Physiological Components of Insect–Plant Interactions

The host-selection process in phytophagous insects is a chain of events (plant stimuli–insect responses) in which each link facilitates the unfolding of the next. Five major steps are defined in this process: (1) host-habitat finding, (2) host finding, (3) host recognition, (4) host acceptance, and (5) host suitability. A scheme of the generalized host-selection process in phytophagous insects is illustrated in Fig. 4.1, using the logic symbology of a flow chart of a computer program.

1. *Host-Habitat Finding*

Dispersing adult populations usually arrive at the general habitat of the host by mechanisms that involve phototaxis, anemotaxis, geotaxis, and probably temperature and humidity preferenda. These mechanisms have important ecological implications and are of interest in pest management, but they have little effect on plant resistance. Quite often agricultural pests stay in the general area where crops are planted, and this phase becomes less important in host selection.

2. *Host Finding*

Long-range sensorial mechanisms, probably visual and olfactory, bring the insect into close contact with a plant. Several aphid and whitefly species tend to alight on yellow surfaces, and larvae of certain beetles are attracted to vertical patterns. When odors of the host plant are present,

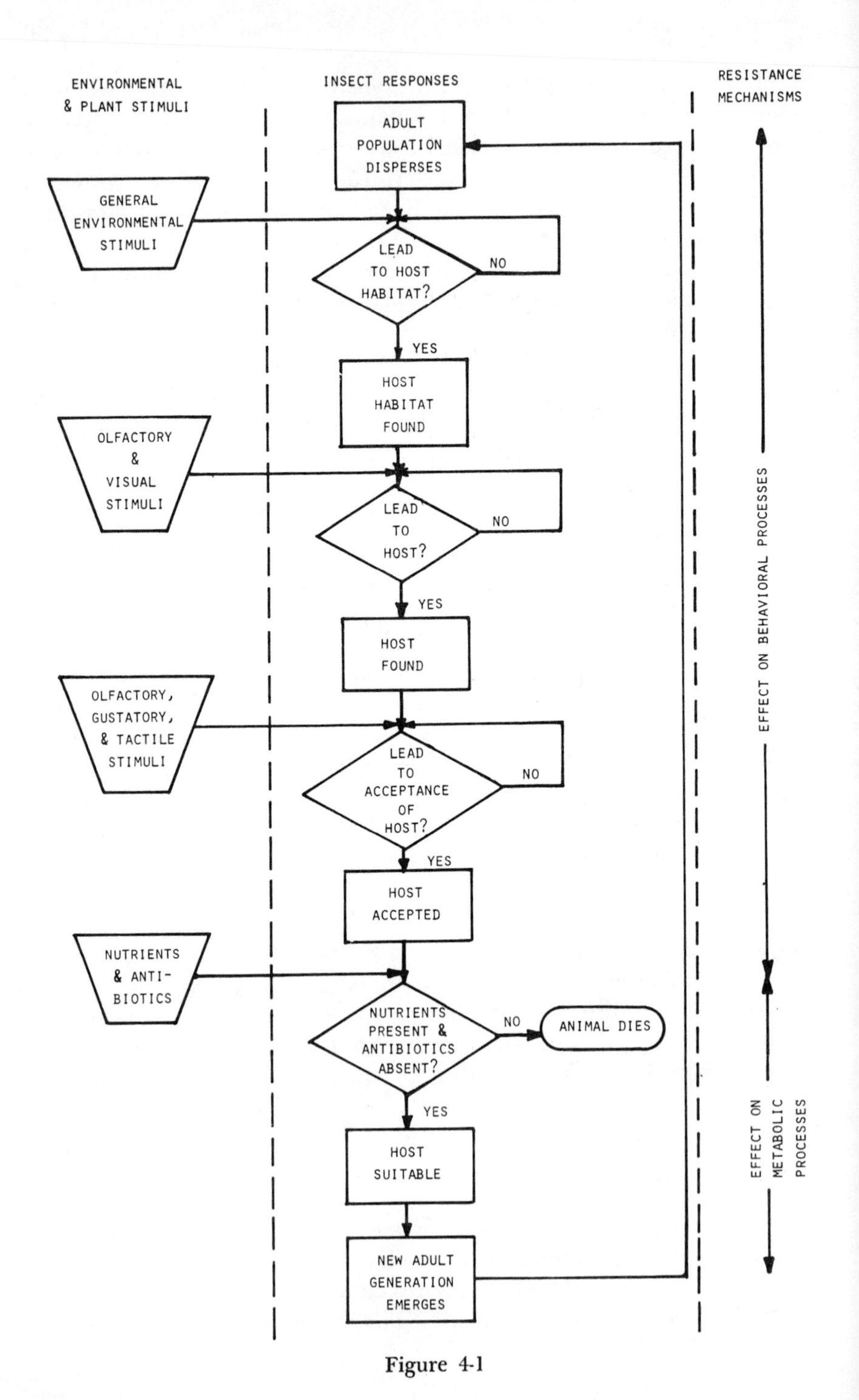
ENVIRONMENTAL
& PLANT STIMULI
INSECT RESPONSES
RESISTANCE
MECHANISMS
ADULT
POPULATION
DISPERSES
GENERAL
ENVIRONMENTAL
STIMULI
LEAD
TO HOST
HABITAT?
NO
YES
HOST
HABITAT
FOUND
OLFACTORY
&
VISUAL
STIMULI
LEAD
TO
HOST?
NO
YES
HOST
FOUND
OLFACTORY,
GUSTATORY,
& TACTILE
STIMULI
LEAD
TO
ACCEPTANCE
OF
HOST?
NO
YES
HOST
ACCEPTED
NUTRIENTS
& ANTI-
BIOTICS
NUTRIENTS
PRESENT &
ANTIBIOTICS
ABSENT?
NO
ANIMAL DIES
YES
HOST
SUITABLE
NEW ADULT
GENERATION
EMERGES
EFFECT ON BEHAVIORAL PROCESSES
EFFECT ON
METABOLIC
PROCESSES

Figure 4-1

grasshoppers and the Colorado potato beetle, *Leptinotarsa decemlineata* (Say) tend to fly upwind (positive anemotaxis), increasing the chances of locating the host (Schoonhoven 1972).

The aggregation of individuals of a given insect species on particular plants is the statistical end result of a higher frequency of landings and a longer duration of visits (Thorsteinson 1960). On contacting the host short-range tactile and olfactory sensorial inputs arrest further movement, causing the insect to remain on the plant. Tarsal and antennal chemoreceptors in contact with a plant receive the stimuli that signal "landing on the right host." Host-finding behavior in aphids has been studied more than in any other phytophagous insect, and it closely follows the mechanisms described above (Kennedy and Fosbrooke 1973). After landing the aphids *Rhopalosiphum incertus* (Walk.) and *Aphis pomi* De Geer apparently perceive a flavonoid (phloricin) which typically occurs in leaves and other organs of apple trees, the preferred host of the aphids (Schoonhoven 1972).

3. Host Recognition

Although larvae are endowed with the sensorial equipment for certain levels of host recognition, quite frequently this phase has been taken care of by the ovipositing female. Certain grasshoppers are known to bite a plant before ovipositing. Caterpillars receiving the proper stimuli test-bite a plant. This first bite causes other chemicals stored intracellularly in the plant to stimulate the gustatory receptors.

4. Host Acceptance

Different chemicals apparently govern the various phases of the feeding process. In the silkworm larvae, *Bombyx mori* (L.), a series of compounds extracted from mulberry leaves was associated with initial biting, swallowing, and continuous feeding (Hamamura 1970). When in the presence of the correct sensorial inputs, caterpillars continue to feed to satiation.

5. Host Suitability

The nutritional value of the plant and the absence of toxic compounds finally determine the adequacy of the food to sustain the various physiological processes related to growth and development of the larvae, and

---

Fig. 4-1 Diagram of the generalized host-selection process in phytophagous insects. The trapezoid symbol is used for inputs (stimuli); the diamond is used for sensorial and integrative processes leading to a "decision"; processes within a stage of the host-selection sequence are shown in rectangles.

longevity and fecundity of the adults. There is little sensorial involvement during this phase, which together with host acceptance is of key importance to the mechanisms of resistance.

C. Plant Components in the Interactive System

Physical and chemical plant components intervene at the various phases of the host-selection process.

1. Physical Factors

Certain morphological characteristics of the host plant such as succulence, toughness, pilosity, and presence of thorns or spines are regarded as *permissive factors* whose presence may act as barriers to normal feeding or oviposition.

Other physical characteristics such as color and shape have an influence in host finding but are considered too general to be critical cues for host recognition. The moth *Autographa precationis* (Guenee) does not oviposit on dandelion, *Taraxacum officinale* Weber, a highly preferred food for the larvae, apparently because the shape of the plant does not elicit the necessary oviposition behavior (Fig. 4.2) (Khalsa and Kogan, unpublished data). In certain cases, however, color may allow aphids to find a host in a more adequate physiological stage (Kennedy 1958).

2. Chemical Factors

Figure 4.3 shows in very general lines the relationships between substances of the primary plant metabolism—sugars, amino acids, purine and pyrimidine bases—and their fundamental polymers with secondary metabolites.

The external environment surrounding the plant is dominated by compounds of the secondary metabolism, which exude from the outer layers of tissues. The compounds generate the olfactory stimuli that prevail in host finding and recognition. The internal plant environment, however, is formed of a complex mixture of compounds, some having nutritional value, some that act as feeding excitants or inhibitants, some toxic, and probably a large number of inert compounds. In general, primary metabolites and their polymers are nutrients that are converted into insect body matter or utilized to produce energy. Secondary metabolites quite often act as token stimuli and have no nutritional value.

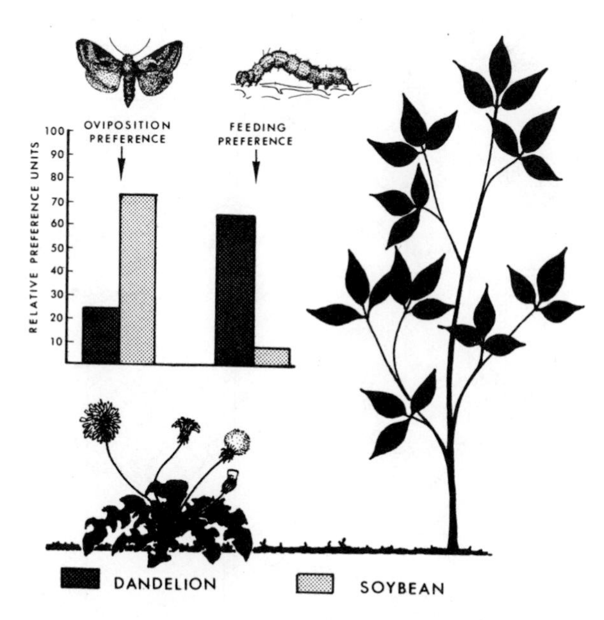

Fig. 4-2 Oviposition (left) and larval feeding (right) preferences of *Autographa precationis* on dandelion, *Taraxacum officinale* Weber, and on soybeans. Adult moths prefer ovipositing on soybeans, but larvae show a marked feeding preference for dandelion. Among other stimuli the erect habit of the soybean plant is believed to influence oviposition preferences.

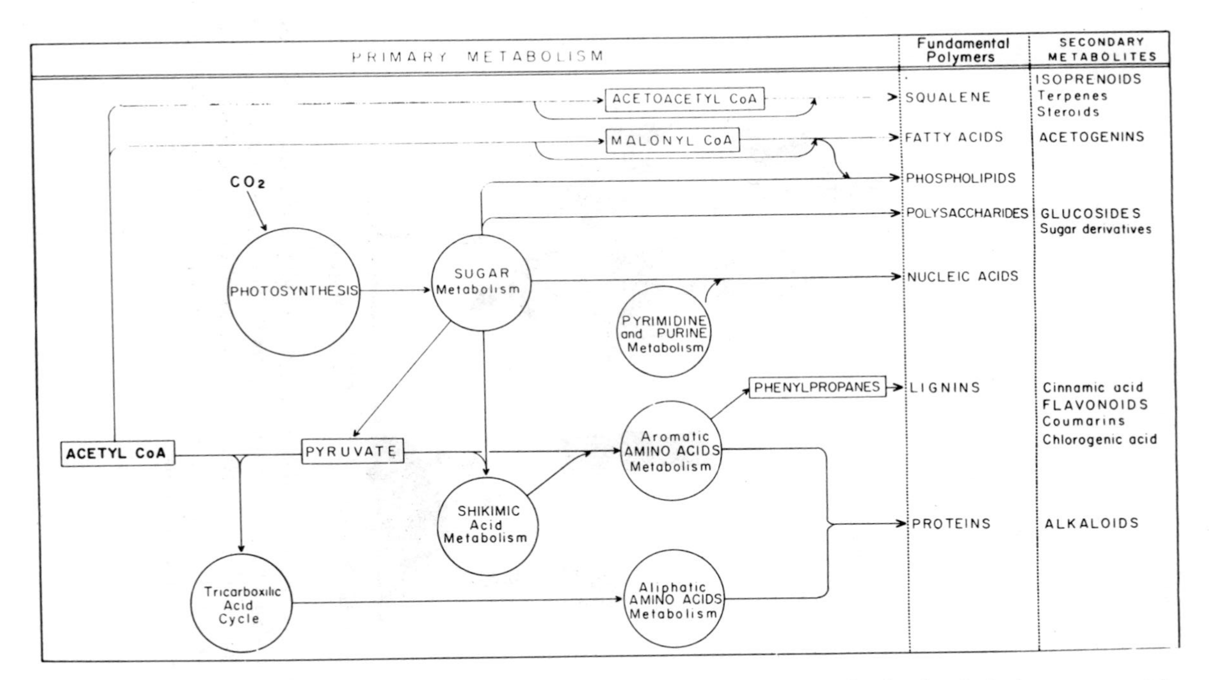

Fig. 4-3 Relationships among various groups of plant components on the basis of their common biosynthetic origins. (Adapted from Bu'Lock 1965, and Hendrickson 1965.)

Theories of host-plant selection have either stressed the role of secondary metabolites (Fraenkel 1969) or ascribed to nutrients an equally important role (Kennedy 1965). It now appears from behavioral and electrophysiological studies that the perception of the host "odor" or "taste" by a phytophagous insect is a holistic phenomenon. Stimuli derived from nutrients and odd compounds are intertwined in complex sensorial inputs. These inputs are decoded by the insect's central nervous system into an expression of "host" or "nonhost," which is the central link in the chain of events of the host-selection process.

D. The Relationships of Plant Stimuli and Insect Responses

Several classifications were proposed to correlate plant stimuli with the various responses observed in insects during the host-selection process. Table 4.1 combines three of the most frequently used systems (Dethier et al. 1960; Beck 1965; Whittaker and Feeney 1971).

Table 4.1 Principal Classes of Chemical Plant Factors (Allelochemics) and the Corresponding Behavioral or Physiological Effect on Insects[a]

Allelochemic Factors	Behavioral or Physiological Effects
Allomones	Give adaptive advantage to the producing organism
Repellents	Orient insects away from plant
Locomotor stimulants	Start or speed movement
Suppressants	Inhibit biting or piercing
Deterrents	Prevent maintenance of feeding or oviposition
Antibiotics	Disrupt normal growth and development of larvae; reduce longevity and fecundity of adults
Kairomones	Give adaptive advantage to the receiving organism
Attractants	Orient insects toward host plant
Arrestants	Slow or stop movement
Excitants	Elicit biting, piercing, or oviposition
Feeding stimulants	Promote continuation of feeding

[a] Adapted from Dethier et al. 1960; Beck 1965; Whittaker and Feeney 1971.

II. HOST-PLANT SELECTION AND MECHANISMS OF RESISTANCE

Every step in the catenary process of host-plant selection is mediated by plant components. All key components must be present at the proper time and in adequate amounts for normal growth and development of the insects. Plant resistance thus may result from disruption of the normal sequence of events either because of a reduction in the level or repression of the action of kairomones, or because of the enhancement of allomones.

Resistance characteristics are under genetic control. Some characters, however, are very labile and fluctuate wildly under the influence of environmental conditions. Accordingly, the mechanisms of resistance may be classified as those that are under the primary control of environmental factors—ecological resistance—and those that are under the primary control of genetic factors—genetic resistance.

A. Ecological Resistance

1. Phenological Asynchrony

Selecting a plant at the proper stage of development is often just as important for an oligophagous insect as is the selection of the right host. A larva that requires fruiting structures for normal growth may starve on the leaves of that same host plant. The phenologies of the plant and of the insects must be synchronized in order for a plant structure to exist when a certain insect stage needs it. Alterations in plant growth patterns that result in asynchronies of insect–host phenologies constitute the modality of resistance called host evasion (Painter 1951).

Asynchronies may be induced by early or late planting of certain plant varieties. The early varieties of soybeans planted in Illinois mature in the beginning of September. When the second generation of the bean leaf beetle, *Cerotoma trifurcata* (Forster), emerges in soybean fields in the early part of September, most plants are already mature and almost ready to harvest. Thus these beetles cannot harm the crop, since the mature pods are not an adequate food. The short life cycle of the crop in central Illinois apparently allows it to escape attacks by the beetles in the latter part of the growing season (Kogan et al. 1974). Rabb (1969) showed that the size, succulence, and maturity of tobacco greatly influenced oviposition by the tobacco hornworm, *Manduca sexta* (L.). In general, most eggs are laid on early-planted tobacco in the initial part of the season. Since early-planted tobacco matures earlier than late-planted

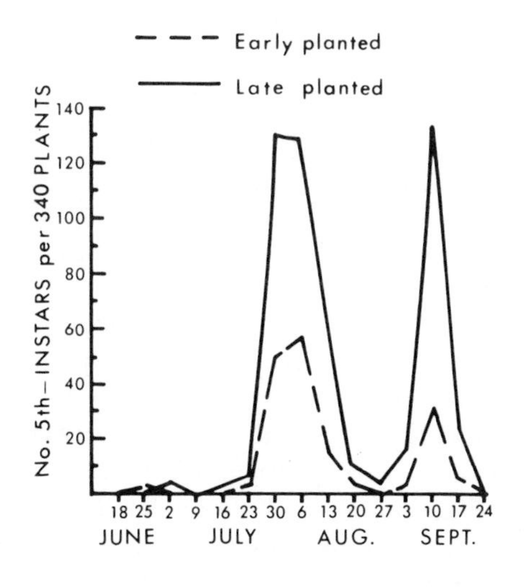

Fig. 4-4 The effect of planting date on infestations of the tobacco hornworm. (Redrawn from Rabb 1969.)

tobacco, the latter is more heavily infested after the end of June. Figure 4.4 clearly shows the differences in fifth-instar hornworm populations on early- and late-planted tobacco plants.

In certain instances, however, multivoltine species are affected in different ways in each generation. Late-planted corn in the Midwest is less infested by the first generation of the European corn borer, *Ostrinia nubilalis* (Hübner), but is more susceptible to attacks by second-generaton larvae (Everett et al. 1958).

Phenological asynchronies are considered a modality of pseudoresistance by Painter, since plants that evade insect attacks by this mechanism may in fact be susceptible if the pest occurs at the right time. Length of life cycle, however, is under genetic control, and breeding for earliness or lateness to enhance asynchronies may be part of a resistance program. In soybeans two independent pairs of genes (E_1e_1 and E_2e_2) affect time of flowering and maturity. When genes were incorporated into a variety 'Clark' background, the normal (e_1E_2) plants matured in 134 days, the early (e_1e_2) ones in 115 days, and the late ones (E_1E_2) in 152 days (Bernard 1971). With a difference of over 1 month in the maturity of the early and the late lines, adjustments can be made to maximize asynchronies if needed to manage bean leaf beetle populations.

2. *Induced Resistance*

Certain environmental conditions may alter the physiology of a plant to the extent that it becomes unsuitable as a host. Under induced resistance are grouped the responses of crop plants to normal cultural practices such as fertilization and irrigation, which may cause drastic quantitative or qualitative changes in the plant.

Phytophagous insects are very sensitive to nutritional changes in the host plant. These changes commonly result from fertilizers absorbed through the roots. Foliar applications of some mineral nutrients and even certain insecticides have also been shown to influence the nutritional value of plants.

Of particular interest is the response to N, P, and K fertilizers, because their incorporation is a routine practice in most crops; they are practically an inherent part of the agroecosystem. It seems that the balance between the three macronutrients in the plant is critical for normal insect development. At a high-N budget insects usually respond with an increase in survival and faster rates of development. Van Emden (1966) indicated that aphids are particularly sensitive to the levels of N in the plant, but respond negatively to the levels of K, even in the presence of high N. He suggested the possibility of using this N–K relationship to induce resistance to aphids. Rodriguez (1960) and Singh (1970) wrote extensive reviews on the effect of plant nutrition on insect pests.

B. Genetic Resistance

In opposition to phenological asynchronies and induced resistance which occur under the strict control of ecological factors, under genetic resistance are grouped those mechanisms based on inherited characters whose expression, although influenced by the environment, is not strictly under environmental control.

The three classic modalities of resistance established by Painter (1951)—*nonpreference, antibiosis,* and *tolerance*—are modalities of genetic resistance. The term nonpreference, although widely used in plant resistance literature, is avoided here. It is semantically confusing to read that a plant shows a nonpreference type of resistance, while preference is in fact displayed by the insect.

1. Resistance Factors Influencing Behavioral Processes

When offered a choice of two or more alternative foods, phytophagous insects usually display a consistent pattern of preferences. Inadequate hosts are often totally rejected, and some insect species starve to death on a diet that lacks the proper stimuli. This type of behavior has been exhaustively tested by workers in the biological control of weeds using the so-called starvation tests (Zwölfer and Harris 1971).

Similar displays of preference are observed in females selecting oviposition sites. Two behavioral responses, feeding preferences of the immature forms and oviposition preferences of the adult female, quite frequently, but not always, coincide. Observations have been made of butterflies ovipositing on plants that are toxic to the larvae, and others rejecting plants that are perfectly adequate food for the immature forms. It is possible that food selection and the selection of oviposition sites are governed by similar plant stimuli, although in some cases different factors may be involved.

Resistance factors influencing feeding preferences have been identified in several cases. Kishaba and Manglitz (1965) measured the resistance to aphids on certain clones of alfalfa and clover. Aphids were offered a choice of leaves from resistant and susceptible clones, and preference was determined by observing the aggregation of insects on plants after a period of time. In the beginning of the experiment, the aphids were uniformly distributed among all plants. After 20 hours, three to four times more aphids had aggregated on the susceptible than on the resistant plants.

The Mexican bean beetle, *Epilachna varivestis* Mulsant, is an oligophagous insect associated with legumes of the genus *Phaseolus*. It is also a serious pest of soybeans in many regions of the United States, and resistance to the beetle has been found in several soybean lines (Van Duyn et al. 1971). Dual-choice experiments were used to study the preference patterns of adults feeding on various lines and varieties of soybeans. The susceptible variety 'Harosoy 63' was used as a standard against which all others were compared. Various levels of preference were recorded among the plants tested. Comparisons of the resistant line PI 229358, the susceptible variety 'Bragg' and the F_1 cross PI 229358 × 'Bragg' clearly showed that the preference was 'Bragg' > F_1 cross > PI 229358, suggesting a semidominant type of the resistance factor (Kogan 1972a). Similar dual-choice experiments using lepidopterous larvae are illustrated in Fig. 4.5, in which the amount of feeding after 12 hours shows definite patterns of preference.

Oviposition preferences are usually measured by counting the number of eggs oviposited by adult females on susceptible or on resistant plants offered simultaneously. The rice stem borer, *Chilo suppressalis* (Walker), deposited 10 to 15 times more egg masses on susceptible than on resistant rice varieties (Pathak 1970). Maxwell et al. (1969) reported the presence of an unidentified oviposition deterrent in *Gossypium barbadense* L.

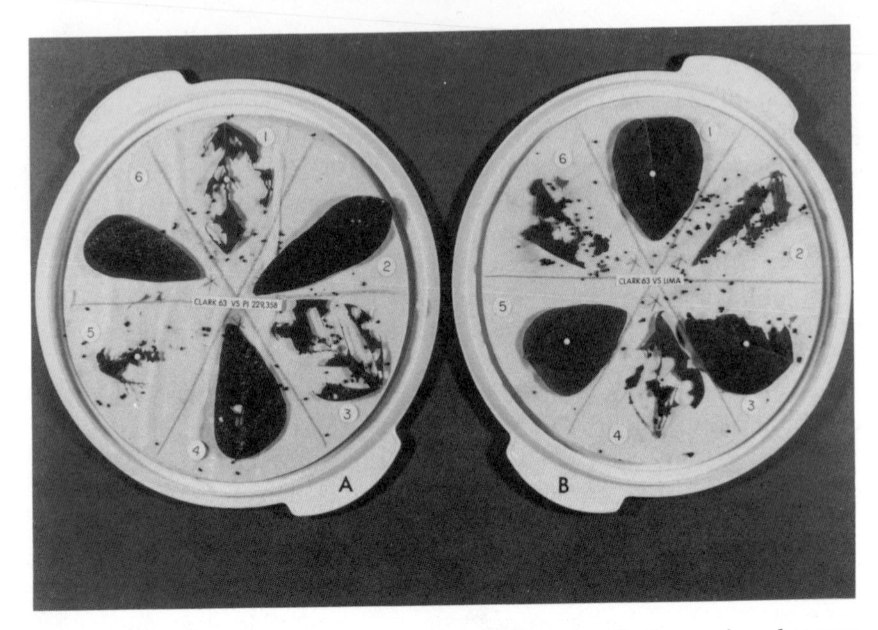

Fig. 4-5 Feeding preferences of the soybean looper. (*A*) Comparison between the susceptible soybean variety 'Clark 63' and the resistant soybean plant introduction PI 229358, showing active feeding on the 'Clark 63' leaflets (positions 1, 3, and 4). (*B*) Comparison between the soybean variety 'Clark 63' and the preferred lima beans showing more active feeding on the lima bean leaflets (position 2, 4, and 6).

cotton. This deterrent factor has been bred into upland cotton germplasm, and several lines have been selected that carry 25 to 40% oviposition reduction in comparison with standard commercial type cottons.

Although there is copious information on feeding excitants in relation to many insect-host associations (see reviews by Fraenkel 1969; Schoonhoven 1968, 1969, 1972; Dethier 1970), little is known about the intimate mechanisms of resistance that affect feeding preferences. Three alternative conditions are conceivably possible: (1) the normal host kairomones are absent or at subthreshold levels in the resistant lines; (2) resistant lines have the normal complement of host kairomones, however, they are inhibited or blocked by antagonistic compounds; (3) only allomones are present in the resistant lines. Evidence for some of these conditions was offered by Maxwell and co-workers, who demonstrated the presence of a feeding suppressant or deterrent in *Hibiscus syriacus* L. When the calyx of *H. syriacus* flowers was removed, the boll weevil fed and oviposited on this plant as well as it did on cotton. On plants with intact flowers, no feeding was observed. The calyxes were extracted, and a nonvolatile, water-soluble material was isolated, which strongly deterred feeding by the weevils. It was concluded that the rejection of *H. syriacus* by the boll weevil in nature results from a low level of attractants, a high level of repellents, and the highly active feeding deterrent extracted from the calyx, which overrides feeding and oviposition excitants which also occur in the plant.

There are no outstanding examples of resistance attributed exclusively to effects on behavioral mechanisms. In field situations insects may feed or oviposit on a nonpreferred host if a better one is not available. Perhaps for this reason resistance factors affecting behavioral mechanisms have been considered of secondary importance. This mechanism, however, may have a more subtle impact on pest populations, and the use of partially nonpreferred or even of more preferred varieties may find use in yet untested methods of cultural control (see Section VIII.B).

2. *Resistance Factors Influencing Metabolic Processes*

The term *antibiosis* encompasses all adverse physiological effects of a temporary or a permanent nature resulting from the ingestion of a plant by an insect. The temporary nature of an antibiotic effect can be demonstrated by transferring the test insect from a resistant to a susceptible plant. On the susceptible host the symptoms of antibiosis disappear, and the insect recovers to its normal physiological state. If, however, antibiosis is due to toxic principles, the symptoms may be irreversible.

Insects fed resistant plants may manifest antibiotic symptoms which range from lethal or acute to very mild or subchronic. The main symptoms commonly observed are (1) death of the larvae in the first few stadia, (2) abnormal growth rates, (3) abnormal conversion of ingested and/or digested food, (4) failure to pupate, (5) failure of adult emergence from the pupae, (6) malformed or subsized adults, (7) failure to concentrate food reserves followed by unsuccessful hibernation, (8) decreased fecundity, (9) reduced fertility, and (10) restlessness and other irregular behavior.

The possible physiological explanations for these symptoms are (1) presence of toxic metabolites (alkaloids, glucosides, quinones), (2) absence or suboptimal amounts of some essential nutrient, (3) unbalanced proportions of nutrients, (4) presence of antimetabolites that render some essential nutrients unavailable to the insects, and (5) presence of enzymes that inhibit normal processes of digestion of food and consequently utilization of nutrients.

Antibiosis is perhaps the most evident mechanism of resistance, and many reports are found in the literature on this subject. A few of these reports are discussed to illustrate the key features of antibiosis.

A convincing demonstration of the nutritional nature of the antibiotic effect on some corn inbreds resistant to the European corn borer was offered by Scott and Guthrie (1966). Resistant corn inbred C.I.31A plants and susceptible WF9 plants were infested with three egg masses of the European corn borer (60 to 70 eggs). Fourteen days after infestation all plants were dissected, and the number of live larvae was recorded. The difference was of the order of 30 live larvae on WF9 to 1 on the resistant C.I.31A. Subsequently, a series of nutrient solutions was pipetted into the whorl of resistant corn plants at a rate of 5 ml solution per plant at 0, 3, 7, and 10 days after initial infestation. Previous laboratory tests had shown that the nutrient solutions contained the ingredients of a completely adequate meridic medium for the borer larvae. The resistant plants thus treated had an average of 31.4 live larvae per plant, whereas the untreated controls had an average of only 1.4 larvae per plant. Similar experiments were repeated under field conditions, and certain nutrients were omitted in different treatments. The most evident effect was obtained when ascorbic acid was deleted from the medium. This result led the investigators to conclude that ascorbic acid, in the presence of one or two more ingredients of the original medium, accounted for the increased larval survival on the treated resistant plants. Indeed, ascorbic acid has been identified as one of the corn leaf factors required for normal larval growth (Chippendale and Beck 1964).

Another nutritionally related antibiotic effect was demonstrated in studies with the rice variety 'Mudgo,' which is resistant to the brown plant hopper, *Nilaparvata lugens* Stal (Pathak 1970). Young females feeding on 'Mudgo' plants had underdeveloped ovaries and contained few mature eggs. The symptoms were ascribed to the reduced asparagine content of the resistant rice.

The effects of nutritional antibiosis are manifested at times by subtle variations in food intake and utilization. A study of these effects on Mexican bean beetle larvae feeding on 21 resistant and susceptible lines and varieties of soybeans and on other legumes revealed a gradual decrease in suitability from the best host, *Phaseolus vulgaris* L., to the worst, the resistant soybean line PI 229358. The range of responses observed could be explained by one or more of the following factors: (1) improper and/or suboptimal levels of feeding excitants, (2) generally reduced levels of the nutritional value of the diet, (3) unbalanced alteration of nutrient proportionality, (4) presence of physical feeding deter-

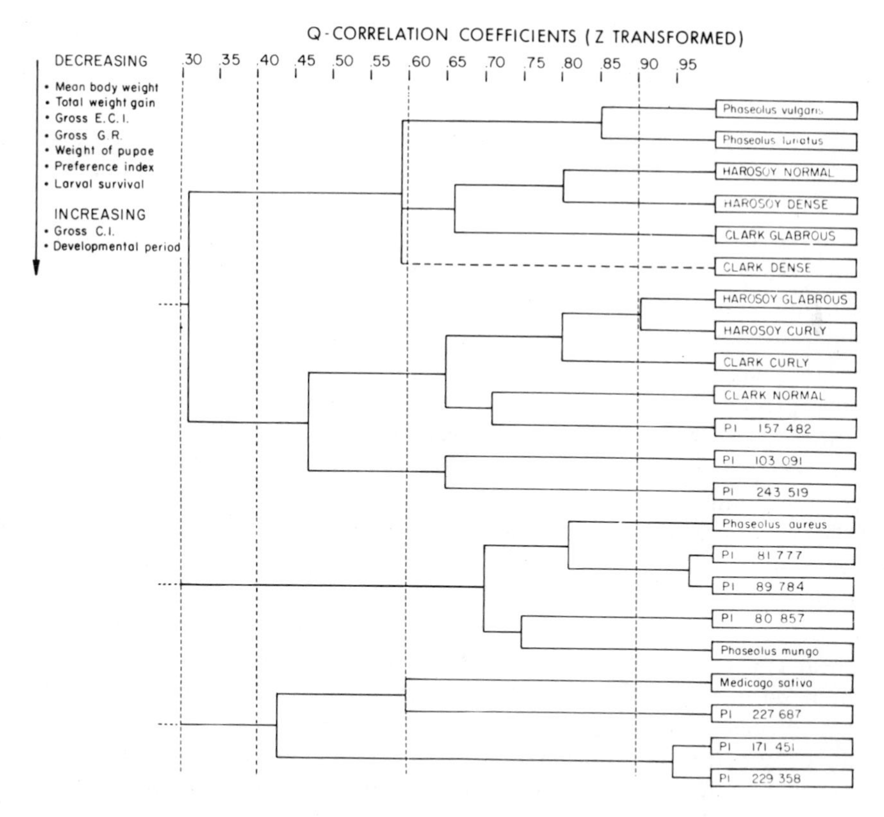

Fig. 4-6 Correlation dendrogram of 21 varieties and lines of soybeans (PIs) and other legumes based on the physiological responses of Mexican bean beetle larvae feeding on these plants. ECI, efficiency of conversion of ingested food; GR, growth rate; CI, consumption index. (From M. Kogan. 1972. Intake and utilization of natural diets by the Mexican bean beetle–*Epilachna varivestis*– a multivariate analysis. Fig. 4, page 119 in J. G. Rodriguez (ed.) Insect and mite nutrition. Copyright North Holland Publishing Company, reproduced by permission.)

rents, and (5) presence of chemical antibiotic factors. Figure 4.6 illustrates the relationships of the 21 cultivars, lines, and species based on 10 types of physiological and behavioral responses recorded with the Mexican bean beetle fed each of these plants (Kogan 1972b).

Some of the most complete studies on antibiosis were made by Beck and co-workers at the University of Wisconsin, and by Brindley, Klun and co-workers at the Ankeny, Iowa, laboratory of the U.S. Department of Agriculture using the corn plant and the European corn borer. The following is a summary of these investigations, based mainly on Beck (1965) and Klun and Brindley (1966).

Very young corn plants are highly resistant to European corn borer establishment and survival. Many lines become susceptible as they mature, but others retain much of their juvenile resistance. Evidence for the presence of borer-toxic compounds was obtained with tissues of both very young corn plants and resistant corn lines. These compounds were temporarily called resistance factor A (RFA), an ether-soluble fraction, and resistance factor B (RFB), an ether-insoluble fraction of plant extracts. Later RFA was shown to contain two distinct compounds, one subsequently identified as 6-methoxybenzoxazolinone (6-MBOA), and the other 2,4-dihydroxy-7-methoxy-1,4-benzoxazine-3-one (DIMBOA). RFB has never been isolated, and it seems to be of secondary importance in borer resistance.

A group of Finnish workers led by Virtanen (Klun and Brindley 1966) demonstrated that 6-MBOA is not an *in vivo* constitutent of corn tissue. They reported that the glucoside 4-O-glucosyl-2,4-dihydroxy-7-methoxy-1,4-benzoxazin-3-one is in fact the *in vivo* precursor of 6-MBOA in both wheat and corn. When the plant tissue is crushed by the mandibles of the borer, for instance, the glucoside is rapidly hydrolyzed to glucose and the aglucone DIMBOA. DIMBOA is then converted to 6-MBOA by heating in an aqueous solution. The *in vivo* conversion of DIMBOA to 6-MBOA probably involves enzymic mechanisms. The process of degradation of the glucoside is as follows.

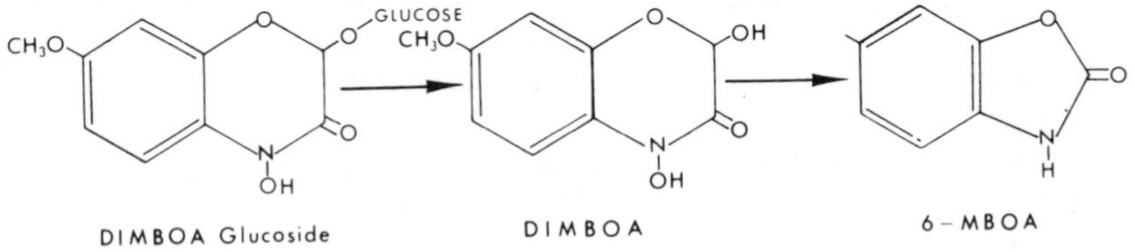

DIMBOA Glucoside → DIMBOA → 6-MBOA

Klun and Robinson (1969) studied the concentration of DIMBOA in dent corn in various tissues at various stages of development. They observed that concentrations were highest in the root and then in decreasing order in the stalk, whorl, and leaves, and that these concentrations differed in each inbred. The high concentration of DIMBOA in seedlings could explain the high level of resistance of young corn plants to the European corn borer. The inbreds that maintained high concentrations of DIMBOA in the whorl at later stages of development were borer-resistant. Resistance to the corn borer was thus dependent on the presence of an effective concentration of the resistance factors in the right tissues at the proper stage of growth of the plant. DIMBOA also has been implicated in corn resistance to the corn earworm, *Heliothis zea* (Boddie) (Reed et al. 1972).

The discovery and identification of a biologically active compound such as DIMBOA is of great scientific interest. It also has practical importance, because it offers plant geneticists a character that can be analyzed and quantified with great precision throughout the stages of a breeding program.

3. Phenetic Resistance

Under this category are grouped the resistance mechanisms related to morphological or structural plant features that impair normal feeding or oviposition by insects, or contribute to the action of other mortality factors.

A direct effect of morphological characters on feeding is the inability of many leafhoppers to reach the parenchyma of plants whose epidermis is covered with a thick layer of long cellulose hairs. Figure 4.7 shows segments of soybean stems of near-isogenic lines of the varieties 'Clark' and 'Harosoy,' and several lines selected for resistance to the Mexican bean beetle. The normal and dense pubescent types are highly resistant to the potato leafhopper, *Empoasca fabae* (Harris) (see Fig. 1.2). A similar effect is observed on cotton in which length and density of hairs are in inverse correlation with leafhopper populations (Fig. 4.8).

Pubescence is only one of many structural characters that affect plant resistance. Characters in corn responsible for increased mortality of corn earworms include husk length, husk tightness, and silk balling. A silk ball is formed when the silk at the apex of the ear, which is the last to elongate, does not grow through the silk channel already filled with silk from the rest of the ear. Instead, the silk at the apex piles up in layers to form an N-shaped ball. Earworms may develop on the silks, but they do not reach the ear (Luckmann et al. 1964). Apparently, the effect of a long husk and a long silk channel is actually due to prolongation of exposure of the worms to the chemical factors present in the silk (Beck 1965).

Agarwal (1969) reviewed the morphological characters related to sugarcane resistance to the sugarcane borer, *Diatraea saccharalis* (Fabricius). Young larvae first feed on leaf and leaf sheath tissues, later entering the stalk. Denticles on the midribs of leaves, the number of vascular bundles, the lignification of the cell walls, and the number of layers of scleren-

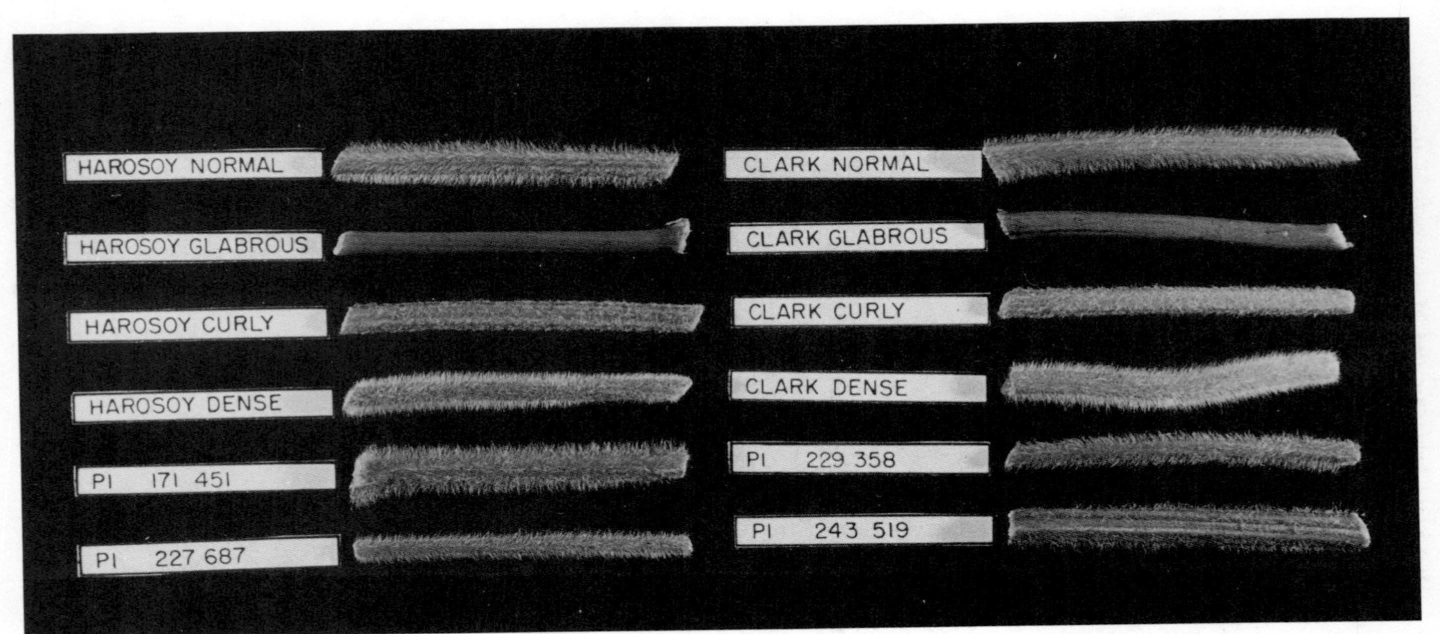

Fig. 4-7 Segments of stems of 12 soybean varieties and lines (PIs) showing various types of pubescence. The glabrous and curly types are most susceptible to potato leafhopper attacks.

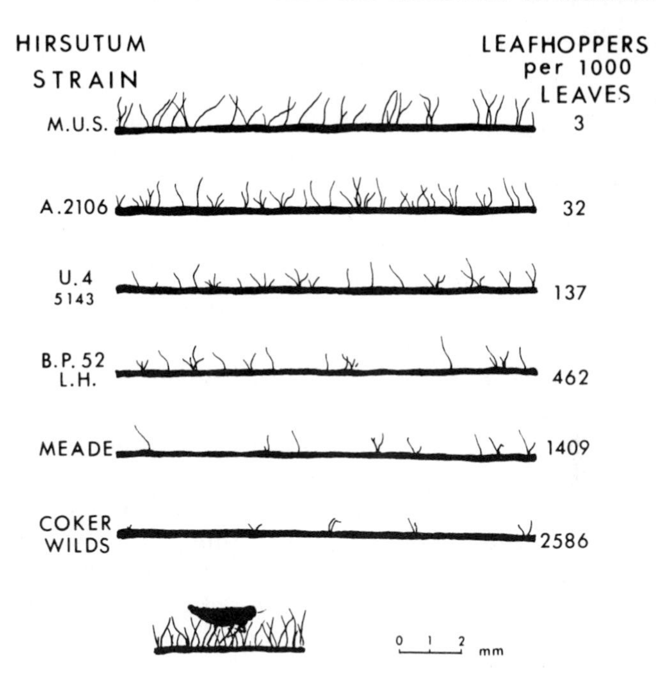

Fig. 4-8 Six typical strips from leaf blades (cross sections) of different varieties of cotton, showing various types of hairiness. On the right are listed the numbers of leafhopper *Empoasca facialis* Jacobi nymphs collected on each cotton variety as an indication of the effect of hairiness on leafhopper populations. The size of the nymph relative to the length of the hairs is shown below. (From R. Painter. 1951 (1968 paperback edition). Insect resistance in crop plants, Fig. 52, p. 233. Copyright by Reginald Painter, University Press of Kansas, Lawrence, Kansas.)

chymatous cells play an important role in the resistance to the first and second instars. As older larvae bore into the stalk, the hardness of the rind and the fiber content of the stalk are the key factors of resistance.

Hard, woody stems of *Cucurbita* spp. plants with closely packed, tough, vascular bundles are the main resistance factors against the squash vine borer, *Melittia cucurbitae* (Harris). Penetration of the stems and feeding are impaired by the structural characteristics of the plants (Howe 1949).

4. *Tolerance*

This modality of resistance refers to the capacity of certain plants to repair injury or grow to produce an adequate yield despite supporting an insect population at a level capable of damaging a more susceptible host. Tolerance usually results from one or more of the following factors: (1) general vigor of plants, (2) regrowth of damaged tissues, (3) strength of stems and resistance to lodging, (4) production of additional branches, (5) efficient utilization by the insect of nonvital plant parts, and (6) lateral compensation by neighboring plants.

In some corn earworm–resistant corn hybrids, the larvae developed normally in the silk channel but never penetrated the ear. The silks have no vital function in the development of the ear after pollination is completed. Similar numbers of earworms were obtained from resistant and from susceptible hybrids, but damage to the resistant hybrids was much less (Wiseman et al. 1972).

Individual plants in a community may be susceptible, but the community as a whole tolerant to insect attacks. Soybeans are a remarkable example of community compensation. If the stand is thinned by attacks on seedlings by early-season pests, the remaining plants will normally grow more branches to cover the spaces opened by the dead plants. Destruction of whole plants at certain stages of development may have little effect on the final yield, as the remaining plants usually compensate for the loss.

III. GENETIC BASIS OF RESISTANCE

Information on the genetic mechanisms of resistance can be of great interest in breeding programs by (1) helping in the identification of various sources of resistance, (2) orienting the program of crossings and progeny selections, (3) expanding the genetic basis of resistant varieties, and (4) developing isogenic lines which can be used in the study of mechanisms of resistance. This information is particularly critical when biotypes of a pest become immune to a resistance character. According to Day (1972), resistance may be oligogenic, polygenic, or cytoplasmic.

A. Oligogenic Resistance

Also called major gene resistance, this is determined by one (monogenic) or a few genes whose individual effects are more or less easy to detect. Examples of monogenic dominant resistance are found in varieties of rye, rice, and sweet potatoes. Monogenic recessive resistance occurs in corn varieties resistant to the western corn rootworm, *Diabrotica virgifera* LeC. (Pathak 1970). Two genes were identified in leafhopper-resistant cotton, resistance being dominant over susceptibility (Painter 1958).

B. Polygenic Resistance

This is determined by many genes, sometimes of individually small effect. The inheritance of polygenic characters is usually complex. They frequently control quantitative characters such as yield and quality of product. Several insect resistance systems are under polygenic control as, for example, European corn borer resistance in corn.

Wheat resistance to the Hessian fly is controlled by a genetic mechanism involving at least five pairs of dominant and partially dominant genes and five recessive factors. There is a complex relationship between the genetics of resistant wheats and the genetics of the various biotypes of Hessian flies that have been identified so far (see Section VI).

C. Cytoplasmic Resistance

Cytoplasmic inheritance is due to self-duplicating, mutable substances found only in the cytoplasm. Because the ovum contributes most of the cytoplasm to the zygote, cytoplasmic inheritance is maternal. This type of inheritance is commonly analyzed by determining the contributions made by each parent in reciprocal crosses. Although quite frequent and extremely important in resistance to pathogens (e.g., susceptibility to southern corn leaf blight is cytoplasmic), it has not been reported in insect resistance (Kindler and Staples 1968).

IV. ONTOGENY OF RESISTANCE

Characteristics of resistance change with age in certain plants. The example of corn has already been considered in the study of antibiosis. Susceptible alfalfa varieties are killed by the spotted alfalfa aphid either at the seedling or at later stages. The resistant 'Lahontan' and 'Moapa' varieties, however, become increasingly resistant with age, and after 41 days become virtually immune to the aphids (Howe and Pesho 1960).

V. EFFECT OF ENVIRONMENTAL FACTORS ON THE EXPRESSION OF RESISTANCE

The effect of environmental factors on the expression of genetically controlled resistance should not be confused with those related to induced resistance. Climatic and edaphic factors can influence to some extent the levels of resistance.

A. Effect of Soil Moisture

Water stresses have an ambiguous effect on resistance. Water-deficient plants become more chlorotic from aphid feeding than do infested leaves on watered plants. However, populations of the bean aphid, *Aphis fabae* Scopoli, on leaves of unwatered plants had a lower reproductive rate than aphids on watered plants. Lower turgor pressure or increased sap viscosity were claimed to reduce sap uptake by the aphids (McMurtry 1962). It seems therefore that stressed plants suffer more from insect attacks, since a given amount of feeding resulted in greater injury to the plants. The pest population in turn is also affected, but apparently not to an extent that could offset the level of damage to the plants.

B. Effect of Soil Fertility

Macro- and micronutrients also have an ambivalent effect. In some instances high levels of nutrients increase resistance, and in others they increase susceptibility. Low levels of nutrients also produce both kinds of effects. As mechanisms of resistance are so diverse, this ambivalence should not be surprising. The metabolic pathways involved in resistance are not necessarily the same in all plants, hence a diversity of effects observed. Unfortunately, little is known of the intimate metabolic processes that could explain these differences. McMurtry (1962) compared the response of susceptible ('Caliverde') and resistant ('Lahontan' C-84 and C-902) alfalfa to various nutrients. Nitrogen, phosphorus, or potassium deficiencies had no effect on the susceptibility of the 'Caliverde' variety, but phosphorus deficiency increased resistance in C-84 and C-902, whereas potassium deficiency reduced resistance in both strains. Nitrogen had no effect on resistance. Kindler and Staples (1970a) replicated these experiments using deficient and excess levels of nitrogen, phosphorus, potassium, calcium, magnesium, and sulfur. None of the treatments made the susceptible clone more resistant. Resistance decreased but was not eliminated when the resistant clone was treated with deficient levels of calcium or potassium, or excess levels of magnesium or nitrogen, but resistance increased in plants receiving deficient levels of phosphorus. Sulfur had no effect.

C. Effect of Temperature

In general, low temperatures have a negative effect on resistance, but certain varieties retain certain levels of resistance even at low temperatures. Isaak et al. (1965) reported that a group of alfalfa clones resistant to the pea aphid and the spotted alfalfa aphid did not differ significantly at 29°C (85°F). At 15.4°C (60°F), however, certain clones suddenly shifted from resistant to susceptible, and other clones progressively lost resistance at the low temperature. All clones regained resistance when the temperature was raised again. When aphids were offered a choice between resistant and susceptible alfalfa shoots at a constant temperature of 10°C, they migrated from the resistant to the susceptible shoots, indicating that resistance was retained even at this low temperature (Kindler and Staples 1970b). The mean temperature over a period of time seems to be more important for the expression of resistance than the maximum or the minimum temperature (McMurtry 1962).

VI. BIOTYPES AND THE EXPRESSION OF RESISTANCE

Biotypes are populations capable of damaging and of surviving on plants heretofore known to be resistant to other populations of the same species. The identification of biotypes has strengthened the evolutionary concept of resistance.

Most biotypes recorded to date are aphids (14 to 20 biotypes in five aphid species). Since most aphid species are parthenogenic, even one individual mutant capable of feeding on a resistant plant can build up a new biotype (Pathak 1970). Dunn and Kempton (1972) used a local (Wellesbourne) strain of the cabbage aphid, *Brevicoryne brassicae* (L.), to select seven resistant clones of Brussels sprouts. The selections were submitted to the attack of aphid populations which originated in seven different localities in England. It was observed that biotypes of the aphid, with differing abilities to colonize respective sprout clones, existed in each area. Of the seven clones originally selected for resistance to the

Table 4.2 Wheat Reactions to Hessian fly Biotype[a]

	Wheat Variety[b]				
Biotype	(T) 'Turkey'	(S) 'Seneca'	(H_3 'Monon'	(H_6) 'Benhur'	(H_5) 'Ribeiro'
GP	S	S	R	R	R
A	S	S	R	R	R
B	S	S	S	R	R
C	S	S	R	S	R
D	S	R	S	S	R
E	S	R	S	R	R
F	S	R	R	S	R
G	S	R	S	S	R

[a] Gallun 1972.
[b] S, susceptible; R, resistant. The dominant genes for avirulence in the biotypes are shown in parentheses.

Wellesbourne aphid, only one always showed at least partial resistance to the other biotypes of *B. brassicae.*

The most complete study on the relationship of resistant lines and biotypes was made with the Hessian fly and wheat. The following account of these studies is based mainly on Gallun (1972). A total of eight races that differ in their ability to survive and develop on wheats having different genes for resistance have been identified. Table 4.2 shows these races and the reaction of the wheats to the races.

The Great Plains biotype of the Hessian fly (GP) is prevalent in the western part of the United States. This race cannot infest wheats having any genes for resistance. Races A, B, C, and D occur mostly in the eastern, soft-wheat region, and their virulence is dependent on specific genes for resistance in the plant. Race E was found almost as a pure race in fields of wheat near Albany, Georgia. The other two races are segregating laboratory populations resulting from crosses between other races (Gallun 1972). From interracial crossings it was concluded that, for every single gene for resistance in the wheat plant, there is a single gene for virulence in the fly and it functions only in the recessive condition. This condition is shown in Table 4.3.

According to Gallun, the Great Plains race has no homozygous recessive genes for virulence, which are required for its survival on resistant wheats; it has only dominant genes for avirulence. F_1 crosses of homo-

Table 4.3 Genotypes of Hessian fly Biotypes[a]

Biotype	Genotype				
GP	tt	SS	H_3H_3	H_6H_6	H_5H_5
A	tt	ss	H_3H_3	H_6H_6	H_5H_5
B	tt	ss	h_3h_3	H_6H_6	H_5H_5
C	tt	ss	H_3H_3	h_6h_6	H_5H_5
D	tt	ss	h_3h_3	h_6h_6	H_5H_5
E	tt	SS	h_3h_3	H_6H_6	H_5H_5
F	tt	SS	H_3H_3	h_6h_6	H_5H_5
G	tt	SS	h_3h_3	h_6h_6	H_5H_5

[a] Gallun 1972. Recessive genes = virulence in fly or susceptibility in wheat; dominant genes = avirulence in fly or resistance in wheat.

zygous GP flies with other biotypes will perform as GP phenotype and not survive on resistant wheats. Releasing 20 GP individuals to every native biotype for three successive generations eliminates the wild-type population at the end of the third release. This fact opens up interesting possibilities for a genetic control program for Hessian flies.

From this evidence on the genetics and characteristics of biotypes, it is apparent that polygenic resistance is less conducive to the development of biotypes. This suggests that in breeding for resistance efforts should be made to locate and combine as many sources of resistance as possible (Pathak 1970).

VII. IMPLEMENTATION OF PROGRAMS IN CROP RESISTANCE

A complete program in crop resistance involves (1) identification of sources of resistance, (2) characterization of resistance mechanisms. (3) breeding of resistance characteristics into agronomic varieties which can economically compete with other established varieties, (4) genetic analysis of resistance traits, and (5) identification of the chemical and/or physical basis of resistance. Such a program necessarily has to be based on a team effort. The implementation of a breeding program requires integration of the input of entomologists, plant geneticists, and plant breeders. The techniques for crossing and hybridization, the genetic analysis of progenies, and the evaluation of the agronomic characteristics of the selections are the contributions of the plant breeder. It is up to the entomologist to identify the sources of resistance, characterize the mechanisms involved, and develop the field and laboratory assays to test the fate of resistance throughout the breeding program. The combined efforts of insect and plant physiologists and of organic chemists are often required to understand the physiology of resistance.

A. Sources of Resistance

The identification of sources of resistance results from (1) the fortuitous discovery of individual plants within a population, or of established varieties that under normal field conditions are less injured or bear a smaller pest population than do other plants or varieties under the same conditions; or (2) the systematic search among lines in variety or in germplasm collections. An example of the former is the observation that 'Lahontan,' an established alfalfa variety, is resistant to the spotted alfalfa aphid (Howe and Smith 1957). An example of the latter is the identification of three soybean lines resistant to the Mexican bean beetle, resulting from the screening of over 300 varieties and lines (Van Duyn et al. 1971).

When mass screening fails to reveal adequate sources of resistance in existing germplasm collections, the search turns to the wild relatives at or close to the centers of origin of the cultivars. These relatives may be plants of the same species, or plants of different species but of the same genus as the cultivar, or even plants belonging to different genera. The use in breeding programs of these noncospecific or congeneric plants is of course conditioned by the compatibility of the genotypes. Radcliffe and Lauer (1970) reported that several lines of Mexican tuber-bearing wild potatoes were resistant to the green peach aphid and to the potato aphid, *Macrosiphum euphorbiae* (Thomas). Crosses between the wild potatoes and *Solanum tuberosum* L. were made possible by the use of colchicine and of induced haploidy in *S. tuberosum*. These techniques permitted the equalization of ploidy in the cultivars and in the wild relatives, thus making the crosses compatible. Wisemen et al. (1967) compared preferences of fall armyworm larvae, *Spodoptera frugiperda* (J. E. Smith), for corn and for *Tripsacum dactyloides* L., stressing the importance of using wild relatives as sources of resistance in corn. This possibility has been raised by Painter (1951), who reported that hybridization with *Tripsacum* spp. may have furnished genes for resistance which entered corn by way of teosinte.

B. Measurement of Resistance

Resistance is normally measured through the effect of the exposure to insects of plants or of plant parts. This effect can be evaluated in terms of the plant as the percentage of damage to the foliage or to fruiting parts, reduction of the stand, percent yield reduction, and general vigor of the plants. It can also be measured in terms of the insects as the number of eggs oviposited, aggregation, food preference, growth rate, food utilization, mortality, and longevity.

In fact, measurement of resistance is a problem of sampling. The same general principles involved in sampling insect populations and in evaluating plant damage that are presented in Chapter 9 also apply to the measurement of resistance.

The required precision of the sampling procedure is a function of the stage of the resistance research. In mass screening of large populations, it is usually enough to rate plants visually on a 1-to-5 scale from resistant to susceptible. A more accurate assessment is necessary in the study of mechanisms of resistance and in investigations of chemical factors related to resistance. In these studies specific bioassays are developed to monitor the effects of particular chemicals on behavioral and metabolic processes related to host-plant finding, acceptance, and suitability.

C. Breeding Sequence

Very seldom does a resistant line have all the desirable agronomic characteristics of a commercial variety. Frequently, resistance is found in plants with many negative traits. Some Mexican bean beetle–resistant soybean lines are viny instead of growing erect like most improved soybean varieties. Breeding programs are therefore set forth to transfer the genes for resistance into the germ plasm of the improved varieties. In this process hybrid plants are produced by convenient crossings. After the F_1 generation the parents' characters segregate, generally following Mendelian laws. The progeny is selected for plants that combine the most desirable traits of both parents. According to the characteristics of the plants and the objectives of the breeding program, the selected plants are selfed (self-pollinated) or backcrossed to the agronomic parent. The level of resistance in each successive generation is monitored using the necessary techniques of measurement.

When desirable characteristics of various parent lines have to be combined, the breeding sequence becomes increasingly complicated. In the development of 'Cody,' a synthetic variety of alfalfa resistant to the spotted alfalfa aphid, 250,000 plants of 'Buffalo' alfalfa were screened. The new variety resulted from combinations of 22 of the selected plants.

D. Recurrent Phenotypic Selection

The blending of different genotypes into one synthetic variety is the objective of a form of mass selection known as *recurrent phenotypic selection*. Such a program was described by Hanson et al. (1972) in the development of alfalfa populations resistant to four diseases (rust, common leafspot, bacterial wilt, and anthracnose) and two insect pests (spotted alfalfa aphid and potato leafhopper). Concurrently, the general vigor of the populations also increased in generations of selection conducted in the field. The plan for recurrent selection in one germplasm pool over a period of 20 years is illustrated in Fig. 4.9. Eighteen sequential cycles (generations) of phenotypic selections are shown in the figure. Parents of

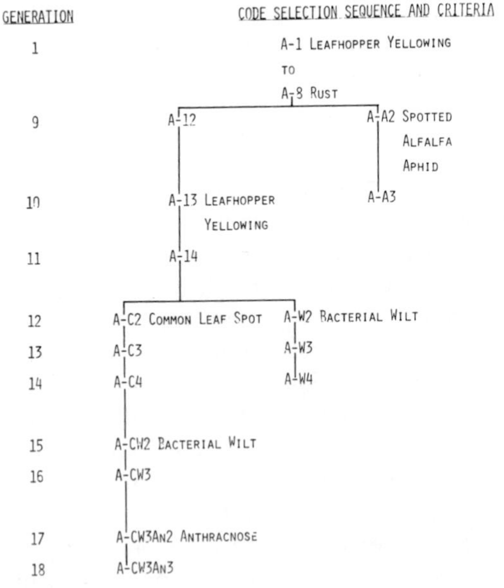

Fig. 4-9 Plan of recurrent phenotypic selection in one alfalfa germplasm pool over a period of 20 years. (Adapted from Hanson et al. 1972. *J. Environ. Qual.* 1(1): 106–111, Fig. 3.)

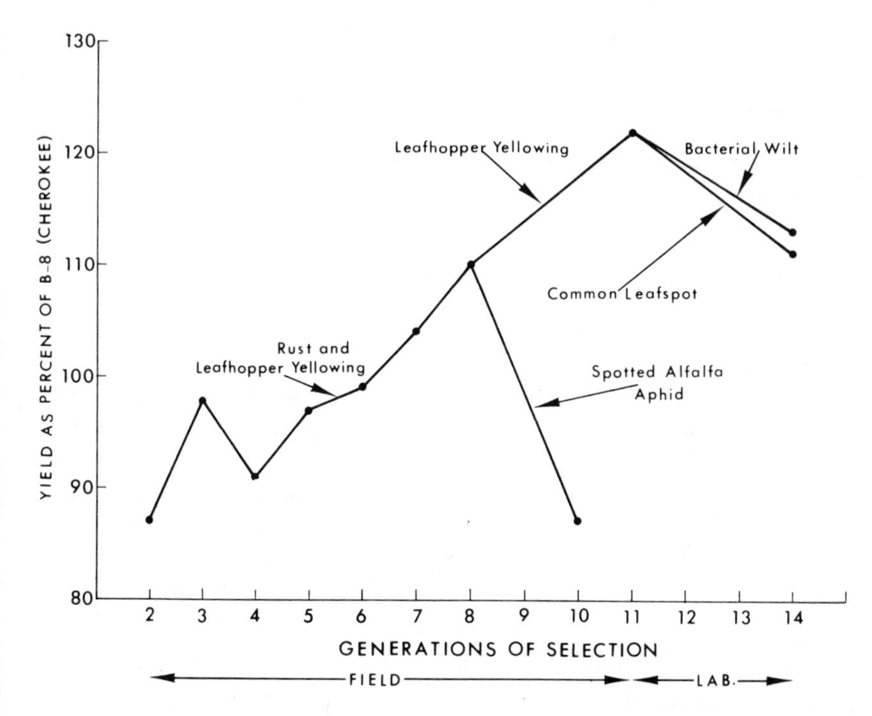

Fig. 4-10 Mean changes in yield in generations 2 through 4 of alfalfa plants selected for resistance against three diseases and one insect pest in successive cycles. (Adapted from Hanson et al. 1972. *J. Environ, Qual.* 1 (1): 106–111, Fig. 3.)

this pool were 300 plants from 'Atlantic,' 33 plants from each of two Kansas synthetics, and 33 plants from a Nebraska synthetic. In each cycle from 1800 to 10,000 plants were evaluated on the basis of the phenotype of the plants. Within each cycle selected plants were intercrossed to produce seed for a new cycle.

The effect of successive cycles of selection on yield of alfalfa is shown in Fig. 4.10. Yields are presented as percentage of yield of 'Cherokee,' a variety released in 1961, resulting from the eighth cycle of selection. Yields generally increased up to the eleventh generation, during which selection for vigor was implemented in the field. Selection in generations 12 to 14 was conducted in the laboratory without regard to vigor. In cycle 8 intense selection pressure was applied for aphid resistance and none for vigor. Yields in this case fell very sharply.

E. Greenhouse versus Field Selection

Mass selection and screening can be performed only under the pressure of insect populations of adequate size. Since natural pest populations fluctuate from year to year, breeding programs have frequently relied on (1) the artificial infestation of plants in the field, (2) the caging of seedlings or of mature plants in the greenhouse with laboratory-reared insects and, (3) measuring insect responses to excised plant parts such as leaves, leaf disks, stems, or fruits.

Particular caution must be exercised in the extrapolation of laboratory or greenhouse data to the field. Based on tests for potato leafhopper resistance using alfalfa seedlings in the greenhouse, Kindler and Kehr (1970) concluded that greenhouse selection was valuable as an initial mass screening tool. Alfalfa seedlings were selected for resistance to the potato leafhopper in the greenhouse, and then the best selections were transplanted to the field. Seedlings of unselected plants of the same pool were transplanted simultaneously. The frequency of resistant plants in the field was not increased by the previous selection of seedlings in the greenhouse. These investigators suggested that this result may be explained by the occurrence of different resistance mechanisms in the seedling and in the mature plant.

F. The Final Product—Outstanding Successes in Plant Resistance

The practical outcome of a breeding program in resistance is the release of improved resistant varieties. A summary of the successful programs in resistance in the last 30 years is found in Painter (1958), Luginbill (1969), Sprague and Dahms (1972), and Maxwell et al. (1973). According to Sprague and Dahms (1972), more than 100 varieties or inbreds of crop plants have been released, carrying resistance factors to more than 25 insect species. In addition to the early success of the grape *Phylloxera* control, practical results have been achieved in the development of insect-resistant varieties of alfalfa, barley, beans, corn, sorghum, rice, sugarcane, and wheat. Examples of some of the most outstanding results are presented below.

Research on insect resistance achieving various degrees of success has involved many other crop plants such as cotton, cucurbits, grasses, oats, onions, peanuts, pecans, peppers, potatoes, soybeans, and tobacco (Maxwell et al. 1973).

Since 1942, 25 Hessian fly-resistant wheat varieties have been released, and in 1969 these were grown on 8.5 million acres in 34 states (Sprague and Dahms 1972). In the past, in years of outbreak the Hessian fly caused losses of $5 to $25 million in the state of Kansas alone. The species is now reduced to the status of a secondary pest, as a result of the widespread use of resistant wheat varieties.

Resistance to the wheat stem sawfly, *Cephus cinctus* Norton, is the most effective way to control this pest. The wheat variety 'Rescue,' introduced into the state of Montana from Canada, was estimated to have saved farmers about $40 million over a period of 10 years.

Most corn hybrids and inbreds now under commercial production incorporate genes for resistance to the European corn borer. Although no direct records are available on the impact of resistance, it is estimated that in 1949 losses due to the European corn borer were of the order of $350 million. In the 1960s, when resistant hybrids were widely used over more than 30 million acres, losses averaged $10 million per year (Horber 1972). Along with resistance, other factors such as diseases, parasites, and new production and harvesting practices are probably responsible for the decline in economic importance of the European corn borer in the Midwest.

Development of resistance in alfalfa to the spotted alfalfa aphid was achieved in a remarkably short time. The variety 'Lahontan' was released in 1954, the same year the aphid was first recorded in the United States. This variety fortunately was partially resistant to the aphid. The resistant variety 'Moapa' was released within a period of 3 years, and since then 17 varieties have been released. Resistant varieties were developed, which were adapted to all ecological conditions where the aphid is a problem (Sprague and Dahms 1972). Savings to growers due to the use of spotted alfalfa aphid–resistant alfalfa varieties were conservatively estimated at $35 million per year (Luginbill 1969). In some areas with epidemic infestations of aphids, resistant alfalfa varieties yielded three to four times more than susceptible varieties (Horber 1972).

VIII. PLANT RESISTANCE IN PEST MANAGEMENT

Resistance used as a single pest-management factor has achieved some outstanding results. Among the most desirable features of plant resistance from an ecological point of view are (1) *specificity*—plant resistance is usually specific to a pest or complex of pest organisms and has no direct detrimental effect on beneficial insects; (2) *cumulative effect*—near-immunity to the insect is not necessary, because the effect on the pest population will be compounded in successive generations; (3) *persistence*—most resistant varieties maintain high levels of resistance for a long time, despite the occasional upsurge of biotypes; (4) *harmony with the environment*—since no unnatural elements are used, there is no danger of contaminating the environment or endangering man or wildlife; (5) *ease of adoption*—once developed, resistant varieties can easily be incorporated into normal farm operations at little or no extra cost, since seed has to be purchased anyway; (5) *compatibility*—plant resistance is compatible with other tactics in pest management, being an ideal adjuvant when resistance alone cannot maintain a pest below the economic threshold.

Although the advantages of plant resistance far outweigh the limitations, the latter nevertheless have to be recognized: (1) *time of development*—because of the long time (3 to 15 years) necessary to identify sources of resistance and breed resistant varieties, the method is not adequate for solving sudden or very localized pest problems; (2) *genetic limitations*—the absence of preadaptive resistance genes among available germ plasm may deter use of the method; induced mutation, although possible, would make development programs even longer and more complex; (3) *biotypes*—the occurrence of biotypes may limit in time and in space the use of certain resistant varieties, but so far breeders have been able to circumvent this problem using polygenic resistance or breeding varieties resistant to certain biotypes; (4) *conflicting resistance traits*—certain plant characteristics may act as resistance factors for some species but induce susceptibility to others; pubescence is a key resistance factor in soybeans against the potato leafhopper, but the soybean pod borer, *Grapholitha glycinivorella* (Matsumura), displays a marked ovipositional preference for pubescent soybean pods (Nishijima 1960); therefore, pubescence is a resistance factor for the potato leafhopper, but a susceptibility factor for the pod borer.

A. Plant Resistance and Chemical Control

Painter (1951) suggested that the use of insecticides on resistant crops may provide a more effective control than either method alone. When a sweetcorn hybrid 'Stowell's Evergreen,' susceptible to the corn earworm, was exposed to natural attacks in Georgia, it had only about 22% damage-free ears, whereas the resistant hybrid 471-U6X81-1 had 78% damage-free ears. However, when both hybrids were sprayed with insecticides, the susceptible hybrid was 86% damage-free and the resistant hybrid about

93% damage-free. These results suggested that resistant hybrids may require less insecticide than susceptible hybrids to achieve an equivalent reduction in pest damage (McMillian et al. 1972).

B. Plant Resistance and Cultural Control

The adjustment of planting dates to achieve phenological asynchronies between host and insect has been discussed before. An interesting integration of resistance, cultural practice, and chemical control is the use in trap crop plantings of resistant and susceptible varieties in combinations.

Outstanding results with trap cropping were obtained by Newsom and co-workers in Louisiana in the control of the bean leaf beetle on soybeans, with a consequent reduction in bean pod mottle virus transmission. In this case an early-maturing variety was planted in a band prior to planting the remainder of the field. The premature plants attracted the colonizing beetle populations which were subsequently killed with insecticides. The bulk of the field did not require treatment for the bean leaf beetle. It is conceivable that a combination of susceptible with resistant, or partially resistant varieties, could also be used in a trap crop scheme.

C. Plant Resistance and Natural Enemies

Classic biological control is usually compatible with plant resistance, except in extreme cases in which the pest is so reduced by resistance that natural enemies cannot find an adequate host population. Van Emden (1966) indicated that even very low levels of resistance may have a dramatic effect on the efficiency of natural enemies. Using an equation proposed by S. Bombasch, Van Emden calculated (Fig. 4.11) that a predator whose voracity fails to exert control (*B*) on a pest population multiplying daily by a factor of 1.2 (*A*), adequately controls a population increasing by a factor of 1.15 (*D*). The limitation on the pest population increase to the 1.15 factor by itself was insufficient to prevent rapid rise of the pest population (*C*). It is therefore apparent that even a mild resistance factor, which reduces pest population increase by 0.05, is enough to turn an inefficient predator into a highly effective one.

A practical example of this effect was offered by Starks et al. (1972), who showed that resistant varieties of barley and sorghum complemented the activity of the parasite *Lysiphlebus testaceipes* (Cresson) in reducing damage to plants and in the production of greenbugs, *Schizaphis graminum* (Rondani) (Fig. 4.12).

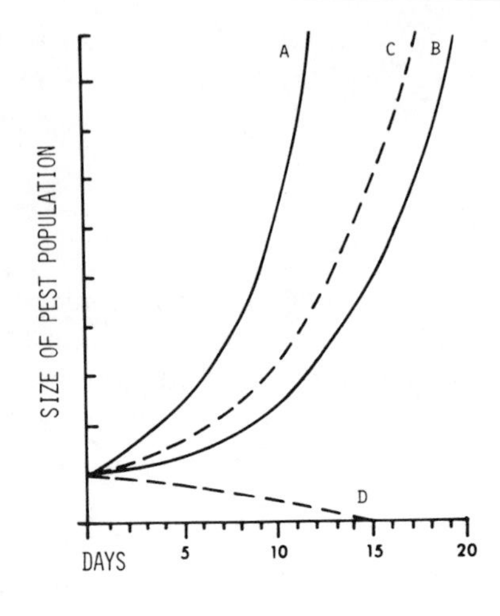

Fig. 4-11 The influence of a low level of plant resistance to insect attack on the effectiveness of natural enemies. Solid line, susceptible plant; dashed line, resistant plant; *A* and *C*, without predators; *B* and *D*, with predators. (Redrawn from van Emden 966. Courtesy of H. F. van Emden, © Fisons Limited, 1966.)

IX. A LOOK INTO THE FUTURE

Since the early successes with grape *Phylloxera* and the woolly aphid, plant resistance has been strongly oriented toward solving problems that defy other more expeditious control methods. The practical orientation of most workers in the field has generated a degree of ambivalence toward the theory and practice of host-plant resistance. Basic work on the genetics of resistance and on the chemical and physical bases of resistance has been recognized as necessary but not essential for progress in the field of plant resistance and the development of resistant varieties.

Although the argument is less a shock of contradictory opinions than a matter of allocation of priorities imposed by budgetary limitations, real progress in plant resistance will come only with broadly based, well-balanced programs. In such programs mass screening, selection, crossing, and further selection go hand in hand with research on the genetics and nature of resistance factors. More than any other field in applied entomology, plant resistance is a team effort involving entomologists, plant pathologists, plant breeders, geneticists, and natural product chemists.

The state of the art does not permit great departures from the classic phases of plant resistance work. It permits us, however, to envision the day when blueprints for the production of a resistant plant will be drawn in a laboratory, combining knowledge of the behavior and metabolism of the target insects and of the genetic control mechanisms of well-defined chemical and physical resistance factors. Models simulating the desired effects will permit testing the various alternatives as they influence the population dynamics of the pest. An optimized plant model will then be

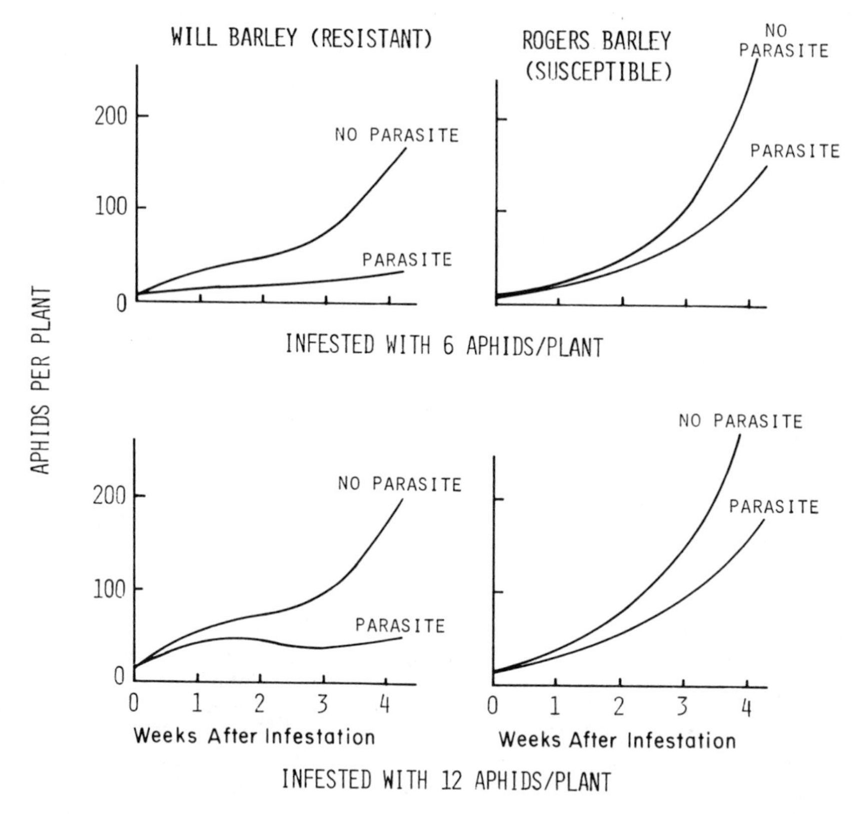

Fig. 4-12 Increase in greenbugs in the absence and presence of one female parasite caged on greenbug-resistant ('Will') and susceptible ('Rogers') barley. (Redrawn from Starks et al. 1972. Courtesy K. J. Starks and the Entomological Society of America.)

passed to geneticists who will produce the prototype of a new variety which is highly productive and highly plastic, and has the genetic make-up needed to resist attacks by insect pests. Field tests will then determine the performance of the new variety under normal field conditions, before final release is made to growers for use in pest-management programs.

REFERENCES

Agarwal, R. A. 1969. Morphological characteristics of sugarcane and insect resistance. *Entomol. Exp. Appl.* **12**:767–776.

Auclair, J. L. 1957. Developments in resistance of plants to insects. *Annu. Rep. Entomol. Soc. Ontario* **88**:7–17.

Baker, H. G. 1972. Human influences on plant evolution. *Econ. Bot.* **26**:32–43.

Balachowsky, A. S. 1951. La lutte contre les insectes Payot, Paris. 380 pp.

Beck, S. D. 1965. Resistance of plants to insects. *Annu. Rev. Entomol.* **10**:207–232.

Bernard, R. L. 1971. Two major genes for time of flowering and maturity in soybeans. *Crop Sci.* **11**:242–244.

Bu'Lock, J. D. 1965. The biosynthesis of natural products. McGraw Hill, New York. 149 pp.

Chippendale, G. M., and S. D. Beck. 1964. Nutrition of the European corn borer, *Ostrinia nubilalis* (Hubn.). V. Ascorbic acid as the corn leaf factor. *Entomol. Exp. Appl.* **7**:241–248.

Day, P. R. 1972. Crop resistance to pests and pathogens. Pages 257–271 *in* Pest control strategies for the future. National Academy of Sciences, Washington, D.C.

de Wilde, J., and L. M. Schoonhoven, eds. 1969. Proceedings of the 2nd international symposium. Insect and host plant. *Entomol. Exp. Appl.* **12**:471–810.

Dethier, V. G. 1954. Evolution of feeding preferences in phytophagous insects. *Evolution* **8**:33–54.

Dethier, V. G. 1970. Chemical interactions between plants and insects. Pages 83–102 *in* E. Sondheimer and J. B. Simeone, eds., Chemical ecology. Academic Press, New York.

Dethier, V. G., L. Barton Brown, and C. N. Smith. 1960. The designation of chemicals in terms of the responses they elicit from insects. *J. Econ. Entomol.* **53**:134–136.

Dunn, Z. A., and D. P. H. Kempton. 1972. Resistance to attack by *Brevicoryne brassicae* among plants of Brussels sprouts. *Ann. Appl. Biol.* **72**:1–11.

Ehrlich, P. R., and P. H. Raven. 1964. Butterflies and plants: A study in coevolution. *Evolution* **18**:586–608.

Emlen, J. M. 1973. Ecology: An evolutionary approach. Addison-Wesley, Reading, Mass. 493 pp.

Everett, T. R., H. C. Chiang, and E. T. Hibbs. 1958. Some factors influencing populations of European corn borer [*Pyrausta nubilalis* (Hbn.)] in the north central states. *Minn. Agr. Exp. Sta. Tech. Bull.* **229**:1–63.

Fraenkel, G. 1959. The raison d'etre of secondary plant substances. *Science* **129**:1466–1470.

Fraenkel, G. 1969. Evaluation of our thoughts on secondary plant substances. *Entomol. Exp. Appl.* **12**:473–486.

Gallun, R. L. 1972. Genetic interrelationships between host plants and insects. *J. Environ. Qual.* **1**:259–265.

Gentry, H. S. 1969. Origin of the common bean, *Phaseolus vulgaris*. *Econ. Bot.* **23**:55–59.

Hamamura, Y. 1970. The substances that control the feeding behavior and the growth of the silkworm *Bombyx mori* L. Pages 55–80 *in* D. L. Wood, R. M. Silverstein, and M. Nakajima, eds., Control of insect behavior by natural products. Academic Press, New York.

Hanson, C. H., J. H. Busbice, R. R. Hill, Jr., O. J. Hunt, and A. J. Oakes. 1972. Directed mass selection for developing multiple pest resistance and conserving germ plasm in alfalfa. *J. Environ. Qual.* **1**:106–111.

Hawkes, J. G. 1970. The origins of agriculture. *Econ. Bot.* **24**:131–133.

Hendrickson, J. B. 1965. The molecules of nature. W. A. Benjamin, New York. 189 pp.

Horber, E. 1972. Plant resistance to insects. *Agr. Sci. Rev.* **10**:1–10, 18.

Howe, W. L. 1949. Factors affecting the resistance of certain cucurbits to the squash borer. *J. Econ. Entomol.* **42:321–326.**

Howe, W. L., and G. R. Pesho. 1960. Influence of plant age on the survival of alfalfa varieties differing in resistance to the spotted alfalfa aphid. *J. Econ. Entomol.* **53**:142–144.

Howe, W. L., and O. F. Smith. 1957. Resistance to the spotted alfalfa aphid in Lahontan alfalfa. *J. Econ. Entomol.* **50:320–324.**

Isaak, A., E. L. Sorenson, and R. H. Painter. 1965. Stability of resistance to pea aphid and spotted alfafa aphid in several alfalfa clones under various temperature regimes. *J. Econ. Entomol.* **58**:140–143.

Kennedy, J. S. 1958. Physiological condition of the host-plant and susceptibility to aphid attack. *Entomol. Exp. Appl.* **1**:50–65.

Kennedy, J. S. 1965. Mechanisms of host plant selection. *Ann. Appl. Biol.* **56**: 317–322.

Kennedy, J. S., and I. H. M. Fosbrooke. 1973. The plant in the life of an aphid. *Symp. Roy. Entomol. Soc. London* **6**:129–140.

Khalsa, M. S. 1973. Host plant selection behavior of *Autographa precationis* (Guenee) (Lepidoptera: Noctuidae). Ph.D. dissertation, University of Illinois, Urbana. 195 pp.

Kindler, S. D., and W. R. Kehr. 1970. Field test of alfalfa selected for resistance to potato leafhopper in the greenhouse. *J. Econ. Entomol.* **63**:1464–1467.

Kindler, S. D., and R. Staples. 1968. Lack of cytoplasmic inheritance of alfalfa resistance to the spotted alfalfa aphid. *J. Econ. Entomol.* **61**:1455–1456.

Kindler, S. D., and R. Staples. 1970a. Nutrients and the reaction of two alfalfa clones to the spotted alfalfa aphid. *J. Econ. Entomol.* **63**:938–940.

Kindler, S. D., and R. Staples. 1970b. The influence of fluctuating and constant temperatures, photoperiod, and soil moisture on the resistance of alfalfa to the spotted alfalfa aphid. *J. Econ. Entomol.* **63**:1198–1201.

Kishaba, A. N., and G. R. Manglitz. 1965. Non-preference as a mechanism of sweet clover and alfalfa resistance to the sweet clover aphid and the spotted alfalfa aphid. *J. Econ. Entomol.* **58**:566–569.

Klun, J. A., and T. A. Brindley. 1966. Role of 6-methoxybenzoxazolinone in inbred resistance of host plant (maize) to first-brood larvae of European corn borer. *J. Econ. Entomol.* **59**:711–718.

Klun, J. A., and J. F. Robinson. 1969. Concentration of two 1,4-benzoxazinones in dent corn at various stages of development of the plant and its relation to resistance of the host plant to the European corn borer. *J. Econ. Entomol.* **62**:214–220.

Kogan, M. 1972a. Feeding and nutrition of insects associated with soybeans. 2. Soybean resistance and host preferences of the Mexican bean beetle, *Epilachna varivestis. Ann. Entomol. Soc. Amer.* **65**:675–683.

Kogan, M. 1972b. Intake and utilization of natural diets by the Mexican bean beetle, *Epilachna varivestis*—A multivariate analysis. Pages 107–126 *in* J. G. Rodriguez, ed., Insect and mite nutrition. North Holland, Amsterdam.

Kogan, M., W. G. Ruesink, and K. McDowell. 1974. Spatial and temporal distribution patterns of the bean leaf beetle, *Cerotoma trifurcata* (Forster) on soybeans in Illinois. *Environ. Entomol.* **3**:607–617.

Luckman, W. H., A. M. Rhodes, and E. V. Wann. 1964. Silk balling and other factors associated with resistance of corn to corn earworm. *J. Econ. Entomol.* **57**:778–779.

Luginbill, P. 1969. Developing resistant plants, ideal method of controlling insects. *USDA Prod. Res. Rep.* 111. 14 p.

Lupton, F. G. H. 1967. Use of resistant varieties in crop production. *World Rev. Pest Control* **6**:47–58.

McMillian, W. W., B. R. Wiseman, N. W. Widstrom, and E. A. Harrell. 1972. Resistant sweet corn hybrid plus insecticide to reduce losses from corn earworms *J. Econ. Entomol.* **65**:229–231.

McMurtry, J. A. 1962. Resistance of alfalfa to spotted alfalfa aphid in relation to environmental factors. *Hilgardia* **32**:501–539.

Maxwell, F. G., J. N. Jenkins, and W. L. Parrott. 1973. Resistance of plants to insects. *Advan. Agron.* **24**:187–265.

Maxwell, F. G., J. N. Jenkins, W. L. Parrott, and W. T. Buford. 1969. Factors contributing to resistance and susceptibility of cotton and other hosts to the boll weevil, *Anthonomus grandis. Entomol. Exp. Appl.* **12**:801–810.

National Academy of Sciences. 1969. Plant and animal resistance to insects. Pages 64–99 *in* Principles of plant and animal pest control. Vol. 3. Washington, D.C.

National Academy of Sciences. 1971. Insect-plant interactions. Report of a work conference. Washington, D.C. 93 pp.

Nishijima, Y. 1960. Host plant preference of the soybean pod borer, *Grapholitha glicinivorella* Matsumura (Lep., Eucosmidae). 1. Oviposition site. *Entomol. Exp. Appl.* **3**:38–47.

Painter, R. H. 1951. Insect resistance in crop plants. Macmillan, New York. 520 pp.

Painter, R. H. 1958. Resistance of plants to insects. *Annu. Rev. Entomol.* **3**:267–290.

Painter, R. H. 1968. Crops that resist insects provide a way to increase world food supply. *Kans. Agr. Exp. Sta. Bull.* 520. 22 pp.

Pathak, M. D. 1970. Genetics of plants in pest management. Pages 138–157 *in* R. L. Rabb and F. E. Guthrie, eds., Concepts of pest management. North Carolina State University, Raleigh.

Rabb, R. L. 1969. Environmental manipulations as influencing populations of tobacco hornworms. *Proc. Tall Timbers Conf. Ecol. Anim. Control Habitat Management.* **1**:175–191.

Radcliffe, E. B., and F. I. Lauer. 1970. Further studies on resistance to green peach aphid and potato aphid in wild tuber-bearing solanum species. *J. Econ. Entomol.* **63**:110–114.

Reed, G. L., T. A. Brindley, and W. B. Showers. 1972. Influence of resistant corn leaf tissue on the biology of the European corn borer *Ann. Entomol. Soc. Amer.* **65**:658–662.

Rees, C. J. C. 1969. Chemoreceptor specificity associated with choice of feeding site by the beetle *Chrysolina brunsvicensis* on its food plant, *Hypericum hirsutum. Entomol. Exp. Appl.* **12**:565–583.

Rodriguez, J. G. 1960. Nutrition of the host and reaction to pests. *Amer. Assoc. Advan. Sci.* **61**:149–167.

Rodriguez, J. G., ed. 1972. Insect and mite nutrition—Significance and implications in ecology and pest management. North Holland, Amsterdam. 702 pp.

Schoonhoven, L. M. 1968. Chemosensory bases of host plant selection. *Annu. Rev. Entomol.* **13**:115–136.

Schoonhoven, L. M. 1969. Gustation and foodplant selection in some lepidopterous larvae. *Entomol. Exp. Appl.* **12**:555–566.

Schoonhoven, L. M. 1972. Some aspects of host selection and feeding in phytophagous insects. Pages 557–566 *in* J. G. Rodriguez, ed., Insect and mite nutrition. North Holland, Amsterdam.

Scott, G. E., and W. D. Guthrie. 1966. Survival of European corn borer larvae on resistant corn treated with nutritional substances. *J. Econ. Entomol.* **59**: 1265–1267.

Singh, P. 1970. Host-plant nutrition and composition: Effects on agricultural pests. *Can. Dep. Agr. Res. Inst. Inf. Bull.* 6. 102 pp.

Snelling, R. O. 1941. Resistance of plants to insect attack. *Bot. Rev.* **7**:543–586.

Sondheimer, E., and J. B. Simeone, eds. 1970. Chemical ecology. Academic Press, New York. 336 pp.

Sprague, G. E., and R. G. Dahms. 1972. Development of crop resistance to insects. *J. Environ. Qual.* **1**:28–34.

Starks, K. J., R. Muniappan, and R. D. Eikenbary. 1972. Interaction between plant resistance and parasitism against greenbug on barleyand sorghum. *Ann. Entomol. Soc. Amer.* **65**:650–655.

Thorsteinson, A. J. 1953. The role of host selection in the ecology of phytophagous insects. *Can. Entomol.* **85**:276–282.

Thorsteinson, A. J. 1960. Host plant selection by phytophagous insects. *Annu. Rev. Entomol.* **5**:193–218.

Van Duyn, J. W., S. G. Turnipseed, and J. D. Maxwell. 1971. Resistance in soybeans to the Mexican bean beetle. I. Sources of resistance. *Crop Sci.* **11**:572–573.

Van Emden, H. F. 1966. Plant insect relationships and pest control. *World Rev. Pest Control* **5**:115–123.

Van Emden, H. F., ed. 1973. Insect/plant relationships. *Symp. Roy. Entomol. Soc. London* 6. 215 pp.

Whittaker, R. H., and P. P. Feeney. 1971. Allelochemics: Chemical interactions between species. *Science* **171**:757–770.

Wiseman, B. R., W. W. McMillian, and N. W. Widstrom. 1972. Tolerance as a mechanism of resistance in corn to the corn earworm. *J. Econ. Entomol.* **65**:835–837.

Wiseman, B. R., R. H. Painter, and C. E. Wassom. 1967. Preference of first-instar fall armyworm larvae for corn compared wtih *Tripsacum dactyloides. J. Econ. Entomol.* **60**:1738–1742.

Zwölfer, H., and G. Harris. 1971. Host specificity determination of insects for biological control of weeds. *Annu. Rev. Entomol.* **16**:157–178.

5

PARASITOIDS AND PREDATORS IN PEST MANAGEMENT

F. W. Stehr

INTRODUCTION

All plant and animal species have natural enemies (parasites, parasitoids, predators, or pathogens) which attack their various life stages. The impact of these natural enemies ranges from a temporary or minor effect to the death of the host or prey.

While this was long recognized, the first dramatic example of deliberate manipulation of insect natural enemies was the importation of the vedalia ladybeetle, *Rodolia cardinalis* (Mulsant), into California in 1888 to control the cottony cushion scale, *Icerya purchasi* Maskell, on citrus. It was an immediate success and provided the impetus and generated support for many other biological control projects. Worldwide, over 120 different pest species subjected to biological control attempts by the *importation* of natural enemies have been partially, substantially, or completely controlled (DeBach 1972) (Table 5.1). More than 133 additional successes have been achieved by the *transfer* of natural enemies from countries having initial successes to different countries with the same pest species (Table 5.1). In addition, innumerable potentially harmful native insects are maintained at noneconomic levels by native natural enemies. Biological control has clearly been successful against many pests, and the potential exists for an even greater role in future pest management systems.

Table 5.1 Worldwide Biological Control Attempts and Successes against Insect Pests by the Importation of Natural Enemies through 1969[a]

Order	Number of Pest Species Involved in Initial Attempts[c]	Number and Degree of Success[b]: Partial		Substantial		Complete		Totals			
		Initial Attempts	Transfers to Other Countries	Initial Attempts	Transfers to Other Countries	Initial Attempts	Transfers to Other Countries	Initial Successes	Successful Transfers	Initial and Transfer Successes	Unsuccessful Initial Attempts against Pest Species
Orthoptera	7	—	—	1	—	2	—	3	—	3	4
Dermaptera	1	—	1	1	—	—	—	1	1	2	—
Thysanoptera	3	—	—	—	—	—	—	—	—	—	3
Hymenoptera	9	1	—	2	1	1	—	4	1	5	5
Diptera	21	7	1	3	2	—	—	10	3	13	11

Coleoptera	37	7	—	4	3	2	—	13	3	16	24
Lepidoptera	51	6	3	14	10	4	—	24	13	37	27
Heteroptera	5	—	—	1	1	1	1	2	2	4	3
Homoptera	89	9	20	22	30	32	60	63	110	173	26
		30	25	48	47	42	61	120	133	253	103
Total	223	55		95		103		253			

[a] Modified from DeBach 1972.

[b] Complete successes refer to complete biological control being obtained and maintained against a major pest of a major crop over a fairly extensive area so that insecticidal treatment becomes rarely, if ever, necessary. Substantial successes include cases in which economic savings are somewhat less pronounced by reason of the pest or crop being less important, by the crop area being restricted (such as on a small island), or by the control being such that occasional insecticidal treatment is indicated. Partial successes are those in which chemical control measures remain commonly necessary but either the intervals between necessary applications are lengthened or results are improved when the same treatments are used or outbreaks occur less frequently. Transfers are additional successes in different countries when natural enemies were moved from a country of previous success.

[c] An initial attempt is only the first project carried out against a pest species.

DeBach (1972) has discussed the number of biological control successes by countries, and gives the following totals (including partial, substantial and complete successes): Hawaii, 23; United States mainland, 43; Canada, 17; Australia, 11; New Zealand, 10; Fiji and Chile, 7 each; South Africa and Peru, 6 each; Israel, Mauritius, USSR, and Tasmania, 5 each; Japan, Seychelles, Mexico, Italy, Puerto Rico, and Greece, 4 each; and 45 other countries with a total of 79. Many of the 79 successes from the other 45 countries were the result of transfers of natural enemies from countries that supported the initial successful effort. One obvious reason for the large number of successes generated by the first 10 countries is the fact that they have had more problems as a result of the introduction of foreign species without their natural enemies by colonists moving from Europe to and between the colonies. Despite this added incentive of very serious problems, countries with the most successes have also been those that have put the most research effort into biological control. Future successes will most surely be directly proportional to the effort put into biological control research.

This chapter introduces some definitions, philosophies, concepts, and problems regarding the use of parasitoids and predators in insect pest management. The use of pathogens in pest management is covered in Chapter 6. Later chapters incorporate more specific examples of their use. General statements are sometimes made, since the conflict between complete accuracy, with its profusion of qualifications, and a more readable style is impossible to resolve.

I. DEFINITIONS AND CONCEPTS

One of the fundamental life styles of the animal kingdom is predation, which is composed of a broad base of animal → plant (herbivore) predation supporting the complex web of animal → animal (carnivore) predation. The terms *predator* and *parasite* have had a long history of usage, and everyone has a general understanding of their meaning. The term *parasitoid* has not been widely used, but there is a useful distinction to be made between a parasite and a parasitoid, although it should be recognized that the life-styles implied by the terms predator, parasite, and parasitoid are all specialized forms of predation. They are not mutually exclusive, and species are found that do not clearly fit into one category.

A *parasite* is an organism that is usually much smaller than its host, and a single individual usually does *not kill* the host. Numerous individuals may irritate, weaken, or otherwise debilitate the host, and occasionally cause its death. Tapeworms, lice, fleas, and even bloodsucking flies such as mosquitoes are examples of parasites. Parasites may complete their entire life cycle on a single host (lice), they may be free-living and nonparasitic during part of their life cycle (mosquitoes and fleas), or they may have complicated life cycles involving several host species (tapeworms). Parasites are generally regarded as pests and are not covered in this chapter.

A *predator* (in the restricted sense) is a free-living organism throughout its life; it kills its prey,[1] is usually larger than its prey, and requires more than one prey to complete its development. Mantids, spiders, and many species of lady beetles are good examples of predators.

Parasitoids have often been included in the parasite category, but a parasitoid is a special kind of predator which is often about the same size as its host, *kills* its host, and requires only one host (prey) for development into a free-living adult. Braconid wasps are good examples of parasitoids.

A. Parasitoids

The biology of parasitoids is among the most complex in the animal kingdom, perhaps being equaled only by that of some of the true parasites with their multiple life forms and hosts. Parasitoid larvae may feed internally (*endoparasitoids*) or externally (*ectoparasitoids*). Hosts of ectoparasitoids are usually paralyzed, presumably so the parasitoid larvae will not be dislodged or lost if the host were to molt. Parasitoids may attack any life stage, but the vast majority attack eggs or larvae, some attacking pupae and relatively few attacking adults. There are such phenomena as egg-larval parasitoids in which the parasitoid egg is laid in the host egg, but the parasitoid larva kills the host larva; or larval-adult parasitism in which the parasitoid egg is laid in the larva and the parasitoid larva finally kills the adult. In some species the adult parasitoid may kill a significant number of hosts by feeding directly on them (predation).

A *solitary* parasitoid species is one in which only a single individual normally completes development per host, all others being killed by the survivor if more than one egg is laid per host. Many ichneumonid wasps and tachinid flies are solitary. A *gregarious* parasitoid species is one in

[1] The terms *prey* and *host* are used interchangeably, unless the distinction is obvious.

Fig. 5-1 Larvae of the gregarious larval parasitoid *Tetrastichus julis* (Walker), which overwinter in the ground in cereal leaf beetle pupal cells. (Photo courtesy Stuart H. Gage, Department of Entomology, Michigan State University.)

which more than one individual of the same species normally completes development in a single host. In some cases more than one individual per host is a necessity, since a single parasitoid larva may not be able to kill the host and complete development. Many braconid and chalcidoid wasps are gregarious (Fig. 5.1), and the number of parasitoids produced per host may be quite large. Some Encyrtidae that parasitize large Lepidoptera and produce more than one individual from a single fertilized egg (polyembryony) may produce several hundred parasitoids of the same sex from a single egg and, if several eggs are laid per host, several thousand parasitoids can be produced (Askew 1971).

Primary parasitoids are those that attack the host directly. They are generally regarded as beneficial, with one major exception (see below).

Parasitoids, like everything else, have enemies. As Jonathan Swift has been paraphrased, "Big fleas have little fleas upon their backs to bite 'em, and little fleas have lesser fleas, and so *ad infinitum*." While hyperparasitism does not proceed *ad infinitum*, parasitoids frequently show such trophic relationships. *Hyperparasitoids* are parasitoids that attack other parasitoids. They are called secondary if they attack primary parasitoids, and tertiary if they attack secondary parasitoids. Secondary parasitoids are generally regarded as harmful, since they reduce the effectiveness of primary parasitoids. Tertiary parasitoids are not common, but they would be regarded as beneficial if we were so unfortunate as to need control of secondary parasitoids. Rare cases of fourth- and even fifth-level parasitism are known, but most parasitism at the tertiary level and above is facultative, that is, the parasitoids are not host-specific and feed on any larvae (including those of their own species) that happen to be present (Askew 1971).

Of course, in the case of biological control of weeds, everything is reversed—the plant is the pest, the phytophagous insect is beneficial since we are trying to reduce the plant, the primary parasitoid is harmful since it reduces the phytophagous insect, the secondary parasitoid is beneficial since it reduces the primary parasitoid, and the tertiary parasitoid is harmful since it reduces the secondary parasitoid. However, the fundamental biological relationships between any phytophagous insect and its parasitoids are similar, whether we are considering biological control of a pest or biological control of a weed. The only difference is whether the plant is regarded as a crop or a weed. For further reading and additional references on biological control of weeds, see Zwölfer and Harris (1971) and Anonymous (1968).

Superparasitism and *multiparasitism* are similar phenomena in that oviposition occurs in hosts that are already parasitized, yet they are significantly different. Females of some species are able to recognize hosts that have already been parasitized by their own or other species, and they do not lay eggs in these hosts. Females of other species cannot recognize a parasitized egg, and apparently all Diptera are unable to do it (Askew 1971). This ability allows superparasitism and multiparasitism to be avoided when hosts are abundant, but when they are scarce it tends to break down, although females of some species are known to resorb their eggs if they cannot find hosts.

Superparasitism occurs when more individuals of the same species are present in a single host than can complete development *in a normal way*. All or some may die, and those that survive may be smaller than normal or weakened, unless the species is a solitary one in which only a single individual survives per host. Abundant superparasitism is not generally regarded as an ideal situation, since much reproductive capacity is wasted if not lost, but it may be an indicator of very high rates of parasitism, since hosts are generally scarce in relation to the number of parasitoids when superparasitism is common. It has also been suggested that superparasitism could be beneficial, since the next generation of parasitoids would be reduced, thereby avoiding intense pressure on the host, which could lead to severe fluctuations between parasitoids and hosts (Anonymous 1969).

Another potential benefit of superparasitism (or multiparasitism) is the thwarting of the encapsulation mechanism of the host, since the host may not be able to encapsulate more than one or a few parasitoids. *Encapsulation* is the surrounding and eventual killing by asphyxiation of parasitoid eggs or larvae by the host's blood cells. Salt (1970) has discussed it and other host defense mechanisms in some detail. Encapsulation can be an important negative factor in a biological control program. Rates are often zero or low for normal hosts, but frequently are higher for unnatural hosts. This encapsulation of foreign bodies by unnatural hosts tends to contradict the hypothesis that parasitoids might be very effective against unnatural hosts because these hosts have not evolved any defenses. In fact, it appears that many hosts have weak defenses against their normal parasitoids and stronger ones against others, because the strangers elicit the "foreign-body reaction" of the host against any foreign object, dead or alive, but see the discussion in Section IV C. In any event the rate of encapsulation in the host or projected host should certainly be checked if at all possible before introductions are made, since encapsulation rates may vary with the local population or subspecies of the host or the parasitoid.

Multiparasitism occurs when more than one species of parasitoid is present in the same host, usually resulting in the death of the less aggressive species of parasitoid(s). It, like superparasitism, is generally regarded as undesirable in most situations.

B. Predators

Predation, in the broadest sense, is one of the major ways of life in the animal kingdom. Ecologists have studied animal predator–prey relationships for many years, but whether they are related to fisheries and wildlife management, food web relationships, classic biological control, or some other mode, the same basic principles apply.

Predation (in the restricted sense) is common among insects, and some of our most successful cases of biological control have been through predators. Both immatures and adults of many species are predaceous,

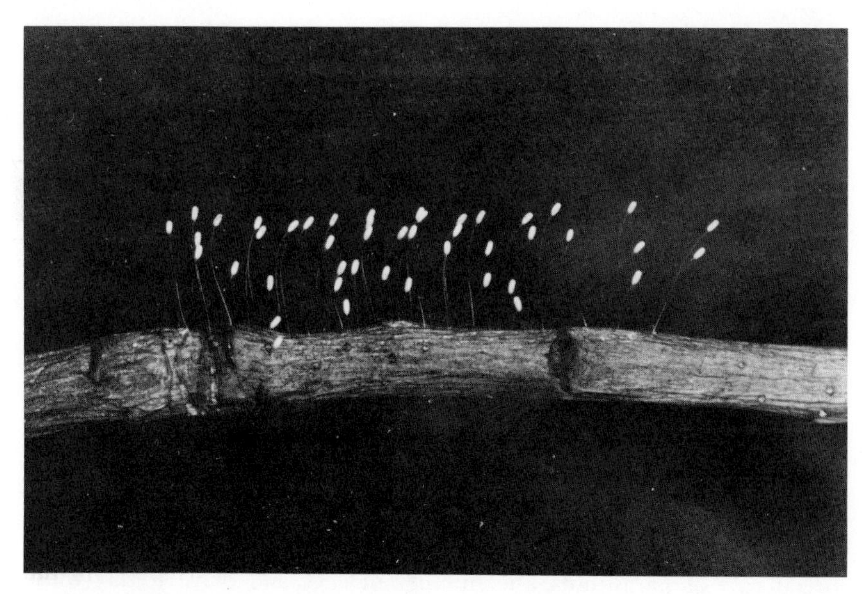

Fig. 5-2 Eggs of *Chrysopa* sp. on silken stalks above the reach of many predators, including *Chrysopa* larvae. (Photo courtesy Ohio Extension Entomology.)

but some, such as the lacewings (*Chrysopa* spp.) are predaceous only as larvae. In fact, lacewing larvae are so voracious that this genus has evolved the habit of laying eggs at the top of silken stalks (Fig. 5.2), where they are presumed to be out of reach of their own larvae and other predators.

Holling (1961) lists five main components of predator–prey relationships as (1) prey density, (2) predator density, (3) characteristics of the environment (e.g., number and variety of alternate foods), (4) characteristics of the prey (e.g., defense mechanisms) and (5) characteristics of the predator (e.g., attack techniques), although he believes the last three are subsidiary factors which affect the magnitude of the first two. He has also reviewed in depth the two fundamental responses of predators to changes in the density of prey or predators. These are the *functional* response, which is a change in the *behavior of the predator* related to a change in the density of the prey and/or the density of the predators, and the *numerical* response, which is a change in the *reproductive rate of the predator* related to changes in the density of the prey (and density of the predators at times), although the movement of predators to prey concentrations and increased survival of predators are sometimes regarded as numerical, too. The same basic principles apply to parasitoids (a specialized kind of predator), although the biological details may be somewhat different, especially in the case of functional (behavioral) responses, since parasitoids attack the prey in a different way, although some parasitoids also kill hosts directly by feeding on them. Stated more concisely, the behavior and reproductive rate of both predators and prey change with changes in the density of either, and with changes in the physical environment.

Predator–prey interactions are complex, and each component may need extensive study and modeling if a full understanding is to be obtained. Only recently has the advent of the computer made it possible to begin to model them with any degree of accuracy that approaches reality. Holling's (1966) monograph on these components gives a good idea of the complexities involved.

Many components interact in predator–prey relationships, including the time the prey are exposed to the predator, the handling time for the predator (recognition time, capture time, and consumption time), the hunger of the predator, learning by both predator and prey and, in the special case of parasitoids, the number of eggs and drive to lay those eggs. Competition between predators can be a major factor at higher densities, as can interference from female parasitoids seeking or attempting to parasitize hosts.

It is difficult to estimate how much detailed knowledge of the components is necessary for intelligent management, but many times this requirement can be met by an empirical knowledge of favorable and unfavorable predator/prey ratios. For example, if it has been established that a favorable ratio is 10 or fewer prey to 1 predator under certain conditions, something must be done to correct the ratio whenever it exceeds 10:1, even though we may not know precisely which components have changed to produce the unfavorable ratio.

II. NATURAL CONTROL AND BIOLOGICAL CONTROL

A. Natural Control

Natural control or the "balance of nature" is perhaps best defined by Huffaker et al. (1971) as "the maintenance of population numbers (or biomass) within certain upper and lower limits by the action of the *whole* environment, necessarily including an element that is density-induced, i.e., *regulating,* in relation to the *conditions* of the environment and the *properties* of the species." It is the combined action of both the biotic and abiotic environment in maintaining the population of any species at a characteristic, yet fluctuating, level. This characteristic level has upper and lower limits which may vary from one geographic area to another, or may change with time in the same geographic area as conditions change.

Natural control affects every living species to some degree sooner or later. The human population has been relatively free from natural control in recent times, but this cannot go on indefinitely as we well know. It is blatantly obvious that in the long run births must equal deaths or any population will overexploit its resources (often with disastrous consequences to both the species and its resources), or the species will become extinct. Overabundance is not quite as disastrous as extinction from the viewpoint of the species, but neither extreme is a common event, or the diversity and abundance of life around us would not exist.

The same holds for pest species—when combinations of conditions are favorable, we have an outbreak, but when conditions become less favorable or man inserts the proper additional mortality factor(s) such as insecticides, new natural enemies, and so on, the pest population reaches a lower level. Whether or not this is below the economic level, and whether or not it stays there are other questions, but natural enemies have the potential to keep it there in many cases, and they help keep it there in even more cases.

The fact that species tend to remain at a characteristic abundance for a given set of biotic and abiotic conditions is generally accepted, but the factors that keep them there and how they operate have been debated for years. Much of the controversy has concerned the meaning of concepts such as density-independent and density-dependent mortality, and the term regulation. A factor that is *density-independent* (nonreactive) has the same effect no matter what the density is, while a factor that is *density-dependent* (reactive) has a different (usually greater) effect at higher densities and a different (usually lesser) effect at lower densities. *Regulation* implies *restricted* fluctuations around some characteristic level, hence must be related to density. Factors such as climate, soil, and photoperiod are generally regarded as acting independent of density, while the action of parasitoids and predators is generally regarded as being related to density. For further discussion of regulation and related subjects, see Huffaker et al. (1971).

B. Biological Control (or Biocontrol)

Traditionally, the term *biological control* has been restricted to the action of natural enemies and phytophagous species used in weed control, although unfortunately it has been broadened by some to include any action of any living organism.

A useful distinction is the recognition of *naturally occurring biological control,* in which man does not actively manipulate natural enemies, and what can be termed *applied biological control,* which involves the use and management of natural enemies by man. However, traditionally, applied biological control has simply been called biological control or classic biological control. This usage is followed here, along with the use of naturally occurring biological control when that distinction is necessary. Biological control should not be broadened to include methods that use biological organisms in some other way, such as host plant resistance (Chapter 4), sterile-male techniques, and genetic manipulation (Chapter 8). To do so confuses the meaning of a term that has a long history of clearly restricted use concerning the management of predators, parasitoids, pathogens, and phytophagous species attacking weeds.

Biological control, be it naturally occurring or applied, has several distinct advantages over many other types of controls, since it is relatively safe, permanent, and economical. One minor disadvantage is that it may take a long time to implement a biological control program, because of the research and other initial effort involved in setting it up. In contrast, pesticides are fast-acting (and broad-spectrum) but, if the developmental time for registration is considered, a lengthy time span can also be involved.

The *safety* of biological control is outstanding, since many natural enemies are host-specific or restricted to a few closely related species. Therefore it is unlikely that nontarget species will be affected, provided of course that the proper taxonomic, biological, and ecological research is carried out to avoid the establishment of undesirable organisms.

Biological control is relatively *permanent,* since it is almost impossible to eradicate any species. Efficient natural enemies often continue to have an effect year after year with little or no assistance from man, provided they are not interfered with in some way. However, since it is nearly impossible to eradicate a species once it is established, it is of the utmost importance that only desirable natural enemies be released. Occasionally, a host population develops immunity to natural enemies by the evolution of a high rate of encapsulation (Muldrew 1953) or some other defense mechanism that reduces the effectiveness of natural enemies. However, this has not been the case for most pest populations, and natural enemies can also evolve, so immunity of hosts has not been a widespread problem, as has resistance to pesticides (pesticides do not evolve).

Biological control is also relatively *economical* since, once efficient natural enemies are present (either native or imported), little may need to be done other than to avoid disruptive practices. In short, biological control has many advantages and no major disadvantages, provided the preliminary basic research is thoroughly and carefully carried out.

Biological control rests on the premise that parasitoids, predators, and pathogens are able to maintain their hosts' populations at lower levels than if they were absent. Hence the objectives of biological control are to maintain the status quo if that is acceptable, or to drive pest populations to lower levels by various manipulations of natural enemies and their environment.

At first glance it appears that natural enemies whose effects increase with increasing host density and decrease with decreasing host density are what we want, since they tend to reduce a population once it reaches a certain density (which may or may not be a damaging density, depending on the tolerance of injury), and relax pressure on the pest at lower levels. However, ideally, we want consistent control without oscillations above economic levels. Hence an ideal natural enemy is one that does *not* permit the host or prey to increase to any degree after it has been reduced to a low level. In other words, the best natural enemy is one that is an exceptional searcher and has a high reproductive capacity, and thus is able to find the host or prey at very low densities and keep it there.

From the pest-management viewpoint, such a natural enemy is ideal, since no management is necessary to keep the pest at a noneconomic level. Unfortunately, we have not yet found such efficient natural enemies for many of our pest species; hence we must rely on the integration of two or more management techniques to keep the pest at noneconomic levels. However, the use of parasitoids and predators should be a *primary consideration* in any pest-management undertaking. It should not be regarded as a second line of defense to be used if other methods fail.

III. IS BIOLOGICAL CONTROL FEASIBLE?

It is impossible to give a "yes" or "no" answer to this question, since each crop must be considered in the light of a complex of pests and other factors which differ from area to area. The following questions are some that can be asked, since the answers may suggest which approaches should be taken. Each of these questions is discussed in more depth as it relates to other management considerations in the following sections.

A. How Much Injury is Tolerable?

This is perhaps the most fundamental consideration because, if little or no injury is tolerated, biological control is almost invariably severely limited. This is discussed further in Section IV.A.

B. What Is the Value of the Crop?

The economics of the situation may dictate the approach to the problem. In the control of pests or weeds on low-value-per-acre crops such as rangeland, biological control using self-regulating systems may be the only economically viable approach, since the use of pesticides or other manipulations by man would be prohibitively expensive. However, in high-value-per-acre crops for which self-regulating systems are not good enough to achieve the desired level of control, natural enemies can often be monitored and manipulated in ways that are not prohibitively expensive.

C. Is the Crop Annual or Perennial?

Historically, our most successful biological control efforts have been for crops that persist for more than 1 year. This does not mean we cannot achieve biological control of pests on annual crops, but it is more difficult because of the inherent instability of annual crop agricultural systems. For example, our annual crops are species of plants that compete poorly with weeds; hence we try to eliminate the weeds which otherwise may add diversity, refuges, and alternate hosts for natural enemies.

Cropping practices also adversely affect annual crop systems as far as biological control is concerned, since the soil is continually disturbed and the crop is often moved to a new site every year to avoid problems with diseases and pests. This may control not only pests, but also natural enemies that overwinter in the soil or crop residue.

Crop rotation may adversely affect biological control in annual crops, but it can also be a positive factor available for management of annual crop systems, which is not available in perennial systems such as forests and rangeland. We also have intermediate crops, such as alfalfa, which persist for a few years but can be cut (Casagrande and Stehr 1973) and otherwise manipulated to the disadvantage of pests and the advantage of parasitoids and predators.

D. Is the Pest Native or Foreign?

This should not be a major consideration, since biological control can be achieved against both native and foreign pests, but the origin of the pest indicates the most likely first approach toward its biological control. If it is foreign, the obvious approach is to investigate the natural enemies that attack it in its homeland. If it is native, ways to better manage existing natural enemies should be investigated along with the possibility of importing exotic ones. However, foreign pests have been regulated by native natural enemies in some cases, so every option should be considered.

E. Are There Any Natural Enemies That Will Provide Control?

The answer to this question is closely related to the first factor; that is, the less injury we tolerate, the smaller the chance of finding natural enemies efficient enough to provide the control required.

We can state with confidence that every pest species has some enemies, and some have many; but whether or not one agent by itself or several used in combination can regulate the pest at established noneconomic levels is difficult to predict. The possibilities of managing existing natural enemies or introducing better ones should be carefully considered.

F. How Many Pests Must be Controlled and What Level of Control Is Required?

Obviously, if pest species are numerous and damage tolerance is low, insecticides will probably be necessary. This reduces the chances for satisfactory management using biological control alone, but does not necessarily preclude the use of natural enemies for control of some pest species. It does mean that the crop system will probably have to be more intensively and carefully managed if potentially useful natural enemies are to work effectively.

G. Are Selective Insecticides or Resistant Natural Enemies Available?

This is intimately related to several of the previous questions, since an insecticide that kills a pest but not the natural enemies that attack it (or more importantly does not kill the natural enemies that control another pest) is an important asset. Unfortunately, we do not have many of these, since most insecticides are also toxic to parasitoids and predators, and most selectivity can be obtained only by careful attention to dosage, timing, formulation, and other manipulations to keep the insecticide away from the natural enemies (Chapter 7).

An equally important factor is the availability of insecticide-resistant natural enemies for use in regulating one or more pests where other pests must be controlled by frequent and/or heavy insecticide applications. These insecticide-resistant natural enemies or the genes for resistance can be introduced into a population in which resistance has not evolved (Croft and Barnes 1971), but it is much better to manage the problem without creating a need for resistant natural enemies if at all possible.

H. Has Biological Control Been Successful against the Same or Ecologically Similar Pests in Other Areas?

This is an obvious consideration, and one that has been used to advantage many times. In fact, the large number of successful transfers in Table 5.1 attests to the validity of the assumption that similar success can be expected against the same pests on the same crops under similar conditions.

Summary

It is important to realize that we are dealing with probabilities in trying to answer these questions. If favorable answers can be arrived at for many of these considerations, the probability of success will be higher than if the answers are largely unfavorable. Nevertheless, there will certainly be a probability greater than zero for any crop system or pest we examine, so we should not look at any crop system or any pest with the view that biological control is impossible. In addition, biological control need not be a complete success, since pest management integrates many tactics to achieve its goal.

IV. MANAGEMENT CONSIDERATIONS

There are three major ways to implement biological control, once the decision has been made to attempt it: (1) the conservation and enhancement of parasitoids and predators already available by manipulation of their environment in some favorable way, (2) the importation and colonization of parasitoids and predators against foreign or native pests, and (3) the mass culture and release of parasitoids and predators. A fourth possibility that can be a part of methods 2 and 3 is the genetic modification of existing populations of natural enemies in the laboratory or in the field so that they are better adapted to the conditions under which they must operate. Laboratory selection has not been very successful to date, although it has been tried (White et al. 1970), probably because of the greatly restricted gene pool available in the laboratory as compared with the wild population, and the very difficult problem of selecting out a laboratory population that will not be inferior in subtle ways once it is released in the field. The most productive approach to date has been to search in other domestic or foreign areas for other populations of natural enemies that possess the desirable traits. This can be done, as in the introduction of insecticide resistance into nonresistant populations (Croft and Barnes 1971) and the introduction of a different and much more effective strain of the walnut aphid parasitoid into central and northern California (Frazer and van den Bosch 1973).

The number and variety of factors that can be managed is almost limitless and many may be employed to manage the pests of a single crop; but the key to any of them is a thorough knowledge of the physical environment and the systematics, biology, ecology, behavior, genetics, and so on, of the pests, their natural enemies, and the plants or animals attacked. However, there is one major judgment which must be made before the potential value of any of them can be objectively estimated—*the setting of realistic economic thresholds* or *economic injury levels.*

A. Economic Injury Levels, Economic Thresholds, and Biological Control

The *economic injury level* has been defined as the lowest pest population that causes economic damage, while the *economic threshold* is a somewhat lower population at which some kind of action must be taken to prevent the pest from reaching the economic injury level (Chapter 1). Both are important considerations in the management of parasitoids and predators, although the primary concern in biological control should be the *adoption of realistic economic injury levels which will permit the effective use of parasitoids and predators.*

Stern (1973) recently reviewed the concepts of economic threshold and economic injury level in relation to the necessity of treatment (usually with insecticides) for crop protection. He concludes that, with few exceptions, there is little usable information relating crop yield to pest density, and this lack of knowledge has led to frequent erroneous judgments and unnecessary controls.

Biological control has the same fundamental objective as any other control measure—avoiding or minimizing economic losses. However, the question of whether or not it should even be attempted is dependent on the use of realistic economic injury levels.

Parasitoids and predators, by their very nature, cannot prevent all damage by a pest, since their relationships with the pest have tended to evolve to the point where both survive and the host is not reduced to zero. Therefore, if no injury is acceptable (which is unrealistic but sometimes the case), some other control must be employed in most situations, and biological control can rarely be attempted. As van Emden and Williams (1974) have pointed out, if no injury is tolerated, a very low density of pests is required, which means we usually must keep a pest *below* its characteristic equilibrium density, often by using insecticides.

It is useful to think of injury as direct or indirect, as suggested by Turnbull and Chant (1961). *Direct injury* is damage to the part of the crop we use, such as the damage caused by the codling moth or the apple maggot to apples. *Indirect injury* is damage to a part of the plant we do not use, such as that caused by scale insects, aphids, or mites sucking sap from apple trees. Some or even a large amount of indirect injury may be tolerable, but in many cases little or no direct injury is tolerable, frequently because of government regulations or because of consumer and processor attitudes toward cosmetic qualities which have little or no relationship to nutritional value.

It is rather obvious that, if the tolerance for insects, insect fragments, or insect damage is zero, and the grower's crop will be rejected or the selling price greatly reduced if it is infested, biological control will usually not be good enough. There is also the unfortunate possibility that biological control may be good enough but the product will be contaminated by beneficial insects, since federal regulations make no distinction between contamination by pests and contamination by beneficial species. Nevertheless, it is obvious that, the more injury or fragments we will tolerate, the better the prospects are for achieving some degree of biological control. This does not mean that biological control cannot be successful against pests causing direct damage—it simply means that the chances for nonchemical control against any pest are better when we are willing to tolerate some injury.

It is not easy to set realistic economic injury levels. In fact, it is one of the most difficult problems of pest management, because of the complex interactions between the severity of the pest infestation and the local climate, season, vigor, variety, and age of the crop when attacked, cropping practices, changing economic conditions, and so on. These complex interactions make it necessary to *use flexible economic injury levels and economic thresholds that will change as conditions change.* This is probably the major reason why realistic economic injury levels and thresholds have not been used for most crops—they are very difficult problems which require the constant monitoring of many variables in order to make the necessary adjustments throughout the growing season. Despite the difficulties inherent in establishing realistic economic injury levels, it is absolutely essential that they be determined if we are to evaluate the chances of achieving them through the use of biological control.

It is worth emphasizing that the *determination* and *use* of realistic economic injury levels are the foundation of good pest management, not just the biological control component. If the economic injury level is relatively high, the range of control options is much greater than if it is low or zero.

B. Conservation and Enhancement

Importation and colonization, mass production and release, and conservation and enhancement can all have important places in a pest-management program, but the conservation and enhancement of natural enemies should be a *first* consideration in every program since, if conservation and enhancement are properly accomplished, the need for other control measures may be greatly reduced or even eliminated for some pests (in effect reducing them to nonpests).

Conservation and enhancement are really different ends of the same continuum, *conservation* being the avoidance of measures that destroy natural enemies, and *enhancement* the use of measures that increase their longevity and reproduction, or the attractiveness of an area to natural enemies. Because of the great variation in climate, crops, pests, natural enemies, and the pesticides used, each problem or potential problem must be examined individually. Hence it is impossible and unnecessary to give detailed examples of the many conservation and enhancement measures that can be employed, but the following are examples of some that have been effectively used. They are not intended to represent either the first use or the latest use or necessarily the best use of the technique; they are merely examples of the kinds of measures that have been used.

1. Protection from Pesticides

Innumerable examples of the interference of pesticides with the control of a pest by natural enemies are known. All kinds and formulations of inorganic and organic materials ranging from stomach poisons to contact poisons to dormant oils, dusts, and so on, can have adverse effects. Perhaps one of the most striking cases of interference is the upset of the long-standing biological control by the vedalia beetle, *Rodolia cardinalis,* of cottony cushion scale, *Icerya purchasi,* on citrus in California by the use of DDT to control other pests in 1946 and 1947 (DeBach 1947). Vedalia beetles were killed by the sprays, while cottony cushion scales were not killed, resulting in serious outbreaks. Adjustment of spray treatments permitted the vedalia beetles to regain control in subsequent years. Another good example is the rise of the red-banded leafroller, *Argyrotaenia velutinana* (Walker), to major pest status after the use of various organic insecticides against the codling moth and other apple pests had reduced the impact of its natural enemies (Paradis 1956).

Perhaps the most widespread interference with natural enemies has been the killing of predaceous mites by various pesticides used for control of plant-feeding mites and other insect pests, which resulted in many species of phytophagous mites which had formerly been minor pests becoming major pests. This has been further aggravated in many crops by the rapid development of resistance. The solution appears to lie in the relatively recent use of predators resistant to some pesticides, the reduction of prey populations through the use of selective pesticides, and the manipulation of prey and predator populations through various habitat management techniques (Chapter 13).

2. Resistant Natural Enemies

There is no inherent reason why parasitoids and predators cannot develop resistance to pesticides just as many pests have, but relatively few have been shown to do so (Croft 1972). However, as Croft has pointed out, it is a more complex situation, since pesticides are aimed at the pests (not the natural enemies) and therefore the pressure is not usually as great on the natural enemies, although it is great enough to disrupt biological control. A second complicating factor is that resistant genotypes of natural enemies cannot be developed if the pesticides are effective against the pests and hosts are scarce. A third factor is the discontinued use of a pesticide as soon as the pest develops resistance, thereby eliminating the selective pressure on the natural enemies at the very time their resistant prey has become abundant enough to support them. Another consideration is the mobility of natural enemies. Most of the natural enemies that have developed resistance to date are predators, perhaps because the mobility of many predators such as predaceous mites is limited; consequently, they are exposed to heavier selective pressure than most parasitoids which can fly into or out of the sprayed area.

The possibility of developing resistant natural enemies is always present, but it is simpler to avoid the necessity by following practices that do not decimate them in the first place. However, in crops in which little or no injury is tolerated, pesticides will be used, and the development and/or detection and use of resistant predators and parasitoids may be an increasingly important tool in the future for the successful management of pests that have developed resistance and pests whose natural enemies are decimated by pesticides applied for the control of other pests.

An important consideration in the use of resistant natural enemies may be the selection of the natural enemy that is best able to control the pest under the spray program in use for other pests, since different natural enemies that prey on the same pest species have been shown to be resistant to different pesticides (Nelson et al. 1973).

3. *Preservation of Inactive Stages*

This is most critical when there is a small reservoir, or none at all, of natural enemies outside the cropped area. At the intial establishment site of the parasitoids of the cereal leaf beetle, *Oulema melanopus* (L.), in Michigan, a high percentage of the major parasitoid, *Tetrastichus julis* (Walker), overwinters in the soil of oat stubble (Fig. 5.1); so a portion of the oat stubble was not plowed until after the parasitoids had emerged in the spring, resulting in a rapid buildup of the parasitoid (Stehr and Haynes 1972).

4. *Avoidance of Harmful Cultural Practices*

Cultural control is a valuable part of managing pests, but plowing, discing, mowing, burning, and so on, can be equally harmful to natural enemies. The effects of such operations on natural enemies must be fully evaluated, and harmful ones eliminated or modified if at all possible. For example, in the irrigated deserts of California where the survival rate of many natural enemies outside of irrigated areas is very low, strip-harvesting of alfalfa so as not to cut the entire field at any time has resulted in much greater survival of natural enemies and the hosts necessary to keep them alive (Schlinger and Dietrick 1960).

5. *Maintenance of Diversity*

This is frequently a necessary part of many other measures, since it may provide alternate hosts, sources of food, overwintering sites, refuges, and so on. Diversity per se is not necessarily essential, but most natural enemies have evolved in communities much more diverse than crop communities; hence it is reasonable to expect the right kinds of diversity to be beneficial.

Hambleton (1944) illustrates the value of crop diversity in the irrigated Canête Valley in Peru where the tobacco budworm, *Heliothis virescens* (Fabricius), reduced cotton yields by 30 to 40% following the planting of nearly the entire valley to cotton. Arsenical insecticides had no effect, so growers were encouraged to plant winter crops such as corn, flax, beans, and sweet potatoes, which supported noneconomic populations of various Lepidoptera (including some *H. virescens*). These species provided the hosts necessary to maintain the populations of predators (especially a nabid) and parasitoids through the winter that controlled *H. virescens* on cotton during the long summer growing season from September to May. He also pointed out that, in other similar Peruvian valleys where diverse crops had always been raised, *H. virescens* had always been there but had never become a problem.

Van Emden and Williams (1974) recently reviewed the subject of insect stability and diversity in agroecosystems. Essentially, they conclude that there is a general relationship between diversity and stability in mature, native, natural systems, but agroecosystems are not mature and not natural, so the same relationships may not hold. They suggest that structural and spatial diversity may be desirable, but species diversity is not necessarily desirable because agroecosystems are unstable and the pests are few and unstable; hence a diversity of natural enemies (which is supposed to promote stability) is not necessarily the best solution in an unstable situation. They do conclude that the maintenance of weeds and other floral diversity outside the crop but within the agroecosystem should be encouraged until we are sure their beneficial capacities can be replaced by planned diversity or other measures.

6. *Alternate Hosts*

The right life stage(s) of the host may not be available at the right time or the right place to ensure the survival of natural enemies throughout the year. Manipulation of the crop, planting of different crops nearby, leaving some weeds, or providing diversity in other ways is usually essential in providing for alternate hosts.

The effectiveness of the egg parasitoid *Anagrus epos* (Girault), an important parasitoid of the grape leafhopper, *Erythroneura elegantula* Osborn, in California vineyards, is linked to the presence of wild blackberry, *Rubus* spp., near vineyards (Doutt and Nakata 1973). These blackberry bushes support populations of a noneconomic leafhopper, *Dikrella cruentata* (Gillette), which breeds throughout the year and provides eggs for *A. epos* throughout the year. *Anagrus epos* dies out in the winter in vineyards far removed from blackberry bushes, because its other host, the grape leafhopper, goes into reproductive diapause and does not produce the eggs necessary for continual breeding of *A. epos,* which does not diapause.

7. Natural Foods (Nectar, Pollen, and Honeydew)

Many parasitoids and predators require foods frequently not available in monocultures. The abundance of the predatory mite *Amblyseius hibisci* (Chant) in avocado orchards in California was related to the availability of pollen, even when prey mites were absent (McMurtry and Johnson 1964). Leius (1967) found much higher parasitism of tent caterpillar eggs and pupae, and of codling moth larvae, in unsprayed apple orchards in Ontario which contained abundant nectar-producing flowers when compared with orchards containing few such flowers.

8. Artificial Food Supplements

In irrigated monoculture desert areas of California, artificial honeydew and pollen in the form of food sprays have induced early oviposition of naturally present aphid lions (*Chrysopa*) and coccinellids in treated fields, resulting in significantly lower populations of aphids in alfalfa and bollworms in cotton (Hagen et al. 1970).

9. Artificial Shelters

In North Carolina substantial reductions of tobacco hornworms were achieved by predacious *Polistes* wasps, following the erection of nesting shelters near field margins (Lawson et al. 1961).

10. Reduction of Undesirable Predators

The recently achieved biological control of the walnut aphid, *Chromaphis juglandicola* (Kaltenbach), by a new ecotype of *Trioxys pallidus* (Haliday) from Iran, is jeopardized in some areas of California by the selective predation of parasitized aphids in preference to unparasitized aphids by the Argentine ant, *Iridomyrmex humilis* (Mayr) (Frazer and van den Bosch 1973).

11. Control of Honeydew-Feeding Ants

Abundant honeydew-feeding ants can make biological control of honeydew-producing species such as scales, mealybugs, and aphids unsatisfactory because of interference with natural enemies. Interference in the control of pest species that do *not* produce honeydew can also occur (DeBach and Huffaker 1971).

12. Favorable Temperatures

Temperature is not easily manipulated in the field, as it is in greenhouses, but the differential effect on hosts and natural enemies can be very striking and should be carefully evaluated, since conditions can be drastically altered by practices such as mowing.

It is well-known that the greenhouse whitefly, *Trialeurodes vaporariorum* Westwood, cannot be controlled below 24°C by the parasitoid *Encarsia formosa* Gahan (Hussey and Bravenboer 1971). The effects are summarized concisely by stating, "At 18°C the fecundity of the whitefly is 10× that of *Encarsia* though the rate of development is equal, while at 26°C the fecundity is equal and the rate of development of *Encarsia* is twice that of the whitefly." Temperature is obviously critical in the use of *Encarsia* to manage the greenhouse whitefly.

13. Avoidance of Dust

Road dust and other dusts have been shown to reduce the effectiveness of some natural enemies in California drastically, while having little effect on the hosts (Bartlett 1951).

C. Importation and Colonization

A person involved in the everyday decisions of pest management is ordinarily not concerned with the importation of new parasitoids or predators, since this is done at the state, university, or federal level, and regulations concerning the importation of plants and animals must be observed and permits obtained. Generally, the pest manager must work with what is already present or can be obtained locally. However, this does not mean that better natural enemies are not available someplace in the world, and efforts to find them should be encouraged by all concerned.

The use of natural enemies from foreign areas is often a major component of a biological control program, especially if the pest(s) are of foreign origin, although sometimes a search is made for more efficient natural enemies of native pests than the native ones already present. Frequently, all or most of the natural enemies are left behind when a pest becomes established in a new area, and in fact quite often the pest is not a pest in its native home. This may be due to natural enemies, differences in climate, differences in host plants, or other factors. However, there is a reasonable chance that natural enemies can be found that will give partial, if not complete, control in the new area.

The importation of natural enemies is not limited to those that attack pests of foreign origin. Any pest (native or introduced) may be a suitable host for natural enemies obtained from related hosts in foreign countries. Pimentel (1963) discusses the subject in some detail and lists many examples. Some of the best examples are in biological control of weeds and the effects of pathogens on plants and animals.

A classic example is the use of the pyralid moth, *Cactoblastis cactorum* (Berg), obtained from the Argentine prickly pear cactus, *Opuntia aurantiaca* Lindl., and introduced into Australia. It failed to control *O. aurantiaca,* but was extremely effective against *O. stricta* (Harv.) from Florida and *O. inermis* (DC.) from Texas, which had infested approximately 60 million acres in Australia (Dodd, 1940). Another example is the decimation of the American chestnut by a fungus native to the Far East, which has little effect on the Oriental chestnut. While we do not regard the American chestnut as a pest, the principle is the same—an exotic natural enemy that has minimal effect on its exotic host has essentially destroyed a species with which it did not evolve.

A good example of the use of a nonadapted parasitoid is the introduction into Fiji of a tachinid fly, *Ptychomyia remota* Aldrich, which controlled the coconut moth, *Levuana iridescens* B-B. This tachinid had been obtained in Malaysia from a related moth, *Artona cotoxantha* Hampson (Tothill et al. 1930).

By now it should be obvious that consideration of the importation of any species of natural enemy from another area (not necessarily foreign) is not to be taken lightly. The intention is to cause a major ecological disruption, and if everything is done properly, it should be beneficial. The first requirement is proper identification of the pest so its native home or its relatives can be found. After the native home is found, the natural enemies must be found, identified, and their biology and ecology studied to determine if they should be introduced, and to identify and exclude any undesirable or potentially undesirable species.

Collection is not simply a matter of obtaining all that can be found in the easiest way. As Flanders (1959) has pointed out, some thought should be given to collecting in areas where the host is never common, since natural enemies found in such areas are more likely to keep the host at a low level, assuming the host is not kept there by some abiotic factor(s). And Force (1974) has suggested searching disturbed and marginal areas for r strategists, since K strategists are most likely to be found in undistributed and optimal areas (see Chapter 2 for discussion of r and K strategists).

Despite the fact that many natural enemies are capable of operating at different levels of pest abundance, some that are able to keep a pest at a low level are not necessarily able to reduce a population to a low level once the pest has reached a high level. Conversely, some of those that are able to bring a population down may not be able to keep it there. Bird and Elgee (1957) report that two introduced ichneumonid parasitoids, *Dahlbominus fuscipennis* (Zett.) and *Exenterus claripennis* (Thom.), of the European spruce sawfly, *Diprion hercyniae* (Hartig), were beginning to build up in 1938 and 1939 but were essentially eliminated during the severe virus epizootic from 1940 to 1942. Following the collapse two different introduced parasitoids, the tachinid *Drino bohemica* Mesnil and the ichneumonid *Exenterus vellicatus* Cushman, were recovered in good numbers and appeared to be responsible (along with the virus) for maintaining the spruce sawfly population at a reasonably low level after 1945.

Material should also be collected from a variety of geographic areas to obtain natural enemies that may come from different host populations, may be differently adapted, and may differ genetically. Remington (1968) has discussed the genetic aspects of introduction.

After natural enemies have been collected and/or reared, they are usually air-shipped to quarantine facilities where they are cultured or stored in environmental chambers until the next growing season. However, the most important functions of quarantine stations are further study of the natural enemies to be sure they have no undesirable qualities, and elimination of hyperparasitoids and anything else that might adversely affect them or other living things in the country where they will be released. Mistakes can be costly and are virtually impossible to correct. The buildings and other facilities at quarantine stations are especially designed and operated to avoid the release of undesirable organisms (Fisher 1964; Fisher and Finney 1964).

Techniques of field release for establishment and subcolonization vary with the species (DeBach and Bartlett 1964) but, once the natural enemy is established, every effort should be made to manage the initial establishment site for the field production of natural enemies to be used for additional subcolonizations, since for many species it is much easier and much cheaper than trying to mass-produce them in the laboratory.

In addition, field collection eliminates the possibility of selecting out laboratory-adapted strains (Ashley et al. 1973), even though this can be greatly reduced by periodic additions of field-collected material to the laboratory culture. Field production of *Tetrastichus julis,* the major parasitoid of the cereal leaf beetle in Michigan, by protection of overwintering sites in the soil, protection from spraying, and provision of additional cereal leaf beetle hosts by planting some oats 2 and 4 weeks later than normal (beetles are attracted to young, succulent grain) has been very successful (Stehr and Haynes 1972). A small portion of this production site has provided sufficient *T. julis* for Michigan State University to subcolonize them throughout the lower peninsula of Michigan and for the Cereal Leaf Beetle Parasite Rearing Laboratory, APHIS, USDA, to subcolonize them throughout the Midwest. This has been done at a fraction of the estimated cost of insectary production, to say nothing of the technical problems that would have had to be overcome for insectary production.

1. *Single-versus Multiple-Species Introductions*

The practical question of whether to introduce one or several species of natural enemies has been debated for years. Most of the available evidence favors multiple-species introductions, but experimental evidence is lacking (and difficult to obtain) for the single-species philosophy, although Force (1974) obtained circumstantial evidence that species with a high reproductive potential and low competitiveness (r species) are better than species that have a lower reproductive potential and higher competitive ability (K species).

The single-species philosophy supported by Turnbull and Chant (1961), Zwolfer (1963), and others suggests that only the single most promising natural enemy should be introduced, in order to avoid any potentially harmful competitive interactions between species. If this species is not satisfactory, another species is introduced, and this procedure is continued until satisfactory control is achieved.

The multiple-species philosophy advocated by DeBach (1966), Hassel and Varley (1969), Huffaker et al. (1971), and others suggests that every reasonable prospect should be brought in, since it is impossible to predict which one(s) will be effective. They believe that more than one parasitoid never does a worse job than one alone. Both philosophies of course require careful screening of the material to be introduced to eliminate hyperparasitoids and other undesirable organisms.

Few will argue against the single-species philosophy in principle, but it has some practical difficulties in implementation, such as determining which agent is the best [a difficult task to say the least, although Force (1974) suggests r species are best], establishing it, differential survival and effectiveness in different environments, and different effectiveness at high or low host densities. Huffaker et al. (1971) discuss this subject thoroughly.

The problem whether or not additional species should be introduced later, with the hope that they may be better, raises the same question of possible interference among natural enemies, which might result in less control, and the related question of competitive displacement by ecological homologs. An ecological homolog is a species that is genetically different, but which occupies an "identical" ecological niche or role in the community. Completely identical niches for two species probably do not exist. However, requirements for food, shelter, hosts, and so on, do not have to be identical in all respects for competitive displacement to operate. If niches are similar, one species will be superior in some crucial way(s) and will tend to replace the other totally or in part of its range. DeBach (1966) discusses the concepts of competitive displacement and coexistence in detail, and is one of the strongest supporters of the idea that the potential of a species to be introduced cannot be predicted and need not be evaluated in depth; he maintains that, if it is a better species, it will displace the existing species and, if not, it will not become established. Not all workers agree, however, and this philosophy should be regarded as one working hypothesis. Another working hypothesis is that of Force (1974), who infers that the species with the greatest reproductive potential (r species) should be the first one introduced, since it can respond faster to increases in the host population, and since it often may be a poor competitor with K species.

One of the best examples of a long-standing failure turned into a success by the introduction of the right natural enemies involves the California red scale, *Aonidiella aurantii* (Maskell), on citrus (DeBach et al. 1971. The first introductions were made in 1889, but *Aphytis lingnanensis* Compere was not introduced until 1948 and *Aphytis melinus* DeBach until 1957. A major reason for this 50-year failure was a series of misidentifications of the parasitoids and the scales, which resulted in searches being made in South America instead of the Far East, introductions of unsuitable parasitoids from similar species of scales, and failure to introduce the effective parasitoid species because they were misidentified as species already present in California. All this serves to emphasize the necessity for sound taxonomy (along with sound biology and ecology) as a fundamental base on which to build successful biological control.

It has also been argued that natural enemies that attack different life stages cannot possibly interfere with each other, since they never come in contact. It is true that they cannot interfere directly in a physical sense, but indirect interference by the killing of one life stage which produces the next life stage needed by another natural enemy may affect the second natural enemy in some way. A case in point is that of *Anaphes flavipes* (Foerster), which parasitizes nearly 100% of the cereal leaf beetle eggs late in the oviposition period. These parasitized eggs do not produce late cereal leaf beetle larvae which are nearly 100% parasitized by the second and overwintering generation of the larval parasitoid *Tetrastichus julis* (Walker). Therefore the rate of buildup of *T. julis* is reduced, since the number of overwintering parasitoids is reduced, even though the final outcome may not be affected (but this is hard to determine, since there is no way of knowing what the outcome would have been if the egg parasitoid had been absent). If possible, such interactions should be investigated in the native country before introductions are made, so those that are potentially undesirable can be avoided.

A related point regarding single- versus multiple-species introductions has been made by Haynes (1973), who suggests that in the concept of modern pest management a relatively poor parasitoid that can be managed may be of considerable value in a pest-management program, even if it does not provide the total solution. He implies that the first introduction should be the species most easily managed to avoid any possible interference from species that may not be manageable.

Whitten (1970) has suggested the genetic "fingerprinting" of enzymes of different populations before release, as a means of determining initial variability for use in assessing which populations become established and how they evolve with the host.

No one has come up with a method to select the best natural enemies for importation, although Force (1972, 1974) and Price (1973) have suggested that natural enemies that are early colonizers or r strategists are more likely to be effective because of their high reproduction rate, high dispersal powers, and tolerance for harsh conditions. These are the conditions encountered in most agricultural cropping systems.

Price (1972) has also made an attempt to develop methods to quantify the selection of natural enemies for introduction, but there is no formally stated national policy on introductions. It is also unlikely there will be one as long as there are such divergent opinions, although the quarantine and importation stations in the United States accept the multiple-species philosophy in practice.

D. Mass Culture and Periodic Release

1. *Mass culture* is the insectary propagation of large numbers (actually millions) of natural enemies for release against selected pests at strategic times. It is relatively expensive and therefore is ordinarily used on limited acreages of high-value crops. It is also restricted to situations in which both the natural enemy and its host (either natural or a suitable substitute) can be easily cultured. Mass culture has usually been accomplished by government laboratories or grower associations, but more recently some private insectaries have been established as a part of pest-management services (Dietrick 1973).

2. Periodic releases have several variations (inoculative, supplementary, or inundative), depending on the purpose and frequency of release, and on the sources of the natural enemies to be released (insectary or field). In some situations two or more kinds of releases may be used to achieve the desired result.

a. *Inoculative releases* are a form in which releases may be made as infrequently as once a year to reestablish a species of natural enemy, which is periodically killed out in an area by unfavorable conditions during part of the year, but operates very effectively the rest of the year. Control is expected from the progeny and subsequent generations, not from the release itself. Oatman et al. (1968) describe the mass release of the predator *Phytoseiulus persimilis* Athias-Henriot in California against the two-spotted spider mite, *Tetranychus urticae* Koch, beginning in January, and the subsequent reduction of both the host and predator to essentially zero by midsummer. Several factors contributed to the need for reestablishment of *P. persimilis* each year, including strawberries being grown as an annual crop, poor dispersal, overexploitation of the prey by the predator, and poor winter survival.

b. *Supplementary releases* of natural enemies can be made when sampling indicates a pest is about to escape control by its natural enemies. Reestablishment of control is expected from the release or the progeny immediately produced by the release. Croft and McMurtry (1972) describe how minimum releases of 128 mites per apple tree of the predaceous mite *Typhlodromus occidentalis* Nesbitt maintained host populations of *Tetranychus mcdanieli* McGregor at less than 15 mites per leaf, while trees with no releases exceeded 120 mites per leaf.

c. *Inundative releases* are the most expensive, since they essentially involve a "biological insecticide" and millions are usually required. The objective is to overwhelm completely the pest with the release, with little or no reliance put on subsequent generations of the natural enemy. They are obviously most economical against pests that have only one generation per year, although they are not limited to such pests.

Inundative releases of egg parasitoids of the genus *Trichogramma* have been tried many times with variable results, often poor (Dolphin et al. 1972), considering the cost and effort involved, although an experimental program of cabbage pest management by Parker (1971) utilizing *Trichogramma* as one of two principal parasitoids is one example of promising results (see below). An outstanding commercial example of mass culture and periodic release given by Doutt (1972) is the release of two species of parasitoids to control black scale and citrus red scale in the Filmore Citrus Protective District in California at a cost of $8 per acre. Only 20 acres were sprayed for black scale and only 10 for citrus red scale out of a total of 8423 acres managed with parasitoids.

Another example is given by Parker (1971), in which densities of *both hosts and parasitoids* were manipulated experimentally in Missouri through periodic releases for control of eight pest species on cabbage. One field was harvested, and only 1 of 1500 heads was unmarketable, although many of the outer leaves showed damage. This clearly demonstrated the feasibility of managing the pests of a short-lived crop which was continually disturbed by farming practices, through the release of natural enemies and their hosts. However, very close supervision was required, because of the necessity for accurate and continuous information on the status of both hosts and natural enemies. Additionally, a necessary requirement for success was the acceptance *by the grower* of visible injury to young plants and to the outer leaves of old plants, which did not affect the quality of the market head.

Mass releases are usually made with insectary stock, but Dietrick (1973) suggests the vacuum harvesting (Stern et al. 1965) of beneficial insects from crops such as sorghum and strip-harvested alfalfa for release in other fields in need of natural enemies. Such methods are most likely to be of direct benefit in irrigated desert areas where a small reservoir or none at all exists in the surrounding nonagricultural land and movement of natural enemies is greatly deterred by the severe environment. However, selected mass-collection and/or mass-production techniques could be very useful in a variety of programs in many areas. For example, Halfhill and Featherston (1973) used field cages in Washington State in attempts to produce enough parasitoids of overwintering pea aphids,

Acyrthosiphon pisum (Harris), in alfalfa to prevent the movement of the pea aphids from the alfalfa to nearby fields of processing peas. Control in alfalfa was achieved until insecticide treatments for *Lygus* disrupted it, but additional production and distribution problems made it impractical to implement the complete program.

V. EVALUATION

Evaluation is one of the most difficult tasks of biological control, since most biological control projects are unique experiments in community relationships, which are virtually impossible to repeat because of the permanent nature of biological control. In other words, there is no way to go back and try different natural enemies or the same natural enemies in a different sequence under the same conditions once some of them have been established. This obviously cannot be done for native insects.

Complete control is the easiest to evaluate since, without setting up experiments or taking samples, it is quite obvious something has reduced the pest to noneconomic levels. In many cases we assume it was the natural enemies, although this is not necessarily true. If characteristic abundance levels for pests are obtained before natural enemies are introduced, this "before-and-after" method of comparing population levels can be quite convincing, even though the proof is only circumstantial. It is, however, not suitable for demonstrating the effects of naturally occurring biological control, and is less convincing if good pest abundance data have not been obtained before the introduction of natural enemies, or if pest populations have not stabilized before the introductions.

"No control" is also relatively easy to evaluate, since it is easy to assume the natural enemies are having a negligible effect if the pest remains a serious pest. Partial successes are difficult to evaluate, but the evaluation of natural enemies that are effective only at low host densities is perhaps the most difficult, since any interference that allows the pest to reach a higher density may make it impossible for the natural enemy to control it. And of course, the demonstration of naturally occurring biological control requires some kind of disruption of the natural enemies before the control becomes apparent.

The available techniques have been separated into three groups: (1) the correlation of pest and natural enemy abundance, (2) the natural enemy exclusion (experimental) method, and (3) the mathematical modeling of population processes. The method to be used depends on the particular system being evaluated, but any combination of methods applicable to a given problem that demonstrates that biological control worked (or did not work), and possibly how it worked (or did not work), is a reasonable approach.

A. Correlation

The correlation method relies largely on life table and density data for pests and their natural enemies, and attempts to show a cause-and-effect relationship between an increase in natural enemies and a decrease in the pest(s), or vice versa. The major deficiencies are the lack of any check or control, the impossibility of demonstrating control when the natural enemy is effective only at low host densities (at high host densities the host increases and decreases without any related change in the density of the natural enemy), and the impossibility of demonstrating control when the natural enemy is so efficient that the host "never" has a chance to increase.

B. Experimental

As DeBach and Huffaker (1971) have pointed out, a clear distinction must be made between the two questions: (1) Does regulation by natural enemies occur? (2) How does it occur? Regulation has been defined by them as the maintenance of an organism's density over an extended period of time between characteristic upper and lower limits. A regulatory factor is defined as one that is wholly or partially responsible for the observed regulation under the given environmental conditions, and whose removal or change in efficiency or degree results in an increase in the average pest population density. This can be tested by experimental techniques.

The experimental techniques all involve the use of some procedure which excludes, eliminates, or drastically reduces the natural enemies in some plots which are then compared to other plots in which such methods have not been used. These are all "present-and-absent" techniques, and they are similar to "before-and-after" methods except that the conditions are more carefully regulated and control plots can be used.

Many different techniques have been used (DeBach and Huffaker 1971). Sometimes identical plots are set up and natural enemies are added to some and not to others. This obviously works best for natural enemies that are slow to disperse, since separation of the plots by any great distance tends to reduce their similarity. Cages are often used to exclude natural enemies, but similar cages which permit entry must be used to eliminate any cage effect. Insecticidal check methods use any materials that selectively kill the natural enemies through differences in toxicity, dosage, formulations, timing, and so on, and have a minimal effect on the pest.

One biological method takes advantage of the aggressive reactions of honeydew-seeking ants toward natural enemies, and has essentially the same effect as selective insecticides, but without any of the complications associated with them or with other methods such as exclusion cages.

Hand removal of natural enemies has been used, and is ideal from the standpoint of avoiding any influence on anything except the natural enemies. The obvious disadvantage is the excessive work required to find and remove most of the natural enemies before they have any impact, and the necessity of working with pests that produce many generations per year or which continually invade the protected area if the pest populations are to build up so results can be obtained in a reasonable period of time.

The trap method is a modification of the insecticidal check method in which an untreated central area is surrounded by a treated area where natural enemies that try to leave or enter the untreated area are killed. It is obviously restricted in use to situations in which the natural enemy is mobile but the pest moves very little (such as scale insects and mealybugs).

The use of experimental techniques in demonstrating the regulatory ability of natural enemies has been clearly successful in many cases, but an examination of the pests they have been used on shows that they have been most effective against pests having limited mobility and many generations per year. Trying to test experimentally the regulating ability of a natural enemy against an insect such as the cereal leaf beetle, *Oulema melanopus* (L.), which has one generation per year, leaves the fields to overwinter, and returns to different fields in different locations the following year is difficult. It is further complicated by the fact that the beetle density per field is affected by the species, age, quality, and acreage of the crop planted. In addition, the larval parasitoids exhibit delayed mortality since they do not kill the host until after the cereal leaf beetle larva pupates and the damage is done. The alfalfa weevil problem is similar, but somewhat less complicated because the crop and the acreages are more stable from year to year. Experimental techniques are needed that can be used on such pests, but these inherent problems are difficult to overcome.

C. Modeling

The advent of computers which can swiftly handle tremendous amounts of data has led to the development of biological systems models which can be used for management and evaluation. These models, combined with adequate input of biological and environmental parameters from the field (Haynes et al. 1973), should make it possible to evaluate the current situation, predict what is likely to happen given certain conditions, and recommend procedures to obtain a desired result.

These models are now being developed, and the necessary biological and environmental data are being collected for many crop systems in many parts of the world. And after the proper models have been developed, still other benefits can be derived, since hypotheses can be tested which could never be tried in the field because of existing conditions, possibly harmful introductions, or time considerations.

D. Other Considerations

Some long-held fundamental assumptions may not be entirely correct. Varley and Gradwell (1971) have discussed the use of models and life tables in assessing the role of natural enemies, and have questioned the common assumption that any form of density-dependent factor regulates a population, and that parasitoids and predators necessarily act as denity-dependent factors. The controversy over single versus multiple-species introductions has not been resolved, and some have questioned the widely held assumptions that superparasitism and hyperparasitism are always harmful, since it is possible that they could help stabilize parasitoid and host numbers in a system in which the pest and parasitoid fluctuate greatly. Other attributes of good natural enemies such as high searching ability, high fecundity, host specificity, and many generations per year may not be necessary or even desirable in some situations.

Many of these options can be investigated by manipulation of paramters in computer-based models, when it would be next to impossible to check them in the field. To do this effectively, realistic models must be developed. Once they are developed, perhaps we will even be able to explain how and why a natural enemy operates. But even if they are not completely successful, they will force us to obtain quantitative data. The prospects for better evaluation and understanding of biological control have never been better, and they should rapidly improve in the future.

VI. CONCLUSION

Are there any principles in the use of parasitoids and predators? The term principle may be too strong, but there are guidelines to follow, some more important than others, depending on the individual crop and location.

1. A thorough study should be made before a decision is reached regarding biological control, no matter how disruptive the cropping practices may be. Biological control is generally regarded as being most effective in very stable systems such as forests, rangeland, orchards, and parks. However, it can still be effective in semistable crops such as alfalfa, and does have a chance in annual crops where the environment is continually disrupted and diversity is minimal if proper management is used.
2. Disruption of naturally occurring biological control by poor choice and/or use of insecticides, harmful cultural practices, and so on, should be avoided. There are usually enough problems without increasing the existing ones or causing new ones.
3. Realistic economic injury levels must be determined and used. Parasitoids and predators do not kill 100% of their hosts or prey. Grower acceptance of injury that does not affect the final market quality of the crop is a closely related consideration.
4. Grower education is also essential for any kind of intensive manipulation of natural enemies, but this is an essential part of the entire pest-management program, not just the biological control component.
5. Avoid killing all the pests if pesticides must be used in conjunction with natural enemies. Parasitoids and predators must have some hosts or prey to survive.
6. Maintain diversity in crop and noncrop areas if necessary to provide refuges, food, and other necessities for both natural enemies and their prey.
7. Improve the chances for the successful use of parasitoids and predators by minimizing the number of pest species that need insecticidal control.
8. If an insecticide must be used to control one or more pests, use one that is selective in some way toward the pests and has minimal effect on the natural enemies, if available. If none is available, look for resistant natural enemies.
9. Know the life cycles and biology of both natural enemies and pests, so all cultural and other control practices can be as beneficial to natural enemies and as harmful to pests as possible.

10. Keep up to date on pest populations in relation to changing economic thresholds and economic injury levels, and on parasitization rates and predator/prey ratios, so the proper measures can be taken to keep them favorable.

11. Every crop is different, and pests and conditions may vary from one region to another, so a careful evaluation must be made for each crop and situation.

12. Management of pest populations in noncrop areas should be part of pest-management efforts, since many pests move from such refuges to crops. The use of natural enemies is one of the best approaches in noncrop areas, since such areas may be very large and the costs of any method that is not self-regulating may be prohibitive.

13. Do not give up—new strains and new species and new management techniques await discovery and use.

14. Biological control is not the answer to everything, but the opportunity exists to do much more than has been done to date with many pests.

Despite what some may think, biological control is more than a trial that succeeds by chance.

REFERENCES

Anonymous. 1968. Weed control. Principles of plant and animal pest control. Vol. 2. Nat. Acad. Sci. Publ. 1597. 471 pp.

Anonymous. 1969. Insect-pest management and control. Principles of plant and animal pest control. Vol. 3, *Nat. Acad. Sci. Publ.* 1695. 508 pp.

Ashley, T. R, D. Gonzalez, and T. F. Leigh. 1973. Reduction in effectiveness of laboratory-reared *Trichogramma*. *Environ. Entomol.* **2**:1069–1073.

Askew, R. R. 1971. Parasitic insects. American Elsevier, New York. 316 pp.

Bartlett, B. R. 1951. The action of certain "inert" dust materials on parasitic Hymenoptera. *J. Econ. Entomol.* 44:891–896.

Bird, F. T., and D. E. Elgee. 1957. A virus disease and introduced parasites as factors controlling the European spruce sawfly. *Can. Entomol.* **89**:371–378.

Casagrande, R. A., and F. W. Stehr. 1973. Evaluating the effects of harvesting alfalfa on alfalfa weevil (Coleoptera: Curculionidae) and parasite populations in Michigan. *Can. Entomol.* **105**:1119–1128.

Croft, B. A. 1972. Resistant natural enemies in pest management systems. *Span* **15**(1):1–4.

Croft, B. A., and M. M. Barnes. 1971. Comparative studies on four strains of *Typhlodromus occidentalis.* III. Evaluations of releases of insecticide resistant strains into an apple orchard ecosystem. *J. Econ. Entomol.* **64**:845–850.

Croft, B. A., and J. A. McMurtry. 1972. Minimum releases of *Typhodromus occidentalis* to control *Tetranychus mcdanieli* on apple. *J. Econ. Entomol.* **65**:188–191.

DeBach, P. 1947. Cottony-cushion scale, vedalia and DDT in central California. *Calif. Citrogr.* **32**(9):406–407.

DeBach, P. 1966. The competitive displacement and coexistence principles. *Ann. Rev. Entomol.* **11**:183–212.

DeBach, P. 1972. The use of imported natural enemies in insect pest management ecology. *Proc. Tall Timbers Conf. Ecol. Control Anim. Habitat Manage.* **3**:211–233.

DeBach, P., and B. R. Bartlett. 1964. Methods of colonization, recovery and evaluation. Pages 402–428 *in* P. DeBach, ed., Biological control of insect pests and weeds. Reinhold, New York. 844 pp.

DeBach, P., and C. B. Huffaker, 1971. Experimental techniques for evaluation of the effectiveness of natural enemies. Pages 113–140 *in* C. B. Huffaker, ed., Biological control. Plenum Press, New York. 511 pp.

Dietrick, E. J. 1973. Private enterprise pest management based on biological controls. *Proc. Tall Timbers Conf. Ecol. Anim. Control Habitat Manage.* **4**:7–20.

Dodd, A. P. 1940. The biological campaign against prickly pear. Commonwealth Prickly Pear Board, Brisbane, Australia. 177 pp.

Dolphin, R. E., M. L. Cleveland, T. E. Mouzin, and R. K. Morrison. 1972. Releases of *Trichogramma minutum* and *T. cacoeciae* in an apple orchard and the effects on populations of codling moths. *Environ. Entomol.* **1**:481–484.

Doutt, R. L. 1972. Biological control: Parasites and predators. Pages 288–297 *in* Pest control strategies for the future. National Academy of Sciences, Washington, D.C.

Doutt, R. L., and J. Nakata. 1973. The *Rubus* leafhopper and its egg parasitoid: An endemic biotic system useful in grape-pest-management. *Environ. Entomol.* **2**(3):381–386.

Fisher, T. W. 1964. Quarantine handling of entomophagous insects. Pages 305–328. *in* P. DeBach, ed., Biological control of insect pests and weeds. Reinhold, New York. 844 pp.

Fisher, T. W., and G. L. Finney. 1964. Insectary facilities and equipment. Pages 381–401 *in* P. DeBach, ed., Biological control of insect pests and weeds. Reinhold, New York, 844 pp.

Flanders, S. E. 1959. The employment of exotic entomophagous insects in pest control. *J. Econ. Entomol.* **52**:71–75.

Force, D. C. 1972. r- and K-strategies in endemic host-parasitoid communities. *Bull. Entomol. Soc. Amer.* **18:**135–137.

Force, D. C. 1974. Ecology of insect host-parasitoid communities. *Science* **184:** 624–632.

Frazer, B. D., and R. van den Bosch. 1973. Biological control of the walnut aphid in California: The interrelationship of the aphid and its parasite. *Environ. Entomol.* **2:**561–568.

Hagen, K. S., E. F. Sawall, Jr., and R. L. Tassan. 1970. The use of food sprays to increase effectiveness of entomophagous insects. *Proc. Tall Timbers Conf. Ecol. Anim. Control Habitat Manage.* **2:**59–82.

Halfhill, E. J., and P. E. Featherston. 1973. Inundative releases of *Aphidius smithi* against *Acyrthosiphon pisum. Environ. Entomol.* **2:**469–472.

Hambleton, E. J. 1944. *Heliothis virescens* as a pest of cotton, with notes on host plants in Peru. *J. Econ. Entomol.* **37:**660–666.

Hassel, M. P., and G. C. Varley. 1969. New inductive population model for insect parasites and its bearing on biological control. *Nature* **223:**1133–1137.

Haynes, D. L. 1973. Population management of the cereal leaf beetle. Pages 232–240 *in* Geier, P. W., L. R. Clark, D. J. Anderson, and H. A. Nix, ed., Insects: Studies in population management. Memoirs of the Ecological Society of Australia. Vol. 1, 295 pp.

Haynes, D. L., R. K. Brandenburg, and P. D. Fisher. 1973. Environmental monitoring network for pest management systems. *Environ. Entomol.* **2:**889–899.

Holling, C. S. 1961. Principles of insect predation. *Ann. Rev. Entomol.* **6:**163–182.

Holling, C. S. 1966. The functional response of invertebrate predators to prey density. *Mem. Entomol. Soc. Can.* 48. 86 pp.

Huffaker, C. B., P. S. Messenger, and P. DeBach. 1971. The natural enemy component in natural control and the theory of biological control. Pages 16–67 *in* C. B. Huffaker, ed., Biological control. Plenum Press, New York. 511 pp.

Hussey, N. W., and L. Bravenboer. 1971. Control of pests in glasshouse culture by the introduction of natural enemies. Pages 195–216 *in* C. B. Huffaker, ed., Biological control. Plenum Press, New York. 511 pp.

Lawson, F. R., R. L. Rabb, F. E. Guthrie, and T. G. Bowery, 1961. Studies of an integrated control system for hornworms on tobacco. *J. Econ. Entomol.* **54:**93–97.

Leius, K. 1967. Influence of wild flowers on parasitism of tent caterpillar and codling moth. *Can. Entomol.* **99:**444–446.

McMurtry, J. A., and H. G. Johnson. 1964. Some factors influencing the abundance of the predaceous mite *Amblyseius hibisci* in southern California. *Ann. Entomol. Soc. Amer.* **58:**49–56.

Muldrew, J. A. 1953. The natural immunity of the larch sawfly (*Pristiphora erichsonii* (Hartig)) to the introduced parasite (*Mesoleius tenthredinis* Morley) in Manitoba and Saskatchewan. *Can. J. Zool.* **31**:313–332.

Nelson, E. E., B. A. Croft, A. J. Howitt, and A. L. Jones. 1973. Toxicity of apple orchard pesticides to *Agistemus fleschneri*. *Environ. Entomol.* **2**:219–222.

Oatman, E. R., J. A. McMurtry, and V. Voth. 1968. Suppression of the two-spotted spider mite on strawberry with mass releases of *Phytoseiulus persimilis*. *J. Econ. Entomol.* **61**:1517–1521.

Paradis, R. O. 1956. Factors in the recent importance of the red-banded leaf-roller, *Argyrotaenia velutinana* (Walker), in Quebec apple orchards. *Que. Soc. Prot. Plants, Rep.* 38, pp. 45–48.

Parker, F. D. 1971. Management of pest populations by manipulating densities of both hosts and parasites through periodic releases. Pages 365–376 *in* C. B. Huffaker, ed., Biological control. Plenum Press, New York. 511 pp.

Pimentel, D. 1963. Introducing parasites and predators to control native pests. *Can. Entomol.* **95**:785–792.

Price, P. W. 1972. Methods of sampling and analysis for predictive results in the introduction of entomophagous insects. *Entomophaga* **17**(2):211–222.

Price, P. W. 1973. Parasitoid strategies and community organization. *Environ. Entomol.* **2**:623–626.

Remington, C. L. 1968. The population genetics of insect introduction. *Ann. Rev. Entomol.* **13**:415–426.

Salt, G. 1970. The cellular defense reactions of insects. Cambridge Univ. Press. Cambridge. 118 pp.

Schlinger, E. I., and E. J. Dietrick. 1960. Biological control of insect pests aided by strip-farming alfalfa in experimental programs. *Calif. Agr.* **14**:8–9, 15.

Stehr, F. W., and D. L. Haynes. 1972. Establishment in the United States of *Diaparsis carinifer*, a larval parasite of the cereal leaf beetle. *J. Econ. Entomol.* **65**:405–417.

Stern, V. M. 1973. Economic thresholds. *Ann. Rev. Entomol.* **18**:259–280.

Stern, V. M., E. J. Dietrick, and A. Mueller. 1965. Improvements on self-propelled equipment for collecting, separating, and tagging mass numbers of insects in the field. *J. Econ. Entomol.* **58**:949–953.

Tothill, J. D., T. H. C. Taylor, and R. W. Paine. 1930. The coconut moth in Fiji, a history of its control by means of parasites. Imp. Bur. Ent., London. 269 pp.

Turnbull, A. L., and D. A. Chant. 1961. The practice and theory of biological control of insects in Canada. *Can. J. Zool.* **39**:697–753.

van Emden, H. F., and G. F. Williams. 1974. Insect stability and diversity in agro-ecosystems. *Ann. Rev. Entomol.* **19**:455–475.

Varley, G. C., and G. R. Gradwell. 1971. The use of models and life tables in assessing the role of natural enemies. Pages 93–112 *in* C. B. Huffaker, ed., Biological control. Plenum Press, New York. 511 pp.

White, E. B., P. DeBach, and M. J. Garber. 1970. Artificial selection for genetic adaptation to temperature extremes in *Aphytis lingnanensis* Compere (Hymenoptera: Aphelinidae). *Hilgardia* **40**:161–192.

Whitten, M. J. 1970. Genetics of pests in their management. Pages 119–135. *in* R. L. Rabb and F. E. Guthrie, eds., Concepts of pest management. North Carolina State University, Raleigh.

Zwolfer, H. 1963. The structure of the parasite complex of some Lepidoptera. *Z. Angew. Entomol.* **51**:346–357.

Zwolfer, H., and P. Harris. 1971. Host specificity determination of insects for biological control of weeds. *Ann. Rev. Entomol.* **16**:157–178.

SELECTED READINGS

Clausen, C. P. 1940. Entomophagous insects. McGraw-Hill Book Co., New York. 688 pp. Reprint ed., Hafner, New York, 1962.

Clausen, C. P. 1956. The biological control of insect pests in the continental United States. *USDA Tech. Bull.* 1139. 151 pp.

Clausen, C. P. 1960. The importance of taxonomy to biological control as illustrated by the cryptic history of *Aphytis holoxanthus* n.sp. (Hymenoptera: Aphelinidae), a parasite of *Chrysomphalus aonidium,* and *Aphytis coheni* n.sp., a parasite of *Aonidiella aurantii. Ann. Entomol. Soc. Amer.* **53**:701–705.

DeBach, P., ed., 1964. Biological control of insect pests and weeds. Reinhold, New York. 844 pp.

DeBach, P. 1965. Some biological and ecological phenomena associated with colonizing entomophagous insects. Pages 287–306 *in* H. G. Baker and G. L. Stevens, ed., The genetics of colonizing species. Academic Press, New York. 588 pp.

DeBach, Paul. 1974. Biological control by natural enemies. Cambridge University Press, Cambridge. 323 pp.

DeBach, P., D. Rosen, and C. E. Kennett. 1971. Biological control of coccids by introduced natural enemies. Pages 165–194 *in* C. B. Huffaker, Biological control. Plenum Press, New York. 511 pp.

den Boer, P. J., and G. R. Gradwell, eds. 1971. Dynamics of populations. Center for Agricultural Publishing and Documentation. Wageningen, Netherlands. 611 pp.

Embree, D. G. 1966. The role of introduced parasites in the control of the winter moth in Nova Scotia. *Can. Entomol.* **98**:1159–1168.

Huffaker, C. B., ed. 1971. Biological control. Plenum Press, New York. 511 pp.

Huffaker, C. B., M. van de Vrie, and J A. McMurtry. 1969. The ecology of tetranychid mites and their natural control. *Ann. Rev. Entomol.* **14**:125–174.

Klomp, H. 1958. On the theories of host-parasite interaction. *Arch. Nied. Zool.* **13**:134–145.

Matthews, R. W. 1974. Biology of Braconidae. *Ann. Rev. Entomol.* **19**:15–32.

Maxwell, F. G., and F. A. Harris. ed. 1974. Proceedings of the summer institute on biological control of plant insects and diseases. University Press of Mississippi, Jackson. 647 pp.

Morris, R. F. 1963. The population dynamics of epidemic spruce budworm populations. *Mem. Entomol. Soc. Can.* 31. 332 pp.

Munroe, E. G. 1971. Status and potential of biological control in Canada. Part 4, Chap. 48 *in* Biological control programmes against insects and weeds in Canada, 1959–1968. Tech. Commun. No. 4, Commonwealth Institute of Biological Control, Trinidad. Commonwealth Agricultural Bureaux, Farnham Royal, Slough SL2 3BN, England.

Nicholson, A. J. 1958. Dynamics of insect populations. *Ann. Rev. Entomol.* **3**: 107–136.

Nickle, W. R. 1972. Nematode parasites of insects. *Proc. Tall Timbers Conf. Ecol. Anim. Control Habitat Manage.* **4**:145–163.

Sabrosky, C. W. 1955. The interrelations of biological control and taxonomy. *J. Econ. Entomol.* **48**:710–714.

Smith, H. S. 1935. The role of biotic factors in determination of population densities. *J. Econ. Entomol.* **28**:873–898.

Sweetman, H. L. 1958. The principles of biological control. W. C. Brown, Dubuque, Iowa. 560 pp.

Thompson, W R. 1956. The fundamental theory of natural control. *Ann. Rev. Entomol.* **1**:379–402.

van den Bosch, R., and P. S. Messenger. 1973. Biological control. Intex Educational Publishers, New York. 180 pp.

Watt, K. E. F. 1965. Community stability and the strategy of biological control. *Can. Entomol.* **97**:887–895.

6

USE OF DISEASES IN PEST MANAGEMENT

J. V. Maddox

INTRODUCTION

Microbial control was defined by Falcon (1971) as "including all aspects of the utilization of microorganisms or their by-products in the control of insect pest species." This definition includes the use of microorganisms as naturally occurring control agents, introduced control agents, and the application of microorganisms and/or their by-products as microbial insecticides.

Insects are infected by viruses, bacteria, protozoa, fungi, rickettsia, and nematodes. Some of these pathogens may be quite common and are frequently the cause of epizootics in natural insect populations, while others may occur so infrequently they are seldom observed. Some of these disease agents may be very pathogenic to their hosts and cause a high rate of mortality, while others may produce only chronic effects.

More than 1000 insect pathogens have been described (Ignoffo and Hink 1971), and this is probably only a small fraction of the total number of pathogens infecting insects. Little is known about many of these pathogens, while others, such as the spore-forming bacterium *Bacillus thuringiensis* Berliner and some of the insect viruses, have been studied extensively.

Insect pathogens can be used in pest management in at least three different ways: (1) by the maximum utilization of naturally occurring diseases, (2) by the introduction of insect pathogens into the insect pest population as permanent mortality factors, and (3) by the repeated application of insect pathogens as microbial insecticides for temporary control of an insect pest.

This chapter briefly covers the types of pathogens that occur in insects and the use of these pathogens in pest-management situations. A chapter of this length obviously cannot do justice to a topic as broad as microbial control of insect pests. For more detailed information on specific insect pathogens and their uses as microbial control agents, the reader is referred to several excellent texts which cover this topic extensively (see Selected Readings).

I. CHARACTERISTICS OF INSECT DISEASES

A. Transmission

The ingestion of an infectious stage of the disease agent is the most common method by which insect diseases are transmitted. A few insect pathogens, such as fungi and some nematodes, can penetrate directly through the cuticle of the insect's body, and ingestion is not necessary. Many insect pathogens can be transmitted from an infected female via the egg to her offspring. When transmission occurs in the ovary and the disease agent is contained inside the egg, it is called transovarian transmission. The insect egg may also be contaminated externally by the disease agent, as a result of secretions from the female accessory glands or fecal material. The newly emerged immature insect then ingests these disease agents and becomes infected. Certain diseases may be transmitted from one insect host to another by hymenopterous parasites which carry disease agents on their ovipositors and in the act of oviposition inject these agents into other insect hosts.

B. Host Range

Some insect pathogens are host-specific and infect only one or very few insect species. Other insect pathogens have a very broad host spectrum and are capable of infecting many different insect species. For example,

Table 6.1 Relative Specificity of Different Types of Insect Pathogenic Viruses Based on Attempted and Successful Cross-Transmission to Various Insect Taxa[a, b]

	Number of Cross-Transmissions[b]				
	To Alien		To Same		
Virus Type	Order	Family	Family	Genus	Total
Granuloses	0/0	38/1	3/2	4/4	56/6
Nuclear polyhedroses	9/3	137/30	43/26	21/21	187/60
Cytoplasmic polyhedroses	7/2	37/17	2/1	1/1	45/29
Noninclusion viruses	19/18	20/17	9/9	9/9	39/35

[a] Courtesy of Ignoffo, 1968, and the Entomological Society of America.

[b] Attempted transmissions/successful transmissions.

the bacterium *Bacillus thuringiensis* Berliner infects the larvae of many species of Lepidoptera, and many species of hyphomycetous fungi infect many insect species from different orders. Some of the insect viruses are relatively host-specific and are restricted to insect species of the same genus, while other insect viruses infect insect species of other families and occasionally of different orders (Ignoffo 1968). The relative specificity of different types of insect viruses is summarized in Table 6.1.

C. Persistence of Insect Pathogens

Under favorable environmental conditions, insect pathogens, such as spore-forming bacteria, are capable of persisting in the environment for many years external to their host. Other disease agents, such as non-inclusion insect viruses, can persist outside their host for periods of only a few weeks at most. Direct sunlight and high temperatures are harmful to most pathogens (Smirnoff 1972; David and Gardiner 1967; Jaques 1972). The spores of bacteria and fungi and insect viruses within their inclusion bodies remain viable for relatively long periods of time in the dark, but are quickly inactivated when exposed to direct sunlight (Fig. 6.1). Polyhedra of the nuclear polyhedrosis virus of the cabbage looper, *Trichoplusia ni* (Hübner), survive in cool, sterile water suspensions for over 15 years and in the soil for at least 5 years, but they survive for less than 1 month on the host plant (Jaques 1967a,b).

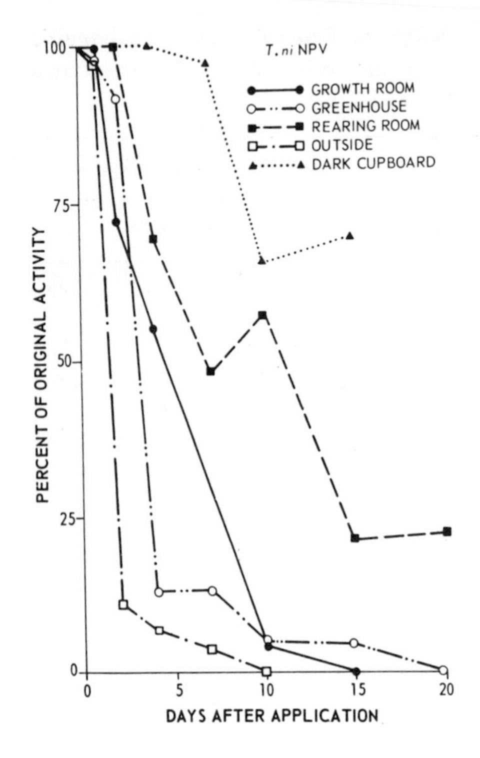

Fig. 6-1 The activity of deposits of the NPV of *Trichoplusia ni* on leaves of collard plants in five locations. Plants treated with 3.6×10^5 polyhedra. Original activity of deposits killed 97% of host larvae. (Courtesy of R. P. Jaques 1972, and Academic Press.)

Temperatures likely to be encountered in the field are not so harmful to insect pathogens as is direct sunlight. A purified granulosis virus of *Pieris brassicae* (L.) was inactivated by 10 minutes at 70°C, 60 minutes at 65°C, or 24 hours at 60°C. It was not entirely inactivated after 5 days at 50°C or 20 days at 40°C, and was not adversely affected after 6 months' storage at −20°C (David and Gardiner 1967).

Relative humidity plays a minor role in the persistence of insect pathogens, except for fungi and certain nematodes (Tanada 1973). High humidity is conducive to germination of fungal spores but is generally unfavorable to spore longevity. For example, if high humidity causes large numbers of fungus spores to germinate when the insect host is absent, the number of viable fungus spores in the environment is reduced (Yendol and Hamlen 1973). Neoaplectanid nematodes survive best at high relative humidity (Fig. 6.2) (Simons and Poinar 1973).

The role of a pathogen as a mortality factor is greatly influenced by its ability to persist for many years in an area under natural conditions. Pathogens that do not persist from year to year must be applied repeat-

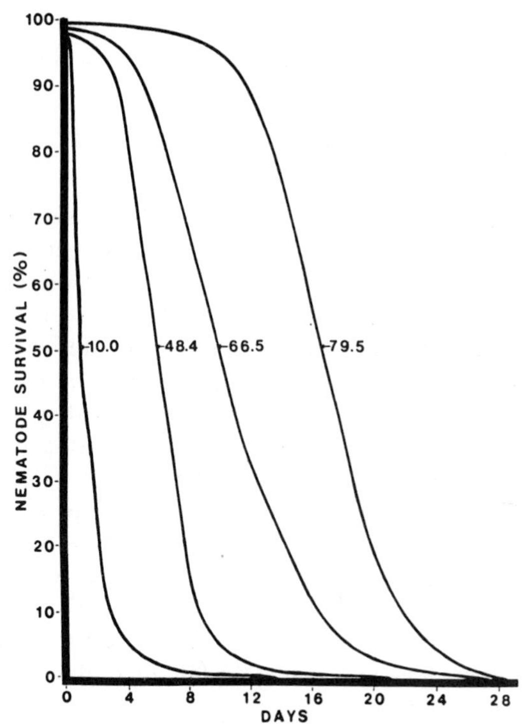

Fig. 6-2 Survival of infective-stage juveniles of *Neoaplectana carpocapsae* maintained at relative humidities of 79.5, 66.5, 48.4, and 10% over a 28-day period. (Courtesy of Simons and Poinar 1973, and Academic Press.)

edly as microbial insecticides for temporary control of an insect pest. In many cases factors other than the longevity of the pathogen are involved in its persistence in the host population. Spatial separation of the pathogen and host may be a factor; the pathogen not only must have a certain resistance to environmental conditions, but must also be able to come into contact with and infect its host. Two of the most important goals in formulating microbial insecticides are (1) to increase the persistence of the pathogens for as long a time as possible, and (2) to place the pathogen in contact with its host in such a way as to achieve maximum infection.

Pathogens often persist in the environment in infected living hosts. Infected insects frequently overwinter successfully and are responsible for initiating the infection the following year. This is one of the most important mechanisms of persistence for many insect pathogens that occur consistently year after year (Neilson and Elgee 1968; Maddox 1973).

D. Pathogenicity

The dosage–mortality relationship for a chemical insecticide is usually expressed as a median lethal dose (LD_{50}). This represents the dose of the chemical per insect that will produce death in half the test subjects. The pathogenicity of an insect disease agent can also be expressed as a LD_{50}, but the LD_{50} alone does not give an accurate picture of the total pathogenic effect. Insects respond to increasing dosages of pathogenic organisms by increased infection and mortality, just as they respond to increased dosages of insecticides. This effect of increasing dosages can be expressed as a dosage–infection or as a dosage–mortality curve. These curves have a sigmoid shape but can be converted to straight lines by plotting the dosages expressed in logarithms against the mortality expressed in probits. Therefore the mortality effect of a pathogen on its host can be expressed as a LD_{50} and characterized by the slope of the log–probit curve. Insect pathogens usually produce curves with very low slopes when compared to insecticides (Table 6.2) (Bucher 1958; Burges and Thomson 1971).

The slope of the dosage–mortality curve indicates the degree of variability in response to treatment within the insect population. The lower the slope, the greater the variability. This means that enormous numbers of a disease agent are usually required to produce 100% mortality. Some mortality occurs at very low dosages and, from the standpoint of insect control, this is not a totally undesirable feature of insect pathogens. When pathogens are introduced into insect populations for long-term control, doses far below the LD_{50} value are often sufficient for successful establishment of the disease in the insect population.

The low slope of dosage–mortality curves for insect pathogens also indicates a more stable host–pathogen relationship. As a hypothetical example, an insect pathogen having a dosage–mortality curve with a high slope and acting as the primary mortality factor in an insect population could cause insect population numbers to fluctuate greatly.

The high LD_{50} values for many pathogens do not necessarily indicate that these pathogens cannot be effective control agents. For example, *Bacillus popilliae* Dutky has a relatively high LD_{50} of 2,000,000 spores per larva (Table 6.2), but this bacterium has been very successful as a control agent for the Japanese beetle. Fields containing Japanese beetles

Table 6.2 The LD_{50} and Slope of Dosage–Mortality Curves of Insect Disease Agents and Insecticides[a]

Disease Agent	Host Insect	Method Administered	LD_{50}[b]	Slope[c]
Nosema locustae Canning	*Melanoplus bivittatus* (Say)	Oral	9,000	0.84
Pseudomonas aeruginosa (Schroeter)	*Melanoplus bivittatus* (Say)	Oral	19,000	0.83
Polyhedrosis virus	*Malacosoma disstria* Hübner	Oral	300,000	0.86
Bacillus popilliae Dutky	*Popillia japonica* Newman	Oral	2,000,000	1.83
Bacillus thuringiensis Berliner	*Pieris rapae* (Linnaeus)	Oral	41,000	2.58
Bacillus thuringiensis Berliner	*Bombyx mori* (Linnaeus)	Oral	2,000,000	2.25
DDT	*Musca vicina* Macquart	Topical	6.2×10^{-3}	5.5

[a] Adapted from Bucher 1958.

[b] Dose is expressed as number of disease agents or milligrams of insecticide.

[c] Slope is expressed as change in value of the ordinate in probits occurring in response to a 10× increase in dose.

can be treated with *Bacillus popilliae* spores at rates much below the LD_{50} value of 2,000,000 spores per larva. Some Japanese beetle larvae become infected, die, and release more spores into the environment. In this manner the spore inoculum gradually builds up in the beetle-infested area until it becomes an important mortality factor. The use of *B. popilliae* as a microbial control agent is discussed later in more detail.

LD_{50} values are also misleading because they do not take into account the debilitating effects of the disease on the host. For example, a cytoplasmic polyhedrosis virus of the fall cankerworm, *Alsophila pometaria* (Harris), has a LD_{50} of 19,000 polyhedra per larva, but infected females do not produce any egg masses, nor do healthy females mated to infected males (Neilson 1965). The microsporidian *Nosema trichoplusiae* Tanabe and Tamashiro has a similar effect on the cabbage looper, *Trichoplusia ni* (Hübner) (Table 6.3). Many other insect diseases have similar debilitating effects which the LD_{50} does not indicate.

Table 6.3 Ovipositional Capacity and Viability of Eggs. *Trichoplusia ni* Adults Infected with *Nosema trichoplusiae*[a]

	Mean Number of Eggs Laid per Female	Percent Hatch
Uninfected female × Uninfected male	1067	85
Uninfected female × heavily infected male	215.5	0
Heavily infected female × uninfected male	130	34
Heavily infected female × heavily infected male	265	0

[a] Adapted from Tanabe and Tamashiro 1967.

Insect diseases transmitted from an infected female to her offspring via the egg (transovarian transmission) also have a pathogenic effect which cannot be expressed in terms of a LD_{50}. For example, Kramer (1959) found a marked difference in the rate of survival between transovarially infected versus disease-free European corn borers, *Ostrinia nubilalis* (Hübner). Among the former, 14% reached adulthood, while among the latter 75% matured to the adult stage.

There is great variability in the pathogenicity of a disease agent toward its host, because two biological variables are involved—the host and the pathogen. The constant pressure of a disease agent exerts a selection pressure on the host population, while at the same time the virulence of the pathogen may also be changing. Repeated passes of a pathogen through the same host species frequently result in increased virulence (Aizawa 1971). Veber (1964) observed a fourfold increase in the virulence of a polyhedrosis of the greater wax moth, *Galleria mellonella* (L.), after eight passes through its host. The mechanisms responsible for these changes in the virulence of pathogens are not fully understood, but they undoubtedly play an important role in the development of epizootics. Virulence must also be preserved in the production of disease agents used in microbial insecticides. The virulence of insect pathogens toward a specific insect host can often be improved by continued passage through that host and selecting virulent strains of the pathogen.

The mortality effects of insect diseases may also be expressed as LT_{50} values (time needed to produce 50% mortality). The LD_{50} must be ex-

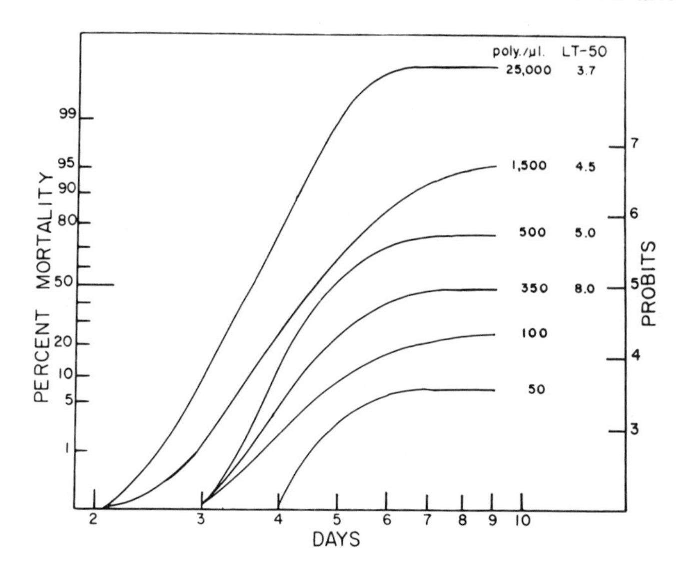

Fig. 6-3 Time–mortality curves for cabbage looper, *Trichoplusia, ni* larvae, fed as 3-day-old larvae media containing various levels of polyhedra of cabbage looper NPV. (Courtesy of C. M. Ignoffo 1964, and Academic Press.)

pressed at a designated interval after administration of the pathogen. For example, if mortality effects of a pathogen on its host are recorded 1 day after administration of the pathogen, the LD_{50} is usually higher (meaning mortality is lower) than if mortality effects are recorded after 5 days. For this reason it is sometimes desirable to express the effect of a pathogen on its host in terms of a LT_{50}. As the dose of the pathogen is increased, the LT_{50}, as expected, decreases. Figure 6.3 shows the dose–LT_{50} relationship of a nuclear polyhedrosis of the cabbage looper, *Trichoplusia ni.*

Insect pathogens often infect one stage of an insect host but not another. For example, *Bacillus thuringiensis* is infectious to many lepidopterous larvae but usually does not infect lepidopterous adults. Some pathogens that can be acquired only during the larval stage of an insect may persist in that host and cause heavy infections in the adult. Other pathogens are capable of infecting any stage of their insect host.

The younger stages of an insect are generally more susceptible to infection than the later stages. First-instar larvae of the armyworm, *Pseudaletia unipuncta* (Haworth), are many times more susceptible to infection by a nuclear polyhedrosis virus than are the later larval instars

(Tanada 1959). Early instar larvae of the cabbage looper are likewise many times more susceptible to the fungus *Nomuraea rileyi* (Farlow) than are later larval instars (Table 6.4). In some cases the last larval instar of an insect host is completely refractory to an insect pathogen that is very infectious to earlier stages of the same insect host. For this reason the timing of microbial applications is very important.

Table 6.4 Mortality of Cabbage Looper Larvae from a Dermal Application of *Nomuraea rileyi* Spores (1×10^6 per ml)[a]

Larval Instar	Test Group Size	Mortality from *Nomuraea rileyi* (%)
First	45	58
Second	24	58
Third	23	13
Fourth	48	6
Fifth	24	0

[a] Adapted from Getzin 1961.

II. TYPES OF INSECT PATHOGENS

Many thousands of insect pathogens have been described. The following is intended to be only a brief introduction to the major groups of insect pathogens. No attempt has been made to cover the classification or identification of insect pathogens. For identification of family or genera, the reader is referred to Weiser and Briggs (1971). Those not familiar with insect diseases can find pictures of many types of disease agents, as well as insects showing symptoms of disease, in *An Atlas of Insect Diseases* (Weiser 1969).

A. Viruses

Insect viruses contain some of the most important of the naturally occurring insect pathogens, as well as promising candidates for microbial insecticides. During the past few years great changes have occurred in the classification of insect viruses. Until recently, viruses described in invertebrates were not compared with vertebrate and plant viruses, and the classification and nomenclature of these invertebrate viruses developed with no obvious link to other fields of virology (Vago et al. 1974). In 1966 the subcommittee of the International Committee on Nomenclature of Viruses (ICNV) began a taxonomic revision of the invertebrate viruses within the broad scope of a unified system of viruses (Vago et al. 1974; Vago 1966). At the present time there is insufficient information on the structure and chemical composition of many invertebrate viruses to assign them to the appropriate genera accurately. For further information on the classification of invertebrate viruses, the reader should consult Bergold et al. (1960), Vago (1966), Wildy (1971), and Vago et al. (1974).

The virus particles of many insect viruses are occluded in protein crystals called inclusion bodies. There are three main types of inclusion viruses—nuclear polyhedrosis viruses, granulosis viruses, and cytoplasmic polyhedrosis viruses—which collectively contain the largest number of viruses known in insects. Both nuclear polyhedrosis viruses and granulosis viruses have been assigned to the genus *Baculovirus* Wildy (1971). Cytoplasmic polyhedroses are closely allied to *Reovirus,* but the ICNV plans to define the position of cytoplasmic polyhedroses more fully (Vago et al. 1974).

1. Nuclear Polyhedrosis Viruses (Arizawa 1963; Bergold 1963; Hughes 1972; Smith 1967; and Smith 1971)

Most nuclear polyhedrosis viruses (NPV) infect lepidopterous larvae, in which case the viruses multiply in the cell nucleus and usually attack tissues of the epidermis, fat body, blood cells, and trachea. Less frequently, the silk glands and epithelial cells of the midgut are attacked by NPV viruses. Hymenopterous larvae of the genera *Viprion, Neodiprion,* and *Gilpinia* are also infected by NPV, but these infections are generally confined to the epithelial cells of the midgut.

NPV virus particles are rod-shaped and vary between 20 and 50 mμ in diameter and between 200 and 400 mμ in length (Smith 1967). The virus rods are encased in an outer envelope which may enclose one or several virus rods, depending on the particular NPV. The viruses enclosed in the envelope are occluded in protein crystals called polyhedra (Fig. 6.4). Polyhedra may exist as dodecahedra, tetrahedra, cubes, or irregular angular forms, and many contain several hundred virus particles. The diameter of these polyhedra varies from 0.5 to 15 μ, depending on the particular NPV, and their size varies considerably, even in an individual insect infected with a specific NPV. Some NPVs have very characteristic shapes which can be used as an aid in identification, but in most cases this is not a reliable diagnostic characteristic.

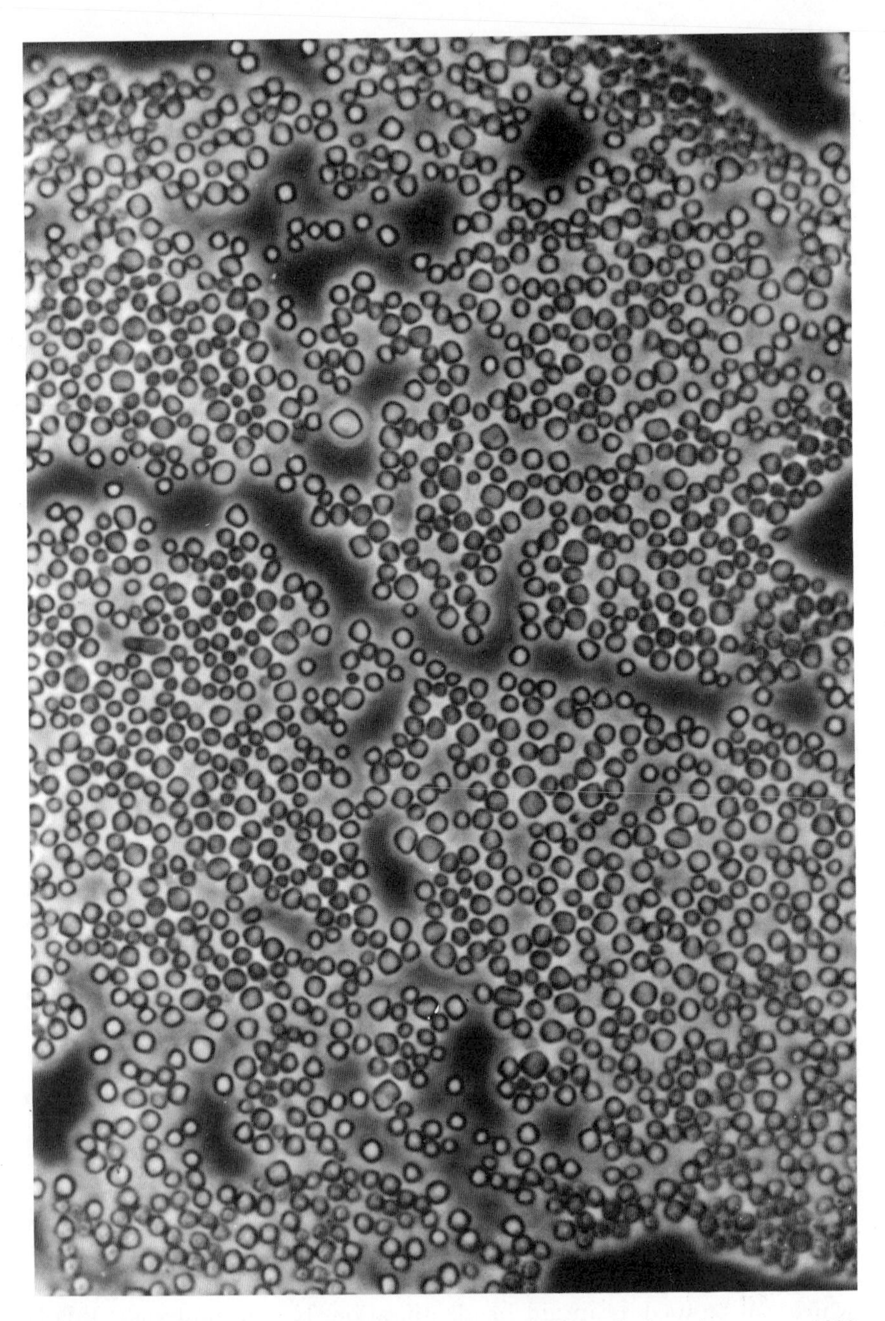

Fig. 6-4 Polyhedra of a NPV of the armyworm, *Pseudaletia unipuncta*. ×2000 phase-contrast.

Fig. 6-5 Larva of the armyworm, *Pseudaletia unipuncta,* dead from a NPV. (Photo by Illinois Natural History Survey Extension staff.)

A NPV is normally transmitted from one insect to another by oral ingestion of polyhedra. When ingested by a susceptible insect, the polyhedra dissolve and release virus rods into the lumen of the midgut. These free virus rods invade susceptible cells, but little is known about the specific method by which this invasion takes place. The period of time from ingestion of the polyhedra to death is from about 4 days to 3 weeks, and varies with different NPVs, insect hosts, the number of polyhedra ingested, the larval instar during which polyhedra are ingested, and the environmental temperature. During most of this period, the infected larvae show no external symptoms at all. A day or two before death occurs, the larval skin sometimes darkens and may have yellow patches or appears oily. The skin becomes very fragile, and the hemolymph becomes turbid; this turbid hemolymph contains large numbers of polyhedra which can easily be identified with a light microscope (Fig. 6.4). Before death the infected larvae often climb to the highest point available, and dead larvae infected with NPV are often found hanging from the tops of their food plants (Fig. 6.5). After death the integument frequently ruptures, releasing millions of polyhedra which often contaminate the food plant of the host.

Most, if not all, NPV infections can also be transmitted from an infected female via the egg to her progeny. For many years this was a controversial subject in insect virology, because the adult moth displays no symptoms and the virus particles often cannot be found in adults; but there is so much evidence that this type of transmission occurs in several insect viruses that it is now generally accepted (Smith 1967).

Latent viral infections, which may be defined as "those phenomena whereby an organism is infected with a virus but yet shows no apparent signs of infection" (Smith 1967), frequently occur in insects. There are many examples of latent virus infections in insects, but little is known about the conditions that determine whether a virus infection is latent or virulent. It is thought that latent infections may result when the initial virus dose is very small and/or if this dosage is received during a late stage of larval development. Several factors can cause these latent infections to develop into active or virulent infections. Steinhaus (1958) collectively calls these factors "stress factors," and they include crowding, temperature extremes, toxic chemicals, food quality, and other diseases. How stresses act on latent viral infections is not fully understood, but latent infections undoubtedly account for many viral epizootics which suddenly appear in insect populations.

2. *Granulosis Viruses* (*Huger 1963; Smith 1967*)

The fat body of lepidopterous larvae is the primary site of infection for granulosis viruses (GV), but the epidermis and tracheal matrix may occasionally be infected. The virus multiplies in both the nucleus and cytoplasm of host cells. GV particles range from 36 to 80 mμ in width to 245 to 411 mμ in length, and are enclosed by an envelope similar to that described for NPVs. GV particles surrounded by these membranes are occluded in a proteinaceous capsule similar to the polyhedral protein that occludes NPVs. The granulosis capsule, however, usually contains only one GV particle rather than the many virus particles contained in the polyhedra of NPVs and CPVs. GV capsules are small, and range in size between 300 and 511 mμ in length and 119 to 350 mμ in width (Fig. 6.6).

GVs are transmitted orally and via the egg. Latent infections also occur. The period between ingestion of the virus and the death of the host generally ranges between 4 and 25 days. External symptoms are not usually apparent in the early stages of infection but, toward the later stages, infected larvae frequently develop a lighter color. The blood of infected larvae is usually turbid and contains large numbers of capsules. The capsules can barely be resolved under the highest power of

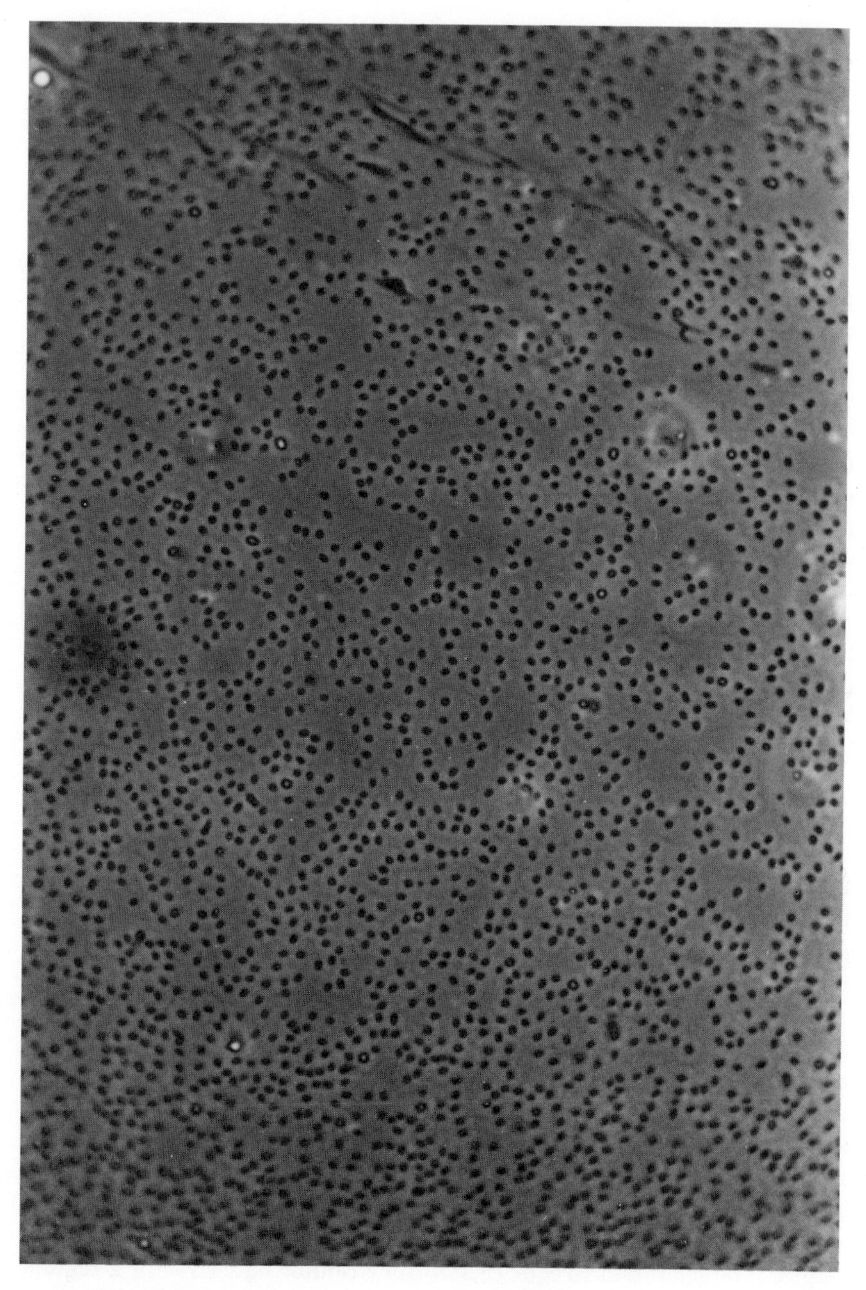

Fig. 6-6 Capsules of a granulosis virus of the armyworm, *Pseudaletia unipuncta*. × 2000 phase-contrast.

a light microscope. GV infections involving the epidermis cause liquefaction of infected larvae similar to NPV infections but, when the epidermis is not involved, liquefaction does not take place.

3. *Cytoplasmic Polyhedrosis Viruses* (*Smith 1963; Smith 1967*)

Cytoplasmic polyhedrosis viruses (CPV) infect the cytoplasm of the midgut epithelium of lepidopterous larvae. CPV particles are icosahedral and vary in size from about 50 to 70 mμ. CPV particles are not enclosed in membranes as are NPVs, but are occluded in protein crystals similar to those described for NPVs. These vary in size from 1 to 7 μ, and some of the larger polyhedra appear almost spherical. Since the midgut is the only tissue involved in CPV infections, the symptoms of CPV infections are quite different from those described for NPV infections. Larvae infected with CPV usually develop very slowly and often appear to have small bodies and large heads. In the later stages of the disease, infected larvae may develop color changes.

CPV infections are transmitted orally and via the egg. Latent infections, as described for NPVs, also occur. CPVs are not as pathogenic as NPVs, and the length of time between ingestion of polyhedra and death is usually much longer than for NPVs. CPVs are not as host-specific as are some other virus diseases, and a CPV from one insect frequently infects several other insect species.

4. *Other Inclusion Viruses*

Several inclusion viruses have been described in addition to those listed above. These viruses are not as common as the other inclusion viruses, and relatively little is known about them. For brief descriptions and further reference material, the reader is referred to Weiser and Briggs (1971).

5. *Noninclusion Viruses*

As the name implies, noninclusion viruses are viruses that are not enclosed within some type of inclusion body. This is a heterogenous group of viruses which have little in common, other than the absence of inclusion bodies. Although the number of noninclusion viruses known to exist in insects is relatively small in comparison with the number of inclusion viruses, this probably does not reflect the total number of noninclusion viruses that actually exist in insects. Since these viruses are not included within a larger proteinaceous crystal, they cannot be seen with a regular light microscope. Many noninclusion viruses probably go undiagnosed for this reason.

B. Bacteria

1. *Nonspore-Forming Bacteria*

The digestive tracts of most insects contain many nonspore-forming bacteria which are often called "potential pathogens" (Bucher 1960). As long as these bacteria remain in the midgut, they are relatively nonpathogenic. However, when introduced into the insect's blood, many of them are extremely pathogenic. "Stress factors" such as temperature extremes, other pathogens, parasites, or poor food quality can cause these bacteria to gain entrance to the hemocoel (Steinhaus 1958). These nonspore-forming bacteria are undoubtedly important mortality factors in many situations but, because they have little invasive power and are cosmopolitan in distribution, little effort has been made to utilize them as microbial control agents.

2. *Spore-Forming Bacteria*

The most important spore-forming bacterial pathogens of insects are *Bacillus popilliae,* which causes milky disease in white grubs, and *Bacillus thuringiensis,* a bacterium that is very pathogenic to many species of lepidopterous larvae.

Bacillus popilliae infects larvae of the Japanese beetle and other scarabs. The disease is transmitted orally by ingestion of spores. The length of time between ingestion of spores and death is greatly influenced by the number of spores ingested and temperature. After ingestion these spores germinate and penetrate the alimentary canal, probably through the Malpighian tubules (Beard 1945). At 30°C vegetative cells can be found in the blood about 30 hours after the initial invasion. At this temperature the vegetative rods multiply rapidly and, when these vegetative forms become very numerous, sporulation occurs. Seven to ten days after the initial infection, the maximum number of spores, between 2 and 5 billion per larva (Dutky 1940; Beard 1945), is reached. At this point the larval blood appears milky, and the larvae usually die shortly thereafter. Many species of white grubs are susceptible to infection by *Bacillus popilliae,* both experimentally and under natural conditions (Fleming 1968). Spores of *Bacillus popilliae* have not been successfully produced in artificial media on a large scale, and spores must be produced in living hosts.

Bacillus thuringiensis is a spore-forming bacterium which is very pathogenic to a large number of lepidopterous larvae. This bacterium can be grown successfully in artificial media, and hundreds of tons are manufactured each year in the United States for use in microbial insecticides.

When *Bacillus thuringiensis* sporulates, it forms a crystal which is toxic when ingested by many lepidopterous larvae (Fig. 6.7). Different Lepidoptera species exhibit varying responses when fed these crystals and/or *Bacillus thuringiensis* spores (Heimpel and Angus 1959; Burgerjon and Martouret 1971). Many lepidopterous larvae are susceptible to the toxic action of the crystal alone, while others are susceptible only to the combined action of the spores and the crystals; a few lepidopterous larvae and several phytophagous hymenopterous larvae are susceptible to the action of *Bacillus thuringiensis* spores alone (Burgerjon and Martouret 1971). Lepidoptera that are susceptible to the crystals alone are divided into types I and II, based on their response to the ingestion of crystals (Heimpel and Angus 1959). Type I and II insects both suffer from midgut paralysis a few minutes after ingesting crystals. Type I insects, consisting of only a few insect species, develop a general paralysis and die after 1 to 7 hours. Type II insects do not develop a general paralysis and die 2 to 4 days after ingestion of crystals. Most susceptible insects are in the type II category. After ingestion of spores the first symptom in both type I and II is cessation of feeding.

The activity of the crystal is dependent on the pH of the larval foregut and midgut and the action of proteolytic enzymes within the gut. Lepidoptera that have a strongly alkaline gut pH (above 8.9) and an enzyme system with selective proteolytic activity in an alkaline medium are susceptible to the crystal. The crystal itself is a protoxin, activated by enzymic hydrolysis, this liberates soluble proteins which are in turn directly toxic (Lecadet and Martouret 1967). The protein crystal toxin of *Bacillus thuringiensis* is thoroughly discussed by Cooksey (1971) and Burgerjon and Martouret (1971).

The number of insects susceptible to only spore–crystal combinations or spores alone is relatively small when compared to the number susceptible to the toxic crystal alone. The spores germinate, enter the hemocoel, and cause general septicemia. The mechanisms involved are not fully understood.

Although many strains of *Bacillus thuringiensis* have been isolated, naturally occurring epizootics due to *Bacillus thuringiensis* are infrequent (Burgerjon and Martouret 1971).

C. Fungi

More than 36 different genera of fungi contain species that cause insect disease (Roberts and Yendol 1971). The classification of many of these

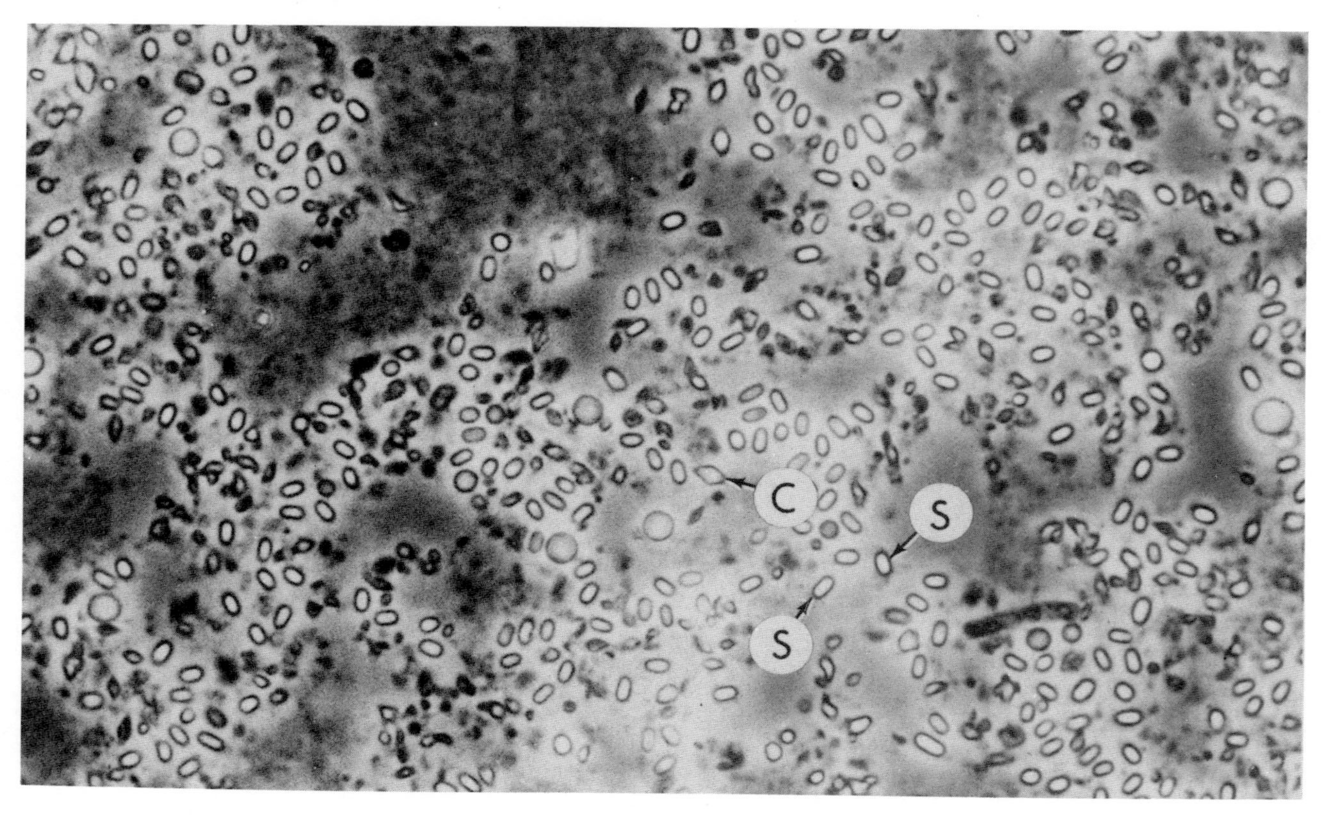

Fig. 6-7 *Bacillus thuringiensis* spores (S) and crystals (C). ×2000 phase-contrast.

entomopathogenic fungi constantly changes, and identification of the fungus species is sometimes quite difficult.

Most fungi are transmitted from one host to another by a spore, usually a conidium. The most common method of infection is through the body wall or cuticle. The conidia germinate and form a special structure which penetrates the cuticle of the insect. The fungus then grows in the insect's body until the insect is filled with mycelia. At this point the insect is usually dead and has a rather firm consistency, somewhat like a small loaf of bread. Under favorable conditions the fungus continues to grow and produces structures which protrude through the cuticle and form spores or conidia (Fig. 6.8). The above description of a fungus infection is very general, and many entomopathogenic fungi produce infections that are quite different.

The development of fungus infections is dependent on environmental conditions. Fungi, perhaps more than any other group of insect pathogens, must have favorable environmental conditions for development of epizootics (MacLeod and Soper 1965). High humidity is considered vital for germination of the fungus spore and transmission of the pathogen from one insect host to another. Many entomopathogenic fungi are not host-specific and infect many different insect species. Many of these fungi can also be grown on artificial media, but the infectivity to insects is sometimes reduced.

D. Protozoa

Certain flagellates, ciliates, amoebas, coccidians, and haplosporidians have pathogenic relationships with insects. However, neogregarines and microsporidians are the most important of the entomopathogenic protozoa (Kudo 1966, Weiser 1961).

Neogregarines and microsporidians are transmitted orally from one insect to another by a resistant spore form (Fig. 6.9). Many species are also transmitted transovarially from an infected female to her progeny. These spore-forming protozoans produce diseases in insects which range from very pathogenic to chronic debilitating infections. They are quite widespread in naturally occurring insect populations, and in many cases are important as naturally occurring mortality factors.

Sporozoan diseases can be diagnosed by the presence of spore forms within the body of the infected host insect. The tissues infected vary with the specific neogregarine or microsporidian. Some species infect a single tissue, while others infect almost every tissue in the insect's body. These

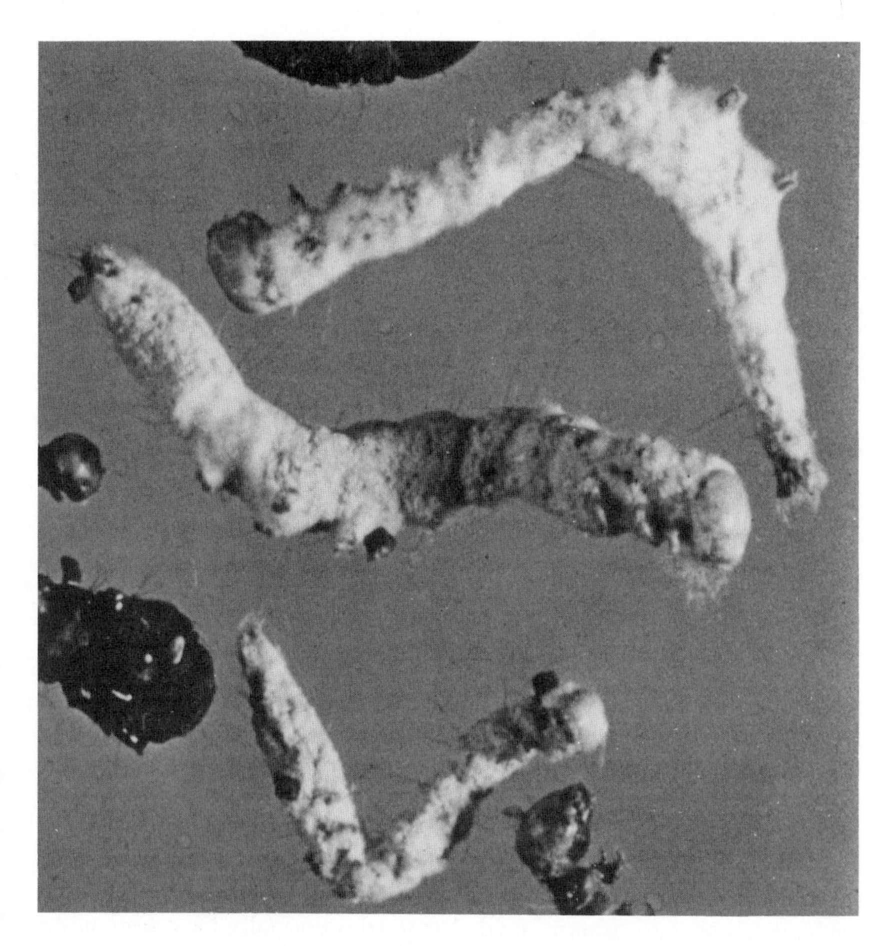

Fig. 6-8 Green cloverworms dead from the fungus *Nomuraea rileyi*. (Photo by Illinois Natural History Extension Staff.)

protozoans often have complex life cycles and are often difficult to identify to species.

Neogregarines, microsporidians, coccidians, and haplosporidians are all obligate parasites and do not complete development in artificial media.

E. Nematodes

Several nematode families contain species that are parasites of insects during at least part of the nematode's development. Well over 100 spe-

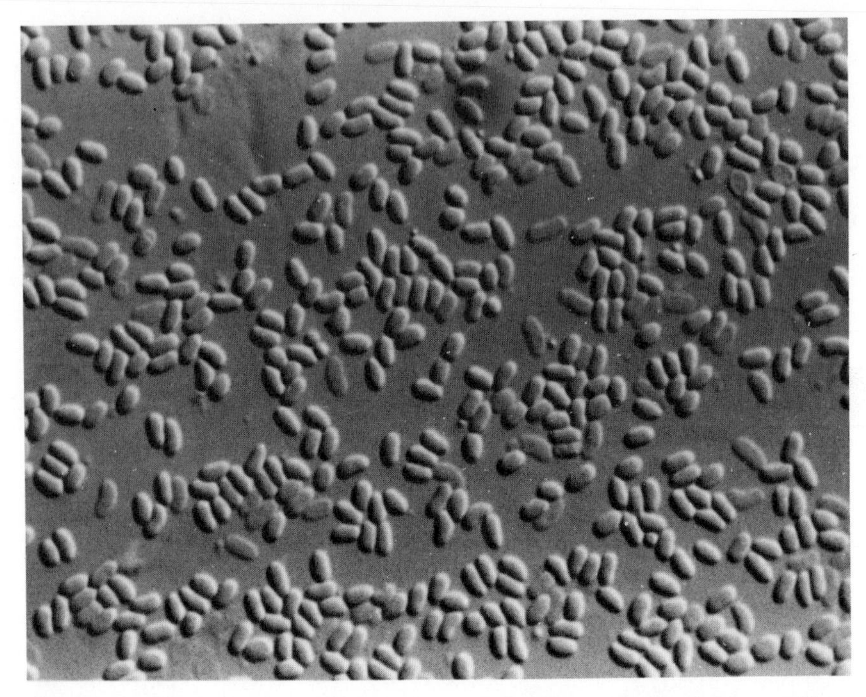

Fig. 6-9 Spores of the microsporidian *Pleistophora schubergi* from midgut tissue of the sod webworm, *Crambus trisectus*. ×1200 Nomarski interference-contrast.

cies of nematodes have been described as having insect associations and, while many of these nematode–insect associations do not appear to harm the insect, others are usually fatal. Nematodes normally have four molts between egg and adult, and between these molts are called juveniles. Most nematodes infect their insect hosts as juveniles which may enter directly through the host cuticle or through the midgut. After entering the hemocoel of its insect host, the nematode juvenile undergoes a period of rapid growth, leaves the host, enters the soil, and molts to form an adult nematode. Mating and oviposition usually occur external to the host (Poinar 1971; Welch 1963).

Some species of nematodes kill their insect host upon leaving to molt into an adult. Other nematodes transport bacteria when they enter the body cavity of their host. The insect dies from bacterial septicemia, usually within several days after being invaded by the nematode, and the nematode feeds on the bacteria in dead host tissue (Poinar and Thomas 1966).

Some nematode species are rather host-specific, while others infect a wide range of insect hosts. Most nematodes are very difficult to culture on artificial media, and no obligate endoparasitic nematode, other than representatives of the genus *Neoaplectana,* has been cultured (Poinar 1971, Dougherty 1960).

III. INSECT DISEASES AS NATURALLY OCCURRING MORTALITY FACTORS

Insect diseases are often important natural regulators of insect populations. Most experienced field entomologists have seen epizootics drastically reduce large insect populations. Unfortunately, these epizootics are usually sporadic, and it is difficult to predict when and where they will occur.

The development and spread of a disease within an insect population is dependent on the interaction of the pathogen, the host, and the environment. These interactions involve many variables and are quite complicated. The characteristics of the pathogen that influence its ability to spread throughout a host population are (1) virulence and infectivity, (2) capacity to survive, and (3) capacity to disperse (Tanada 1963). Density and susceptibility to the insect disease are important characteristics of the host population. The most important environmental factors include temperature, humidity, physiochemical conditions of the soil, microorganisms, parasites, predators, and other animals and plants (Tanada 1964).

One might well ask, "If naturally occurring insect diseases are so sporadic in nature, how can they be used in a pest-management program?" Even though the study of most insect disease has not yet reached the state of sophistication that allows consistent prediction of a naturally occurring epizootic, such epizootics can be used in pest-management programs if two important factors are carefully considered. First, the damage threshold of the insect population for a particular crop must be known (Chapter 1). Second, the entomologist must be familiar with the most important diseases in the insect population. If, for example, one is aware that a certain insect pest is frequently controlled by a naturally occurring insect disease, he can watch that population very carefully and avoid treatment with insecticides for as long as possible, and never before the insect population exceeds the damage threshold. The insect population should be constantly monitored for the initial appearance of insect disease. The appearance of diseases in dense insect populations frequently signals the onset of an epizootic and the subsequent collapse of the insect population. In some cases chemical control applications are applied to insect populations just at the onset, or even after disease epizootics have occurred. Many of these applications could be prevented if the economic threshold of the pest population and the possibility of a disease epizootic were carefully evaluated before applying control measures.

Viruses and fungi are the two groups of insect pathogens most often credited with causing epizootics. Bacteria, protozoans, and nematodes are thought to cause epizootics less frequently than viruses and fungi, but this is no doubt influenced by the fact that viral and fungal infections are more easily recognized than are protozoan, nematode, and bacterial infections.

Populations of the armyworm, *Pseudaletia unipuncta* (Haworth), are frequently controlled by the natural occurrence of NPV and GV infections which, in Hawaii, usually maintain the insect population below the economic threshold (Tanada 1968). In Illinois these viruses are also important in the natural control of the armyworm. The extent of virus epizootics usually varies from area to area but, by carefully observing the armyworm population, chemical treatments can often be avoided.

Many species of aphids are controlled by the natural occurrence of the fungus genus *Entomophthora*. In a seven-year study *Entomophthora* infections were probably the major factors in controlling the green peach aphid, *Myzus persicae* (Sulzer), in Maine (Shands et al. 1972). Although the factors that initiate the occurrence of *Entomophthora* infections in aphids are largely environmental and unpredictable, once *Entomophthora* infections appear in dense aphid populations, an epizootic usually occurs, making further control measures unnecessary. *Entomophtora* sp. also accounted for a high percentage of the natural control of the soybean looper, *Pseudoplusia includens* (Walker), in Alabama (Harper and Carner 1973) (Fig. 6.10). Without the natural control provided by the fungus, chemical control would be necessary to prevent severe defoliation and crop loss (Harper and Carner 1973).

The fungus *Nomuraea rileyi* is responsible for epizootics in populations of many phytophagous lepidopterous larvae (Rabb 1971). In Illinois, large populations of the green cloverworm, *Plathypena scabra* (Fabricius), may occur in soybean fields, but they are usually controlled by the natural occurrence of *Nomuraea rileyi*. Unfortunately, the onset of this epizootic is not predictable, and chemical treatments are often applied to green cloverworm populations, many of which would have been controlled naturally by *Nomuraea rileyi* if treatments had not been applied. Accurate prediction of naturally occurring insect diseases must be pos-

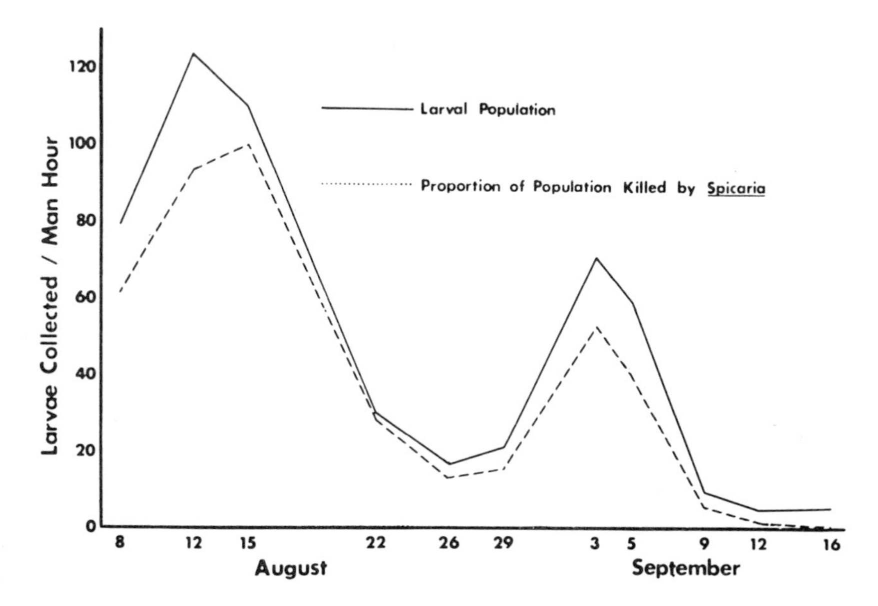

Fig. 6-10 Contributions of *Nomuraea* sp. to larval mortality in populations of *Pseudoplusia includens* at Tallahassee, Alabama, in 1969. (Adapted from Harper and Carver 1973. Academic Press.)

sible if they are to be utilized to their fullest extent in pest-management systems. We must continue our efforts to better understand the factors that are responsible for the development of epizootics in insect populations.

IV. INTRODUCTION AND APPLICATION OF INSECT DISEASES FOR LONG-TERM SUPPRESSION

According to Steinhaus (1954), the introduction of an insect pathogen into an insect population can bring about more-or-less permanent control if the economic level of host density is higher than the threshold level of the disease. Conversely, if the economic threshold of host density is lower than the disease threshold level, temporary control similar to that produced by chemical insecticides will result.

As one might expect, the introduction of pathogens into insect populations for long-term suppression has been most successful on crops that tolerate a relatively high level of host density. This is especially true of forest insect pests, and several insect pathogens introduced into populations of forest pests have produced satisfactory control. Two of the most successful introductions have involved NPVs of two sawfly species which were accidentally introduced into North America from Europe. The European spruce sawfly, *Gilpinia hercyniae* (Hartig), and the European pine sawfly, *Neodiprion sertifer* (Geoffroy), have both been at least partially controlled by the introduction of these viruses. It is not surprising that these pathogen introductions were particularly successful in insects of foreign origin, which were introduced into an environment where their natural parasitoids, predators, and diseases were largely absent. The NPV of the European spruce sawfly was probably introduced by accident with imported European parasites (Balch and Bird 1944). This sawfly has since been kept under control by the virus disease and the imported parasites (Bird and Elgee 1957).

One of the most successful introduction programs for the control of an insect pest involved the use of the bacterium *Bacillus popilliae,* which infects larvae of the Japanese beetle, *Popillia japonica.* Although the Japanese beetle was introduced into the United States from Japan, the bacterium *Bacillus popilliae* is probably indigenous to the United States. The pathogen is propagated by injecting spores into live larvae of the Japanese beetle. Bacterial spores are removed from the injected larvae, dried, and mixed with a sufficient amount of talc to give a final mixture containing 100 million spores/g (Fleming 1968). Twenty-three grubs, each containing 2 billion spores, produce about 1 lb of spore dust (Fleming 1968). The application of this spore dust at intervals of 10 ft over established turf, using a hand rotary-type spreader ($1\frac{3}{4}$ lb spore dust per acre), was thought to be the most practical method for colonizing the pathogen (White and Dutky 1942). The density of the grub population was usually a more important factor in the establishment and buildup of the pathogen to an effective level than was the dosage of spores applied (Fleming 1968; White and Dutky 1942). These larvae become infected and further inoculate the soil, causing the pathogen to build up to a level sufficient for consistent control of Japanese beetle larvae. *Bacillus popilliae* has spread naturally into uninoculated areas and is presently one of the most important biological factors in the regulation of Japanese beetle populations in the eastern United States (Fleming 1968).

Insect pathogens may also be periodically introduced into insect populations in order to initiate epizootics at an earlier time than they would normally occur. For example, under natural conditions a NPV of the forest tent caterpillar, *Malacosoma disstria* Hübner, does not reach epizootic levels until 4 to 6 years after the first year of high caterpillar populations (Sippell 1952). Stairs (1965) initiated epizootics in the early stages of a forest tent caterpillar outbreak by spraying with a NPV. The virus was also carried over into the next generation and spread from the point of introduction.

Several factors should be considered when introducing a pathogen into an insect population: (1) the concentration of the insect pathogen must be high enough to produce infection in at least some of the insect hosts; (2) the host population in which the pathogen is to be applied should have a relatively high density in order to assure propagation of the pathogen and its survival from one generation to the next; and (3) the pathogen should be applied when a susceptible stage of the host insect is present.

V. MICROBIAL INSECTICIDES

If the economic level of host density is lower than the disease threshold level, temporary control similar to that produced by chemical insecticides will result (Steinhaus 1954). When repeated applications of the insect pathogen are used in a manner similar to chemical insecticides for temporary suppression of an insect pest, the pathogen can be considered a microbial insecticide. The difference between microbial insecticides and insect pathogens applied for long-term control is not always distinct. Microbial insecticides sometimes have more than a short-term effect on the insect pest population and, as earlier examples have illustrated, pathogens introduced into insect populations for long-term control may be reapplied at intervals to promote the onset of an epizootic.

The principal advantages in using microbial insecticides are safety and host specificity. The two groups of insect pathogens that have received the greatest attention as microbial insecticides, inclusion viruses and the spore-forming bacterium *Bacillus thuringiensis,* do not affect warm-blooded animals. The relative specificity of most insect pathogens allows them to be used without the danger of destroying beneficial parasitoids and predators. Host specificity, however, may be a disadvantage in cases in which control of several insect pests is desired.

Methods used for production, formulation, storage, application, and standardization of chemical insecticides are different from those for microbial insecticides.

A. Production

Some insect pathogens that are not obligate parasites can be mass-produced in artificial media. *Bacillus thuringiensis* is produced commercially by a large-scale fermentation process. The most common method for producing *B. thuringiensis* is by submerged fermentation (Dulmage and Rhodes 1971) (Fig. 6.11). Various modifications of this method are currently used to produce *B. thuringiensis* spores. This process must be carefully controlled, as the strain of *B. thuringiensis* used and the fermentation conditions greatly influence the insecticidal activity of the product.

Obligate insect pathogens, such as viruses, must be produced in living insect tissue; this means the use of either tissue culture or the whole insect. Because tissue culture is not at the present time economically feasible for virus production, the living insect host is used for the production of large quantities of virus. Figure 6.12 shows a large colony of the insect

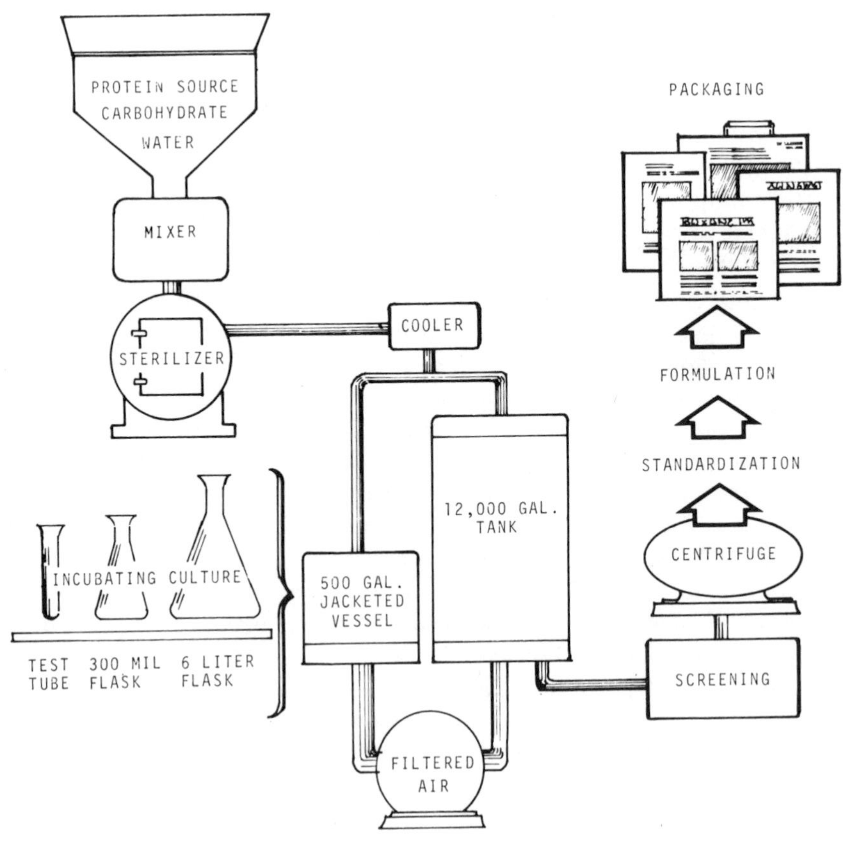

Fig. 6-11 Submerged fermentation of *B. thuringiensis*. (Ignoffo 1967. Redrawn by L. LeMere.)

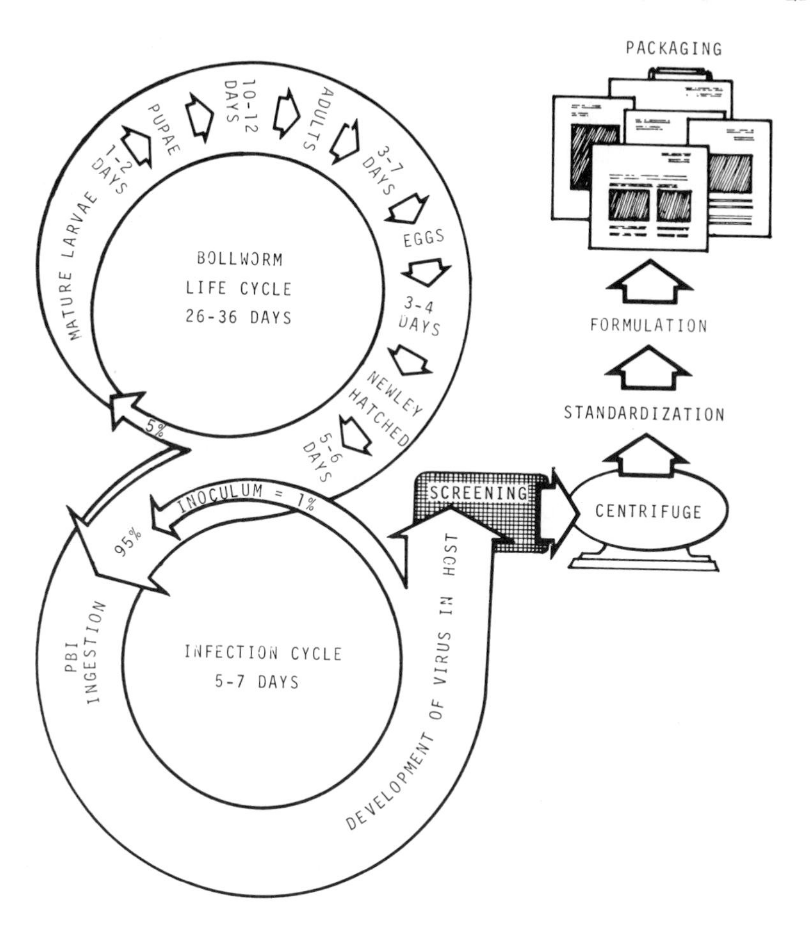

Fig. 6-12 Schematic representation of the production of a NPV in a living host system. (Drawing by L. LeMere from Ignoffo and Hink 1971.)

host maintained on an artificial medium. Young larvae are infected by adding polyhedra to the surface of the diet. After an incubation period of about 1 week, the infected larvae are blended with water and the polyhedra removed by filtration and differential centrifugation (Ignoffo and Hink 1971). This material is then standardized and formulated.

B. Standardization

Burges and Thomson (1971) define standardization as "adoption of the most appropriate units for the measurement of the insecticidal potency of a pathogenic preparation." This enables a manufacturer to reproduce a product of consistent potency and compare its potency with products of other manufacturers. In the production of microbial control agents, both the yield and infectivity of the pathogen vary to some degree with each batch of pathogen produced. Therefore for each batch the potency or the amount of active ingredient per resulting formulation must be measured and expressed in terms that allow comparison between formulations prepared from different batches by the same company, as well as those produced by other companies. Standardization has been achieved with *B. thuringiensis* by establishment of an international unit (IU). A strain of *B. thuringiensis*, E-61, has been designated the primary reference standard and arbitrarily assigned a potency of 1000 IU/mg. Bioassays of each *B. thuringiensis* formulation are compared with those of the E-61 standard and expressed in international units per milligram relative to this standard. Therefore, when the potency of *B. thuringiensis* formulations is expressed in international units, it is comparable to the potency of every other formulation of *B. thuringiensis* for which potency is also expressed in international units.

The proper standardization of microbial insecticides is very important, since application rates for the control of various insect pests are based on the potency of the microbial insecticide formulation being used. If standardization is not achieved and the potency of a microbial insecticide varies with each formulation, it is impossible to establish recommendations on the quantity of insecticide necessary to attain a given level of control for a specific insect pest.

International units have not yet been firmly established for microbial insecticides containing insect viruses, but some type of system similar to that used for *B. thuringiensis* will undoubtedly be established for those viruses that are commercially developed as microbial insecticides (Dulmage 1973).

C. Formulation

In formulating chemical insecticides one usually deals with compounds that are soluble. Insect pathogens, however, are physical particles which must remain intact in order to retain their infectivity. The pathogens cannot be dissolved, and so must be uniformly distributed as a suspension if the formulation is a liquid, or as dry particles if the formulation is a dust. A formulation may contain many additives to enhance the effectiveness of the pathogen. Wetting agents are often added to improve coverage of the sprayed surface; stickers or adhesives cause the particles to stick to the surface area. Protectants are usually added to a microbial insecticide formulation to help protect the pathogen from sunlight and ultraviolet radiation. The formulation of microbial insecticides has not received the attention it deserves. Most of the methods used for formulating microbial insecticides are only modifications of methods previously used for formulating chemical insecticides. The principles of uniformly delivering a soluble chemical may be completely different from those of uniformly delivering an intact disease agent. The improvement of methods for formulating microbial insecticides will undoubtedly improve their effectiveness.

D. Application

Microbial insecticides can be applied as sprays, dusts, baits, or in combination with other pathogens, many chemical insecticides, chemosterilants, and pheromones. The spray droplet or dust particle size, deposit, coverage, and drift of microbial insecticides, and the properties of the application equipment that influence these characteristics, are very important but have not been thoroughly investigated (Falcon 1971). Pathogens are presently applied with spray equipment designed for the application of chemical insecticides. There is a great need for the development of application equipment specifically designed for delivering uniform quantities of the pathogen in a manner that ensures ingestion of a maximum number of these pathogens by the host insect.

Best results are usually attained if microbial insecticides are applied when early stages of the insect pest are present; late larval instars are much more difficult to control with microbial insecticides than are earlier larval instars. This is generally true of chemical insecticides as well, but this effect is much greater with microbial insecticides.

Microbial insecticides are generally more effective at higher temperatures. This is usually the result of a dual effect. The disease agent develops more rapidly at higher temperatures, up to the point where it is inactivated, and at higher temperatures insects consume a larger amount of food and therefore ingest larger quantities of the pathogen. Microbial insecticides should be applied when temperatures are likely to remain above the activity threshold of the host for at least several hours during the day following treatment (Franz 1971).

Direct sunlight is harmful to most insect pathogens. Spray equipment should be adjusted to give good coverage on the undersides of the leaves of the host plant and, if the insect pest is active at night, applications of microbial insecticides should be applied in the late afternoon or early evening, allowing ingestion of the disease agent before it has been exposed to direct sunlight.

Obviously, the behavioral characteristics of the specific insect pest must be considered when applying microbial insecticides. The location of the insect pest and its period of peak activity, feeding behavior, and mobility dictate how and when a microbial insecticide should be applied. For example, insects like the European corn borer, *Ostrinia nubilalis* (Hübner), and codling moth, *Laspeyresia pomonella* (L.), which tunnel into their host plants, are vulnerable to control by microbial insecticides for only a short and specific period of time during larval development. Applications of microbial insecticides must be timed to cover the plant surface during those vulnerable periods.

E. Registration of Insect Pathogens for Use as Microbial Insecticides

For any pesticide, including microbial insecticides, to move in interstate commerce in the United States, it must be registered by the U.S. Environmental Protection Agency (EPA). Any proposed use of a pesticide that may result in residues on the food products of humans or animals must have an established tolerance level or an exemption from tolerance requirements. Food crops cannot have residues in excess of the tolerance level. A pesticide having an exemption from tolerance requirements does not have a residual tolerance level and may be present on the food crop at any level. Obviously, only pesticides that demonstrate absolute safety for man can be expected to obtain an exemption from tolerance requirements. Before the EPA registers a pesticide, it must be provided with acceptable evidence of ingredients, efficacy in achieving the results intended, safety to man and his environment, analytical methods suitable for determination of residues, and a suitable label describing the ingredients, methods of use, and any necessary precautionary measures (Upholt et al. 1973). One of the major problems in registering microbial insecticides has been deciding what types of studies are necessary for evaluating the safety of insect pathogens. It was obvious that the protocols used for evaluating the safety of chemical insecticides were not adequate for microbial insecticides. The tests used to evaluate the safety of insect pathogens for vertebrates are extensive and have been thoroughly reviewed by Ignoffo (1973).

At the present time only one microbial insecticide, *Bacillus thuringiensis*, is registered for use on food crops in the United States and is being produced commercially. In 1970, the *Heliothis* NPV was officially granted the status of temporary exemption from the requirement of a tolerance for residues in or on cottonseed (Ignoffo 1973). This was the first exemption granted for a viral insecticide. The granting of exemption from tolerance requirements to the *Heliothis* NPV was based on a series of safety tests (Table 6.5). The events leading to the registration of the *Heliothis* NPV have been covered by Ignoffo (1973). Conditions for establishing an exemption from tolerance required that production of virus material meet the following standards: "(1) Bacterial contaminants must not exceed 10^7 colonies/g product; (2) human bacterial pathogens, e.g., *Salmonella, Shigella, Vibrio,* must not be present; (3) the virus-product must be safe when injected (intraperitoneally) or fed (21-day test) to mice; (4) virus used in the product must be serologically identical to the virus used in production" (Ignoffo 1973). In 1973, the *Heliothis* NPV produced by one company was granted permanent exemption from tolerance in or on cottonseed by the EPA (*Federal Register* 1973).

Table 6.5 Series of Tests Used to Establish That the *Heliothis* NPV Was Safe and Could Be Granted a Temporary Exemption from the Requirement of Tolerance[a]

Type of Test	Animal System
Acute toxicity-pathogenicity	
Per os diet	Rat or mouse, birds, fishes, oyster, shrimp
Inhalation	Rat
Dermal-topical sensitivity	Man, rabbit, guinea pig
Eye sensitivity	Rabbit
Intraperitoneal injection	Rat or mouse
Subcutaneous injection	Rat
Subacute toxicity-pathogenicity	
Per os diet	Monkey, dog, rat or mouse
Inhalation	Monkey, dog, rat or mouse
Subcutaneous injection-sensitization	Monkey, dog, rat or mouse
Teratogenicity	Rat or mouse
Carcinogenicity	Rat or mouse
Replication potential	Man, primates, tissue culture
Phytotoxicity	Agricultural crops
Invertebrate specificity	Beneficial and other anthropods

[a] Courtesy of Ignoffo 1973, and *Annals of the New York Academy of Science.*

The EPA is developing guidelines for the type of data they can accept for establishing tolerances or exemptions from tolerance requirements for microbial insecticides (Upholt et al. 1973). These guidelines should reduce the time required for the portion of registration dealing with safety for future microbial insecticides.

Another important aspect of registration of a microbial insecticide is proof of efficacy, or the ability of the microbial insecticide to control the target pest. Again, the methods used to evaluate chemical insecticides are often not appropriate for evaluating microbial insecticides. The primary criterion for chemical control, immediate mortality, often does not measure the total effect of a microbial insecticide. Mortality may occur over an extended period of time, and sublethal infections are often transmitted to the next generation. Microbial insecticides are often compared with chemical insecticides on small experimental plots in which there is no possibility of a return of naturally occurring parasites and predators. Microbial insecticides produce the best results when used over large areas for several consecutive years. It is difficult to undertake preliminary efficacy tests on such a large scale.

F. Economics

The development of a chemical insecticide from synthesis to commercial production is estimated to cost about $5.5 million (Rogoff 1973). There is no reason to believe that the average cost for developing a microbial insecticide would be less. Industry is undoubtedly reluctant to invest such large sums of money in the development of many of the insect pathogens that have a limited host range. Many insect pathogens that could be very useful in insect control programs if they were available commercially are relatively host-specific. If these pathogens were developed as microbial insecticides, they would not command a large sales volume, even though the cost of development would be no less than that of a broad-spectrum pesticide that would command a large sales volume. Therefore from a business standpoint the development of host-specific microbial insecticides is usually a poor investment. State or federal investments may be necessary if host-specific microbial insecticides are to be developed by industry (Falcon 1971; Dulmage 1971).

G. Examples of Insect Control with Microbial Insecticides

Experimental and commercial microbial insecticides have been used for the control of many insect pests of fruit, forest, vegetable, and field crops, as well as ornamental plants, stored products, and medically important insects. Most of these microbial insecticides contain either spore-forming bacteria or insect viruses.

The use of the bacterium *Bacillus popilliae* for control of the Japanese beetle has already been cited. Preparations of this bacterium are available commercially and have been extremely successful in controlling Japanese beetle populations.

In the United States, the only other insect pathogen currently available to the general public as a microbial insecticide, *Bacillus thuringiensis,* is used extensively for the control of many different insect species. *Bacillus thuringiensis* preparations have been especially useful in controlling foliage-chewing species of Lepidoptera on vegetable crops and tobacco where the use of chemical insecticides is often restricted because of harmful residues. *Bacillus thuringiensis* preparations also provide the homeowner a safe means of controlling many lepidopterous pests of ornamentals. Excellent control of the bagworm, *Thyridopteryx ephemeraeformis* (Haworth), was obtained with *B. thuringiensis* (Bishop et al. 1973) (Table 6.6). Even older larvae, which are difficult to control with chemical insecticides, were killed by *B. thuringiensis* applications. Falcon (1971) cites numerous examples of the uses of *B. thuringiensis* for insect control.

There are several interesting uses for a by-product of *Bacillus thuringiensis,* the thermostable exotoxin released into the media in which *B. thuringiensis* is grown. This exotoxin kills several dipterous pests and

Table 6.6 Field Evaluations of *Bacillus thuringiensis* var. *kurstaki* Applied as Foliar Sprays on Eastern Red Cedar for Bagworm Control[a]

Pounds per 100 gal water (3.4×10^9 IU/lb)	Mean Percent Reduction Following Treatments		
	June 12	July 3	July 24
4	99.0 *a*[b]	98.5 *a*	96.5 *a*
2	100.0 *a*	100.0 *a*	86.0 *ab*
1	98.5 *a*	96.5 *a*	72.5 *bc*
0.5	72.5 *a*	75.0 *b*	71.5 *c*
Untreated	0.0 *b*	0.0 *c*	2.0 *d*

[a] Adapted from Bishop et al. 1973.

[b] Means followed by same letter not significantly different at 5% level as determined by Duncan's multiple range test.

small amounts can be incorporated into the food of domestic animals and enough material is passed in the droppings of these animals to kill fly larvae (Gingrich and Eschle 1971).

There are many published accounts of the successful use of viral insecticides for the control of insect pests (Stairs 1971, Falcon 1973). The *Heliothis* NPV has been tested more extensively for insect control than any other viral insecticide. This NPV infects six species of *Heliothis* and *Helicoverpa* (Allen 1968). These insects are important pests of corn, cotton, soybean, tomato, and tobacco plants. This presents a wide range of conditions under which this NPV can be used. The degree of control obtained by the use of this NPV has not been consistent, but in most cases the larval population was significantly reduced and the crop yield significantly increased over the untreated check. In many cases yield and larval control compared favorably with insecticide treatments. In experiments by Chapman and Ignoffo (1972), the *Heliothis* NPV reduced larval corn earworm populations and increased cotton yields by as much as 92% over the untreated check. Experiments by many other workers have been equally successful.

Noninclusion viruses have not been used as microbial control agents as extensively as have inclusion viruses. A noninclusion virus of the mite *Panonychus citri* (McGregor) produced effective control when applied as sprays, but the best method of applying the virus was by the introduction of infected mites (Gilmore 1965; Gilmore and Tashiro 1966).

The microsporidian *Nosema locustae* significantly reduced grasshopper populations when spores were applied in a wheat bran bait mixture (Henry et al. 1973). The average reduction in grasshopper density attributed to 30 spores/in.^{2} of bait was 46%, and that attributed to 900 spores/in.^{2} was 73%.

VI. THE ROLE OF MICROBIAL CONTROL IN PEST-MANAGEMENT PROGRAMS

We consider the various methods of insect control as tools to be used in pest-management programs. One control method is usually not sufficient for an entire pest-management program, and no single control method works in every pest-management program. Microbial control is a good example. Microbial control will play an important role in many pest-management programs, while in others it may be totally impractical, either because it does not give the degree of control desired, or because it is simply not economically feasible. However, where microbial control can be used as a tool in pest-management programs, it conforms nicely with the objectives of pest management: (1) microbial control agents are relatively host-specific and do not upset other biotic systems, thus causing an upsurge of previously unimportant insect pests; (2) microbial control agents are safe for humans and do not cause environmental contamination; and (3) microbial control is compatible with most other control methods.

When naturally occurring insect diseases frequently act as controlling factors in the insect pest population, they should be considered an integral part of the pest-management program and utilized to the fullest extent possible. Even though naturally occurring epizootics are not controlled by man, the recognition of an epizootic in an insect population as a signal for discontinuing chemical control methods can be just as valuable as the more direct control tactics available in pest management.

The introduction and colonization of insect pathogens for long-term control should be considered a pest-management tactic in such situations as forests or perennial crops for which the economic threshold of the insect pest is likely to be high and the ecosystem rather stable. Periodic introduction of insect pathogens to initiate epizootics prior to the probable natural occurrence of such epizootics should also be considered.

Microbial control is often most effective when combined with other methods of control, such as the use of resistant varieties of cultivated plants. These combinations may produce acceptable levels of control, when the use of either of these two methods independently does not. Applications of *Heliothis* NPV were more effective against the bollworm and the tobacco budworm when applied to a resistant variety of cotton (Fernandez et al. 1969).

Because of their host specificity, microbial insecticides are often useful for controlling a specific pest problem without upsetting the ecological relationships of other animals in the area. Microbial insecticides can be safely used in situations in which broad-spectrum pesticides cause an upsurge in previously unimportant insect pests, or when large areas containing much wildlife must be treated. Microbial insecticides may be especially useful in pest-management programs for some vegetables when the residues of chemical insecticides often present a problem. A vegetable crop may be treated with *Bacillus thuringiensis* immediately before harvest with no medical or legal dangers.

VII. THE FUTURE

Even though pathogens are often very important in the regulation of insect populations, we know little about the ecology of insect pathogens.

There are very few quantitative data on the specific factors that initiate epizootics of insect populations (Tanada 1964). The population models so useful in assessing the role of insect parasitoids and predators have not been developed for insect pathogens, because the corresponding relationships for pathogens necessary for the construction of these population models have not been determined. If insect pathogens are to be utilized to their fullest extent, a better understanding of their ecology is necessary. Although there are many documented cases of disease being an important regulating mechanism in insect populations, few comprehensive field studies on these epizootics have been made. Such studies are very complicated, because in the field the pathogen, the host, and the environment are interacting and changing, all at the same time. An understanding of why and how natural epizootics occur will not only allow accurate predictions of epizootics, but will also allow the pest-management program to be manipulated to induce the occurrence of epizootics at times prior to their probable natural occurrence.

Relatively few attempts have been made to introduce or colonize pathogens for the long-term control of insect pests. As in the introduction of parasitoids and predators, the proportion of successful pathogen introductions has been and probably will continue to be low (Burges and Hussey 1971) but, even with the low proportion of successes, this method of microbial control remains one of the most economical and efficient ways of using insect pathogens. The discovery of new insect pathogens and a better understanding of their ecology should produce more successful introductions.

In many ways the future looks bright for microbial insecticides, even though at the present time only two insect pathogens, *Bacillus thuringiensis* and *Bacillus popilliae,* are registered and have labels for use in the United States. *Bacillus thuringiensis* is established as a bona fide insect control agent and is used on a worldwide basis. The potency and field effectiveness of *B. thuringiensis* formulations are constantly being improved, and its usage is increasing yearly.

Many of the insect viruses have shown excellent promise as microbial insecticides. Proof of safety to nontarget organisms has been a major obstacle in the development of microbial insecticides. The approval of an exemption from tolerance registration for the *Heliothis* NPV and efforts on the part of the EPA to develop guidelines for the tests necessary to establish the safety of other pathogens are very encouraging.

The development of new methods of formulation and application designed specifically for microbial insecticides is needed. This will undoubtedly improve the effectiveness of microbial insecticides. The selection of more virulent strains of insect pathogens, various combinations of different insect pathogens, the combination of pathogens with bait and attractants, and the combination of pathogens with chemical insecticides are areas of research that should enhance the usefulness of microbial insecticides.

The costs of developing a chemical insecticide from synthesis to commercial product are astounding. There is no reason to believe that the development of a microbial insecticide would cost less. Therefore industry may be reluctant to invest large sums of money in developing a microbial insecticide which has a limited host range and thus a limited sales potential. The gross sales of such a product could not be expected to approach those of a broad-spectrum insecticide (Rogoff 1973). Since industry must operate on a profit margin, the limited sales volume of many microbial insecticides may be a limiting factor in their commercial production.

Even though many insect pathogens have been successfully used in insect-control programs, there are many problems yet to be solved if microbial control is to reach its full potential. As these problems are solved, insect pathogens will play an increasingly important role in pest-management programs.

REFERENCES

Aizawa, K. 1963. The nature of infections caused by nuclear-polyhedrosis viruses. Pages 381–412 *in* E. A. Steinhaus, ed., Insect pathology, an advanced treatise. Vol. 1. Academic Press, New York.

Aizawa, K. 1971. Strain improvements and preservation of virulence. Pages 655–672 *in* H. D. Burges and N. W. Hussey, eds., Microbial control of insects and mites. Academic Press, New York.

Allen, G. E. 1968. Evaluation of the *Heliothis* nuclear polyhedrosis virus as a microbial insecticide. *Proc. Joint U.S.–Jap. Comm. Sci. Coop. Panel* 8. *Fukuoka,* pp 37–41.

Balch, R. E., and F. T. Bird. 1944. A virus disease of the European spruce sawfly, *Gilpinia hercynide* (Htg.), and its place in natural control. *Sci. Agr.* **25**:65–80.

Beard, R. L. 1945. Studies on the milky disease of Japanese beetle larvae. *Conn. Agr. Exp. Sta. Bull.* **491**:505–582.

Bergold, G. H. 1963. The nature of nuclear-polyhedrosis viruses. Pages 413–456 *in* E. A. Steinhaus, ed., Insect pathology, an advanced treatise. Vol. 1. Academic Press, New York.

Bergold, G., K. Aizawa, K. Smith, E. A. Steinhaus, and C. Vago. 1960. The present status of insect virus nomenclature and classification. *Int. Bull. Bacteriol. Nomencl. Taxon.* **10**:259–262.

Bird, F. T., and D. E. Elgee. 1957. Virus disease and introduced parasites as factors controlling the European spruce sawfly, *Diprion hercyniae* (Htg.), in Central New Brunswick. *Can. Entomol.* **89**:371–378.

Bishop, E. J., T. J. Helms, and K. A. Ludwig. 1973. Control of bagworm with *Bacillus thuringiensis. J. Econ. Entomol.* **66**:675–676.

Bucher, G. E. 1958. General summary and review of utilization of disease to control insects. *Proc. Xth Int. Congr. Entomol. Montreal, 1956.* **4**:695–701.

Bucher, G. E. 1960. Potential bacterial pathogens of insects and their characteristics. *J. Insect Pathol.* **2**:172–195.

Burgerjon, A., and D. Martouret. 1971. Determination and significance of the host spectrum of *Bacillus thuringiensis.* Pages 305–325 *in* H. D. Burges and N. W. Hussey, eds., Microbial control of insects and mites. Academic Press, New York.

Burges, H. D., and N. W. Hussey. 1971. Past achievements and future prospects. Pages 687–709 *in* H. D. Burges and N. W. Hussey, eds., Microbial control of insects and mites. Academic Press, New York.

Burges, H. D., and E. M. Thomson. 1971. Standardization and assay of microbial insecticides. Pages 591–622 *in* H. D. Burges and N. W. Hussey, eds., Microbial control of insects and mites. Academic Press, New York.

Chapman, A. J., and C. M. Ignoffo. 1971. Influence of rate and spray volume of a nucleopolyhedrosis virus on control of *Heliothis* in cotton. *J. Invertebr. Path*ol. **20**:183–186.

Cooksey, K. E. 1971. The protein crystal toxin of *Bacillus thuringiensis:* biochemistry and mode of action. Pages 247–274 *in* H. D. Burgess and N. W. Hussey, eds., Microbial control of insects and mites. Academic Press, New York.

David, W. A. L., and B. O. C. Gardiner. 1967. The effect of heat, cold, and prolonged storage on a granulosis virus of *Pieris brassicae. J. Invertebr. Pathol.* **9**:555–562.

Dougherty, C. E. 1960. Cultivation of nematodes. Pages 297–318 *in* J. N. Sasser and W. R. Jenkins, eds., Nematology. University of North Carolina Press, Chapel Hill.

Dulmage, H. T. 1971. Economics of microbial control. Pages 581–590 *in* H. D. Burges and N. W. Hussey, eds., Microbial control of insects and mites. Academic Press, New York.

Dulmage, H. T. 1973. Assay and standardization of microbial insecticides. *Ann. N.Y. Acad. Sci.* **217**:187–199.

Dulmage, H. T., and R. A. Rhodes. 1971. Production of pathogens in artificial media. Pages 507–540 *in* H. D. Burges and N. W. Hussey, eds., Microbial control of insects and mites. Academic Press, New York.

Dutky, S. R. 1940. Two new spore forming bacteria causing milky disease of Japanese beetle larvae. *J. Agr. Res.* **61**:57–68.

Falcon, L. A. 1971. Microbial control as a tool in integrated control programs. Pages 346–364 *in* C. B. Huffaker, ed., Biological control. Plenum Press, New York.

Falcon, L. A. 1973. Biological factors that affect the success of microbial insecticides: Development of integrated control. *Ann. N.Y. Acad. Sci.* **217**:173–186.

Federal Register. 1973. Vol. 38, p. 14169–14170. Government Printing Office, Washington, D.C.

Fernandez, A. T., H. M. Graham, M. J. Lukefahr, H. R. Bullock, and N. S. Hernandez, Jr. 1969. A field test comparing resistant varieties plus applications of polyhedral virus with insecticides for control of *Heliothis* spp. and other pests of cotton, 1967. *J. Econ. Entomol.* **62**:173–177.

Fleming, W. E. 1968. Biological control of the Japanese beetle. *U.S. Dep. Agr. Tech. Bull.* 1383. 78 pp.

Franz, J. M. 1971. Influence of environment and modern trends in crop management on microbial control. Pages 407–444 *in* H. D. Burges and N. W. Hussey, eds., Microbial control of insects and mites. Academic Press, New York.

Getzin, L. W. 1961. *Spicaria rileyi* (Farlow) Charles, an enomogenous fungus of *Trichoplusia ni* (Hübner). *J. Insect Pathol.* **3**:2–10.

Gilmore, J. E. 1965. Preliminary field evaluation of a noninclusion virus for control of the citrus red mite. *J. Econ. Entomol.* **58**:1136–1140.

Gilmore, J. E., and H. Tashiro. 1966. Fecundity, longevity, and transinfectivity of citrus red mites (*Panonychus citri*) infected with a noninclusion virus. *J. Invertebr. Pathol.* **8**:334–339.

Gingrich, R. E., and J. L. Eschle. 1971. Susceptbility of immature house flies to toxins of *Bacillus thuringiensis*. *J. Econ. Entomol.* **64**:1183–1188.

Harper, J. D., and G. R. Carner. 1973. Incidence of *Entomophthora* sp. and other natural control agents in populations of *Pseudoplusia includens* and *Trichoplusia ni*. *J. Invertebr. Pathol.* **22**:80–85.

Heimpel, A. M., and T. A. Angus. 1959. The site of action of crystalliferous bacteria in lepidopterous larvae. *J. Invertebr. Pathol.* **1**:152–170.

Henry, J. E., K. Tiahrt, and E. A. Oma. 1973. Importance of timing, spore concentrations, and levels of spore carrier in applications of *Nosema locustae* (Microsporida: Nosematidae) for control of grasshoppers. *J. Invertebr. Pathol.* **21**:263–272.

Huger, A. 1963. Granuloses of insects. Pages 531–575 *in* E. A. Steinhaus, ed., Insect pathology, an advanced treatise. Vol. 1. Academic Press, New York.

Hughes, K. M. 1972. Fine structure and development of two polyhedrosis viruses. *J. Invertebr. Pathol.* **19**:198–207.

Ignoffo, C. M. 1964. Bioassay technique and pathogenecity of a nuclear polyhedrosis virus of the cabbage looper, *Trichoplusia ni* (Hübner). *J. Insect Pathol.* **6**:237–245.

Ignoffo, C. M. 1967. Possibilities of mass-producing insect pathogens. Pages 91–117 *in* P. A. van der Lan, ed., Insect pathology and microbial control. North-Holland, Amsterdam.

Ignoffo, C. M. 1968. Specificity of insect viruses. *Bull. Entomol. Soc. Amer.* **14**:265–276.

Ignoffo, C. M. 1973. Effects of entomopathogens on vertebrates. *Ann. N.Y. Acad. Sci.* **217**:141–164.

Ignoffo, C. M., and W. F. Hink. 1971. Propagation of arthropod pathogens in living systems. Pages 541–580 *in* H. D. Burges and N. W. Hussey, eds., Microbial control of insects and mites. Academic Press, New York.

Jaques, R. P. 1967a. The persistence of a nuclear polyhedrosis virus in the habitat of the host, *Trichoplusia ni.* I. Polyhedra deposited on foliage. *Can. Entomol.* **99**:785–794.

Jaques, R. P. 1967b. The persistence of a nuclear polyhedrosis virus in the habitat of the host, *Trichoplusia ni.* II. Polyhedra in soil. *Can. Entomol.* **99**:820–829.

Jaques, R. P. 1972. The inactivation of foliar deposits of viruses of *Trichoplusia ni* (Lepidoptera: Noctuidae) and *Pieris rapae* (Lepidoptera: Pieridae) and tests on protective additives. *Can. Entomol.* **104**:1985–1994.

Kramer, J. P. 1959. Some relationships between *Perezia pyraustae* Paillot (Sporozoa, Nosematidae) and *Pyrausta nubilalis* (Hübner) (Lepidoptera, Pyralidae). *J. Insect Pathol.* **1**:25–33.

Kudo, R. R. 1966. Protozoology. Charles C Thomas, Springfield, Ill. 1174 pp.

Lacadet, M. M., and D. Martouret. 1967. Enzymatic hydrolysis of the crystal of *Bacillus thuringiensis* by the proteases of *Pieris brassicae.* II. Toxicity of the different fractions of the hydrolysate for larvae of *Pieris brassicae. J. Invertebr. Pathol.* **9**:322–330.

MacLeod, D. M., and R. S. Soper. 1965. The influence of environmental conditions on epizootics caused by entomogenous fungi. *Proc. 12th Int. Congr. Ent. London,* pp. 724–726.

Maddox, J. V. 1973. The persistence of the microsporida in the environment. *Misc. Publ. Entomol. Soc. Amer.* **9**:99–104.

Neilson, M. M. 1965. Effects of a cytoplasmic polyhedrosis on adult Lepidoptera. *J. Invertebr. Pathol.* **7**:306–314.

Neilson, M. M., and D. E. Elgee. 1968. The method and role of vertical transmission of a nucleopolyhedrosis virus in the European spruce sawfly, *Diprion hercyniae. J. Invertebr. Pathol.* **12**:132–139.

Poinar, G. O. 1971. Use of nematodes for microbial control of insects. Pages 181–203 *in* H. D. Burges and N. W. Hussey, eds., Microbial control of insects and mites. Academic Press, New York.

Poinar, G. O., Jr., and G. M. Thomas. 1966. Significance of *Achromobacter nematophilus* Poinar & Thomas (Achromobacteraceae; Eubacteriales) in the development of the nematode DD-136 (*Neoaplectana* sp. Steinernematidae). *Parasitology* **56:**385–390.

Rabb, R. L. 1971. Naturally occurring biological control in the eastern United States, with particular reference to tobacco insects. Pages 294–311 *in* C. B. Huffaker, ed., Biological control. Plenum Press, New York.

Roberts, D. W., and W. G. Yendol. 1971. Use of fungi for microbial control of insects. Pages 125–149 *in* H. D. Burges and N. W. Hussey eds., Microbial control of insects and mites. Academic Press, New York.

Rogoff, M. H. 1973. Industrialization. *Ann. N.Y. Acad. Sci.* **217:**200–210.

Shands, W. A., G. W. Simpson, I. M. Hall, and C. C. Gordon. 1972. Further evaluation of entomogenous fungi as a biological agent of aphid control in northeastern Maine. *Maine Life Sci. Agr. Exp. Sta. USDA Tech. Bull.* 58. 32 pp.

Sippell, W. L. 1952. Winter rearing of forest tent caterpillar, *Malcosoma disstria* (Hbn.). *Bi-mon. Rep. For. Biol. Div., Can. Dep. Agr.* **8:**1–2.

Simons, W. R., and G. O. Poinar, Jr. 1973. The ability of *Neoaplectana carpocapsae* (Steinernematidae: Nematoda) to survive extended periods of desiccation. *J. Invertebr. Pathol.* **22:**228–230.

Smirnoff, W. A. 1972. The effect of sunlight on the nuclear-polyhedrosis virus of *Neodiprion swainei* with measurement of solar energy received J. *Invertebr. Pathol.* **19:**179–188.

Smith, K. M. 1963. The cytoplasmic virus diseases. Pages 457–497 *in* Insect pathology, an advanced treatise. Vol. 1. Academic Press, New York.

Smith, K. M. 1967. Insect virology. Academic Press, New York. 256 pp.

Smith, K. M. 1971. The viruses causing the polyhedroses and granuloses of insects Pages 479-507 *in* K. Maramorosch and E. Kurstak, eds., Comparative virology. Academic Press, New York. 584 pp.

Stairs, G. R. 1965. Artificial initiation of virus epizootics in forest tent caterpillar populations. *Can. Entomol.* **97:**1059–1062.

Stairs, G. R. 1971. Use of viruses for microbial control of insects. Pages 97–124 *in* H. D. Burges and N. W. Hussey, eds. Microbial control of insects and mites. Academic Press, New York.

Steinhaus, E. A. 1954. The effects of disease on insect populations. *Hilgardia* **23:**197–261.

Steinhaus, E. A. 1958. Stress as a factor in insect disease. *Proc. 10th Int. Congr. Entomol., Montreal 1956* **4:**725–730.

Tanabe, A. M., and M. Tamashiro. 1967. The biology and pathogenicity of a microsporidian (*Nosema trichoplusiae* n. sp.) of the cabbage looper, *Trichoplusia ni* (Hübner) (Lepidoptera: Noctuidae). *J. Invertebr. Pathol.* **9:**188–195.

Tanada, Y. 1959. Descriptions and characteristics of a nuclear polyhedrosis virus and a granulosis virus of the armyworm, *Pseudaletia unipuncta* (Haworth) (Lepidoptera: Noctuidae). *J. Invertebr. Pathol.* **1**:197–214.

Tanada, Y. 1963. Epizootiology of infectious diseases. Pages 423–475 *in* E. A. Steinhaus, ed., Insect pathology, an advanced treatise. Vol. II. Academic Press, New York.

Tanada, Y. 1964. Epizootiology of insect disease. Pages 548–578 *in* P. DeBach, ed., Biological control of insect pests and weeds. Reinhold, New York.

Tanada, Y. 1968. The role of viruses in the regulation of the population of the armyworm, *Pseudaletia unipuncta* (Haworth). *Proc. Joint U.S.–Jap. Semin. Microb. Control Insect Pests. U.S.–Japan Comm. Sci. Coop. Panel* **8**. *Fukuoka,* pp. 25–31.

Upholt, W. M., R. A. Engler, and L. E. Terbush. 1973. Regulation of microbial pesticides. *Ann. N.Y. Acad. Sci.* **217**:234–237.

Vago, C. 1966. Perspectives sur la classification des virus d'invértebrés. *Entomophaga* **11**:347–353.

Vago, C., K. Aizawa, C. M. Ignoffo, M. E. Martignoni, L. Taresevitch, and T. W. Tinsley. 1974. Present status of the nomenclature and classification of invertebrate viruses. *J. Invertebr. Pathol.* **23**:133–134.

Veber, J. 1964. Virulence of an insect virus increased by repeated passages. *Entomophaga Mém. Sér.* **2**:403–405.

Weiser, J. 1961. Die mikrosporidien als parasiten der insekten *Monogr. Angew. Entomol. 17*. 149 pp.

Weiser, J., and J. D. Briggs. 1971. Identification of pathogens. Pages 13–66 *in* H. D. Burges and N. W. Hussey, eds., Microbial control of insects and mites. Academic Press, New York.

Welch, H. E. 1963. Nematode infections. Pages 363–392 *in* E. A. Steinhaus, ed., Insect pathology, an advanced treatise. Vol. II. Academic Press, New York.

White, R. T., and S. R. Dutky. 1942. Cooperative distribution of organisms causing milky disease of Japanese beetle grubs *J. Econ. Entomol.* **35**:679–682.

Wildy, P. 1971. Classification and nomenclature of viruses. Monographs in virology. Karger, Basel. Vol. 5, 81 pp.

Yendol, W. G., and R. A. Hamlen. 1973. Ecology of entomogenous viruses and fungi *Ann. N.Y. Acad. Sci.* **217**:18–30.

SELECTED READINGS

Bulla, L. A., ed. 1973. Regulation of insect populations by microorganisms. *Ann. N.Y. Acad. Sci.* **217**.

Burges, H. D., and N. W. Hussey, eds. 1971. Microbial control of insects and mites. Academic Press, New York. 861 pp.

DeBach, P., ed. 1964. The biological control of insect pests and weeds. Reinhold, New York. 844 pp.

Steinhaus, E. A., ed. 1963. Insect pathology, an advanced treatise. Academic Press, New York. Vol. I, 661 pp., Vol. II, 689 pp.

Weiser, J. 1969.. An atlas of insect diseases. Irish University Press, Shannon. 292 pp.

7

INSECTICIDES IN PEST MANAGEMENT

Robert L. Metcalf

Insecticides are the most powerful tool available for use in pest management. They are highly effective, rapid in curative action, adaptable to most situations, flexible in meeting changing agronomic and ecological conditions, and economical. Insecticides are the only tool for pest management that is reliable for emergency action when insect pest populations approach or exceed the economic threshold. In the words of the Committee on Plant and Animal Pests, National Academy of Sciences, ". . . a major technique such as the use of pesticides can be the very heart and core of integrated systems. Chemical pesticides will continue to be one of the most dependable weapons of the entomologist for the foreseeable future. . . . There are many pest problems for which the use of chemicals provides the only acceptable solution. Contrary to the thinking of some people, the use of pesticides for pest control is not an ecological sin. When their use is approached from sound ecological principles, chemical pesticides provide dependable and valuable tools for the biologist. Their use is indispensable to modern society" (NAS 1969).

Despite these impressive credentials, much use of insecticides, as related in Chapter 1, has been ecologically unsound, leading to such disadvantages as insect pest resistance, outbreaks of secondary pests, adverse effects on nontarget organisms, objectionable pesticide residues, and direct hazards to the user (Smith 1970). "The prime difficulty . . . lies in man's tendency to substitute pesticides for effective bioenvironmental controls. Pesticides should be employed primarily as 'stop-gap' or 'fire-fighting' tools and sound bioenvironmental controls relied upon as the primary control method" (PSAC 1965). The misuse, overuse, and unnecessary use of insecticides (von Rumker and Horay 1972) have been the most important factors in the growth of interest in pest management, and indeed this concept seeks to maximize the advantages in their use and minimize the disadvantages. It should always be remembered that the application of insecticides represents purposeful environmental contamination and can be justified only where benefit/risk ratios are clearly tilted in favor of insecticide use.

The past 30 years has seen an amazing growth of insecticide use in the United States, from less than 30 to more than 200 chemical compounds (Kenaga and Allison 1969), and from less than 100 million to more than 560 million pounds in annual production (*Pesticide Review* 1972). About 50% of the total use is in agriculture and, of this, three-quarters is applied to five crops: cotton 47%, corn 17%, apples 6%, tobacco 3%, and soybeans 2% (Pimentel 1973). Of the remaining use, 25% is industrial, 15% home and garden, and 10% governmental. Thus one-half of all insecticide use in United States agriculture is on two nonfood crops, cotton and tobacco. It is estimated (ACS 1969) that 40% of all insecticide use in the United States is directed at the control of the cotton boll weevil, the cotton bollworm, and the codling moth. About 12% of United States cropland (excluding pastures) is treated annually with insecticides, and the rate is much higher in major agricultural regions, for example, 30% of the total crop acreage of Illinois is treated annually with insecticides (Illinois State Department of Agriculture 1972).

I. ADVANTAGES OF INSECTICIDES FOR PEST MANAGEMENT

The very widespread use of insecticides for pest control is largely a result of their convenience, simplicity, effectiveness, flexibility, and economy. These properties represent virtues for use in pest-management programs.

A. Insecticides Afford the Only Practical Control Measure for Insect Pest Populations Approaching or at the Economic Threshold

Insecticides are equally useful in a typical chemical control program or in a pest-management program when other carefully planned control measures have failed and emergency intervention is necessary.

B. Insecticides Have Rapid Curative Action in Preventing Economic Damage

Lethal action is rapid and high mortality of the pest population is usually obtained within a few hours to a day or two. Thus it is possible to spray lettuce, broccoli, asparagus, cabbage, and so on, within a day or two of harvest with a rapidly degradable insecticide such as tetraethyl pyrophosphate (TEPP), mevinphos, or nicotine and to obtain insect-free marketable produce of high quality, free from residue hazard. Swarms of migratory locusts, *Schistocerca gregaria* or *Locusta migratoria,* have been sprayed by aircraft with dinitro-*o*-cresol (DNOC) and destroyed while in flight, thus protecting entire agricultural regions. Ultralow-volume applications of malathion to millions of acres in the United States have been used for the total destruction of adult *Aedes nigromaculis* infected with Venezuelan equine encephalomyelitis to avert destructive epidemics in horses. Similar applications have been made to entire cities to control epidemics of yellow fever transmitted by *Aedes aegypti* (L.) mosquitoes. Perhaps the most compelling example is the use of the familiar "bug bomb" with a pyrethrins aerosol to "shoot down" mosquitoes or bedbugs disturbing a night's slumber.

C. Insecticides Offer a Wide Range of Properties, Uses, and Methods of Application to Pest Situations

The 200-odd chemicals registered for insecticidal use in the United States have a very wide range of properties from fumigant gases, for example, hydrogen cyanide and methyl bromide; through short-lived contact action, for example, nicotine, tetraethyl pyrophosphate, and dichlorvos; to long-term residual persistence, for example, methoxychlor and endosulfan. These insecticides are applied as fumigants, smokes, aerosols, sprays, dusts, granulars to soil, residual fumigants, baits, and seed treatments; impregnated into cloth, timbers, and paper; and administered as systemics to plants and animals. Insecticidal compounds such as carbaryl, diazinon, endosulfan, malathion, methoxychlor, methyl parathion, and toxaphene are registered for the protection of 50 to 100 or more raw agricultural commodities. Combinations of several insecticides can be used together to achieve a desired range of properties, for example, the use of methoxychlor, pyrethrins, and piperonyl butoxide in a household aerosol to control flies, mosquitoes, cockroaches, ants, and bedbugs; or an orchard tank mix spray of carbaryl, azinphosmethyl, and dicofol to control apple pests.

D. The Use of Insecticides Is Low in Cost and Often Results in Substantial Financial Returns

Insecticidal chemicals are produced in relatively large volumes and are low in cost compared to many other synthetic chemicals. Present costs per pound for insecticide, which is usually adequate to treat an acre or more of agricultural land or crop, are (*Pesticide Review* 1972):

Aldrin	$1.05	Lindane	$1.37
Carbaryl	$0.80	Malathion	$0.79
Chlordane	$0.59	Methoxychlor	$0.66
DDT	$0.22	Methyl parathion	$0.46
Dichlorvos	$3.75	Toxaphene	$0.24
Heptachlor	$1.03		

Insecticide application usually requires a minimum of labor, as in seed dressings, granular applications at time of planting, aircraft sprays, and incorporation of baits and attractants drawing the insect to the insecticide. Estimated costs per acre for typical agricultural operations range from $1 to $3.

Costs of insecticide applications in public health programs are very low. The annual cost of malaria control by DDT residual house spraying against *Anopheles* mosquitoes at 1 g/m^2 has been estimated at 10¢ to 12¢ per capita (Soper et al. 1961). Control of the human louse vector of typhus, *Pediculus humanus humanus* L., by DDT, lindane, or malathion powder dusted into clothing is about 25¢ per capita. Ultralow-volume aerial spraying with malathion at 3 to 6 oz per acre gives 90 to 95% control of adult mosquitoes at about 90¢ per acre.

Benefit/Cost Ratios

Benefit/cost ratios for the use of insecticides are generally considered to return $4 to $5 to the user for every $1 invested (PSAC 1965). Pimentel (1973) estimated that $2.1 billion (1960) in additional losses would occur in United States agriculture without pesticide use, and calculated a return of $2.82 for each $1 invested. Benefit/cost ratios for specific individual pest control operations seem even higher. The use of aldrin in Illinois at 1 lb per acre to control susceptible corn rootworms, *Diabrotica* spp., returned 8.5 bu of corn per acre for a ratio of $4 to $1 (Petty 1970); the treatment of sugar beets in California with granular phorate at 1 lb per acre to control the green peach aphid, *Myzus persicae* (Sulzer), the southern garden leafhopper, *Empoasca solana* De Long, the carmine spider mite, *Tetranychus telarius* L., and the virus diseases they transmit increased yields of sugar beets by 3 to 4 tons per acre and sugar to 1500 lb per acre, for a ratio of about $18 to $1 (Reynolds et al. 1967); and the use of DDT in Wisconsin at 2 lb per acre to control the Colorado potato beetle, *Leptinotarsa decemlineata* (Say), and the potato leafhopper, *Empoasca fabae* (Harris), increased yields 68%, or 65 bu per acre, for a ratio of about $29 to $1 (Metcalf 1968).

World Health Organization data indicate that in East Pakistan, from 1963 to 1966, every dollar invested in malaria control saved an average of $1.84 in labor and improved health standards (WHO 1968). In addition, of course, no monetary value can be applied for deaths prevented.

II. LIMITATIONS IN THE USE OF INSECTICIDES FOR PEST MANAGEMENT

Despite the manifold advantages of insecticides in pest control, their use often results in major disadvantages, especially to pest-management systems. Consideration of these will provide guidance for the optimum use of insecticides in pest-management programs.

A. Insect Resistance to Insecticides

The genetically acquired resistance of insects to insecticide toxicity continues to be the most serious barrier to the successful use of insecticides. The importance of this phenomenon first became evident with the worldwide development of DDT resistance by the housefly, *Musca domestica* L., as DDT-resistant strains appeared almost simultaneously in 1946–1947 in Sweden, Denmark, Italy, and the United States within 2 years after the first employment of DDT residual sprays. Recent surveys indicate that there are more than 119 species of insects of agricultural importance resistant to insecticides (Brown 1970).

Insect resistance to insecticides is believed to develop largely if not entirely as a result of selection of preadaptive mutants possessing genetically controlled mechanisms for detoxication, target site insensitivity, or other means of survival in the presence of the insecticide, for example, the enzyme DDT-ase of the resistant housefly, which converts DDT to the inactive DDE. These genetic factors may be present in very low frequency in the population before insecticidal treatment; for example, *Anopheles gambiae* in northern Nigeria was found to contain heterozygotes for the single gene of dieldrin resistance in 0.4 to 6% of the population. Intensive selection by dieldrin or BHC residual house spraying increased the gene frequency for resistance to 90% after 1 to 3 years (Brown and Pal 1971).

Insecticide resistance has had nearly disastrous effects on many control programs. The control of floodwater *Aedes* mosquitoes in California has successively required and exhausted the effectiveness of all available organochlorine and organophosphorus insecticides. Similarly, the European red mite, *Panonychus ulmi* (Koch), has successfully become resistant to almost the entire gamut of available acaricides used in fruit tree protection.

The development of resistance in the *Anopheles* vectors of malaria has seriously disrupted the global control and eradication program of the World Health Organization. The 36 species of *Anopheles* that have developed dieldrin resistance and the 14 species with DDT resistance provide an enormous challenge to the program, which is already forced to use malathion and propoxur as substitute insecticides in Central America, the Middle East, and in India and Pakistan.

The regular and frequent use of insecticides has almost inevitably produced target pest resistance. The only reasonable hope for delaying or avoiding pest resistance lies in pest-management practices that decrease the frequency and intensity of selection by reduced use of insecticides, and encourage control by natural enemies and cultural practices which may destroy most of the insecticide-selected resistant mutants before they can breed resistant progeny.

Case History

The resistant strain of the western corn rootworm, *Diabrotica virgifera* L., a major pest of corn in the Midwest, spread rapidly from a single locus in southeastern Nebraska in 1961 until it had encompassed most of the corn-growing area in eight midwestern states by 1973 (Fig. 7.1). The resistant strain of the western corn rootworm is almost completely nonsusceptible to the annual treatments of aldrin and heptachlor that were customary for several years (Kuhlman and Petty 1973), and the use of these materials in Illinois declined from 6,648,000 acres treated in 1966 to 1,813,300 acres in 1973. Correspondingly, the acreage treated with organophosphorus and carbamate insecticides, to which the rootworm is still susceptible, has increased from 174,800 acres in 1966 to 3,496,900 acres in 1973 (Kuhlman et al. 1974).

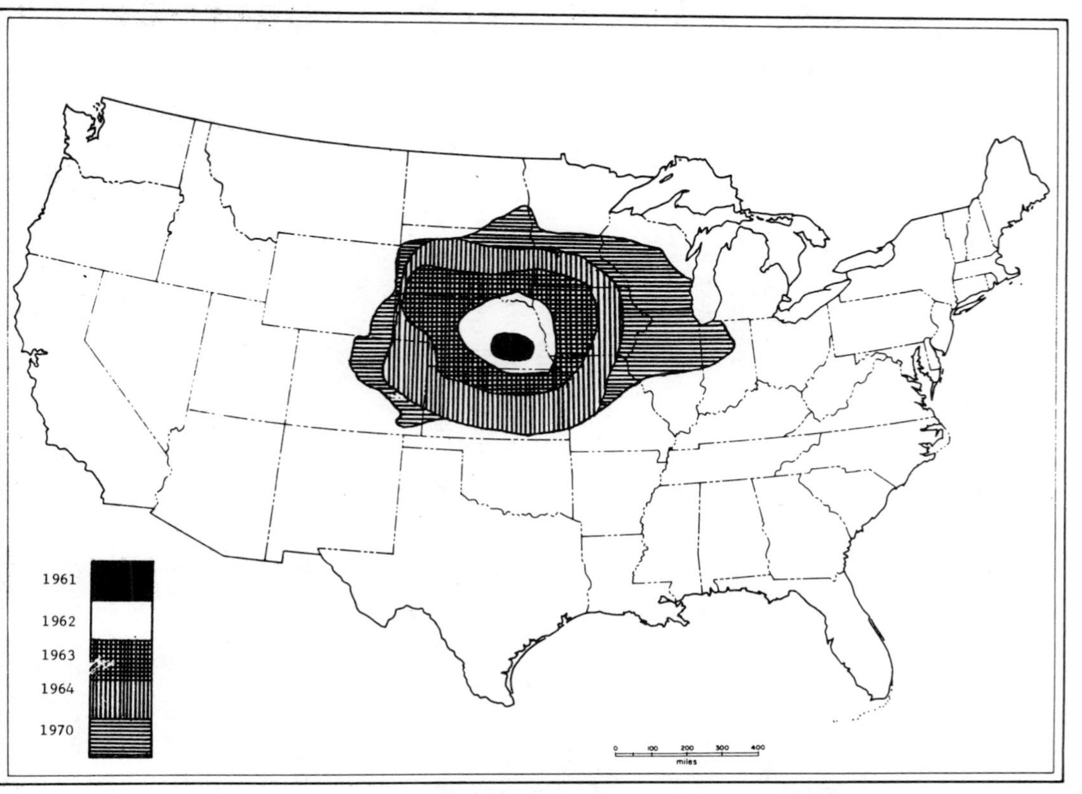

Fig. 7-1 Spread of cyclodiene-resistant western corn rootworm, *Diabrotica virgifera*, in the corn belt. (Data from Brown 1969 and Chiang 1973.)

B. Outbreaks of Secondary Pests

The use of broad-spectrum, persistent insecticides has had important consequences on the nature of pest complexes attacking many crops. In 1947, the employment of DDT sprays for citrus pest control was followed by a dramatic resurgence of the cottony cushion scale, *Icerya purchasi* Maskell, which had been under excellent biological control since the introduction of the predatory beetle *Rodolia cardinalis* (Mulsant) into California in 1888. The outbreak was clearly the result of the highly toxic effect of DDT on the predator *Rodolia*. The European red mite, *Panonychus ulmi* (Koch), became a major pest in most deciduous orchards because it was totally resistant to DDT applied for codling moth control, although its predators were severely affected. In the Imperial Valley of California an obscure insect, the cotton leaf perforator, *Bucculatrix thurberiella* Busck, became a major cotton pest following the widespread use of DDT. Brought under reasonable control by substitution of the stomach poison trichlorfon for DDT, the cotton leaf perforator again became an epidemic pest following heavy applications of carbaryl to eradicate the pink bollworm, *Pectinophora gossypiella* (Saunders), which virtually eliminated the leaf perforator's natural enemies (Smith 1970). The red-banded leaf roller, *Argyrotaenia velutinana* (Walker), was a casual or rare pest attacking apples in the eastern United States and Canada until complex spray programs using DDT, parathion, TDE, and malathion disrupted its complex of natural enemies and the leaf roller became a major apple pest. Many other examples of the role of insecticides in altering pest complexes could be cited (NAS 1969).

Case History

The dramatic effects of overtreatment with broad-spectrum insecticides resulting in insecticide resistance and the development of secondary pests have been demonstrated in the cotton-growing area of the Lower Rio Grande Valley in northeastern Mexico (Adkisson 1971). The original primary pest, the cotton boll weevil, *Anthonomus grandis* Boheman, was controlled by intensive spraying and dusting with dieldrin, aldrin, endrin, and BHC. Resistance to these insecticides developed in the late 1950s, and methyl parathion was substituted. DDT was added to the treatment schedule to control the secondary pests, the cotton bollworm, *Heliothis zea* (Boddie), and the tobacco budworm, *Heliothis virescens* (Fabricius). By 1962, these two insects had developed resistance and could not be controlled by DDT, other organochlorines, or carbamates, and had become the major cotton pests, while the boll weevil had faded into insignificance. Temporary control was obtained by increasing the dosages of methyl parathion two to four times and increasing the number of applications. This resulted in acceptable control of the bollworms, but caused a great increase in the number of cases of human poisoning by insecticides. Drift of the heavy insecticide applications into citrus orchards caused major upsets in the biological control of citrus pests and resulted in heavier pesticide applications to citrus trees. By 1968, the budworms were becoming resistant to methyl parathion and producers were treating 15 to 18 times and still suffering heavy losses in yields. By 1969, the budworm was almost totally resistant to all available insecticides, and cotton production was almost totally destroyed. By 1970, cotton production was abandoned in this area where it was no longer profitable because of increased treatment costs and severe crop losses. Adkisson (1971) records the following catastrophic decline in a once flourishing industry.

	Cotton	
Year	Acres	Bales
1960	710,715	370,382
1962	499,790	362,197
1964	191,780	54,066
1965	102,605	50,207
1966	43,515	18,183
1967	24,178	15,957
1970	1,200	

C. Adverse Effects on Nontarget Species

The insecticides most generally applied, for example, carbaryl, toxaphene, methyl parathion, malathion, chlordane, and aldrin, have broad-spectrum activity and are toxic to a great variety of animals. Moreover, in conventional applications less than 1% of the total dosage contacts the pest. It is important to remember that such insecticide use in the United States is aimed at the control of about 1000 insect pest species, but may affect adversely many of the remaining 200,000 species of plants and animals that may be essential for human survival (Pimentel 1971).

1. *Natural Enemies*

The repeated applications of insecticides to cotton, deciduous fruits, alfalfa, tobacco, and other crops obviously has a destructive effect on beneficial insect populations. As a result, fields and orchards under heavy insecticide treatment schedules may become veritable "biological deserts" (van den Bosch and Stern 1962). This wholesale destruction of natural parasitoids and predators by fixed treatment schedules is contrary to the principles of pest management and has two major consequences: (1) the rapid resurgence of the target pest species uncurbed by any natural enemies, and (2) the selection of secondary pests formerly under control by natural enemies. Both of these conditions demand still further insecticide treatments and eventually give rise to a situation such as the one on cotton described above (Chapter 1, Section II.B), in which treatments spiral, and resistant secondary pests become almost uncontrollable. The reestablishment of adequate complexes of natural enemies in such biological deserts may require 2 to 4 years (Smith 1970).

2. *Honeybees and Other Pollinators*

The honeybee, *Apis mellifera* L., is an important domestic animal, producing annually in the United States honey and beeswax valued at about $50 million. More importantly, the honeybee is estimated to pollinate 80% of deciduous fruits, vegetables, legumes, and oil seed crops. The annual value of these bee-pollinated crops is more than $1 billion (USDA 1967). Poisoning of honeybees by agricultural applications of insecticides is a major problem of beekeepers, especially where applications are made to crops during the bloom period. The honeybee is genetically a very specialized insect, and because of its sheltered larval life has not developed the wide variety of detoxication enzymes found in many insects. As a result, the honeybee is extremely susceptible to many pesticides in common use. Applications of carbaryl in southern California to eradiate the pink bollworm are stated to have obliterated more than 30,000 colonies of honeybees used for alfalfa pollination. Wild bees such as bumblebees, *Bombus* spp., the alfalfa leaf-cutting bee, *Megachile rotundata* (Fabricius), and the alkali bee, *Nomia melanderi* Cockerell, are also very important in pollination and are highly susceptible to broad-spectrum insecticides.

In pest-management programs every effort should be made to avoid spraying crops when they are in bloom, and wherever possible when treating deciduous fruits, vegetables, and forage crops to use insecticides with low intrinsic toxicity to bees (Table 7.1).

3. *Effects on Wildlife*

Many insecticides are highly toxic to vertebrates as well as invertebrates, and their use has caused adverse effects in some wildlife populations (Cope 1971; Newsom 1967). Fish are especially susceptible to the organochlorine insecticides such as endrin, dieldrin, aldrin, heptachlor, toxaphene, chlordane, DDT, and methoxychlor, which are toxic in the parts-per-billion range (Table 7.1).

Fish are particularly susceptible to continuous exposure to trace contamination by persistent organochlorine insecticides and may concentrate and store these materials in thousand- to millionfold amounts. DDT aerial spraying of the Lake George region of New York for control of the gypsy moth, *Porthetria dispar* (L.), was responsible for the total loss of lake trout fry in 1955–1958. The affected fry were killed immediately after absorption of the yolk sac which contained 4.75 ppm DDT, enough to poison the newly emerged fish (Burdick et al. 1964). DDT spraying of the Yellowstone River system for control of the spruce budworm, *Choristoneura fumiferana* (Clemens), greatly reduced the populations of important fish food organisms, stone flies (Plecoptera), mayflies (Ephemeroptera), caddis flies (Trichoptera), and midge larvae (Chironomidae). These effects persisted for as long as a year and were responsible for widespread mortality of brown trout, mountain whitefish, and longnose sucker (Cope 1961).

The adverse effects of insecticides on nontarget species can be minimized by (1) proper selection of insecticides, (2) avoiding overtreatment, (3) proper timing for applications using light and pheromone traps, and (4) making applications as selective as possible, using seed and granular treatments, systemic insecticides, bait sprays, and so on. These are discussed in Section III.3.

D. Hazards of Pesticide Residues

Most of the present usage of insecticides is based on the deposition of persistent residues which protect against insect infestation or attack for periods of weeks to months after application. This is the strategy for spraying apples against the codling moth, *Laspeyresia pomonella* (L.), and dwellings against the *Anopheles* spp. vectors of malaria, and treating soil against the corn rootworms, *Diabrotica* spp., and woolens against fabric pests. Thus insecticide persistence per se is a valuable attribute. However, in most examples of insecticide usage, it is impossible to utilize persistence against the pest without contaminating the environment and

Table 7.1 Pest Management Rating of Widely Used Insecticides[a]

Insecticide	Mammalian Toxicity	Nontarget Toxicity				Environmental Persistence	Overall Rating
		Fish	Pheasant	Bee	Average		
Aldicarb	5	3	5	5	4.3	3	12.3[b]
Aldrin	4	4	4	4	4.0	5	13.0
Azinphos-methyl	4	3	2	4	3.0	3	10.0
Carbaryl	2	1	1	4	2.0	2	6.0
Carbofuran	5	2	5	5	4.0	3	12.0[b]
Carbophenothion	4	2	4	4	3.3	2	9.3
Chlordane	2	3	2	2	2.3	3	7.3
DDT	3	4	2	2	2.7	5	10.7
Demeton	5	2	5	2	3.0	2	10.0
Diazinon	3	2	5	4	3.7	3	9.7
Dicofol	2	1	2	1	1.3	4	7.3
Dieldrin	4	4	3	4	3.7	5	12.7
Dimethoate	3	1	4	5	3.3	2	8.3
Disulfoton	5	3	5	2	3.3	3	11.3
Chlorpyrifos	3	3	3	5	3.7	3	9.7[b]
Endosulfan	4	4	2	2	2.7	3	9.7
Endrin	5	5	5	2	4.0	5	14.0
Ethion	3	2	3			2	7.0[b]
EPN	4	2	3	4	3.0	4	11.0
Gardona	1	4	1	4	3.0	1	5.0[b]

(Continued)

producing unwanted effects against nontarget organisms. Applications of pesticides often drift long distances from the target site, and indeed it appears that less than half the total amount of insecticides applied as sprays to crops is deposited on plant surfaces for effective insect control and less than 1% is applied to the pests themselves (PSAC 1965). Thus insecticide residues are nearly everywhere, and highly persistent nondegradable compounds such as DDT, DDE, dieldrin, heptachlor epoxide, lindane, endrin, toxaphene, dicofol, and so on, are detectable in soil, air, water, and in human adipose tissues as well as those of wildlife (Table 7.2). Although authorities differ about the long-term significance of these persistent residues to human health, their presence is clearly undesirable.

Table 7.1 (*continued*)

Insecticide	Mammalian Toxicity	Nontarget Toxicity				Environmental Persistence	Overall Rating
		Fish	Pheasant	Bee	Average		
Heptachlor	4	3	4	4	3.7	5	12.7
Lindane	3	3	2	4	3.0	4	10.0
Malathion	2	2	1	4	2.3	1	5.3
Methoxychlor	1	3	1	1	2.3	2	5.3
Methyl parathion	4	1	5	5	3.7	1	9.7
Mevinphos	5	3	5	4	4.0	1	10.0
Naled	2	2	3	4	3.0	1	6.0
Oxydemeton-methyl	3	2	4	2	2.7	2	7.7[b]
Parathion	5	2	4	4	4.0	2	11.0
Phorate	5	4	5	2	3.7	3	11.7
Phosphamidon	4	1	5	3	3.0	2	9.0
Tetraethyl pyrophosphate	5	2	5	3	4.0	1	10.0
Toxaphene	3	4	4	1	3.0	4	10.0
Trichlorfon	2	1	2	1	1.3	1	4.3
Zectran	4	1	5	5	3.7	2	9.7

[a] For explanation of ratings, see Section IV.B.

[b] Insufficient data for accurate rating.

From R. L. Metcalf, "Selective Use of Insecticides in Pest Management," page 200 in *Proceedings of the Summer Institute on Biological Control of Plant Insects and Diseases,* edited by Fowden G. Maxwell and F. A. Harris, Jackson: University Press of Mississippi, 1974.

Moreover, through biomagnification they may be concentrated 10^3- to 10^7-fold in the bodies of fish and birds, particularly at the upper ends of food chains.

About 50% of all pesticide production is used for the protection of agricultural commodities. Much of the initial pesticide load is lost by "weathering," through the action of rain and dew and by photochemical oxidation; by enzymic destruction in tissues of plants or animals; and through losses in harvesting and food processing. The processes of degradation and persistence of pesticide residues usually follow first-order chemical kinetics, and residue persistence curves which are independent of concentration can be determined for each insecticide on each crop

Table 7.2 Average Insecticide Residues in the United States Environment

	Surface Waters (ppb)[a]	Cropland soil (ppm)[b]	Dietary intake (mg per day)[c]	Human Fat (ppm)[d]
BHC	0.003–0.022	—	0.007	0.20–0.60
Chlordane	—	0.04	—	—
DDT-T[e]	0.008–0.144	0.31	0.077	10.3–11.1
Dieldrin	0.008–0.122	0.03	0.006	0.15–0.31
Endrin	0.008–0.214	0.01	Trace	0.03
Heptachlor epoxide	0.001–0.008	0.01	0.003	0.10–0.24
Toxaphene	—	0.07	—	—
Dicofol	—	0.01	0.008	—

[a] Breidenback et al. 1967.

[b] Wiersma et al. 1972.

[c] Duggan 1969.

[d] Durham 1969.

[e] DDT + DDE + DDD (TDE).

substrate. From such data harvest-time residues can be estimated and a time interval established between application and harvest. Based on animal feeding studies and incorporating a safety factor, ideally 100-fold, to allow for human sensitivities, together with potential human daily dietary intake, the U.S. Food and Drug Administration (FDA) has established pesticide residue tolerances for each registered pesticide on each food commodity.

Typical FDA tolerances for the most widely used insecticides are:

Carbaryl	5–100 ppm	Aldrin	0.1–0.25 ppm
Toxaphene	3–7 ppm	Methoxychlor	1–100 ppm
DDT[1]	1–7 ppm	Diazinon	0.75–40 ppm
Methyl parathion	1 ppm	Phorate	0.1–1.0 ppm
Malathion	0.3 ppm	Azinphosmethyl	0.3–5 ppm

It should be evident that a major aim of pest management must be to decrease the total amounts of pesticide residues and especially to substitute readily degradable or "soft" pesticides, such as organophosphorus or carbamate insecticides, for persistent, nondegradable, or "hard" organochlorine insecticides.

[1] DDT tolerances were canceled by the U.S. Environmental Protection Agency (EPA) on January 1, 1973.

Case History

Lake Michigan is the fourth largest freshwater lake in the world, with an area of 58,016 km^2 and a volume of 4871 km^3. It has an average water retention time of 30.8 years (Rainey 1967). Lake Michigan is particularly vulnerable to pesticide pollution, since its watershed drains heavily treated agricultural regions and its south shore is surrounded by one of the largest metropolitan areas of the United States. The open waters of Lake Michigan have an average contamination of about 6 ppt DDT (plus DDE and DDD) and 2 ppt dieldrin. However, important game and commercial fish in the lake have concentrated these residues enormously, so that their total body burden in parts-per-million averages: chubs–DDT 10.19, dieldrin 0.19; coho salmon–DDT 14.09, dieldrin 0.12; and lake trout–DDT 18.80, dieldrin 0.27 (EPA 1972). The residues in individual fish approach or exceed 15 ppm DDT and 0.5 ppm dieldrin for bioconcentration values of as much as 2.5×10^6 for DDT and 2.5×10^5 for dieldrin.

The levels of DDT in sport and commercially valuable species of fish exceed the 5-ppm actionable level established by FDA, and the values for dieldrin approach or exceed the 0.3-ppm actionable level. Thus the commercial sale of Lake Michigan fish is essentially prohibited. As Rainey (1967) points out, the slow water turnover of Lake Michigan makes such contamination by slowly degradable pesticides a natural disaster for which there is no apparent solution.

E. Direct Hazards from Insecticide Use

Many insecticides are highly toxic to a wide range of animals including man himself. It is not widely recognized that insecticides such as aldicarb, azinphos, carbofuran, fensulfothion, demeton, disulfoton, endrin, mevinphos, parathion and methyl parathion, phorate, and tetraethyl pyrophosphate, all with rat oral LD_{50} values of <15 mg/kg (Table 7.1) are among the most toxic chemicals known. Man is almost universally more susceptible than the rat, often by a factor of 10-fold or more. The high toxicity ot these insecticides leaves no margin for careless application or improper storage, and spray operators and pilots, field workers and fruit pickers, and especially children can become victims of severe or fatal poisoning. It is estimated that there are about 200 deaths annually in the United States from insecticide poisoning (Blackbourne 1970). Hayes (1969) estimated that an average of 100 nonfatal poisonings occur for each fatality, and enumerates 58 insecticides known to have caused poisoning in man. The safety record in the United States is relatively good, considering the large volumes of toxic materials handled. In developing countries the toll is far higher.

Aerial spray applications of highly toxic insecticides have caused poisoning of persons in the area contaminated by spray drift, and workers picking fruit in orchards and vineyards have suffered a substantial number of incidents of parathion poisoning from toxic residues on foliage and fruit (Jegier 1969). Epidemics of poisoning by insecticides, especially through accidental contamination of flour with parathion and endrin, have poisoned hundreds of persons and caused many deaths in India, Malaya, Arabia, Egypt, Columbia, and Mexico (Mrak 1969).

Human safety must be of paramount importance in the use of insecticides in all pest-management programs. Insecticides highly toxic to humans, such as those listed above (see Table 7.1), should not be chosen for the vast majority of pest-management operations. If such highly toxic materials must be used, they should be applied only under very restricted situations by trained personnel. Much attention should be given to proper storage and disposal of insecticide containers to keep them from children and the illiterate. Workers exposed to all phases of the application of toxic insecticides should wear protective gloves, masks, and clothing, and should observe strict hygienic precautions. Field workers should scrupulously observe "safe reentry periods," as specified under the Occupational Health and Safety Act, U. S. Department of Labor. In all instances in which there is any doubt about the safety of a particular insecticide application, it should be eliminated from pest-management programs and an alternative measure developed.

III. PROPER USE OF INSECTICIDES IN PEST MANAGEMENT

It is of the utmost importance to the pest-management philosophy to determine how insecticides can be used most effectively and harmoniously in pest-management programs. Two major principles will promote this objective (PSAC 1965):

1. Substitute treat-when-necessary insecticide applications for presently employed routine-treatment insecticide schedules.
2. Recognize that 100% control of most insect pests is not required to prevent economic loss.

Adherence to these principles could reduce insecticide use on United States crops by as much as 50% (PSAC 1965; Pimentel 1973). Further guidance can be gained from Geier's (1966) definition of the practice of pest management:

1. Determine how the life system of the pest needs to be modified to reduce its numbers to tolerable levels, that is, below the *economic threshold.*
2. Apply biological knowledge and current technology to achieve the desirable modification, that is, *applied ecology.*
3. Devise procedures for insect pest control both suited to current technology and compatible with economic and environmental quality aspects, that is, *economic and social acceptance.*

As the United States moves into an era of pest management, wholesale changes must be made in our strategies for the employment of chemicals. In the pest-management era the use of insecticides can be categorized in three ways: (1) carefully timed suppressive applications aimed at a weak point in the insect's life cycle, (2) emergency applications reserved for epidemic situations in which all other control measures are inadequate and the insect population approaches or exceeds the economic threshold, and (3) preventive treatments of highly selective impact made with the least dosage of insecticide and calculated to provide the least disturbance of the environment. To implement pest-management practices on a broad scale demands more sophisticated entomological knowledge and techniques, including a detailed knowledge of the properties of the insecticides themselves in relation to effects on target and nontarget organisms, on human and public health, and on the total quality of the environment.

These premises then suggest that in future we shall use substantially smaller quantities of insecticides for specific programs of pest control, that the chemicals chosen may be quite different from those heavily relied on today, and that our standards of efficacy of control, efficiency in crop production, and economics of pesticide use are likely to be significantly different from those of today.

A. Selective Use of Insecticides

Selectivity of pest-control intervention that endeavors to curb the numbers of pest species with minimal effects on all other components of the environment is the keynote of pest management. Man has produced few if any less ecologically selective operations than the aircraft spraying or dusting of large acreages with a broad-spectrum insecticide such as parathion or methyl parathion. Yet this has become a common operation in cotton insect pest control, although the ultimate futility of exclusive reliance on this strategy has been described previously (Section II.B). The insecticides in greatest use in the United States today (yearly production >10 million lb each annually, Lawless et al. 1972)—carbaryl, 55; toxaphene, 50; DDT, 45; methyl parathion, 45; parathion, 15; malathion, 30; chlordane, 25; and aldrin, methoxychlor, and diazinon, 10 million pounds—are generally effective against a wide range of insect life, both destructive and beneficial, and are toxic to many other organisms in the environment. DDT and the cyclodienes are of low biodegradability and subject to biomagnification in food chain organisms. The parathions and diazinon, although biodegradable, are of general toxicity to all forms of life. Only carbaryl, methoxychlor, and malathion combine suitable biodegradability with adequate safety to higher animals. For many operations in pest management, however, even carbaryl and malathion are not of sufficiently narrow spectrum for invertebrates. However, under the stress of increasing insect resistance and environmental quality problems, and aided by the Federal Environmental Pesticide Control Act of 1972, which requires the proper application of pesticides to ensure greater protection of man and the environment, changes will be made more rapidly, both in the development of physiologically selective insecticides and in the ecological and behavioral selectivity of their use.

1. Physiological Selectivity

Insecticides that are intrinsically selective for the phylum Arthropoda or the class Insecta alone or—even more desirable—for a few related insect species are few in number, but interest in their development is increasing. Such compounds have as their target site some specific developmental pattern characteristic and are specific to arthropods alone, or are biological toxins whose evolutionary attack has become centered on the Insecta. Juvenile hormone mimics or juvigens such as isopropyl 11-methoxy-3,7,11-trimethyldodeca-2-4-dienoate (methoprene) (Diekman 1972) and 1-(4′-ethylphenoxy)-6,7-epoxy-3,7-dimethyl-2-octene (R-20458) (Pallos and Menn 1972) mimic the action of the insect juvenile hormone in preventing metamorphosis during immature life and consequent development to the adult stage (Bowers 1971; Robbins 1972). They are most effectively applied to late-instar larvae immediately before pupation.

The complex processes of cuticle formation are virtually unique to the Insecta and related arthropods, and several types of new insecticides are being developed which interfere with specific biochemical reactions there. Antioxidant chemicals such as 2,6-di-*tert*-butyl-4-(α,α-dimethylbenzyl)-phenol (MON-0585) are effective against mosquito larvae in which they prevent metamorphosis and adult emergence (Sacher 1971). The compound 1-(4-chlorophenyl)-3-(2,6-difluorobenzoyl)-urea (TH6040) is a highly active representative of a large group of benzyl-phenyl ureas that are highly effective stomach poisons in insect larvae. These compounds act during molting to prevent deposition of new cuticle, probably by preventing the acetylation of glucose to form glucoseamine which is a primary constituent of chitin (Van Daalen et al. 1972; Wellinga et al. 1973).

The insect bacterial toxin of *Bacillus thuringiensis* Berliner (BT) has an almost specific effect against a small group of lepidopterous larvae, cabbage loopers, gypsy moths, cankerworms, tent caterpillars, and so on, and is being widely used as a specific insecticide. The nuclear polyhedrosis virus (NPV) of the bollworm, *Heliothis zea* (Boddie), is being produced commercially as the first viral insecticide (Viron/H) (Greer et al. 1971) and is effective orally against only five species of the genus *Heliothis*. These two biological insecticides seem to be entirely devoid of toxicity to any other groups of organisms, (Chapter 6).

All these newer approaches to selective insect control seem to have an interesting future in pest management.

It should be remembered that certain conventional pesticides also have this property to some degree. Specific acaricides, dicofol, ovex, tetrasul, tetradifon, omite, and others, are almost entirely specific to the Acarina (mites and ticks) and are essentially nontoxic to insects, wildlife, and humans and higher animals. They have an important role to play in pest management, especially when used to minimize resistance, selection, and environmental persistence.

2. *Ecological Selectivity*

Efforts to make the use of insecticides totally compatible with the aims of pest management must seek to reduce both the frequency of application and the dosage applied. Authorities agree that insecticide treatments are in general excessive. Adkisson (1971) declares that "sufficient knowledge is available . . . to . . . reduce insecticidal use by 50%, or more, on several major crops."

To begin to accomplish these reductions on important crops, the pest-management specialist must move toward more selective means of application of the insecticide and to the replacement of preventive routine-treatment schedules with treat-when-necessary schedules, based on sound knowledge of pest ecology.

a. *Use of Life Tables.* Traditionally, pest-control interventions were aimed at weak points in the life cycle of the insect pest. With the development of modern insecticides and their use in preventive applications, much of this lore was ignored or forgotten. These weak points must be exploited in pest-management strategies. For example, apples infected with the larvae of the plum curculio, *Conotrachelus nenuphar* (Herbst), drop, thus permitting collection and disposal of most of the infestation; larvae of the European corn borer, *Ostrinia nubilalis* (Hübner), overwinter in cornstalks just above the ground surface, where they are vulnerable to clean plowing and destruction of crop refuse; and the 10- to 14-day developmental cycle of *Plasmodia,* from gametocytes to sporozoite formation, in the stomach of the female *Anopheles,* allows DDT residual house spraying to act as an effective malaria control measure.

The development of the life table concept (Le Roux 1963) provides keys to the practical understanding of pest problems in agroecosystems, and to the intelligent and practical manipulation of control factors in crop habitats (NAS 1969). As a first step in understanding practical pest management, life tables should be assembled for the host crop plant (Chapter 1) and for the insect pests concerned. The life table for the host crop provides essential information about critical areas of pest damage and about the benefit/risk ratios of proposed interventions.

Case History

The cabbage crop is attacked by numerous important insect enemies of which several species of cutworms are the most serious enemies of newly set plants, cutting them off shortly after transplant. The cabbage maggot, *Hylemya brassicae* (Bouché), tunnels through and eats off the roots of young plants. As the cabbage matures, three species of cabbage caterpillars, the imported cabbageworm, *Pieris rapae* (L.), the cabbage looper, *Trichoplusia ni* (Hübner), and the diamondback, *Plutella xylostella* (L.), chew holes in the leaves and ruin the appearance of the heads. A *crop life table* for the cabbage plant in eastern Canada is presented in Chapter 1. It shows that cutworms are the major mortality factor during the establishment of the plants, and that cabbage caterpillars cause major damage during heading and harvest. Pest-control interventions must be directed at these two groups of pests in order to produce a marketable and profitable crop. From the operating statistics in Chapter 1, it is evident that the losses are great enough to require substantial applied control. From these data a minimal pest-management program would include:

1. Application of poison bait to control cutworms, for example, 1% carbaryl bran
2. Application of sprays at heading to control caterpillars

These should be timed by the appearance of large numbers of moths in black-light or pheromone traps and could include BT toxin or a rapidly degradable insecticide such as mevinphos.

b. *Pest Life Tables.* These tables describe in critical detail the stages in the life cycle that contribute most to the pest population, and the key factors—parasitism, predation, diseases, food supply, migration, and weather—that are responsible for the density-dependent or density-independent regulation of population trends. This information is critical to planning the most effective insecticide interventions, their timing, and their integration with other key factors.

Case History

The fruit tree leaf roller, *Archips argyrospilus* (Walker), is a pest of apples in the eastern United States and Canada, and in peak years may web over and eat cavities in 80 to 90% of the fruit if uncontrolled. The eggs are laid in the fall in masses of 30 to 100, covered with a smooth varnishlike material, and plastered on twigs and branches. A life table for this insect is given in Table 7.3. It shows that the egg stage has a very low mortality from natural factors. Consequently, a substantial increase in egg mortality through application of a selective ovicide should provide a decisive check on the larval population, as there is only a single generation. This vulnerability of the egg stage has been recognized for many years, and spring spraying with a dormant oil ovicide is a traditional control measure.

3. *Selectivity through Improved Application*

Broadcast application of sprays and dusts for insect control is a very inefficient process. Estimates suggest that only about 10 to 20% of the insecticides applied as dusts and 25 to 50% of those applied as sprays are deposited on plant surfaces for insect control; less than 1% is applied to the insect pests themselves. These data suggest that, even under optimum conditions, 50 to 75% of sprayed or dusted insecticide is useless for pest

Table 7.3 Life Table for Fruit Tree Leaf Roller *Archips argyrospilus* on Apple[a]

Growth Period, *x*	Mean Number Living per 100 Leaf Clusters, *lx*	Mortality Factor, *dxF*	Mean Number Dying *dx*	Percent Mortality 100*rx*	Range in 100*rx*
Egg	97	Desiccation	8	8	Small
		Physiological	2	2	Small
L, 1–2	83	Dispersion	57	69	58 to 81
L, 3–5	26	Parasites[b]	1.0	4	Small
		Predators[c]	0.7	3	Small
		Birds, etc.[d]	19.0	73	35 to 87
Pupa	5.3	Parasites[e]	1.0	19	Small
		Predators	0.3	28	Small
		Physiological	0.2	6	Small
		Birds, etc.	1.5	4	Small
Moth	2.3	Migration	0.4	17	−138 to +30

Generation survival, 1.9–1.96%; mortality, 95.1–98.04%.
Index of population trend, 103%.
Constant mortality rate, 98.2%.

[a] After LeRoux, 1963.

[b] *Eumea, Compsilura, Phytodictus, Horogenes, Exochus, Gravenhorstia, Microgaster, Apanteles, Ceratochaeta.*

[c] *Podisus, Zelus, Chrysopa, Hemerobius.*

[d] Redwing, cowbird, bronzed grackle, Baltimore oriole.

[e] *Itoplectis, Gravenhorstia.*

control and falls to the ground or drifts away from the treatment area and becomes an undesirable environmental contaminant (PSAC 1965). This waste also causes substantial and unnecessary economic loss to the grower.

There are several relatively simple ways in which overtreatments can be minimized or avoided. Some of them become obvious when we discard preconceived ideas that routine fixed-schedule treatments must be the backbone of insect pest-control practices.

a. *Reduced-Dosage Schedules.* Spray schedules and label directions may prescribe inflated dosages of insecticide. Experimentation with decreased dosages has often brought about agreeable results in pest control with minimal adverse effects. Thus Adkisson (1971) showed that disulfoton used against the greenbug. *Schizaphis graminum* (Rondani), in Texas was fully as effective over a 2-week period at 0.1 lb per acre as at 0.25 lb per acre, and parathion at 0.1 lb per acre gave control almost as effective as that at the recommended dosage of 0.5 lb per acre.

Azinphosmethyl is recommended as the standard spray for the control of codling moth, *Laspeyresia pomonella* (L.), on apples at a dosage of about 5 lb 25% wettable per acre. Madsen and Williams (1968) obtained nearly as effective control at one-half the dosage, 2.5 lb per acre, and reasonably good control at one-quarter the dosage, 1.25 lb per acre. Use of reduced dosages is an important first step in beginning the transition program toward successful pest management of many insects.

With decreased dosages it is of course necessary to evaluate carefully the benefit/risk ratio, including savings in insecticide and effects on other members of the pest complex. Lowered dosages of broad-spectrum insecticides may often increase insecticide selectivity by decreasing damage to parasites and predators (van den Bosch and Stern 1962).

b. *Selectivity through Nonpersistence.* Nonpersistent, rapidly degradable insecticides such as nicotine, tetraethyl pyrophosphate, mevinphos, and trichlorfon can be used to achieve substantial selectivity if applications are properly timed. Thus beneficial insects may survive in pupal stages, protected locations, or in untreated reservoirs. Stern and van den Bosch (1959) found that nonpersistent mevinphos was highly destructive to the spotted alfalfa aphid, *Therioaphis maculata* (Buckton), but produced no mortality in parasites emerging from mummified aphids or coccinellid larvae hatching from eggs; malathion and parathion were highly destructive to the beneficial species and were less effective in overall control.

An excellent example of this principle is the pest-management program for the alfalfa weevil, *Hypera postica* (Gyllenhal) (Wilson and Armbrust 1970). A single spray of 4 oz of methyl parathion or 1 lb of malathion per acre is applied in late March to control young larvae developing from the eggs of overwintering weevils, while the important parasite *Bathyplectes curculionis* (Thomson) is still hibernating in its protective cocoon. The parasite is unharmed by the rapidly degradable insecticides and controls the low populations of alfalfa weevil surviving the insecticide treatment (Chapter 12).

c. *Selectivity with Systemic Insecticides.* These usually show pronounced selectivity, especially against sucking plant pests such as aphids, mites, thrips, leafhoppers, and sometimes chewing insects. Insecticides such as schradan, demeton, and oxydemeton-methyl applied to plant foliage rapidly penetrate the leaf cuticle and are translocated throughout the xylem tissue; they act as stomach poisons to sucking insects with little or no harm to parasites, predators, and pollinators. Stern and van den Bosch (1959) found demeton at the very low dosage of 2.0 oz per acre was most effective in controlling the spotted alfalfa aphid, *Therioaphis maculata,* without serious damage to beneficial parasites and predators. Demeton proved to be the most suitable material for an integrated control program for this pest, and produced substantially better control over several weeks than either parathion or malathion, which were very destructive to the beneficial insects as well as to the spotted alfalfa aphid. More persistent systemics such as phorate, disulfoton, aldicarb, and carbofuran are best applied as granulars to the soil at the time of planting, or as seed dressings. They are translocated to the above-ground portions of the plant, concentrating in the most rapidly growing areas such as new leaves and fruits. Aldicarb in particular has pronounced systemic effects on the adult cotton boll weevil, *Anthonomus grandis* Boheman, and when applied to soil at 2 lb per acre as a side dressing (Bariola et al. 1971) reduced the population of adult weevils from 94 to 96%. However, this high dosage of aldicarb almost eliminated the natural enemies of the bollworms, *Heliothis* spp., which produced very severe damage to the treated plots (Timmons et al. 1973).

d. *Selectivity with Seed Treatments.* Application of insecticides to seeds before or at time of planting offers the most efficient and concentrated means of protecting the germinating seed and the seedling plant. Such applications are minimal in dosage and least disturbing to the environment. The savings in application costs and in the total amount of pesticide are striking.

1. Treatment of seeds of a wide variety of field and vegetable crops with lindane, aldrin, heptachlor, dieldrin, or endrin at about 0.25 oz (7 g) per acre has given 70 to 95% mortality of wireworms and ensured the production of satisfactory stands at a reduction of over 99% in the conventional dosage of 2 to 3 lb per acre (Reynolds 1958).
2. Use of planter-box treatment of corn seeds with diazinon at 1.3 oz (37 g) per 100 lb of seed provides adequate wireworm protection at 0.4 oz (11 g) per acre, a reduction of 98% in the conventional dosage of 24 oz per acre applied as a 7-in. band at planting.
3. Treatment of oat seed with propoxur at 4 oz/100 lb of seed or 2.8 oz (80 g) per acre provided high kill of the cereal leaf beetle, *Oulema melanopus* (L.), over 40 to 50 days after planting and gave a reduction

of 85% in the conventional application of 1 lb of carbaryl per acre (Ruppel et al. 1970).

4. Application of the systemic insecticides phorate or disulfoton to alfalfa, sugar beet, or cotton seed at 4 to 8 oz (110 to 220 g) per acre gave control of aphids, thrips, and leafhoppers attacking the seedling plants for several weeks (Reynolds et al. 1957). The principal disadvantages in seed treatments are possible injury to germination; damage to wildlife, especially game birds, by ingestion of seeds; and possible consumption by humans or domestic animals of treated seeds which may be illegally used for food.

e. *Granular Applications at Planting.* Insecticides impregnated at 5 to 20% on granulated clay, bentonite, or diatomaceous earth in which the particles range from 30 to 60 mesh offer significant advantages as selective soil treatments. Granular applications can be made as a continuous band treatment along with the seed at planting, and the rate of release of the insecticide can be controlled by the moisture-absorptive properties of the granular carrier. This technique is substantially less economical than seed treatment, but is much less likely to injure the seed and less hazardous to seed-eating animals. Granular insecticides are easy to use, and they can be applied with precision by ground equipment and with a minimum of drift by aerial equipment. The larger dosages employed can extend the useful protective period of the soil application over an entire growing season. This technique is well suited to the control of pests attacking plant roots, and soil applications of granular carbofuran, fensulfothion, phorate, and fonofos are widely used at about 1 lb per acre to control the larvae of corn rootworms, *Diabrotica* spp.

The granular seed or side dressing technique is also very well suited to the slow release of systemic insecticides such as phorate, disulfoton, or aldicarb (Section III.A.3.c), which may control pests feeding on aerial portions of the crop or ornamental for relatively long periods of time.

f. *Ear or Fruit Treatment.* Treatment of sweet-corn silks with an injection of 5% DDT in mineral oil to kill the young larvae of the corn earworm, *Heliothis zea* (Boddie), soon after they hatch from eggs laid on the silks was an early and classic demonstration of ecological selectivity. Anderson and Reynolds (1960) simplified the operation and brought it to acceptable pest-management status by using a stencil-type paint brush to apply 5% carbaryl dust, which is biodegradable, to the corn silks.

4. *Behavioral Selectivity*

Insecticide selectivity can be substantially enhanced and application rates reduced by specific timing or placement of the insecticide in relation to insect behavior. The few good examples that exist are tantalizing in terms of the broader applicability of this principle to pest management.

Case Histories

1. Insecticides such as methyl parathion, azinphosmethyl, and carbaryl are very toxic to the honeybee, *Apis mellifera* L. Therefore it is best to apply them to fruit trees after bloom is completed, or at least in the evening when bees are not visiting blossoms. This simple modification of spraying practice has successfully prevented devastation of bee colonies in fruit-growing areas.

2. The gravid melon fly, *Dacus cucurbitae* Coquillett, enters tomato fields to oviposit but leaves these fields at dusk and spends the evening on adjacent vegetation. Application of DDT or parathion to such vegetation adjacent to crop areas between sunset and 7 AM reduced the average infestation of tomato fruits from 65 to 3% and avoided residue problems and hazards to pickers (Nishida and Bess 1950, Nishida 1954).

3. The use of DDT as an interior house spray at 1 g/m^2 for malaria control is an outstanding example of behavioral selectivity, which is effective because the female *Anopheles* mosquito enters human habitations in search of a blood meal and during her digestion process prefers to rest on walls or ceilings, where she receives a lethal exposure to DDT. This process could be refined still further through spot treatment of dark corners, as discussed in Chapter 15.

4. An astonishing example of behavioral selectivity in the use of insecticides against the tsetse fly, *Glossina swynnertoni,* has been described by Chadwick et al. (1965). The tsetse fly was eradicated over a 35 mi^2 area of Africa by selective treatment of resting places with 3% endosulfan or dieldrin applied on the underside of tree branches 1 to 4 in. in diameter, 4 to 9 ft above the ground, and inclined less than 35° from the horizontal.

5. Baits of mirex applied at 1.7 g per acre with corncob grits and soybean oil to control the imported fire ant, *Solenopsis saevissima richteri* Forel, are effective because the oil-loving worker ants carry it back to the nest where the slow-acting stomach poison kills the queen and ultimately destroys the colony (*Science* 1971).

a. *Use of Attractants.* Many residue problems could be avoided if it were not necessary to apply insecticides to food crops. With increasing knowledge of sensory communication among insects, especially in the elucidation of the chemical pheromones regulating mating behavior and the chemicals responsible for attraction to sites for feeding and oviposition (Chapter 8), there is abundant opportunity to use these chemical messengers to attract insect pests away from food crops to alternate sites treated with insecticides. Several outstanding examples can be cited.

The Mediterranean fruit fly, *Ceratitis capitata* (Wiedemann), has been eradicated from Florida citrus by using aqueous bait sprays of yeast protein with malathion at 1.2 lb of toxicant per acre (Steiner et al. 1961).

The oriental fruit fly, *Dacus dorsalis* Hendel, is a pest of more than 150 species of citrus and deciduous fruits and melons. The eggs are laid in "stings" in the fruit, and the maggots tunnel through the flesh to complete their growth. The males of this species are specifically attracted by methyl eugenol, an ingredient of several essential oils found in host plants of the fly. The oriental fruit fly was eradicated from the Island of Rota by dropping fiberboard squares impregnated with 28 g of a mixture of 3% naled insecticide and 97% methyl eugenol, at the rate of 125/mi^2 or 3.4 g of naled per acre (Steiner et al., 1965). Both the insecticide and the boards are degradable and offer little if any environmental hazard.

b. *Timing Applications by Light or Pheromone Traps.* It has been emphasized that a major goal in moving into pest-management strategies is the substitution of treat-when-necessary insecticide applications for routine-treatment insecticide schedules. This change would substantially decrease the number of treatments and the volume of insecticide necessary for the satisfactory control of many important pests. The benefits would include decreased costs to the grower, lessened opportunity for the development of insecticide resistance, better integration of biological and chemical control, and improved environmental quality and safety. To implement this enlightened approach to pest management, it is necessary to have simple and effective means of measuring insect populations. Problems and progress in the quantitaive sampling of insect populations are discussed in detail in Chapter 9. Two important techniques deserve mention here as being especially suited to guiding insecticide interventions for the control of many destructive lepidopterous pests of farm, garden, and orchard.

1. *Ultraviolet or black-light traps.* Lepidoptera in particular are selectively attracted to black-light traps emitting ultraviolet light. Use of these traps at three per square mile to trap the moths of the tobacco hornworm, *Manduca sexta* (L.), and the tomato hornworm, *M. quinquemaculata* (Haworth), together with tobacco stalk destruction to kill

overwintering insects, reduced insecticide treatments necessary for tobacco production by over 90% in North Carolina (Gentry et al. 1967).

2. *Sex pheromone traps.* Sex pheromones of important insect pests are rapidly being isolated and identified, and many are available commercially.[1] A listing of the chemical structures of the known sex pheromones of important insect pests is given in Chapter 8. Sex pheromone traps are being used to guide reduced chemical spray programs in apple orchards by monitoring not only the periods of insect pest activity but also the presence or absence of the pest in a particular orchard, consequently indicating whether insecticidal interventions may be omitted safely, with possible reductions of 40 to 50% in the amount of insecticide needed (Glass and Hoyt 1972). Similar pheromone monitoring of the cabbage looper, *Trichoplusia ni,* has resulted in effective commercial control of this pest on lettuce in New Jersey, where two timed insecticide interventions replaced a standard spray schedule of up to eight treatments.

Case History

The codling moth, *Laspeyresia pomonella,* is the most important pest of deciduous fruits, attacking apple, pear, quince, and also English walnut. It is a particularly severe pest of apples and pears, because of the high standards of consumer acceptance which require unblemished fruits. Serious crop losses are prevented only by regular seasonal spray programs beginning soon after bloom and continuing until harvest. These spray schedules were originally based on lead arsenate, then on DDT, and presently on azinphosmethyl. Four or more cover sprays may be required to give total protection in heavily infested orchards. The intensive spray programs to control the codling moth have affected the population regulation of other orchard pests, reducing some to insignificance and causing others such as the red-banded leaf roller, *Argyrotaenia velutinana,* the pear psylla, *Psylla pyricola* Foerster, the European red mite, *Panonychus ulmi* (Koch), and the two-spotted spider mite, *Tetranychus urticae* Koch, to become major pests. It has been very difficult to develop pest-management programs in apple and pear orchards because of the tyranny of the regular codling moth spray schedules.

On pears (Batiste 1970; Batiste et al. 1973) and on apples (Madsen and Vakenti 1973), very promising efforts have been made to de-

[1] Zoecon Corporation, Palo Alto, California.

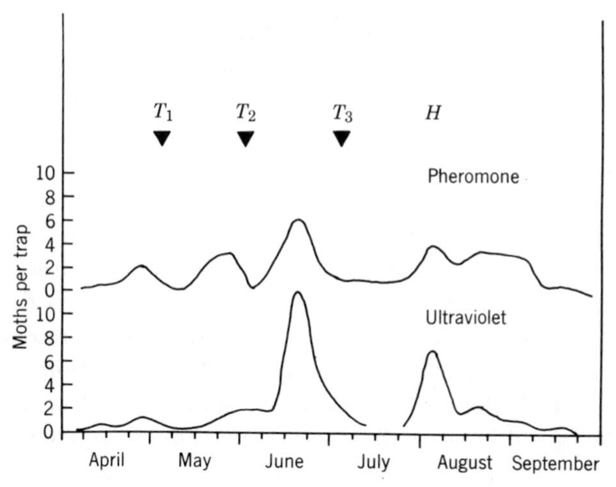

Fig. 7-2 Use of pheromone traps and ultraviolet light traps for timing spray applications for codling moth control in California pear orchards. T = spray treatment, H = harvest. (Data from Batiste et al. 1973.)

velop integrated pest-management programs based on the timed application of reduced dosages of azinphosmethyl, using codling moth traps baited with the synthetic female sex pheromone 8,10-dodecadien-1-ol (Codlemone), together with convenient commercially available pheromone traps (Fig. 9.13). As shown in Fig. 7.2, the peaks of male catches in the pheromone traps, corresponding to the emergence of the three generations or broods of the codling moth, can be used to predict egg laying and the consequent need for a protective spray. Using this means of timing, together with observations on the appearance of the first fruit entries, Madsen and Vakenti (1973) reduced a conventional schedule of three cover sprays to one, or at most two, timed sprays per season and still produced commercial-quality apples. Such a program not only provides the basis for a total pest-management program on deciduous fruits, but also results in substantial savings to the grower.

IV. SELECTING INSECTICIDES FOR PEST-MANAGEMENT PROGRAMS

Once it has been decided that insecticide intervention is essential for a pest-management operation, how is the choice of an individual insecticide to be made? As we have seen, the criteria for using insecticides in pest management are complex, yet the ultimate success of the program may be determined by the insecticide selected as well as by the method of application and by the timing of the treatment. The informed choice of a specific insecticide for any program is complex, involving knowledge of the chemical properties of the compound, its biological activity on the target insect pest, its toxicity to humans and domestic animals, its effects on nontarget organisms—crops, parasites, predators, pollinators, and wildlife—and its environmental fate in air, water, soil, and food. Thus we have an enormous responsibility when we prescribe or recommend an insecticide for a pest-management program. In this section we attempt to develop a methodology for insecticide selection that will both simplify the choice and improve the overall aspects of environmental quality.

A. Federal Registration of Pesticides

The Federal Environmental Pesticide Control Act of 1972 requires registration with the EPA of all pesticides shipped in interstate commerce. Before registration is granted, the manufacturer is required to provide scientific evidence that the product (1) is effective against the pest or pests listed on the label, (2) does not injure humans, crops, livestock, wildlife and/or damage the total environment when used as directed, and (3) when used as directed does not result in illegal residues on food or feed. This law requires the proper application of pesticides to ensure greater protection of man and the environment. It provides severe penalties for knowing misuse of pesticides by farmers, applicators, dealers, and distributors, and therefore is a strong legislative force toward proper pest management.

Under the 1972 act, more than 200 chemical insecticide compounds and thousands of formulated products are registered with the EPA. Therefore to a considerable extent the choice of a specific insecticide for any purpose is biased by federal pesticide registration and by legal residue tolerances. Yet these permit wide latitude on most important crops. For example, the approximate number of individual insecticide chemicals registered on important United States crops is: alfalfa, 39; apple, 72; citrus, 47; corn, 66; cotton, 50; potato, 37; soybean, 23; tobacco, 22; and tomato, 49. Few entomologists can claim to be entirely familiar with the important properties, efficacy, safety, and environmental-quality aspects of this relatively large number of insecticides.

B. Pest-Management Rating of Insecticides

To provide further rationale for intelligent choice of insecticidal tools by the pest-management specialist, an overall evaluation of the suitability of each compound seems to be both desirable and useful. Such ratings have been attempted for various types of pesticides (Metcalf 1972; Petty 1971).

A pest-management rating (Table 7.1) has been devised for the common insecticides registered for and widely used on important agricultural crops in the United States, in regard to their safety and overall effects on environmental quality. Ratings were made on the basis of the average performance in (1) acute toxicity to man and domestic animals, (2) overall toxicity to three important environmental indicator organisms, and (3) environmental persistence. Each category was assigned a rating of 1 to 5 with increasing hazard:

1. *Mammalian toxicity* was rated from rat oral LD_{50} in milligrams per kilogram (Hayes 1963):

1 = >1000
2 = 200–1000
3 = 50–200
4 = 10–50
5 = <10

2. *Nontarget toxicity* was rated as the *average* of individual ratings for:

Pheasant Oral LD_{50} (mg/kg)	Rainbow Trout 48-hour/LC_{50} (ppm)	Honeybee Topical LD_{50} (mg/kg)
1 = >1000	1 = >1.0	1 = >100
2 = 200–1000	2 = 0.1–1.0	2 = 20–100
3 = 50–200	3 = 0.01–0.1	3 = 5–20
4 = 10–50	4 = 0.001–0.01	4 = 1–5
5 = <10	5 = <0.001	5 = <1

These organisms are important environmental indicators for which a wide range of toxicity values has been measured (Pimentel 1971).

3. Environmental persistence was rated as the approximate average soil half-life: 1 = <1 month, 2 = 1–4 months, 3 = 4–12 months, 4 = 1–3 years, 5 = 3–10 years.

The combined ratings for the individual insecticides are presented in Table 7.1. Insecticides with the lowest ratings are those that seem least likely to cause overall disturbances in environmental quality and *ceteris paribus* should be favored for employment in pest-management programs. The individual ratings for toxicity to the various organisms and for persistence should be a useful guide to specific applications.

In summary, the insecticides can be segregated into four classes:

1. Suitable for general use in pest-management programs (rating 4 to 7): carbaryl, Gardona, malathion, methoxychlor, naled, trichlorfon, BT toxin, ovex, cryolite
2. To be used in pest-management programs only under skillful supervision (rating 8 to 10): azinphosmethyl, chlorpyrifos, demeton, diazinon, dicofol, dimethoate, endosulfan, lindane, mevinphos, methyl parathion, phosphamidon, oxydemeton methyl, tetraethyl pyrophosphate, toxaphene, nicotine
3. To be used in pest management only under restricted conditions (rating 11 to 13): such as seed and soil treatment with aldicarb, carbofuran, disulfoton, phorate, parathion, EPN; or indoor treatment with DDT
4. Little if any place in pest management except structural pest control (rating 13 to 15): aldrin, dieldrin, endrin, heptachlor.

V. MODEL PEST-MANAGEMENT PROGRAMS USING INSECTICIDES

Successful integrated pest-management approaches to the use of insecticides are discussed elsewhere in this book (Chapters 1, 8, and 11 through 15). Two outstanding programs are summarized here as conforming to our ideas of how insecticides can best be used in a complex program involving most of the elements of pest management. Both programs were developed in response to the attack of a single imported insect pest on an important field crop, and both programs have been carefully and soundly constructed from a large amount of research dealing with the total crop-insect pest–natural enemy ecosystem. The pest-management program for the spotted alfalfa aphid successfully suppressed a nearly disastrous invasion of the United States. The pest-management program for the cereal leaf beetle is much newer and still under revision. It needs additional input to both the parasite complex and to the development of a really selective insecticide for use in emergency control.

Case Histories

1. The spotted alfalfa aphid, *Therioaphis maculata* (Buckton), was introduced into central New Mexico in early 1954, probably

as a single agamic female imported by aircraft from the insect's native home in the Mediterranean area. In its new home, free from the biological restraints of specific natural enemies, the insect multiplied at an unbelievable rate, entering Arizona, California, Nevada, Texas, and Oklahoma in 1955; by 1957, it had spread from coast and coast and from Mexico to central Wisconsin. Because of the enormous numbers of aphids present, up to 1 billion per acre, feeding injury from a toxic saliva that often killed young plants, and the secretion of profuse amounts of honeydew which supports the growth of sooty mold fungus, it became virtually impossible to grow marketable alfalfa hay in many of the infested areas. Yield and protein and carotene content of the hay were often reduced by half. Damage in California alone was estimated at $42 million in 1956 (Smith 1959).

Control in the early stages of the epidemic was possible only through frequent aerial spraying with parathion at 4 oz or malathion at 10 oz per acre. These broad-spectrum treatments caused widespread destruction of beneficial insects and were clearly undesirable on a general forage crop fed to milk cows.

Pest-Management Program: A practical pest-management program for the spotted alfalfa aphid in California was developed by Stern et al. (1959) and Stern and van den Bosch (1959). This program was organized about newly developed varieties of alfalfa highly resistant to the aphid attack and consisted of the following specific components.

a. Plant only alfalfa varieties tolerant to damage by aphid feeding, namely, 'Cody,' 'Moapa,' 'Lahontan.'

b. Establish and maintain a vigorous stand through proper water management, fertilization, and weed control.

c. Encourage predators such as lady beetles, syrphid flies, lacewings, and the hemipterans *Nabis* and *Geocoris.*

d. Establish a complex of imported hymenopterous parasites—*Praon pollitans, Trioxys utilis,* and *Aphelinus semiflavus.*

e. Encourage *Entomophthora* fungi.

f. Cut alfalfa so that there is little green leaf remaining on stubble, to kill aphids by exposure to the sun.

g. If the aphid exceeds the economic threshold of 0.5 to 1 per seedling or 20 to 40 per stem, spray, using the selective systemics demeton or mevinphos at 1 to 2 oz per acre. These insecticides are rapidly absorbed by the plant, producing sap toxic to the aphids, but do not drastically damage populations of natural enemies.

h. Treat alfalfa seed with disulfoton or phorate using 1.5 lb of charcoal powder per 100 lb of seed at planting. These systemic insecticides effectively protect young seedlings from being killed by aphid attack and ensure a good stand.

2. The cereal leaf beetle, *Oulema melanopus* (L.), a Eurasian pest of small grains, was introduced into southwest Michigan and adjacent Indiana in July–August 1962 and has become a threat to the entire grain-growing region of the United States. The adult beetles severely damage oats and wheat in the early spring, and the larvae continue feeding on these crops throughout the summer. Control has generally been by aerial application to grain fields of carbaryl or malathion at 1 lb per acre (Ruppel and Wilson 1964). However, these broad-spectrum insecticides are deleterious to beneficial insects, and the treatment is expensive for field crops.

Pest-Management Program: As developed in Michigan by Ruppel et al. (1970), Stehr (1970), and Maltby et al. (1971), an integrated program consists of:

a. Plant tolerant varieties of oats or wheat.

b. Encourage natural enemies such as the spotted ladybird, *Coleomegilla maculata* (De Geer).

c. Establish the eulophid parasite *Tetrastichus julis* (Walker) and the mymarid parasite *Anaphes flavipes* (Foerster).

d. Treat cereal seeds with propoxur or carbofuran at 4 oz/100 lb of seed at planting, to control adult beetles early in the season.

e. If the beetle population exceeds the economic threshold, spray the field with malathion or carbaryl at 1 lb per acre. (A more selective insecticide would be better here, according to Ruppel 1972).

VI. SUMMARY AND CONCLUSIONS

Problems of insecticide resistance and concern for the quality of the environment and for the balanced ecology of host–pest–natural enemy complexes will dictate major changes in the use of pesticides in the immediate future.

Pest-management practices will generally involve the logical, precise, and careful use of insecticides.

Methods must be developed to use insecticides specifically and selectively and with particular concern for the economic thresholds of pests attacked, so that unnecessary treatments are avoided and insecticidal intervention at the target site involves maximum ecological selectivity.

Insecticides must be selected for use in specific pest-management programs on the basis of overall safety to humans and domestic animals, to nontarget organisms, and to the overall quality of the environment, as well as for specific effectiveness against the target species.

The development of successful pest-management programs will provide an enormous challenge to entomological practitioners who must be equipped with knowledge about host–pest–natural enemy complexes, about the toxicological and environmental properties of insecticides, about the practical economics of risk versus benefit, and with a code of ethics to match the responsibilities they will assume.

REFERENCES

Adkisson, P. L. 1971. Objective uses of insecticides in agriculture. *Agr. Chem. Symp.*, pp. 43–51.

American Chemical Society (ACS). 1969. Cleansing our environment; the chemical basis for action. Washington, D.C. 250 pp.

Anderson, L. D., and H. T. Reynolds. 1960. A comparison of the toxicity of insecticides for the control of corn earworm on sweet corn. *J. Econ. Entomol.* **53**:22–24.

Bariola, L. A., R. L. Ridgway, and J. R. Coppege. 1971. Large-scale field tests of soil applications of aldicarb for suppression of populations of boll weevils. *J. Econ. Entomol.* **64**:1280–1284.

Batiste, W. C. 1970. Timing sex-pheromone traps with special reference to codling moth collection. *J. Econ. Entomol.* **63**:915–918.

Batiste, W. C., A. Berlowitz, W. H. Olson, J. E. DeTar, and J. L. Loos. 1973. Codling moth: Estimating time of fruit egg hatch in the field—A supplement to sex-attractant traps in integrated control. *Envir. Entomol.* **2**:387–391.

Blackbourne, B. D. 1970. Pesticide poisoning—A medical examiner's view. *Proc. 22nd Ill. Custom Spray Oper. Train. Sch.*, pp. 16–20.

Bowers, W. S. 1971. Insect hormones and their derivatives. *Bull. World Health Organ.* **44**:381–389.

Breidenback, A. W., C. G. Gunnerson, F. K. Kawahara, J. J. Lichtenberg, and R. S. Green. 1967. Chlorinated hydrocarbon pesticides in major river basins, 1957-65. *U.S. Pub. Health Rep.* **82**:139–156.

Brown, A. W. A. 1969. Insect resistance. *Farm Chem.* September-November 1969, Feb. 1970.

Brown, A. W. A., and R. Pal. 1971. Insecticide resistance in arthropods. World Health Organization, Geneva. 491 pp.

Burdick, G. E., E. J. Harris, H. J. Dean, T. M. Walker, J. Shed, and D. Colby. 1964. The accumulation of DDT in lake trout and the effect on reproduction. *Trans. Amer. Fish. Soc.* **93:**127–136.

Chadwick, P. R., J. S. S. Beesley, P. J. White, and H. T. Matechi. 1965. An experiment on the eradication of *Glossini swynnertoni* Aust. by insecticidal treatment of its resting sites. *Bull. Entomol. Res.* **55:**411–419.

Chiang, H. C. 1973. Bionomics of the northern and western corn rootworms. *Annu. Rev. Entomol.* **18:**47–72.

Cope, O. B. 1961. Effects of DDT-spraying for spruce budworm on fish in the Yellowstone River system. *Trans. Amer. Fish. Soc.* **90:**239–251.

Cope, O. B. 1971. Interactions between pesticides and wildlife. *Annu. Rev. Entomol.* **16:**325–364.

Diekman, J. 1972. Use of insect hormones in pest control. Pages 69–73 *in* Implementing practical pest management strategies. Proceedings of a national extension insect pest-management workshop. Purdue University, Lafayette, Indiana.

Duggan, R. E. 1969. Pesticide residues in foods. *Ann. N.Y. Acad. Sci.* **160:**173–182.

Durham, W. F. 1969. Body burden of pesticides in man. *Ann. N.Y. Acad. Sci.* **160:**183–195.

Environmental Protection Agency (EPA). 1972. An evaluation of DDT and dieldrin in Lake Michigan. EPA-R3-72-003.

Geier, P. W. 1966. Management of insect pests. *Annu. Rev. Entomol.* **11:**471–490.

Gentry, C. R., F. R. Lawson, C. M. Knott, J. M. Stanley, and J. M. Lam, Jr. 1967. Control of hornworm by trapping with blacklight and stalk cutting in North Carolina. *J. Econ. Entomol.* **60:**1437–1442.

Glass, E. H., and S. C. Hoyt. 1972. Insect and mite pest management on apples. Pages 98–106 *in* Implementing practical pest management strategies. Proceedings of a national extension insect pest-management workshop. Purdue University, Lafayette, Indiana.

Greer, F., C. M. Ignoffo, and R. F. Anderson. 1971. The first viral pesticide. A case history. *Chem. Tech.,* pp. 342–347.

Hayes, W. J., Jr. 1963. Clinical handbook on economic poisons. U.S. Public Health Service. Atlanta, Ga. 144 pp.

Hayes, W J., Jr. 1969. Pesticides and human toxicity. *Ann. N.Y. Acad. Sci.* **160:**40–54.

Illinois State Department of Agriculture. 1972. Illinois pesticide use, Springfield, Illinois.

Jegier, Z. 1969. Pesticide residues in the atmosphere. *Ann. N.Y. Acad. Sci.* **160:** 143–182.

Kenaga, E. E., and W. E. Allison. 1969. Commercial and experimental organic insecticides. *Bull. Entomol. Soc. Amer.* **15**(2):85–148.

Kuhlman, D. E., T. A. Cooley, and J. Walt. 1974. Adult corn rootworm populations in Illinois, August 1973. *Proc. 26 Ill. Custom Spray Sch., pp.* 159–164.

Kuhlman, D. E., and H. B. Petty. 1973. Summary of corn rootworm insecticide demonstrations, 1968–1972. *Proc. 25th Ill. Custom Spray Sch.*, pp. 140–4.

Lawless, E. W., R. von Rumker, and T. L. Ferguson. 1972. The pollution potential in pesticide manufacturing. *EPA Tech. Stud. Rep.* TS-00-72-04.

LeRoux, E. J., ed. 1963. Population dynamics of agricultural and forest insect pests. *Mem. Entomol. Soc. Can.* 32. 103 pp.

Madsen, H. F., and J. M Vakenti. 1973. Codling moth: Use of Codlemone®-baited traps and visual detection of entries to determine need of sprays. *Environ. Entomol.* **2:**677–679.

Madsen, H. F., and K. Williams. 1968. Effectiveness and persistence of low dosages of azinphosmethyl for control of the codling moth. *J. Econ. Entomol.* **61:**878–879.

Maltby, H. L., F. W. Stehr, R. C. Anderson, G. E. Moorehead, L. C. Barton, and J. D. Paschke. 1971. Establishment in the United States of *Anaphes flavipes,* an egg parasite of the cereal leaf beetle. *J. Econ. Entomol.* **64:**693–697.

Metcalf, R. L. 1968. Methods of estimating effects. Pages 17–29 *in* Chichester C. O., ed., Research on pesticides. Academic Press, New York.

Metcalf, R. L. 1972. Selective use of insecticides in pest management. Pages 74–97 *in* Implementing practical pest management strategies. Proceedings of a national extension insect pest management workshop. Purdue University, Lafayette, Indiana.

Mrak, E. (ch.). 1969. Report of the Sec. Commission on pesticides and their relationship to environmental health. U.S. Department of Health, Education, and Welfare, Washington, D.C.

National Academy of Sciences (NAS). 1969. Insect-pest management and control. Pub. 1695. Washington, D.C.

Newsom, L. D. 1967. Consequences of insecticide use on non-target organisms. *Annu. Rev. Entomol.* **12:**257–286.

Nishida, T. 1954. Further studies on the treatment of border vegetation for melon fly control. *J. Econ. Entomol.* **47:**226–229.

Nishida, T., and H. A. Bess. 1950. Applied ecology in melon fly control. *J. Econ. Entomol.* **43:**877–883.

Pallos, F. M., and J. J. Menn. 1972. Pages 303–16 *in* J. J. Menn and M. Beroza, eds., Insect juvenile hormones: Chemistry and action, Academic Press, New York

Pesticide Review. 1972. U.S. Department of Agriculture, Washington, D.C.

Petty, H. B. 1970. DDT, other persistent insecticides, and our environment. *Proc. 22nd Ill. Custom Spray Oper. Train. Sch.*, pp. 169–189.

Petty, H. B. 1971. *Proc. 23rd Ill. Custom Spray Oper Train. Sch.*, p. 274.

Pimentel, D. 1971. Ecological effects of pesticides on non-target species. Executive Office of the President, Office of Science and Technology. U.S. Government Printing Office. Washington, D.C.

Pimentel, D. 1973. Extent of pesticide use, food supply, and pollution. *J. N.Y. Entomol. Soc.* **81**:13–37.

President's Science Advisory Committee (PSAC). 1965. Restoring the quality of our environment. The White House. Washington, D.C. 317 pp.

Rainey, R. H. 1967. Natural displacement of pollution in Great Lakes. *Science* **155**:1242–1243.

Reynolds, H. T. 1958. Research advances in seed and soil treatment with systemic and non-systemic insecticides. *Adv. Pest Control Res.* II, pp. 135–182.

Reynolds, H. T., R. C. Dickson, R. M. Hannibal, and E. F. Laird, Jr. 1967. Effects of the green peach aphid, southern garden leafhopper, and carmine spider mite populations upon yield of sugar beets in the Imperial Valley, California. *J. Econ. Entomol.* **60**:1–7.

Reynolds, H. T., T. R. Fukuto, R. L. Metcalf, and R. B. March. 1957. Seed treatment of field crops with systemic insecticides. *J. Econ. Entomol.* **50**: 527–539.

Robbins, W. E. 1972. Hormonal chemicals for invertebrate pest control. Pages 172–196 *in* Pest control strategies for the future. National Academy of Sciences, Washington, D.C.

Ruppel, R. F. 1972. Tests of chemical control of the cereal leaf beetle. *J. Econ. Entomol.* **65**:824–827.

Ruppel, R. F., J. Valarde, and S. L. Taylor. 1970. Integrated control of the cereal leaf beetle. *Mich. State Univ. Res. Rep.* **122**:5–6.

Ruppel, R. H., and M. C. Wilson. 1964. Aerial application of insecticides to control spring infestations of the cereal leaf beetle on small grains. *J. Econ. Entomol.* **57**:899–903.

Sacher, R. 1971. *Chem. Eng. News.* November 28, p. 9.

Science. 1971. Mirex and the fire ant: Decline in fortunes of "perfect" pesticide, **172**:358–359.

Smith, R. F. 1959. The spread of the spotted alfalfa aphid, *Therioaphis maculata* (Buckton), in California. *Hilgardia* **28**:647–685.

Smith, R. F. 1970. Pesticides: Their use and limitations in pest management, Pages 103–113 *in* R. L. Rabb, and F. E. Guthrie, eds., Concepts of pest management. North Carolina State University, Raleigh.

Soper, F. L., J. A. Andrews, K. F. Bode, G. R. Coatney, W. C. Earle, S. M. Keeny, E. F. Knipling, J. A. Logan, R. L. Metcalf, K. D. Quarterman, P. F. Russell, and L. L. Williams. 1961. Report and recommendations on malaria: A summary. *Amer. J. Trop. Med. Hyg.* **10**:451–502.

Stehr, F. W. 1970. Establishment in the United States of *Tetrastichus julis,* a larval parasite of the cereal leaf beetle. *J. Econ. Entomol.* **63**:1968–1969.

Steiner, L. F., W. C. Mitchell, E. J. Harris, T. T. Kozuma, and M. S. Fugimoto. 1965. Oriental fruit fly eradication by male annihilation. *J. Econ. Entomol.* **58**:961–964.

Steiner, L. F., G. G. Rohwer, E. L. Ayers, and L. D. Christianson. 1961. The role of attractants in the recent Mediterranean fruit fly eradication program in Florida *J. Econ. Entomol.* **54**:30–35.

Stern, V. M., and R. van den Bosch. 1959. The integration of chemical and biological control of the spotted alfalfa aphid. II. Field experiments on the effects of insecticides. *Hilgardia* **29**:103–130.

Stern, V. M., R. F. Smith, R. van den Bosch, and K. S. Hagen. 1959. The integration of chemical and biological control of the spotted alfalfa aphid. I. The integrated control concept. *Hilgardia* **29**:81–101.

Timmons, F. D., T. S. Brook, and F. A. Harris. 1973. Effects of aldicarb applied side-dress to cotton on some arthropods in the Monroe County, Mississippi, boll weevil diapause control area in 1969. *J. Econ. Entomol.* **66**:151–153.

U.S. Department Agriculture (USDA). 1967. Beekeeping in the United States. Agriculture Handbook no. 335.

Van Daalen, J., J. Meltzer, R. Mulder, and K. Wellinga. 1972. A selective insecticide with a novel mode of action. *Naturwissenschaften* **59**:312–313.

van den Bosch, R., and V. M. Stern. 1962. The integration of chemical and biological control in arthropod pests. *Annu. Rev. Entomol.* **7**:367–386.

von Rumker, R., and F. Horay. 1972. Pesticide manual. U.S. Agency for Interational Development.

Wellinga, K., R. Mulder, and J. Van Daalen. 1973. A synthesis and laboratory evaluation of 1-(2,6-disubstituted benzoyl)-3-phenylureas, a new class of insecticides. *J. Agr. Food Chem.* **21**:348–354.

Wilson, M. C., and E. J. Armbrust. 1970. Approach to integrated control of the alfalfa weevil. *J. Econ. Entomol.* **63**:554–557.

Wiersma, G. B., H. Tai, and P. F. Sand. 1972. Pesticide residue levels in soils. FY 1969-National Soils Monitoring Program. *Pestic. Monito. J.* **6**:194–288.

World Health Organization (WHO). 1968. *World Health Mag.* April, pp. 10–11.

8

ATTRACTANTS, REPELLENTS, AND GENETIC CONTROL IN PEST MANAGEMENT

Robert L. Metcalf and Robert A. Metcalf

Pest-management strategies can utilize the instinctual behavior of the pest insects themselves for the manipulation and regulation of their populations. Such strategies are sophisticated in concept and development, but may be relatively simple in routine execution. Familiar examples are the use of repellents for biting flies to protect livestock and increase weight gain and milk production, and the use of attractants in sampling insect populations (Chapter 9) and in timing insecticide applications in orchards (Chapter 13). In this chapter we consider ways to direct or modify insect pest behavior in regard to searching for food, shelter, oviposition sites, or mates; to decrease population levels; and to improve the disparity between the general equilibrium position and the economic threshold (Chapter 1). Useful discussions of these techniques can be found in National Academy of Sciences Publications, *Insect Pest Management and Control* (NAS 1969) and *Pest Control Strategies for the Future* (NAS 1972).

I. ATTRACTANTS

Attractants are the glamour tools of insect pest management and have been described as "new, imaginative, and creative approaches to the problem of sharing our earth with other creatures" (Carson 1962). Their use in insect pest management is precise, specific, and ecologically sound. It is now well established that many phases of insect behavior in searching for food, oviposition sites, and sexual partners are stimulated and controlled by chemicals. For the most part these are natural products, but some useful synthetic chemical attractants are known, either discovered serendipitously or through structural optimization of the essential features of natural attractant substances.

Chemical substances that deliver behavioral messages have been termed *semiochemicals*; they act either intraspecifically between individuals of the same species or interspecifically between members of different species (Law and Regnier 1971). Interspecific semiochemicals that favor the producer are called *allomones,* while those that favor the receiver are called *kairomones* (Brown et al. 1970). *Pheromones* are semiochemicals used for intraspecific communication between individuals of a single species (Karlson and Butenandt 1959).

A. Pheromones

These substances are exocrine secretions of animals, which cause a specific reaction in the receiving individual of the same species, that is, in insect alarm, sexual attraction, aggregation, or tracking; or specific changes in physiological development, for example, sexual determination or maturation (Karlson and Butenandt 1959). Knowledge of specific pheromone substances in insects has developed very rapidly since the chemical characterization of the sex pheromone of the female silkworm moth, *Bombyx mori* (L.), as *trans, cis*-10,12-hexadecadien-1-ol (Butenandt et al. 1959). For utility in pest management, the sex pheromones that initiate and control mating behavior appear to be the most promising, and those of important insect pests are listed in Table 8.1.

1. Sex Pheromones

These substances are widely if not universally distributed in the Insecta and appear to have reached their evolutionary peak in the Lepidoptera where they have already been demonstrated in more than 170 species (MacConnell and Silverstein 1973). In most Lepidoptera studied these sex pheromones are typically elaborated by eversible glands on the terminalia of the female insect, although the male of the greater wax moth, *Galleria mellonella* (L.), produces a sex pheromone, stated to be undecanal, in wing glands and, in Danainae butterflies (Nymphalidae),

Table 8.1 Important Insect Pheromones[a]

Species	Pheromone	Structure
Silkworm, *Bombyx mori*	*Trans*-10-*cis*-12-hexadecadien-1-ol (bombycol)	$CH_3(CH_2)_2(CH{=}CH)_2(CH_2)_8CH_2OH$
Gypsy moth, *Porthetria dispar*	*Cis*-7,8-epoxy-2-methyloctadecane (disparlure)	$CH_3(CH_2)_9CH{-}CH(CH_2)_4CH(CH_3)_2$ (epoxide O bridging the two CH)
Pink bollworm, *Pectinophora gossypiella*	*Cis*, *cis*- and *cis*-*trans*-7,11-hexadecadienyl acetate (gossyplure)	$CH_3(CH_2)_3CH{=}CHCH_2CH_2CH{=}CH(CH_2)_6OC(O)CH_3$
Codling moth, *Laspeyresia pomonella*	*Trans*, *trans*-8,10-dodecadien-1-ol	$CH_3CH{=}CH{-}CH{=}CH(CH_2)_6CH_2OH$
European corn borer, *Ostrinia nubilalis*	*Trans*-11-tetradecenyl acetate[b]	$CH_3CH_2CH{=}CH(CH_2)_{10}OC(O)CH_3$
Red-banded leaf roller, *Argyrotaenia velutinana*	*Cis*-11-tetradecenyl acetate	$CH_3CH_2CH{=}CH(CH_2)_{10}OC(O)CH_3$
Oriental fruit moth, *Grapholitha molesta*	*Cis*-8-dodecenyl acetate	$CH_3(CH_2)_2CH{=}CH(CH_2)_7OC(O)CH_3$
Southern armyworm, *Spodoptera eridania*	*Cis*-9-tetradecenyl acetate	$CH_3(CH_2)_3CH{=}CH(CH_2)_8OC(O)CH_3$
Cabbage looper, *Trichoplusia ni*	*Cis*-7-dodecenyl acetate	$CH_3(CH_2)_3CH{=}CH(CH_2)_6OC(O)CH_3$

(Continued)

Table 8.1 (*continued*)

Species	Pheromone	Structure
Eastern spruce budworm, *Choristoneura fumiferana*	*Trans*-11-tetradecenal	$CH_3CH_2CH{=}CH(CH_2)_9CHO$
Black carpet beetle, *Attagenus megatoma*	*Trans*-3-*cis*-5-tetradodecadienoic acid	$CH_3(CH_2)_7CH{=}CHCH{=}CHC(O)OH$
Cotton boll weevil, *Anthonomus grandis*	*Cis*-2-isopropenyl-1-methyl-cyclobutane ethanol	CH_3; CH_2CH_2OH; $C{=}CH_2$; CH_3
	Cis-3,3-dimethyl-Δ^1-β-cyclo-hexane ethanol	$CHCH_2OH$; CH_3; CH_3

	Cis- and *trans*-3,3-dimethyl-Δ^1-α-cyclohexaneacetaldehyde	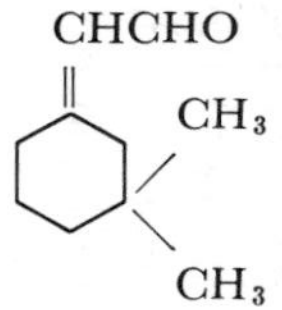 CHCHO, CH_3, CH_3
Southern pine beetle, *Dendroctonus frontalis*	1,5-Dimethyl-6,8-dioxabicyclo-[3.2.1.]-octane (frontalin)	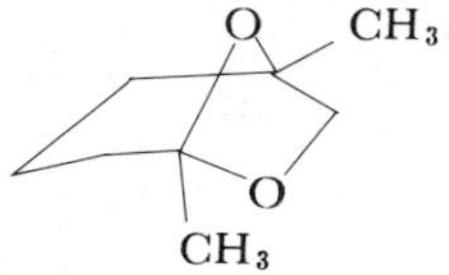 O, CH_3, O, CH_3
Western pine beetle, *Dendroctonus brevicomis*	*Exo*-7-ethyl-5-methyl-6,8-dioxabicyclo-[3.2.1.]-octane (brevicomin)	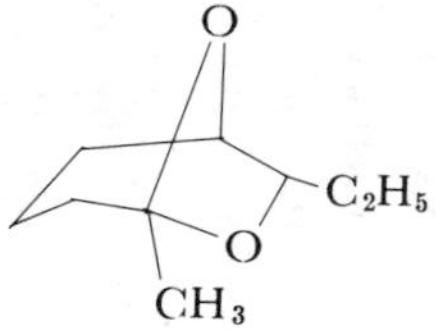 O, C_2H_5, O, CH_3
Ips and *Dendroctonus* beetles	*Cis*-(+)-verbenol	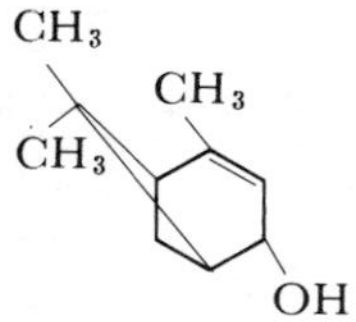 CH_3, CH_3, CH_3, OH

[a] See Law and Regnier 1971, Beroza 1972, and MacConnell and Silverstein 1973.

[b] A New York race responds to mostly trans isomer and an Iowa race responds to mostly cis isomer.

the male produces the sex pheromones, for example, *trans, trans*-10-hydroxy-3,7-dimethyl deca-2,6-dienoic acid in the monarch butterfly, *Danaus plexippus* (L.), from extrusible glands on the abdomen called hair pencils (Brower et al. 1965; Meinwald et al. 1968). The male cotton boll weevil, *Anthonomus grandis* Boheman, produces a mixture of at least four pheromones: *cis*-2-isopropenyl-1-methylcyclobutane ethanol, *cis*-3,3-dimethyl-Δ′-β-cyclohexane ethanol, and *cis*- and *trans*-3,3-dimethyl-Δ′-α-cyclohexane acetaldehyde (Tumlinson et al. 1969). The release of sex pheromones is a complex process correlated with sexual maturity of the virgin female and with photoperiod and light intensity; for example, in the Noctuidae it typically occurs at minimum light intensity (Shorey et al. 1968).

Female sex pheromones are typically received by specialized sensory sensilla of the male antennae, accounting for the hyperdevelopment of these organs in most male moths. These sensilla are exceedingly sensitive to the pheromone, and in *Bombyx mori* it appears that the threshold for male response is about 10,000 molecules per cubic centimeter of air, suggesting that the threshold is approximately one molecule of pheromone per antennal receptor site (Law and Regnier 1971). Thus the simultaneous depolarization of several antennal receptors is sufficient to elicit the male searching response which takes him upwind into the odor corridor (Fig. 8.1) emanating from the female moth. The male is believed to follow upwind, orienting himself to the odor corridor by random searching whenever he strays into a region of subthreshold odor level,

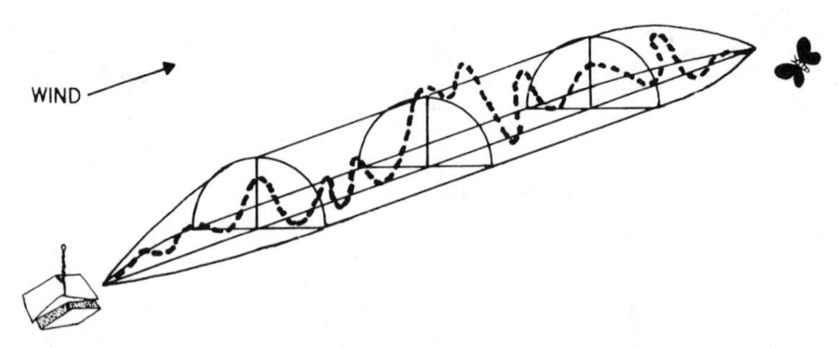

Fig. 8-1. Odor corridor from pheromone trap. Its dimensions depend on rate of pheromone release and wind velocity. (After Bossart and Wilson 1963.) The dotted line shows hypothetical flight path of male moth orienting toward the trap by flying upwind through the corridor where the concentration of pheromone is above the threshold for response. (After Farkas and Shorey 1972.)

until he reaches the region of steep pheromone odor gradient near the female, where osmotactic orientation brings him quickly to the female. Secondary pheromones may be involved in the final fixation of the mating behavioral pattern (Law and Regnier 1971).

Sex pheromones have an extraordinarily high ratio of emission rate Q to threshold concentration K, for example, 10^{10} to 10^{12} cm^3 per second, and may be perceived for hundreds of yards downwind (Wilson and Bossert 1963).

2. *Alarm Pheromones*

These substances are elaborated by mandibular glands, anal glands, or the sting apparatus and typically produce either flight or aggression. A typical alarm pheromone of Dolichoderine ants is 2-methylheptanone, which is responsible for the fruity odor of crushed workers and produces immediate confused and erratic behavior of all the workers in the immediate vicinity. When discharged through the mandibles onto an intruder ant, the latter becomes marked as an aggressor. Alarm pheromones have a high ratio of emission rate Q to threshold concentration K, for example, 10^3 to 10^5 cm^3 per second in *Acanthomyops claviger* (Roger) (Wilson 1970), giving a maximum transmission radius of about 10 cm over which warning signals persist for about 8 minutes. *Trans*-β-farnesene has recently been identified as an interspecific alarm pheromone of aphids (Bowers et al. 1972).

3. *Trail-Marking Pheromones*

These are substances of low persistence elaborated by foraging ants and termites. The ant *Formica rufa* L. appears to use formic acid as a trail marker. The major trail-marking pheromone of the Texas leaf-cutting ant, *Atta texana* (Buckley), is methyl 4-methylpyrrole-2-carboxylate (Tumlinson et al. 1972).

Various substances have been described as trail-marking pheromones in Isoptera, for example, hexanoic acid in *Zootermopsis nevadensis* (Hagen) (Karlson et al. 1968) and *cis, cis, trans*-3,6,8-dodecatrien-1-ol in *Reticulitermes* (Matsumura et al. 1968). Trail pheromones have a relatively low Q/K ratio, ca. 1.0 cm^3 per second, and thus are transmitted over a short range (Wilson 1970).

4. *Aggregation Pheromones or Arrestants*

These are chemicals or chemical combinations that cause insects to aggregate or congregate. The aggregation pheromone of the khapra beetle, *Trogoderma granarium* Everts, is reported to be a mixture of fatty acid

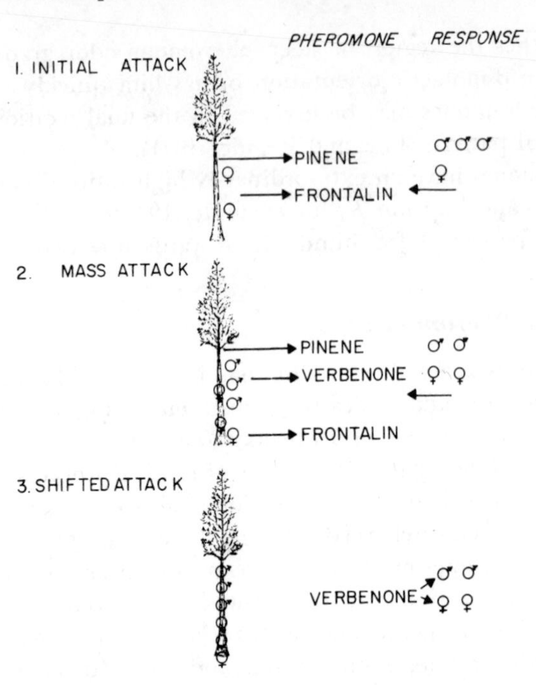

Fig. 8-2. Pheromone control of attack and colonization of pines by southern pine beetle, *Dendroctonus frontalis* Zimmerman. (After Renwick and Vité 1970.)

esters, methyl and ethyl oleate, ethyl palmitate, ethyl stearate, and ethyl linoleate (Ikan et al. 1969). Bark beetles, *Ips* spp. and *Dendroctonus* spp., employ groups of natural and metabolically altered pine terpenes as aggregation pheromones, as shown in Table 8.1 and Fig. 8.2 (Silverstein et al. 1968; Renwick and Vité 1971).

B. Food Lures

These are natural chemical substances present in many plant or animal hosts, which direct the insect pest toward suitable sites for feeding. Food lures may function as olfactory stimulants, producing orientation behavior in which the insect travels upwind to the source in a manner similar to the search for a source of sex pheromone. The response of the male oriental fruit fly, *Dacus dorsalis* Hendel, to methyl eugenol is of this nature (Fig. 8.3), and is directed by olfactory receptors on the antenna, which can detect 1×10^{-8} g or less of the lure (Steiner 1952, and

Fig. 8-3 Wild male oriental fruit flies, *Dacus dorsalis* Hendel, attracted to a small line of methyl eugenol applied around the edge of a windowpane.

unpublished data). Food lures or odor stimuli act as token representatives of the nutritive components of suitable food, for example, floral scents for nectar-feeding insects, essential oils for phytophagous insects, decomposition products for scavengers, and carbon dioxide, water, and lactic acid for bloodsucking insects (NAS 1969). Specific examples of food lures include sugar and propiononitrile for the housefly, *Musca domestica* L., coumarin for the sweet clover weevil, *Sitona cylindricollis* Fahraeus, sinigrin for the diamondback moth, *Plutella xylostella* (L.), and 3-hexen-1-ol and 2-hexene-1-ol for the silkworm, *Bombyx mori* (L), (Beck 1965). Natural and synthetic substances effective as practical insect food lures are shown in Table 8.2.

C. Oviposition Lures

These are natural chemical substances that control the selection by the adult female of sites for oviposition, for example, *p*-methylacetophenone for the rice stem borer (Saito and Munakata 1970). This area has considerable potential for exploitation, as it has been shown, for example, that corn earworm moths, *Heliothis armigera* (Hbn.), oviposit on twine impregnated with the juice from corn silk.

Table 8.2 Important Insect Food Lures[a]

Species	Lure	Structure
Oriental fruit fly, *Dacus dorsalis*	Methyl eugenol	$(CH_3O)_2C_6H_3CH_2CH{=}CH_2$
Melon fly, *Dacus cucurbitae*	*p*-Acetoxyphenethyl methyl ketone (Cue-lure)	$CH_3C(O)OC_6H_4CH_2CH_2C(O)CH_3$
Mediterranean fruit fly, *Ceratitis capitata*	*t*-Butyl-2-methyl-4-chlorocyclohexanecarboxylate (Trimedlure)	$Cl{-}C_6H_9(CH_3){-}C(O)OC(CH_3)_3$
Japanese beetle, *Popillia japonica*	Eugenol and phenethyl propionate	$HO(CH_3O)C_6H_3CH_2CH{=}CH_2$; $C_6H_5CH_2CH_2O(O)CC_2H_5$
European chafer, *Amphimallon majalis*	Propyl 1,4-benzodioxan-2-carboxylate (Amlure)	$C_6H_4(O{-}CH_2{-}CH(C(O)OC_3H_7){-}O)$
Coconut rhinoceros beetle, *Oryctes rhinoceros*	Ethyl 3-isobutyl-2,2-dimethyl cyclopropane carboxylate	$(CH_3)_2CHCH_2CH{-}CHC(O)OC_2H_5$, both CH joined to $C(CH_3)_2$ (cyclopropane)
Yellow jacket, *Vespula* spp.	Heptyl butyrate	$CH_3(CH_2)_5CH_2O(O)CCH_2CH_2CH_3$
Green June beetle, *Cotinis nitida*	Caproic acid	$CH_3(CH_2)_4C(O)OH$

D. Use of Attractants in Pest Management

Broadly speaking, effective insect attractants—sex pheromones, aggregation pheromones, or food lures—can be utilized in insect control in three major ways: (1) in sampling or monitoring pest insect populations (see Chapter 9), (2) in attracting insects to destruction by traps or poison baits, and (3) in confusing insects and distracting them from normal mating, aggregating, feeding, or oviposition behavior.

1. *Attractants in Sampling and Monitoring Insect Pest Populations*

This area has great promise in pest management. Insect surveys to determine the presence of responsive pest species have been built around the use of attractants. Thus a mixture of geraniol and eugenol, and later of phenethyl propionate and eugenol (7:3), has been used for many years to survey areas for infestation by the Japanese beetle, *Popillia japonica* Newman (McGovern et al. 1970). Methyl eugenol has been regularly employed in California to monitor infestation by the oriental fruit fly, *Dacus dorsalis,* and cue-lure, or *p*-acetoxyphenylethyl methyl ketone for the melon fly, *Dacus cucurbitae* Coquillett, (Alexander et al. 1962). In Florida, trimedlure has been used to monitor reinfestation by the Mediterranean fruit fly, *Ceratitis capitata* (Wiedemann), (Beroza et al. 1961) and to determine the precise areas of infestation to be treated.

Disparlure, the sex pheromone of the gypsy moth, *Porthetria dispar* (L.), is used in survey trapping to detect peripheral infestations of this pest in eastern hardwood forests. About 60,000 traps are in use, baited with 1.0 μg of disparlure and 5.0 mg of octanoin to promote persistence and slow release. The astounding potency of disparlure is shown by the following data (Beroza 1971):

	Male Moths Trapped after		
	1 week	6 weeks	12 weeks
Natural extract 10 ♀ terminalia	11	6	0
Disparlure, 0.001 μg	69	20	7
Disparlure, 0.01 μg	88	90	40
Disparlure, 0.1 μg	155	109	126
Disparlure, 1.0 μg	127	146	138

The use of the codling moth sex pheromone, *trans, trans*-8,10-dodecadien-1-ol (Table 8.1), for determining early spring emergence of male

Fig. 8-4 Commercially developed pheromone trap in apple tree for monitoring and controlling lepidopterous pests of apple.

Laspeyresia pomonella (L.), has resulted in improving the timing of apple spray programs and in decreasing the frequency of sprays as discussed in Chapter 7 (see Fig. 7.2) (Batiste 1970). Reports from New Jersey indicate that the use of pheromone trapping of the cabbage looper, *Trichoplusia ni* (Hübner), with *cis*-7-dodecenyl acetate to time a spray program for the protection of lettuce, decreased the number of spray applications required with no loss in marketability (Kiner, personal communication).

2. *Attractants in Trapping Insect Pests*

The classic example of this approach is the use of methyl eugenol to lure male oriental fruit flies, *Dacus dorsalis* Hendel, to toxic baits. This insect was eradicated from the island of Rota by dropping fiberboard squares impregnated with methyl eugenol and a contact insecticide, naled, at 3.4 g of toxicant per acre as described in Chapter 7 (Steiner et al. 1965). Bait sprays of yeast, protein, and malathion applied at 1.2

lb of toxicant per acre were used as the chief weapon in the successful campaign to eradicate the Mediterranean fruit fly, *Ceratitis capitata* (Wiedemann) from Florida (Steiner et al. 1961). The sex pheromone of the red-banded leaf roller, *Argyrotaenia velutinana* (Walker), *cis*-11-tetradecenyl acetate, has been used for 3 years in New York apple orchards in sticky traps for male moth control. One commercial trap (Fig. 8.4) was used per tree, each containing 20 μl of a 60:40 mixture of dodecenyl acetate and sex pheromone; the traps were rebaited every 6 weeks (Trammel et al. 1974). Fruit injury in trapped orchards averaged about 0.6% over the 3-year period, described as acceptable commercial control. For other examples see Table 8.1.

Older entomologists are familiar with a variety of poison baits using bran, molasses, and suitable insecticide to control grasshoppers, cutworms, armyworms, crickets, and earwigs (Metcalf et al. 1962). Poison baits for ants, incorporating syrup for sweet-eating ants and fat or vegetable oil for fat-eating species have been standard control measures for many years, for example, the peanut butter-mirex bait is used to control the imported fire ant, *Solenopsis saevissima richteri* Forel (Chapter 7, section III.A.4).

The complex pheromone and tree exudate mixtures that control the behavior of the bark beetles have provided interesting and productive opportunities for new techniques of pest management, as illustrated in the following case history.

Case History

Bark beetles of the genus *Dendroctonus* (Scolytidae) are economically the group of insects most destructive to coniferous forests. In the United States alone they are estimated to destroy several billion board feet of standing timber annually. They leave large areas of forest covered with dead trees, slowing regeneration and providing a fire hazard. These small beetles, which have a greatly enlarged prothorax and clubbed antennae, mine beneath the bark, feeding on the cambium and adjacent tissues. In the case of the western pine beetle, *D. brevicomis* LeConte, the female lays an average of 50 eggs in a gallery mined under the bark. On hatching, the larvae tunnel first in the cambium and then in the bark, requiring 30 to 60 days before pupation. There are two full generations and a partial third generation per year (Metcalf et al. 1962).

Recent studies on several *Dendroctonus* species have greatly advanced our knowledge of the mechanisms controlling selection of trees for attack, mating, and aggregation behavior. These have been shown to be primarily under the control of a complex of pheromones (McNew 1971). These pheromones have evolved so as to coordinate the mass attack on individual trees, which is necessary to diminish the sap flow that provides the trees' major defense against attack. They further act to adjust the sex ratio.

As shown in Fig. 8.2 for *Dendroctonus frontalis* Zimmerman, the attack may be divided into three stages (Renwick and Vité 1971). First, one or a few females find a suitable tree, preferably one with a weak sap flow which may have been the result of a variety of factors such as flood, drought, lightning, mechanical injury, diverse agents, and so on. These females release the pheromone frontalin on landing. Frontalin itself is only weakly attractive but, as the females bore into the tree, resin is exuded which contains the terpene α-pinene. The combination of frontalin and α-pinene is very attractive to females, but even more so to males. Males are attracted in a 3:1 ratio over females. As the mass attack progresses, the ratio of males to females attracted decreases as a result of the release of verbenone by males, which inhibits the male response to α-pinene and frontalin at low concentrations. This causes the sex ratio of the mass attack to approach 1:1. As the mass attack becomes successful, the flow of sap, hence of α-pinene, is slowed. Further successfully feeding females do not produce frontalin. At the same time, the verbenone concentration reaches high levels which inhibit both male and female responses to α-pinene and frontalin. Thus recruitment ceases.

Although the chemical ecology of *Dendroctonus, Ips,* and other Scolytidae is far from being completely understood, progress already made in identifying the pheromones and associated tree terpenes has made possible several imaginative approaches to bark beetle control in forests (McNew 1971). These include (1) application of the pheromone complex to baited trees to trap the beetles, and subsequently preventing brood development by increasing the phloem moisture level by injecting cacodylic acid into cuts near the base of the tree; (2) application of the pheromone complex to baited trees to attract and aggregate predaceous beetles of the families Cleridae and Ostomidae, which will destroy the bark beetle aggregations before they can enter the bark; (3) treatment of selected bait trees with the pheromone complex followed by logging of the baited trees so as to harvest bark beetles with the timber; (4) application of the pheromone complex in a slow-release formulation to trunks of trees sprayed with benzene hexachloride insecticide to destroy the aggregating beetles; and (5) baiting cardboard cylinders or panels covered with an adhesive such as Tanglefoot with the pheromone complex to trap the aggregating beetles.

The selection of methods for the pest management of the bark beetles depends on their compatibility with sound forestry practices (Chapter 14) and on local conditions of forest ecology and of bark beetle–predator complexes. For example, insecticide-pheromone applications should be used only where predatory beetles are of little importance, as in areas with short growing seasons.

A somewhat similar use of the pheromone of the male boll weevil, *Anthonomus grandis* Boheman, grandlure (Table 8.1,) has been described in Chapter 11. This chemical is the aggregating pheromone for weevils of both sexes, and attracted boll weevils into cottonfields treated with the systemic insecticide aldicarb which subsequently killed the weevils as they fed on the treated cotton (Lloyd et al. 1972).

3. *Pheromones as Mating Confusants*

The most elegant use of pheromones in pest management is their employment in suppressing mating by confusing male insects. The use of this procedure is still relatively unproven but rests on the theory that, if many sources of synthetic sex pheromone are introduced into the native environment of the insect, there will be a greatly reduced probability of the male (or female) finding a member of the opposite sex. The discovery of disparlure, the sex pheromone of the gypsy moth, *Porthetria dispar* (L.) (Table 8.1), provided a very stable compound active at as little as 5 g/ha. When 25 g of disparlure per hectare was dispensed in Pennsylvania on a ground cork base, the mating success of the gypsy moth was decreased from 81.3% in control areas to 35.0% (Cameron 1973). Similar tests with an improved microencapsulated formulation of disparlure at 15 g/ha disrupted mating sufficiently so that population increases of the gypsy moth were precluded, that is, below 10% (Cameron et al. 1974).

Experiments with the natural sex pheromone of the pink bollworm, *Pectinophora gossypiella* (Saunders), gossyplure, and a less active synthetic analog, hexalure, continuously evaporated into the air of cottonfields showed that the consequent disruption of mating produced a reduction in larval boll infestation comparable to commercial insecticide applications (Chapter 11) (Shorey et al. 1974). With rapidly increasing knowledge of the sex pheromones of important pest insects, the mating confusant technique is certain to be widely studied, refined, and exploited for pest management.

II. REPELLENTS

Repellents are chemicals that prevent insect damage to plants or animals by rendering them unattractive, unpalatable, or offensive. Thus repellents include a wide range of chemicals from volatile substances active in the vapor phase to protect man against biting flies and mosquitoes, for example, dimethyl phthalate and 2-ethyl-1-,3-hexanediol, to persistent chemicals such as Bordeaux mixture and tetramethylthiuram disulfide, which act as *feeding deterrents* to foliage feeders. Repellents have been defined *sensu strictu* as chemicals that cause insects to make oriented movements away from their source (Dethier 1963). However, this definition does not include deterrents or irritating insecticides like pyrethrins and DDT, which act as *excitorepellents* on insect tarsi, resulting in acute disturbances of the sensory nervous system and uncoordinated behavior and causing the insect to escape the treated area.

A. Repellents for Foliage Feeders

This area of entomology has not progressed very far in the past 100 years, as the most successful foliage repellent is still Bordeaux mixture which was devised in France in 1882. Bordeaux mixture is produced from copper sulfate, hydrated lime, and water in a 6-10-100 mixture and acts as a repellent to flea beetles, leafhoppers, and the potato psyllid *Paratrioza cockerelli* (Sulc) (Metcalf et al. 1962). Newer foliage repellents include tetramethylthiuram disulfide, which repels feeding by the Japanese beetle, *Popillia japonica* Newman, and 4′-(dimethyltriazeno)-acetanilide, which is a feeding repellent or deterrent to a wider range of foliage-feeding insects including the cabbage looper, *Trichoplusia ni* (Hübner); the cotton leafworm, *Alabama argillacea* (Hübner); the cotton boll weevil, *Anthonomus grandis* Boheman; and the spotted cucumber beetle, *Diabrotica undecimpunctata howardi* Barber (NAS 1969).

Naturally occurring feeding repellents or deterrents exist for many insects and undoubtedly play a role in host selection and specificity. The best studied example is 6-methoxybenzoxazolinone (6-MBOA) which is a major factor in corn varieties resistant to feeding by the European corn borer, *Ostrinia nubilalis* (Hübner) (Chapter 4, Section II.B.2).

The use of foliage repellents offers few advantages in pest-management programs. These chemicals need very thorough coverage to prevent insect damage, and over a large area merely direct insect attack elsewhere to untreated crops. The chemicals themselves provide about the same degree of environmental hazard as conventional insecticides without providing any decreases in the pest population.

B. Repellents for Crawling Insects

The use of creosote or 4,6-dinitro-*o*-cresol as soil barriers against chinch bug, *Blissus leucopterus leucopterus* (Say), migration from small grain to corn is a practical example of excitorepellency which provides good crop protection with minimal environmental contamination or disturbance (Metcalf et al. 1962).

C. Repellents for Wood-Feeding Insects

Wooden timbers and structures placed in or near the soil suffer enormous damage from the feeding of termites (order Isoptera), and lesser damage from powder-post beetles, *Lyctus* and *Anobium* spp. Damage in the United States is estimated as about $500 million annually. Various wood species have substantial resistance to termite attack, which is largely due to specific chemicals present acting as feeding deterrents. Examples include pinosylvin monomethyl ether in *Pinus sylvestris* L.; taxifolin in Douglas fir, *Pseudotsuga menziesii* (Mirb.) Franco; and β-methylanthraquinone from East Indian teak, *Tectona grandis* L. (NAS 1969). Redwood *Sequoia sempervirens* (D. Don) Endl. is also highly resistant to termite attack.

Synthetic substances such as pentachlorophenol and its sodium salt also act as feeding deterrents to termite attack and are widely used as surface treatments or pressure impregnants for telephone poles and foundation timbers. Soil treatments around foundations of buildings with 0.5% aldrin and dieldrin, 0.8% lindane, 1% chlordane, or 8% DDT in oil solution or water emulsion have been approved by the Federal Housing Authority as giving protection of above-ground timber against termite attack for 5 years or more. These insecticides (Chapter 7) presumably act both as excitorepellents and as soil poisons.

D. Repellents to Fabric-Eating Insects

Clothes moths *Tinea pellionella* (Linnaeus) and *Tineola bisselliella* (Hummel) (family Tineidae) and carpet beetles, *Anthrenus* spp. and

Attagenus piceus (Olivier) (family Dermestidae), cause damage to woolen goods, rugs, furniture, and so on, estimated annually in the United States at about $200 million. The use of feeding repellents offers almost complete protection to various types of woolen goods when properly applied. The colorless dyestuffs Eulan CN and Mitin FF can be firmly affixed to wool during the dyeing operation and provide protection over the lifetime of the article when applied at 1 to 3% of the weight of the cloth. Sodium aluminum fluosilicate in a 0.5 to 0.7% water solution with an equivalent amount of wetting agent is an effective repellent for fabric pests when applied to wool by spraying or dipping and cannot be removed by dry cleaning. Fabrics treated with 0.25 to 0.75% DDT or 0.05% dieldrin based on cloth weight are protected against clothes moths and carpet beetles for years (Metcalf et al. 1962). The wide use of the last-mentioned two poisons in dry cleaning processes for moth proofing has resulted in undesirable environmental pollution.

E. Repellents to Bloodsucking Insects

1. Repellents for Protection of Man

A spectacular example of insect repellency is the use of a variety of slowly volatile chemicals on skin or clothing to prevent the feeding of bloodsucking mosquitoes, flies, mites, and ticks. Where these creatures are also vectors of human and animal diseases, the use of such repellents can become a component of pest-management programs (Chapter 15). For example, the widespread use of dimethyl phthalate and 2-ethyl-1,3-hexanediol by the U.S. Armed Forces in the South Pacific in World War II as repellents for *Anopheles* mosquitoes was an important part of the armed forces antimalaria program (Smith 1970). Similarly, dibutyl phthalate and benzyl benzoate were used for clothing impregnation to prevent the attack of *Trombicula* spp. mites, vectors of scrub typhus (Chapter 15).

The use of newer repellents such as deet or diethyl-*m*-toluamide and benzil as described in Table 8.3 has provided outdoor workers, sportsmen, and so on, with comparative comfort and tranquillity during biting seasons of pest mosquitoes (*Aedes, Culex, Psorophora* spp.), blackflies (*Simulium* spp.), other biting flies [*Tabanus, Chrysops, Stomoxys calcitrans* (L.), *Leptoconops* spp., etc.], mites, and ticks.

Case History

The following quote from Agassiz's *Lake Superior* (Agassiz and Cabot 1850) will serve as a reminder of the severity of such attacks

Table 8.3 Important Insect Repellents[a]

Species	Repellent	Structure
Mosquitoes, *Anopheles*, *Aedes*, *Culex* spp.	Dimethyl phthalate	$C_6H_4(COOCH_3)_2$
Mosquitoes, flies, fleas	2-Ethyl-1,3-hexanediol (612)	$CH_3(CH_2)_2CHOHCH(C_2H_5)CH_2OH$
Mosquitoes, flies, fleas	*N*,*N*-Diethyl *m*-toluamide (deet)	$CH_3C_6H_4C(O)N(C_2H_5)_2$
Mosquitoes, flies, fleas	2-Ethyl-2-butyl-1,3-propanediol	$HOCH_2C(C_2H_5)(C_4H_9)CH_2OH$
Ticks, fleas	*N*-Butylacetanilide	$C_6H_5N(C_4H_9)C(O)CH_3$
Mites (chiggers), *Trombicula* spp.	Benzyl benzoate	$C_6H_5C(O)OCH_2C_6H_5$
Mites (chiggers)	Benzil	$C_6H_5C(=O)-C(=O)C_6H_5$
Mites (chiggers)	Dibutyl phthalate	$C_6H_4(COOC_4H_9)_2$
Flies on cattle	Dibutyl succinate	$C_4H_9OOCH_2CH_2COOC_4H_9$
Flies on cattle	Dipropylpyridine-2,5-dicarboxylate	$C_3H_7OOC(C_5H_3N)COOC_3H_7$
Flies on cattle	Butoxypolypropylene glycol	$C_4H_9O[CH_2CH(CH_3)O]_nCH_2CH(CH_3)OH$
Flies on cattle	2-hydroxyethyl octyl sulfide	$HOCH_2CH_2SC_8H_{17}$

[a] Metcalf et al. 1962.

by biting flies on the human person: "Nothing could tempt us into the woods so terrible were the blackflies. One, whom scientific ardor tempted a little way up the river in a canoe, after water plants, came back a frightful spectacle, with blood-red rings around his eyes, his face bloody, and covered with punctures. The next morning his head and neck were swollen as if from an attack of erysipelas." Today the outdoorsman can obtain protection from blackfly attack (*Simulium*) for up to 5 hours by judicious application of dimethyl phthalate, 2-ethyl-1,3-hexanediol, or diethyl-*m*-toluamide to the exposed skin or to clothing or a veil. Such personal protection can become an important component of a pest-management program for onchocerciasis, a filarial disease of man, in which the blackfly is the vector (Chapter 15).

Individual repellents for biting insects are generally applied directly to the skin, and deet (Table 8.3) is perhaps the most generally suitable for this purpose, giving 3 to 8 hours protection against mosquitoes, deerflies, sandflies, biting gnats, fleas, mites (chiggers), ticks, and land leeches (Beroza 1972). Application of repellents to clothing or to wide-meshed netting as a space repellent can extend the length of protection. The formulation most resistant to leaching by rain and laundering is the U.S. Armed Forces formulation M-1960: 3 parts benzyl benzoate, 3 parts *N*-butylacetanilide, 3 parts 2-ethyl-2-butyl-1,3-propanediol, and 1 part Tween-80 emulsifier. This mixture applied to clothing at 2 g/ft^2 protects against mites, ticks, and mosquitoes for several weeks (Hall et al. 1957).

1. Repellents for Protection of Domestic Animals

These have been of limited value in pest management, because rapid skin absorption limits their effectiveness to 1 to 2 days, and because of the problems of obtaining thorough coverage. Materials that have had limited use include butoxypolypropylene glycol, di-*n*-butylsuccinate, dipropylpyridine-2,5-dicarboxylate, and 2-hydroxyethyl octyl sulfide (NAS 1969) (Table 8.3). Sprays of synergized pyrethrins, which act as excito-repellents, can extend practical protection to livestock against all biting flies when applied twice daily. Practical repellent protection for cattle against the attacks of the stable fly, *Stomoxys calcitrans* (L.); the horn fly, *Haematobia irritans* (L.); the horsefly; deer flies (*Tabanidae*); mosquitoes (Culicidae); the face fly, *Musca autumnalis* de Geer, and the housefly, *Musca domestica* L., would doubtless be profitable to beef producers and dairymen in preventing substantial weight loss and decrease in milk production because of the constant annoyance and blood loss (NAS 1969; Bruce and Decker 1958).

III. GENETIC CONTROL OF PEST POPULATIONS

Genetic control refers to a variety of methods by which a pest population can in theory or practice be controlled through the manipulation of its genetic component or other mechanism of inheritance (NAS 1969; Knipling 1972; Smith and von Borstel 1972).

Genetic control mechanisms have not at this time proved to be widely useful. Their single major success has been in control of the screwworm, *Cochliomyia hominivorax* (Coquerel), a major pest to livestock in the southern United States. In this case control is effected through the competition for matings between sterile mass-reared males and the natural population of males. At present the screwworm has been eradicated from the southeastern United States. Additional details are given in Section III.C.

Despite problems that have been encountered in attempts to use genetic methods of control on other pests, their potential importance to the field of pest management makes an understanding of the parameters involved necessary. This is particularly true when one considers the continued advances in theory and practice of genetic control mechanisms, together with the decline in effectiveness and realization of adverse effects on the environment of some chemical methods of control.

A. The Sterile-Male Technique

As has been stated, the sterile-male technique has been the most successfully applied of genetic control methods. Further, most potentially useful methods of genetic control are constrained by the same parameters. Therefore it is instructive to examine in some depth the workings of this method. Toward this end, a modified version of a demonstrative model of the sterile-male technique presented by Knipling (1967) is used.

In general, populations show exponential growth where resources are not limiting and other factors are held constant. The results, over three generations, for such a hypothetical population where the average reproductive success per female is five is presented in Table 8.4. Note that the sex ratio is assumed to be 1:1 and no differential mortality occurs, that is, a population of 10 females implies a population of 10 males.

Table 8.4 Rate of Increase for a Population in Which the Reproductive Success is Five Daughters per Female per Generation

Generation	Number of Females per Unit Area
Parental	1,000,000
F1	5,000,000
F2	25,000,000
F3	125,000,000

Table 8.5 shows the expected population size per generation where 90% mortality is produced in each parental population by application of a pesticide.

Table 8.5 Rate of Decrease of a Population Subjected to 90% Control Each Generation[a]

Generation	Number of Females per Unit Area
Parental	1,000,000
F1	500,000
F2	250,000
F3	125,000
F4	62,000

[a] An increase rate of 5× females is assumed for the females surviving each generation.

It should be noted that the amount, hence the cost, of a certain level of control with a pesticide is in general proportional to the area treated, not to the size of the pest population. The damage done by a pest is in general proportional to the number of individuals per unit area. Thus the cost/benefit ratio of an insecticide control program increases as the size of the pest population decreases.

Table 8.6 shows the theoretical levels of control attained when the ratio of sterile males released to wild males present is 9:1 in the parental generation and the same absolute number of sterile males is released in each successive generation. It is assumed that each female chooses a single mate at random from the entire pool of males. In the parental

Table 8.6 Rate of Decrease of a Population Subjected to Sustained Release of the Number of Sterile Males Necessary to Produce a 9:1 Ratio of Sterile to Wild Males in the Parental Generation

Generation	Females (Number)	Sterile Males (Number)	Ratio of Sterile to Fertile Males	Reproducing Females (Number)
Parental	1,000,000	9,000,000	9:1	100,000
F1	500,000	9,000,000	18:1	26,316
F2	131,580	9,000,000	68:1	1907
F3	9,535	9,000,000	942:1	10
F4	50	9,000,000	180,000:1	0

generation 1 out of 10 matings are made with a fertile male and produce offspring. Thus, of the initial 1 million females, 100,000 mate with fertile males and produce 500,000 females in the F_1 generation.

If the ratio of sterile males to wild males had been less than 4:1, the pest population would have continued to increase.

The important point to emphasize is that while the level of control with a pesticide is proportional to the area infested, control by release of sterile males is proportional to the size of the pest population. Thus the cost/benefit ratio decreases with a decrease in the size of the pest population.

In general, this model suggests that the potential exists to integrate the use of the two methods so as to exploit the different shapes of their respective cost/benefit curves (Chapter 3). An insecticide might be used initially to reduce high population numbers, and then genetic control substituted to either maintain an acceptable population level or reduce it further, as demonstrated in Table 8.7.

B. Sterility Versus Competition among Gametes

To understand how this model's predictions may be made more general, it is necessary to clarify the confusion engendered by common synonymous usage of the terms *genetic control* and *sterile-male technique.*

A sterile male is one that produces either inactive sperm or none at all. To effect sterility it is usually necessary to kill the primary germ cells in order to keep the males from producing new normal sperm after irradia-

Table 8.7 Rate of Decrease of a Population Subjected to the Combined Use of Sterile Males and a Pesticide (at 90% Control Level) for one Generation Followed by the Release of Sterile Males Alone

Generation	Normal Trend (Number)	Original Population Reduced by 90% (Number)	Females Remaining (Number)	Sterile Males Released (Number)	Females Reproducing (Number)
Parental	1,000,000	100,000	100,000	900,000	10,000
F1	5,000,000	—	50,000	900,000	2,632
F2	25,000,000	—	13,160	900,000	189
F3	125,000,000	—	945	900,000	0

tion. Thus competition for mates can be between fertile wild and sterile released males. Here the efficiency of the technique is greatly reduced if the female mates more than once, given that multiple matings provide excess sperm.

In many other mechanisms of genetic control, a deleterious mutation is introduced into the natural population by a sperm that actually fertilizes an egg. In this case competition for eggs is between normal sperm and those carrying the deleterious mutation. The effectiveness of the techniques is independent of the number of matings. Thus the predictions of the model hold when multiple matings occur and when competition is between mutant and normal sperm instead of between sterile and normal males (Smith and von Borstel 1972).

Because of the frequent occurrence of multiple matings and problems associated with the irradiation of every individual to be released, as in the sterile-male technique, other methods of genetic control have been examined. Theoreticians and experimentalists have had to satisfy the often opposing requirements of isolating a mutant that will be deleterious when mixed with the pest population and at the same time can be viable so as to be raised in large numbers. As examples, two of the theoretical solutions that have shown the most promise are described.

1. Conditional Lethals

An allele that is lethal only when and if a certain environmental condition occurs is said to be conditionally lethal. A conditionally lethal mutation for which some experimental success has been recorded is temperature sensitivity. The mutant allele causes a carrier's death when a certain temperature is experienced. It is apparent that large numbers of carriers of dominant conditionally lethal mutations could be raised by controlling the critical parameter under breeding conditions.

2. *Chromosomal Translocations*

A chromosomal translocation is the exchange of a chromosomal section by two nonhomologous chromosomes. A strain that is homozygous for the translocation may suffer no loss in viability and thus can be raised in large numbers. When the translocated strain is bred to the wild population, the gametes of individuals that are heterozygous for the normal and translocated chromosomes often have substantial deletions and duplications of chromosome arms, which produce abnormal gametes. Between 35 and 85% lethal gametes are produced for a single chromosomal translocation. If more than one chromosome is involved, lethality can approach 100% (Pal and LaChance 1974; Foster et al. 1972).

C. General Problems of Genetic Control Techniques

Ten years experience with the idiosyncracies of genetic control methods has shown that they work, but that the problems of application and cost are far greater than was initially realized. One set of problems may be characterized as technological or relating to the cost of applying our present techniques. Here might be included genetic engineering, that is, our ability to induce the mutation desired and maintain it without reducing the competitive ability of the strain in the wild. For example, just being realized is the extent to which laboratory strains of animals may differ from the populations they were derived from, for example, the Norway rat and the common laboratory rat. The laboratory rat cannot compete for mates with the wild rat and therefore cannot act as a carrier for a deleterious mutation. Another technological problem is the rearing of numbers of individuals at orders of magnitude greater than the size of the pest population (and often segregating them by sex) at what is considered an acceptable cost.

A second set of problems may be characterized as intrinsic properties of animal populations. The importance of knowledge of such fundamental population parameters as the number of individuals, the population's rate of growth, and the rate and distance of immigration is obvious in relation to the model of the sterile-male technique and its application to the screwworm. For example, consider the importance of population density to a population's rate of growth. In one experiment with *Culex pipiens quinquefasciatus* Say (*Culex fatigans*), 70% survival from egg to adult was recorded at 3100 eggs/m^2 versus less than 1% survival at 387,000 eggs/m^2 (Pal and LaChance 1974). This result is probably due in part to the virtual elimination of intraspecific competition at low population levels. Another undesirable effect of decimating a population is the concomitant damage to predator populations, which further accelerates the rebuildup of the pest population size.

Another population property of considerable interest is the rate at which resistance to control can be expected to appear. At present the paucity of successfully run control programs using genetic properties leaves us without empirical data on this point. We can theorize that the rate at which a pest population, such as the screwworm, becomes resistant is proportional to the selection of females that mate assortatively with wild males. Differences that might lead to resistance could be as simple as a change in the time of day at which mating occurs. Any similar characteristic represented unequally in the released population and the wild population tends to have its frequency increased with the selective pressures of control. The successful screwworm control case in Florida is potentially the least likely to have developed resistance, since the parents for the next generation of males to be sterilized can be continually renewed from the wild population and will thus mirror its changing characteristics. At present this is not being done, which may partly explain the recent decrease in control observed in the southwestern United States (Table 8.8).

This situation cannot be said to be generally true of other genetic control mechanisms and other pests. Particularly, mechanisms such as the use of translocations or temperature-sensitive mutations use individuals derived from a single mating, and thus do not represent the dif-

Table 8.8 Screwworm Infestations in the Southwestern United States Following Eradication Efforts[a]

	1962	1963	1964	1965	1966	1967	1968	1969	1970	1971	1972
Confirmed cases	49,484	4916	223	446	1203	835	9268	161	92	444	92,198
Counties infested	280	224	83	75	112	74	157	60	31	61	348

[a] USDA data.

ferences within the pest population. Nor is it easy with present techniques to go back to the wild population and isolate a new mutant. It should be noted that, when we can isolate deleterious mutants as fast as differing populations emerge, there is no theoretical end to the effectiveness of the control procedures.

Case History

The female screwworm, *Cochliomyia hominivorax* (Coquerel), is a dark, shiny, blue-green blowfly with three black stripes on the prothorax. It is about twice as large as the housefly and lays shinglelike masses of 200 to 400 eggs, cemented to the dry skin near wounds on animals. The eggs hatch within a day, and small maggots with screwlike spinose ridges enter the wound and feed on pieces of healthy flesh. Their body secretions prevent healing of the wounds, and resultant contamination produces foul-smelling, pus-discharging sores. After 4 to 10 days, when the larvae become full grown, they drop to the ground and pupate over 3 to 14 days in the soil. There are commonly 8 to 10 generations a year in a warm climate such as Florida (Metcalf et al. 1962). Untreated screwworm infested wounds may kill cattle, and the wounds severely damage the quality of the cowhide, so that annual damage in the southeastern United States has been estimated as high as $20 million and that in the southwestern United States at $50 to $100 million. During 1935 a severe outbreak is reported to have caused 1,200,000 cases of attack in domestic animals and 55 cases in man (Baumhover 1966).

The screwworm is primarily an inhabitant of tropical and subtropical areas of North and South America, but may disperse as much as 500 mi from its normal overwintering area. Extensive studies by Knipling, Bushland, and others demonstrated that the female screwworm fly mates only once in her lifetime and this, together with the relatively small natural population size, about two per square mile, made this destructive insect an appropriate candidate for an eradication trial through the release of sterile males, as proposed by Knipling (1955). It was found that male screwworms could be sterilized by 5 kV of gamma radiation from ^{60}Co, administered to the 5-day-old pupa. The males so sterilized could compete normally with wild males for mating opportunities.

The island of Curacao, 40 mi north of Venezuela, where the screwworm severely attacked the goat population, was selected for the first field trial, and approximately 800 sterile flies were dropped by air over each of the 170 mi² of Curacao. Collection of egg masses from various host animals showed that sterility was 69% after 1 week, 79% at 4 weeks, 88% at 6 weeks, and 100% at 7 weeks, closely following the theoretical calculations (Table 8.6) (Baumhover 1966). This successful eradication in 1953 set the stage for the eradication attempt in Florida. A large rearing and sterilization facility which could produce 2 million flies weekly was established at Sebring, Florida, where the flies were reared in trays on a mixture of 2 parts water, 2 parts lean ground beef, 1 part citrated blood, and 0.24% formalin (Smith 1960). Pupation took place in sand-filled trays, and the sifted pupae were exposed to 8000 r of gamma radiation from ^{60}Co. The sterile pupae were packaged at 440 per carton and released from aircraft at the rate of 200/mi², and as many as 80 million flies, were released weekly over about 85,000 mi² in the winter of 1958–59. The final infestation of screwworm was recorded on February 19, 1959. The eradication program was estimated to cost $10 million and to have saved southeastern ranchers $140 million since 1958 (Baumhover 1966).

The enormous success of genetic control spurred demands for a similar eradication project in the southwestern United States. The difficulties were much greater because of the large area involved and the constant reinfestation from Central and South America. Nevertheless, with advanced methods of rearing and sterilization, a plant was constructed at Mission, Texas, to produce up to 150 million sterile screwworm flies per week. Beginning in February 1962, again after a severe winter which limited the area of infestation, sterile flies were dropped over an area which has now expanded to more than 300,000 mi² including a sterile-fly barrier zone extending 2000 mi along the United States–Mexican border and from 300 to 500 mi in width. At first the results were uniformly encouraging, and the cases of screwworm fly infestation in the United States dropped rapidly from 50,000 in 1962 to 92 in 1970 (Table 8.8). The cost of the program was approximately $5 million annually against savings estimated of $100 million (Knipling 1967). However, that was the low point and infestations have risen steadily, reaching 92,198 in 1972 (Table 8.8) (Cuellar and Brinklow 1973). The reasons for this astonishing reversal are not well understood, but may relate to a behavioristic selection for assortative mating which has diminished the competitivity of the factory-produced sterile males (Cuellar and Brinklow 1973).

Under discussion is a plan to push the area of infestation back to the Isthmus of Panama. There the cost of maintaining the barrier would be reduced in proportion to the reduction in the width of the zone of contact with the infested area, and the benefits would accrue to the entire North American continent.

D. Chemosterilization

A large variety of chemicals is known that interrupt the mitotic apparatus of the cell and prevent its reproduction. Many of these have been discovered through programs for cancer chemotherapy. Such compounds have proved highly effective in causing sterility in insects and have been proposed as substitutes for gamma radiation in the production of sterile males (Borkovic 1966). Such sterilization could be produced in the field by luring insects to traps containing the chemosterilant. Examples include 0.5% tepa in sugar bait for sterilizing the housefly, *Musca domestica,* and a 0.025% tepa solution in drinking water in a protein hydrolyzate–baited trap for sterilizing the Mexican fruit fly, *Anastrepha ludens* (Loew).

Compounds evaluated as insect chemosterilants include alkylating agents, aziridines such as tris-(1-aziridinyl)phosphine oxide (tepa) and 2,2,4,4,6,6-hexahydro-2,2,4,4,6,6-hexakis(1-aziridinyl)-1,3,5,2,4,6-triazatriphosphorine (apholate), tetramethylene-bis-1,4-methylsulfonate (busulfan), antimetabolites such as 5-fluorouracil and 5-fluoroorotic acid and the folic acid analogs aminopterine and methotrexate, *s*-triazines such as tris-2,4,6-dimethyl-*s*-triazine (hemel), triphenyl tins, and ureas such as 2-imidazolidinone and ethylene thiourea (Borkovic 1966).

Unfortunately, effective chemosterilants are all compounds that display strong mutagenic action in man and higher animals, and their use, even under carefully supervised conditions represents a definite hazard to man and the environment. Their use in pest management cannot be recommended.

REFERENCES

Alexander, B. H., M. Beroza, T. A. Oda, L. F. Steiner, D. H. Miyashita, and W. C. Mitchell. 1962. The development of male melon fly attractants. *J. Agr. Food Chem.* **10:**270–276.

Batiste, W. C. 1970. A timing sex-pheromone trap with special reference to codling moth collection. *J. Econ. Entomol.* **63:**915–1918.

Baumhover, A. H. 1966. Eradication of the screwworm fly. *J. Amer. Med. Assoc.* **196:**240–248.

Beck, S. D. 1965. Resistance of plants to insects. *Annu. Rev. Entomol.* **10**:207–232.

Beroza, M., N. Green, S. I. Gertler, L. F. Steiner, and D. H. Miyashita. 1961. New attractants for the Mediterranean fruit fly. *J. Agr. Food Chem.* **9**:361–365.

Beroza, M. Insect sex attractants. 1971. *Amer. Sci.* **59**:320–325.

Beroza, M. 1972. Attractants and repellents for insect pest control. Pages 226–253 *in* Pest control strategies for the future, National Academy of Sciences, Washington, D.C.

Borkovic, A. B. 1966. Insert chemosterilants Vol. VII. Advances in Pest Control Research. R. L. Metcalf, ed. Wiley-Interscience, New York. 140 pp.

Bowers, W. S., L. R. Nault, R. E. Webb, and S. R. Dutky. 1972. Aphid alarm pheromone: Isolation, identification, synthesis. *Science* **171**:1121–1122.

Brower, L. P. S., van Zandt Brower, and F. P. Cranston. 1965. Detailed study of hair pencil courtship of *Danaus gilippus berenice*. *Zoologica* **50**:1–39.

Brown, W. L. Jr., T. Eisner, and R. H. Whitlake. 1970. Allomones and kairomones: Transpecific chemical messengers. *Bioscience* **20**:21–22.

Bruce, W. N., and G. C. Decker. 1958. The relationship of stable fly abundance to milk production in dairy cattle. *J. Econ. Entomol.* **51**:269–274.

Butenandt, A., R. Beckmann, D. Stamm, and E. Hecker. 1959. Über den Sexual Lockstoff des Seidenspinners *Bombyx mori*, Reindarstellung und Konstitution. *Z. Naturforsch.* **B14**:283–284.

Cameron, E. A. 1973. Disparlure: A potential tool for gypsy moth population manipulation. *Bull. Entomol. Soc. Amer.* **19**:15–19.

Cameron, E. A., C. P. Schwalbe, M. Beroza, and E. F. Knipling. 1974. Disruption of gypsy moth mating with microencapsulated disparlure. *Science* **18**: 972–973.

Carson, R. 1962. Silent spring. Houghton-Mifflin, Boston, 368 pp.

Cueller, C. B. and D. M. Brinklow. 1973. The screwworm strikes back. *Nature* **242**, 493.

Dethier, V. 1963. The physiology of insect senses. John Wiley, New York. 266 pp.

Farkas, S. R., and H. H. Shorey. 1972. Chemical trail-following by flying insects: a mechanism for orientation to a distant odor source. *Science* **178**:67–68.

Foster, G. G., M. J. Whitten, T. Prout and R. Gill. 1972. Chromosome rearrangements for the control of insect pests, *Science* **176**:875–880.

Hall, S. A., N. Green, and M. Beroza. 1957. Insect repellents and attractants. *J. Agr. Food Chem.* **5**:663.

Ikan, R., E. D. Bergmann, U. Yinon, and A. Shulov. 1969. Identification, synthesis and biological activity of an "assembly scent" from the beetle *Trogoderma granarium*. *Nature* **223**:317.

Karlson, P. and A. Butenandt. 1959. Pheromones (ectohormones) in insects. *Annu. Rev. Entomol.* **4**:39–58.

Karlson, P., M. Luscher, and H. Hummel. 1968. Extraktion und biologische Auswertung des Spurpheromons der Termite *Zootermopsis nevadensis J. Insect Physiol.* **14**:1763–1769.

Knipling, E. F. 1955. Possibilities of insect control or eradication through the use of sexually sterile males. *J. Econ. Entomol.* **48**:459–462.

Knipling, E. F. 1967. Sterile technique-principles involved, current application, limitations, and future application. Pages 587–616 *in* J. W. Wright and R. Pal, eds., Genetics of insect vectors of disease. Elsevier, Amsterdam.

Knipling, E. F. 1972. Sterilization and other genetic techniques. Pages 272–287 *in* National Academy of Sciences, Pest control for the future. Washington, D.C.

Law, J. H. and F. E. Regnier. 1971. Pheromones. *Annu. Rev. Biochem.* **40**:533–548.

Lloyd, E. P., W. P. Scott, K. K. Shaumak, F. C. Tingle, and T. B. Davich. 1972. A modified trapping system for suppressing low-density populations of overwintering boll weevils. *J. Econ. Entomol.* **65**:1144–1147.

MacConnell, J. G. and R. M. Silverstein. 1973. Recent results in insect pheromone chemistry *Angew. Chem., Int. Ed.* **12**:644–654.

Matsumura, F., H. C. Coppl, and A. Tai. 1968. Isolation and identification of termite trail-following pheromone. *Nature,* **219**:963–964.

McGovern, T. P., M. Beroza, T. L. Ladd, Jr., J. C. Ingange, and J. P. Purimos. 1970. Phenethyl propionate, a potent new attractant for Japanese beetles. *J. Econ. Entomol.* **63**:1727–1729.

McNew, G. L. 1971. The Boyce Thompson Institute program in forest entomology that led to the discovery of pheromones in bark beetles. *Contrib. Boyce Thompson Inst.* **24**:251–262.

Meinwald, J., A. M. Chalmers, T. E. Pliske, and T. Eisner. 1968. Pheromones. III. Identification of *trans, trans*-10-hydroxy-3,7-dimethyl-2,6-decadienoic acid as a major component in "hairpencil" secretion of the male monarch butterfly. *Tetrahedron* Lett. **47**:4893–4896.

Metcalf, C. L., W. P. Flint, and R. L. Metcalf. 1962. Destructive and useful insects. 4th ed. McGraw-Hill, New York, 1087 pp.

National Academy of Sciences. 1969. Insect pest management and control. Principles of plant and animal pest control. Vol. 3, Pub. 1695. Washington, D.C.

National Academy of Sciences. 1972. Pest control strategies for the future. Washington, D.C.

Pal, R. and L. E. LaChance. 1974. The operational feasibility of genetic methods for control of insects of medical and veterinary importance. *Annu. Rev. Entomol.* **19**:269–292.

Renwick, J. A. A. and J. P. Vité. 1971. Systems of chemical communication in *Dendroctonus. Contrib. Boyce Thompson Inst.* **24**:283–292.

Saito, T. and K. Munakata. 1970. Insect attractants of vegetable origin, with special reference to the rice stem borer and fruit-piercing moths. Pages 225–236 *in* D. L. Wood, R. M. Silverstein, and M. Nakajima, eds., Control of insect behavior by natural products. Academic Press, New York. 345 pp.

Shorey, H. H., L. K. Gaston, and R. N. Jefferson. 1968. Insect sex pheromones. Pages 57–126 *in* Advances in pest control Research. R. L. Metcalf, ed., Vol. VIII. Wiley-Interscience, New York.

Shorey, H. H., R. S. Kaae, and L. K. Gaston. 1974. Sex pheromones of Lepidoptera. Development of a method for pheromonal control of *Pectinophora gossypiella* in cotton. *J. Econ. Entomol.* **64:**347–350.

Silverstein, R. M., R. G. Brownlee, T. E. Bellas, D. L. Wood, and L. E. Browne. 1968. Brevicomin: Principal sex attractant in the frass of the female western pine beetle. *Science* **159:**889–891.

Smith, C. L. 1960. Mass production of screwworm (*Callitroga hominovorax*) for the eradication program in the southeastern states. *J. Econ. Entomol.* **53:** 1110–1116.

Smith, C. N. 1970. Repellents for anopheline mosquitoes. *Misc. Pub. Entomol. Soc. Amer.* **7:**99–115.

Smith, R. H. and R. C. von Borstel. 1972. Genetic control of insect populations. Science 178:1164–1174.

Steiner, L. F. 1952. Methyl eugenol as an attractant for the oriental fruit fly. *J. Econ. Entomol.* **45:**241–248.

Steiner, L. F., W. C. Mitchell, E. J. Harris, T. T. Kozuma, and M. S. Fujimoto. 1965. Oriental fruit fly eradication by male annihilation. *J. Econ. Entomol.* **58:**961–964.

Steiner, L. F., G. G. Rohwer, E. L. Ayers, and L. D. Christenson. 1961. The role of attractants in the recent Mediterranean fruit fly eradication program in Florida. *J. Econ. Entomol.* **54:**30–35.

Trammel, K., W. L. Roelofs, and E. H. Glass. 1974. Sex-pheromone trapping of males for control of redbanded leafroller in apple orchards. *J. Econ. Entomol.* **67:**159–170.

Tumlinson, J. H., J. C. Moser, R. M. Silverstein, R. G. Brownlee, and J. M. Ruth. 1972. A volatile trail pheromone of the leaf cutting ant, *Atta texana. J. Insect Physiol.* **18:**809–814.

Tumlinson, J. H., D. D. Hardee, R. C. Gueldner. A. C. Thompson, P. A. Hedin, and J. P. Minyard. 1969. Sex pheromones produced by male boll weevil: Isolation, identification, and synthesis. *Science* **166:**1010–1012.

Wilson, E. O. and W. H. Bossert. 1963. Chemical communication among animals. *Rec. Prog. Horm. Res.* **19:**673–716.

Wilson, E. O. 1970. Chemical communication within animal species. Pages 133–55 *in* E. Sondheimer and J. B. Simeone, eds., Chemical ecology. Academic Press, New York.

STRATEGY

9

THE QUANTITATIVE BASIS OF PEST MANAGEMENT: SAMPLING AND MEASURING

William G. Ruesink and Marcos Kogan

Modern pest management cannot operate without accurate estimates of pest and natural enemy population densities, or without reliable assessments of plant damage and its effect on yield. Acquiring quantitative information about the agroecosystem is a preliminary phase of any basic or applied work on insect–plant interactions.

Data collecting for research purposes differs considerably from that needed for making management decisions. Research requires precise estimates of parameter values, whereas management requires the rapid classification of situations into a decision category (e.g., spray or do not spray). Researchers can accept long delays in data processing, while managers want immediate answers.

Chapter 10 goes into the details of how acquired data are used to design a good management strategy. The goal of this chapter is to explain how to sample insect populations adequately and how to assess plant damage. No analysis is discussed beyond preliminary data reduction.

Measurements taken to estimate population density fall into these groups: absolute methods, relative methods, and population indices. Absolute methods yield estimates as density per unit of land area in the habitat. Relative methods yield density per some unit other than land area and cannot be converted to absolute estimates without a major effort to correct for the insect's behavior and/or the effect of habitat. The sweep net is an example. Population indices do not count insects at all, but rather they are measures of insect products (e.g., frass or nests) or effects (e.g., plant damage).

Some relative methods of population measurement are used in qualitative monitoring programs. For instance, in the establishment of quarantines, it often makes little difference how many individuals are caught in a trap. The information sought is just the presence or absence of a species in a given area [for example, the gypsy moth *Porthetria dispar* (L.), in Illinois, or screwworm flies *Cochliomyia hominivorax* (Coquerel), in areas under an eradication program]. In this case perhaps the key feature is the ability of the trap to detect minute or even transient populations.

Evaluating damage depends, to a large extent, on whether the insect is a direct or an indirect pest. Direct pests affect the desired product of a crop, for example, the corn earworm, *Heliothis zea* (Boddie), destroying corn kernels, the boll weevil, *Anthonomus grandis* Boheman, feeding on cotton bolls, and the velvet bean caterpillar, *Anticarsia gemmatalis* Hübner, feeding on soybean pods. Indirect pests damage plant parts that may or may not affect yield, according to the role of the part in the physiology of the plant vis-á-vis the development of the desired product, for example, the velvet bean caterpillar feeding on soybean leaves, and rootworms feeding on corn roots. The effect of indirect pests on yield is often more difficult to assess, and quite frequently this difficulty is the cause of unwarranted decisions in control programs.

The duality that exists in the relationship between accuracy and ease of acquisition of data on pest populations also occurs with regard to damage estimation. In this case, however, accurate estimates may disagree substantially with the more expedient ones which depend heavily on subjective judgments.

I. ABSOLUTE METHODS

Absolute methods provide data of the type most often desired by population ecologists. Successive estimates of the number of insects per unit of land area are necessary for constructing life tables and for nearly all other studies of population dynamics of uncaged field populations. They can be used to calculate oviposition and mortality rates, and in validation studies of descriptive population models. Four distinct approaches have been used in entomology to measure absolute population densities: (1) distance to nearest neighbor, (2) sampling a unit of habitat, (3) recapture of marked individuals, and (4) removal trapping.

A. Distance to Nearest Neighbor

This approach has been applied much more by botanists than by entomologists. The method is based on randomly selecting an individual, and then measuring the distance to its nearest neighbor. The mobility of insects and the risk that one will fail to find the nearest neighbor have limited its application in entomology. Furthermore, the theory apparently has been developed only for species that are randomly distributed. Nevertheless, the approach can be used in entomology, especially for sessile insects such as scales, or for well-marked nests or colonies (e.g., tent caterpillars).

The simplest formula, attributed to Clark and Evans (1954), is:

$$m = (pr)^{-2} \tag{1}$$

where m is density per unit area, r is the mean distance between nearest neighbors, and p is an index of aggregation which takes the value of 2 for randomness and exceeds 2 for aggregated distributions (Clark and Evans 1954; Waloff and Blackith 1962).

A modification of the above formula, which uses the sum of r^2 rather than the mean r squared, is reputed to be more accurate when p differs greatly from 2. Other modifications are available which utilize the information contained in the distances to the second, third, . . . , and nth nearest neighbors (Southwood 1966).

B. Sampling a Unit of Habitat

Whereas the nearest-neighbor approach is useful only for individuals inhabiting a substrate and works best for those on the soil surface, the method of sampling a unit of habitat applies equally well to a broad range of habitats including soil, litter, soil surface, on or in the vegetation, and even airborne. The main advantage of this approach is its broad range of applicability, for it makes no assumptions about the spatial distribution of the species nor about its activity or behavior. In general, there is no concern about the efficiency of the sampling method, for it is assumed that, by bringing in an entire unit of habitat, no individuals are left behind.

The disadvantages of this approach lie largely in the processing of the samples. Almost always the species being sampled represents a small portion of the total matter per sample, and somehow it must be separated from the soil, plant matter, and other insects. Later we discuss some of the methods used to extract insects from samples, but first we look at how the samples are acquired from air, vegetation, soil, and litter.

1. *Sampling from Air*

The techniques used for sampling air are better mechanized than for other habitats; suction traps and rotary traps are the two basic methods used (Fig. 9.1*a* and *b*). The advantage of the rotary trap is that catch is independent of wind speed; however, Taylor (1961) has calculated efficiency factors for suction traps, which correct for wind speed, trap design, and insect size. Efficiency factors are not available for rotary traps, but it is known that even the best ones miss about 15% of the population (Taylor 1961).

Since the catch with these traps is almost exclusively insects, there is no problem in counting the number per sample. There is a problem in converting the catch, typically in insects per hours, into aerial density.

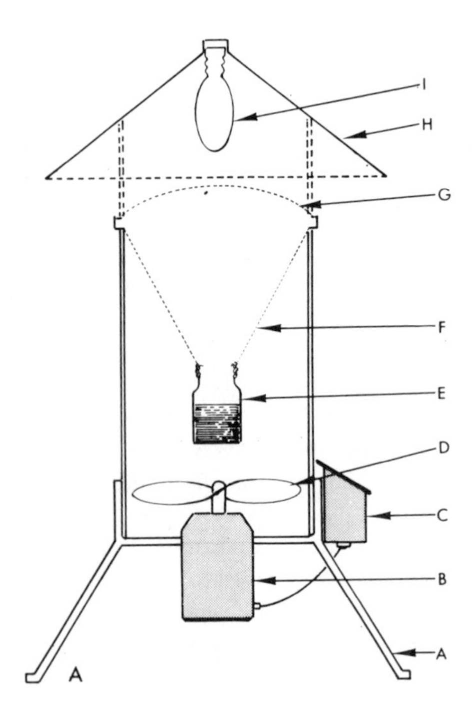

Fig. 9-1*A* Combination light and suction trap. *A*, Metal stand; *B*, fan motor; *C*, main electrical connection box; *D*, fan; *E*, Killing jar; *F*, funnel; *G*, metal screen cover—variable mesh according to size of insects to be screened out; *H*, Metal or wooden roof; *I*, light source.

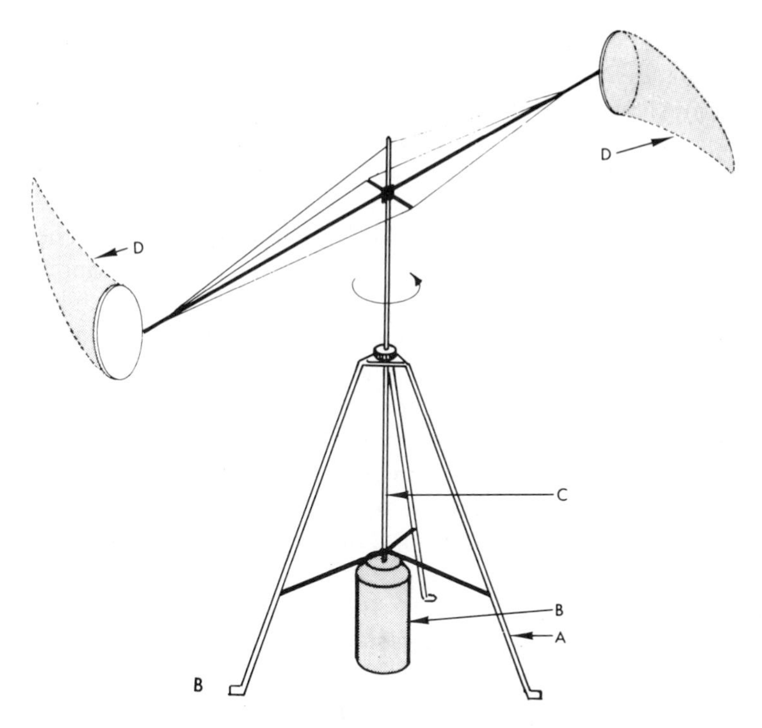

Fig. 9-1*B* Rotary net. *A*, Metal tripod; *B*, electric motor; *C*, vertical axle; *D*, revolving nets.

Even after correcting for the efficiency of the trap as explained above, one must divide by the volume of air sampled per hour to estimate density at the trap height. Total aerial population *P* can be calculated, if density estimates are available for at least three heights, from the equation

$$P = \frac{C}{(\lambda - 1)\, Z_e^{\lambda - 1}} \qquad (2)$$

where C, λ, and Z_e are calculated from the densities at various heights following the technique of Johnson (1957).

2. *Sampling from Vegetation*

Two essentially distinct approaches are used for estimating absolute density from vegetation. The most direct is to sample all plant matter over a unit area of soil surface. The difficulty in this approach is that in some habitats it is difficult to determine just what vegetation is inside the sample. For example, in a crop that has lodged, should parts be taken from plants lying across the sample area? In orchard or forest situations, it is practically impossible to sample this way.

The alternate approach is to use a plant or a plant part, such as a branch or leaf, as the sample unit. It then becomes necessary to determine the density of these sample units per unit of land area. The absolute population density is then the product of insects per sample unit and sample units per unit land area. In alfalfa, for example, insects per square meter can be estimated from insects per alfalfa stem multiplied by the number of stems per square meter.

A technique often used in sampling insects from vegetation involves placing a collecting sheet on the ground beneath the plant or area being sampled, and then knocking the insects off the plants onto the sheet by physically beating the vegetation or by using a chemical knockdown agent such as an aerosol pyrethrin. This technique has been much used in forest entomology and to some extent with row crops such as soybeans. The main drawback is that some species apparently resist being dislodged from the vegetation, even when exposed to vigorous beating and/or pyrethrins. Densities of these species are underestimated when sampled by this technique.

3. *Sampling from Soil and Litter*

Only one technique is used for sampling from soil and litter, namely, digging up a unit area to a predetermined depth. The only real problem that arises is deciding the proper depth to dig. It is not safe to assume that shallow samples are adequate, since eggs of the western corn rootworm have been found more than 8 in. deep, and Colorado potato beetles overwinter as deep as 4 ft in sandy soil (Mail and Salt 1933). However, bean leaf beetle eggs in soybeans have not been found below a 3-in. depth (Waldbauer and Kogan 1973).

If the insect being sampled is fairly abundant, quite small, and possibly several centimeters deep in the soil, the best sampling device is a core sampler (Fig. 9.2), such as the one used for sampling bean leaf beetle eggs from the soil. If the insect is larger, or less dense, and is found in the litter or top 1 cm of soil, then it is better that the sample unit be of larger area and not so deep in the soil. Often a square meter or a square foot of litter plus the top 1 cm of soil is a reasonable sample unit. This method was used by Ruesink (1972) in sampling for overwintering adults of the cereal leaf beetle.

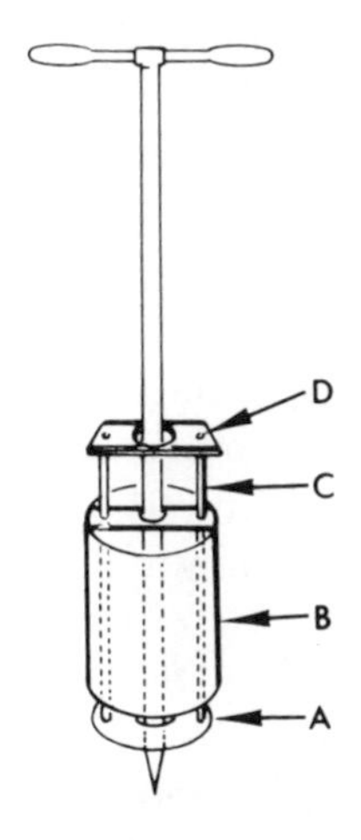

Fig. 9-2 Soil core sampler used in surveys of soil arthropods. *A*, Sliding ejection plate; *B*, cylindrical core borer; *C*, ejection plate holder; *D*, depth regulator.

4. *Extracting the Insect from the Sample*

Separating the insect being studied from the other material in the sample may involve many hours of hand sorting and sifting. Although hand sorting is very time-consuming and tedious work, it is still the method most entomologists prefer for obtaining the best possible data. Even this method can result in large errors, however, for it depends on workers remaining highly efficient through hours of boring labor. It is common for efficiency to drop sharply toward the end of a long day.

An often used alternative to hand sorting is placing the samples in a Berlese funnel (Fig. 9.3) and expelling the insects, using a heat source such as an incandescent light, into a vial of alcohol. This method depends on insect behavior and obviously does not account for dead individuals or for immobile forms. The rate of drying is very important. If it is too slow, individuals may molt (or even pupate) before being extracted, and each sample will tie up a funnel unnecessarily long. If it is too fast, individuals will be killed by the heat or low humidity before they are able to move through the sample into the vial.

Washing the samples, followed by screening and/or floating, is another common alternative to hand sorting. This method is particularly useful for separating insects from soil, but does not work as well for litter samples or soil samples containing much vegetation. Even eggs less than 1 mm in length can be retrieved with a high degree of reliability using a machine such as the one illustrated in Fig. 9.4.

Regardless of the extraction technique used, an effort must be made to check its efficiency. It is not uncommon for Berlese funnels to be less

Fig. 9-3 Berlese funnels. (*A*) Battery of funnels used in the alfalfa weevil project, Illinois Natural History Survey. (*B*) Funnel open to show internal arrangement of the sample and heat source. (*C*) Funnel closed for operation.

than 50% efficient, and hand picking can be equally inefficient. There is no best method for all insects; rather the method of extraction must be selected to best suit the insect, the needs, and the resources at hand.

C. Recapture Marked Individuals

When this method is used to estimate the absolute population size, the most serious limitation is satisfying the assumption that the marked individuals become thoroughly mixed into the population.

The simplest way to use recapture data was proposed by Lincoln (1930) and is today called the *Lincoln index*. This method requires the additional assumption that no births, deaths, immigration, or emigration occur between the time of releasing the marked animals and the subsequent recapture sampling. This assumption applies to both the natural population and the marked individuals. Total population size P can be computed from the number marked and released a, the total number of individuals taken in the recapture sampling n, and the number of marked individuals recaptured r from the equation

$$P = \frac{an}{r} \tag{3}$$

The best estimate of P is obtained when a and n are approximately equal.

Modifications of the simple Lincoln index include methods based on (1) the successive release of groups of marked animals with interspersed recapture sampling, and (2) a single release with multiple recaptures. Many of these modifications are designed to eliminate the assumption of a stable population without migrations. Southwood (1966) presents details on these modified versions of the Lincoln index.

D. Removal Trapping

This method is based on the fact that, as the population is reduced by removing members from it, the catch per trap per unit time decreases. Initial population size can be estimated from the rate the catch decreases, using one of three mathematical techniques described by Southwood (1966).

The simplest method is to plot the number caught in the nth sampling against the previous total catch and to draw a straight line through these points. An estimate of initial population size is then obtained by reading from the graph the value at which the line intersects the total catch axis.

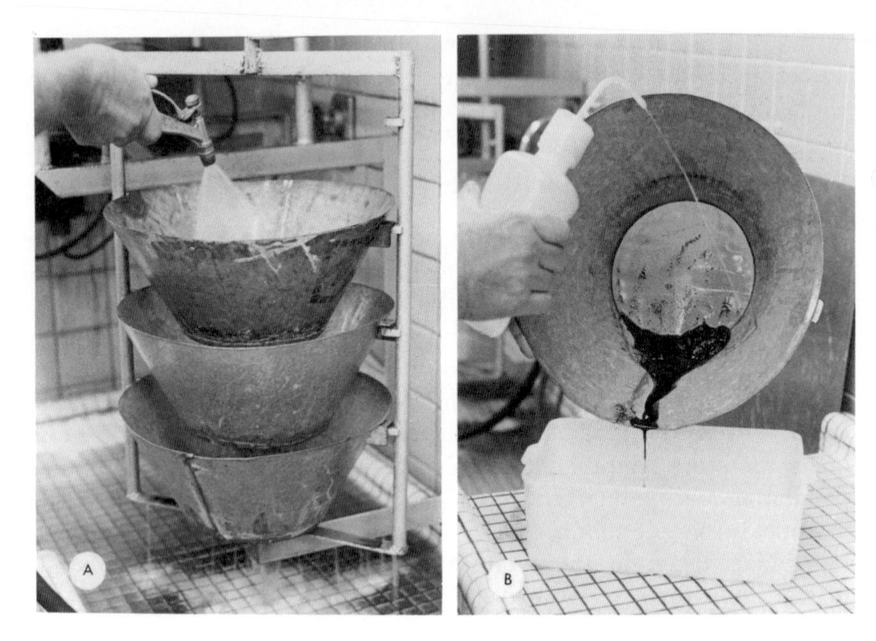

Fig. 9-4 Illinois egg separator. (Adapted from Horsfall 1956, see Waldbauer and Kogan 1973.) (*A*) Superimposed sieves in a rotating rack. (*B*) Transferring the washed sample collected on the bottom 100-mesh screen. (*C*) Washing the sample from a transfer sieve into a separatory funnel containing a saturated

Removal trapping is very rarely used in entomology because of the assumptions that must be satisfied and the great number of individuals that must be trapped and removed from the population. The assumptions of the method are:

1. No births, deaths, or migration may occur during the sampling program.
2. The probability of being caught must be identical for each member of the population and cannot change during the sampling program.

Zippin (1956) has shown that both the initial population size and the proportion of the population caught influence the accuracy of the population estimate. Table 9.1 indicates the percentage of the population to be captured for several population sizes and three levels of accuracy.

As an example, consider using a sweep net as the removal trapping technique for cereal leaf beetle adults. Assume we are working in a 10-

sodium chloride solution; a bubbling stirrer is kept inside the funnel. (*D*) Washing eggs and debris which float in the salt solution into a transfer sieve. Final separation of eggs from debris is made in distilled water under a stereomicroscope.

acre field in which the density is expected to be between 0.1 and 1.0 beetles per square foot. Since 1 acre contains 43,560 ft^2, we expect that the population is between 43,560 and 435,600 beetles. If we require that the population estimate have a standard error within 20% of the mean, Table 9.1 indicates that at least 20% of the population of 43,560 must be removed, or 10% of the 435,600. Available data on the sweep net (Ruesink and Haynes 1973) imply that with reasonably good sweeping conditions (e.g., a conversion factor of 0.33) this translates to 28,000 and 14,000 sweeps, respectively.

For the sake of completing the example, we assume that 10,000 sweeps were taken at 3-hour intervals and that the catches were 6500, 6400, 5400, and 5000 (see Fig. 9.5). Linear regression analysis on these four points yields a line which intersects the total catch axis at about 75,000. Since our total catch was 23,300, or about 30% of the initial population, we can estimate from Table 9.1 that 75,000 is probably within 10% of the true initial population.

Table 9.1 Percentage of the Population That Must Be Caught for Certain Accuracies of Estimation

Initial Population Size	Accuracy of the Estimate[a]		
	±40%	±20%	±10%
1,000	33	45	60
10,000	18	26	36
100,000	9	13	18
1,000,000	5	7	9
10,000,000	2	3	5

[a] Accuracy is defined as 100 × standard error/mean.

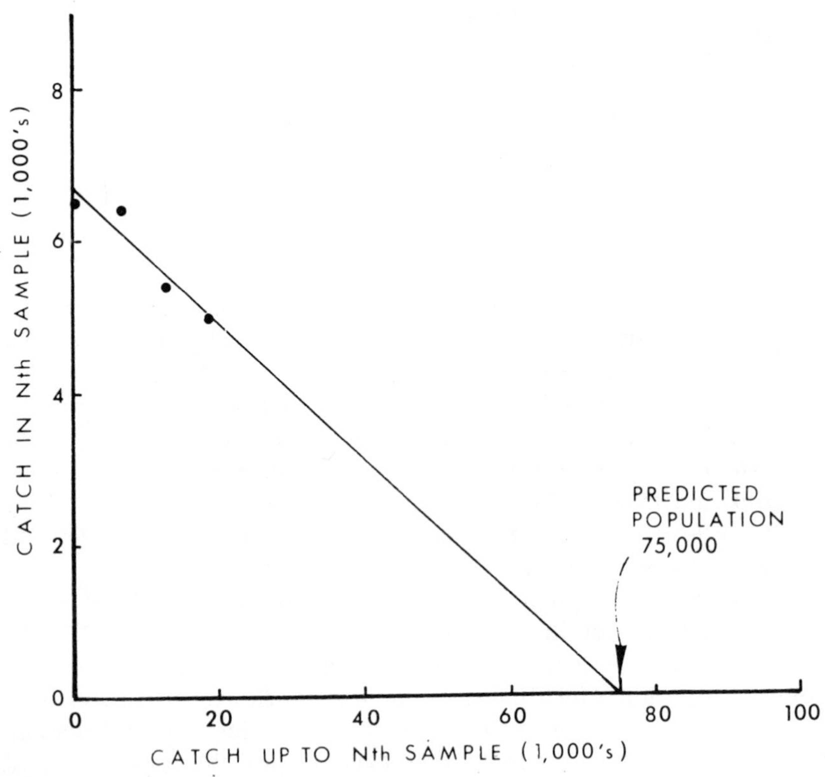

Fig. 9-5 Computation of population size by linear regression.

Another important consideration is what to do about those members of the population that are caught. In life-table work removal of 10% of the population represents a significant mortality factor. In some cases it may be feasible to store the captured individuals until the removal trapping is complete, and then release them back into the population.

II. RELATIVE METHODS

Whereas with absolute methods the intention is to capture and count every insect over a unit area of earth surface, the intention with relative methods is to capture a more-or-less consistent, if unknown, portion of those present. The sweep net and the black-light trap are two of the most common methods used to sample insect population density.

Relative methods have one distinct advantage over absolute methods; a given amount of labor and equipment yields much more data. By wisely choosing which relative method to use, a person can usually capture at least 100 times as many insects as with the same effort using an absolute method. On statistical grounds alone, this is often the reason an absolute method cannot be used.

Southwood (1966) lists five factors that may simultaneously affect the catch of any relative method:

1. Actual density or population size
2. The number of animals in a particular phase
3. Level of activity
4. Efficiency of the relative method being used
5. Responsiveness of the particular sex and species to the trap stimulus (includes attraction and avoidance).

In general, all five of these affect the catch of a relative method, while the intention is to measure only the actual density or population size. Some attempts have been made to calibrate relative methods, to allow conversion of catches to estimates of absolute density; recent examples include the sweep net (Menhinick 1963; Pedigo et al. 1972; Ruesink and Haynes 1973) and pitfall traps (Gist and Crossley 1973). The task of developing conversion methods is time-consuming and not particularly rewarding, hence it remains a relatively open area for additional research.

The *phase* of a population refers to differences in behavior principally due to age. For example, Geier (1960) showed that prereproductive females of the codling moth are more attracted to light traps, while mature females are more attracted to bait traps.

Level of activity is largely a weather response; activity generally increases as temperature increases until an optimal condition is reached, after which activity decreases. Some insects are most active in the subdued light of dawn and dusk and less active both at night and at midday. Most insects have a preferred humidity range and actively seek optimal conditions. All these and more weather conditions interact to affect the activity level of the insect being sampled.

The efficiency of the sampling method is also affected by weather and by the habitat being sampled. Sweep nets are known to be affected by the height and density of the crop and by the vertical distribution of the insects in the crop. Light traps are less efficient when competing with a full moon.

The response of an insect to light traps, bait traps, and pheromone traps must be positive for the trap to catch anything. Different species and different sexes respond differently. Even the same sex of the same species responds differently when environmental conditions change. Sweep nets and pitfall traps may be intentionally avoided by some insects (ever try catching a tiger swallowtail?), thus reducing the catch. Some insects drop to the ground when a shadow passes over; these species cannot be caught in a sweep net unless the sweeper walks toward the sun.

Even without being able to convert relative estimates to absolute densities, the data obtained by these methods can be highly useful. For instance, a sweep net catch can provide estimates of annual population changes in univoltine species by comparing catch per sweep for 2 years. There is no need to know density per unit area.

Relative methods can be lumped into two broad classes—catch per unit effort and trapping. The first includes (1) visually searching for a fixed time or area, (2) the sweep net, (3) the D-Vac, and (4) shaking and beating methods. The second includes (1) Malaise traps, (2) windowpane traps, (3) pitfalls traps, (4) sticky traps, (5) visual traps, and (6) traps using attractants. Selecting the best method for a specific problem requires thorough consideration of all options. Southwood (1966) has compiled a very helpful summary of techniques that have been used; some of them are discussed briefly.

A. Visual Searches

Visually searching an area may be considered an absolute measure of population size, but in general less than 100% of the insects present are found. The proportion found may vary with conditions, thus making the relationship between absolute population and catch unpredictable.

A variation on this method is called *flushing*; for example, a vehicle driven across the grasslands of Africa has been used to flush locusts (Scheepers and Gunn 1958), and the number flushed per unit distance can be used as a relative measure of population density.

A 5-minute visual search has been used in field corn in Illinois to obtain a relative estimate of adult corn rootworm density. The use of these estimates must be approached with caution, because of changes in behavior with the weather and the age of the insects and because of differences among observers in their ability to spot and identify the insect several feet away.

B. Sweep-net Catch

The sweep net has been much criticized, yet it remains as the most often used method for sampling insects from small grain, forage, and many row crops. The reason is simple; no other method can capture as many insects from vegetation per man hour without both costing more for the equipment and doing more damage to the crop.

The efficiency of the sweep net usually varies with:

1. Different species
2. Different habitats, particularly the height of the vegetation
3. Different weather, particularly wind speed, air temperature, and intensity of solar radiation
4. Different time of day, reflecting diel cycles of behavior of the species
5. Different styles of sweeping.

The usefulness of sweep net data can be greatly improved by standardizing items and recording items 1 to 4 for each sample taken. The difference in catch among several persons can be minimized if the style of sweeping is restricted as follows:

1. Use a pendulum swing, as if you were sweeping the sidewalk with a broom.
2. In short vegetation swing the net as deep as possible without taking too much dirt into the net. In tall vegetation sweep only deep enough to keep the upper edge of the sweep net opening even with the top of the vegetation.
3. Sweep one stroke per step while walking at a casual pace.
4. Use a sweep net having a 15-in. diameter opening.

Fig. 9-6 D-Vac mechanical sucking net. (*A*) Backpack model. (*B*) Hand model. (By permission of Dietrick, Riverside, Calif.)

C. Shaking and Beating

This method is most widely used in row crops, such as soybeans or sorghum, but it is also adequate for sampling the insect fauna of bushy weeds. The method consists of spreading a piece of heavy cloth on the ground along the stems of the plants to be sampled. The plants are bent over the cloth and vigorously shaken or beaten with a wooden stick. The insects dislodged from the plant fall on the cloth and are collected and counted. In sampling for fast-moving insects, it is recommended that an aspirator be used for collecting.

There are few comparative studies on the efficiency of shaking and beating and other sampling methods. Kogan et al. (1974) found no correlation between the numbers of bean leaf beetles sampled by beating and by sweep net.

D. Vacuum Trapping

A device with the tradename D-Vac captures insects by sucking them into a fine mesh net held open inside a rigid enclosure. A portable gasoline

Fig. 9-7 Metal-framed trap in a soybean field.

motor propels a blower which generates the suction. Two models are available—a smaller one which is held in front of the operator, and a heavier, backpack model (Fig. 9.6). The D-Vac was described by Dietrick et al. (1959).

E. Malaise Traps

A Malaise trap (Fig. 9.7) essentially consists of a tent made of netting with one open side into which insects either fly or crawl. Since most insects automatically crawl up the netting once inside, they can be trapped in vials of preservative placed in the upper corners or peak of the tent. Since these traps depend upon the insect to enter them accidentally, they work best for highly active species such as adult Diptera and Hymenoptera.

F. Windowpane Traps

Flying Coleoptera can be sampled using windowpane traps (Fig. 9.8), which consist simply of a vertical pane of glass or Plexiglas with a trough

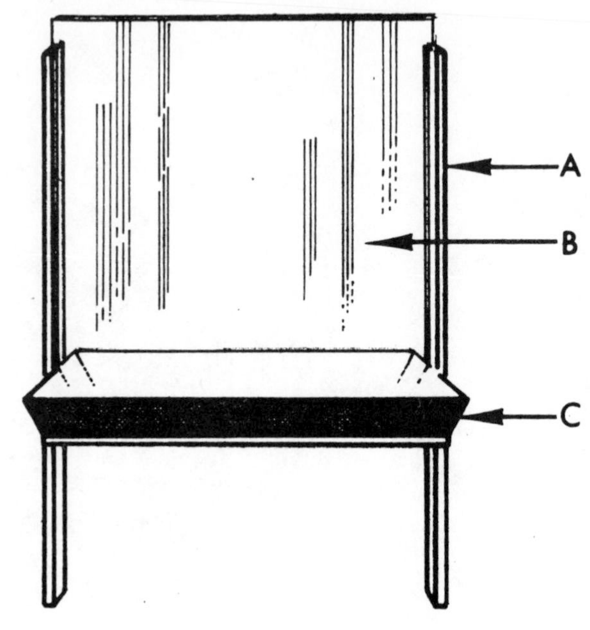

Fig. 9-8 Windowpane trap. *A*, Wooden or metal support; *B*, glass or Plexiglas pane; *C*, collecting pan usually filled with detergent solution. (Modified from Southwood 1968.)

of preservative beneath it. Any insect that hits the glass and reacts by falling is caught. These traps are particularly useful for determining direction of flight, and can also provide data as to when dispersal flights occur.

G. Sticky Traps

Sticky traps (Fig. 9.9) are conveniently thought of as a variation on windowpane traps, in which glass, screening wire or other surface is covered with some substance sticky enough to hold the insect where it hits, rather than allowing it to fall into the preservative. Sticky traps can be used for many species that cannot be taken with windowpane traps.

Since the insects are caught in place, the sticky substance can be put on almost anything and anywhere. Often these traps are combined with baits or attractants to enhance the catch of certain species.

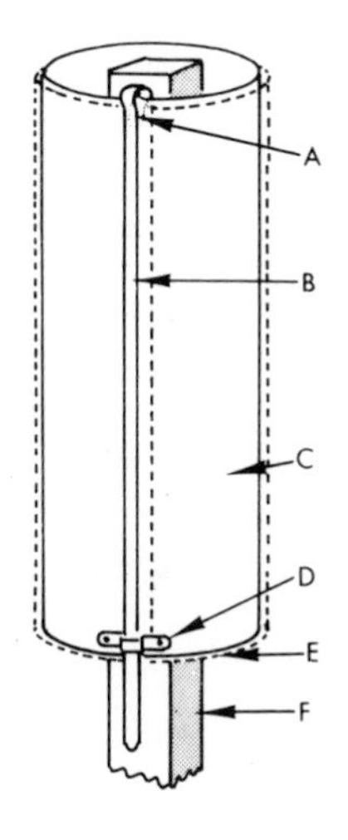

Fig. 9-9 Cylindrical sticky trap. *A* and *B*, Brass strip to hold the trapping material; *C*, Rigid cylindrical support; *D*, socket for the brass strip holder; *E*, sheet of fabric, paper, or plastic coated with a sticky grease and wrapped around support *C*; *F*, wooden post. (Adapted from Southwood 1968.)

H. Pitfall Traps

Pitfall traps (Fig. 9.10) are used for species that roam the soil surface, such as ground beetles, spiders, and Collembola. When used without baits, they catch whatever accidentally falls in; when baited, they draw individuals from considerable distances. Traps baited with beer are an often recommended control practice for slugs in the home garden.

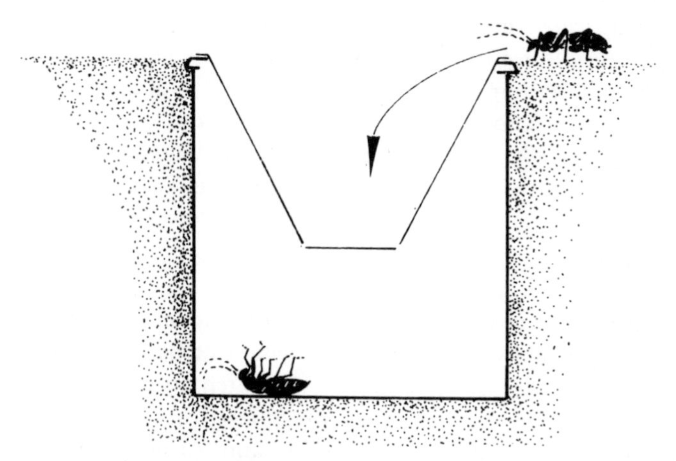

Fig. 9-10 Pitfall trap.

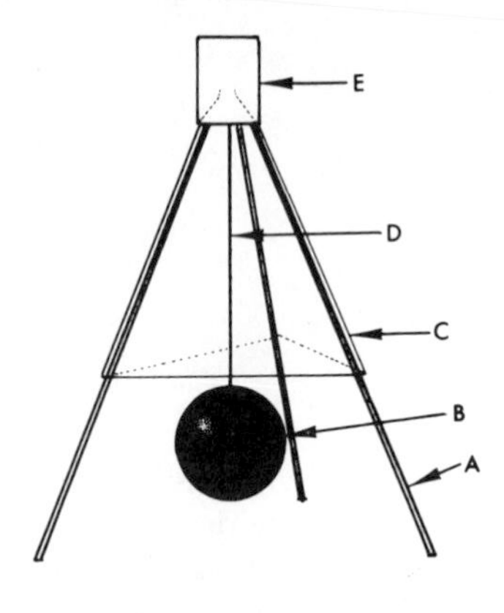

Fig. 9-11 Manitoba horsefly trap. *A*, Metal tripod; *B*, suspended black sphere; (*C*, transparent plastic cover; *D*, sphere holding string; *E*, collecting jar. (Adapted from Southwood 1968.)

I. Visual Traps

Phototactic reactions are used to attract insects into a suction trap, a Malaise trap, or some other type. The visual stimulus can be generated by shape and color as, for example, in the Manitoba horsefly trap (Fig. 9.11).

The most common devices using visual attraction are light traps. A typical light trap (Fig. 9.12) uses an ultraviolet fluorescent tube above a collecting jar containing a killing agent. As with other relative sampling methods, black-light traps do not yield data that allow the comparison of densities among species. The results are useful for comparing densities of a species from one year to the next and for monitoring the emergence dates of many pests.

A light-trap catch must be interpreted with the same caution as any other relative measure of abundance. With sufficient corroborative data on absolute population size, it will someday be possible to convert from light-trap catch to absolute density.

These traps have been mostly used for moths and mosquitoes and of course are restricted to species that fly at night or at twilight.

J. Traps Using Attractants

Species that are not attracted to light may be trapped instead by using a sticky board in combination with an attractant or bait. Pheromones (i.e., chemical sex attractants) (Chapter 8) are often used for this purpose (Fig. 9.13). Again, the data must be interpreted cautiously, but the fact that some pheromones are highly specific and attract individuals from great distances makes pheromone traps very useful sources of data.

Fig. 9-12 Illinois Natural History Survey light trap.

Fig. 9-13 Pheromone trap. (By permission of Zoecon.)

III. INSECT PRODUCTS

In some cases a species that is difficult to sample directly creates products that are easily sampled by absolute methods. The insect product most often sampled is the frass, or excrement, of lepidopterous defoliators of forests. The rate that frass is being produced can be estimated from the amount falling into a box or funnel placed under the trees. The size and shape of the frass pellets are rather constant for a given species and instar; this allows one to identify the species and age composition of defoliators without seeing one. Density estimates are more difficult, but work done by Waldbauer (1964), Gosswald (1935), Green and DeFreitas (1955), Pond (1961), and many others helps to convert frass catch to absolute population density.

Insect nests are another product that can be sampled rather easily. The nests of caterpillars can be counted to obtain a fast, yet reasonably accurate, estimate of larval density over large areas (Morris 1964).

IV. ASSESSING PLANT DAMAGE

The amount of damage caused by arthropods to crop plants is a function of the pest density, the characteristic feeding or oviposition behavior of the arthropod species, and the biological characteristics of the plants. Each of these factors is differentially affected by environmental and other biotic factors, and correlation between population levels and plant damage is frequently difficult to establish. Chapters 3 and 10 discuss some of the concepts and theoretical methods used in defining these correlations. The practical assessment of damage is critical in pest-management programs, and this assessment, with and without regard to its correlation with population levels, is the object of this section.

The main reasons for making damage evaluations in pest management are (1) to define the economic status of a given pest species, (2) to establish economic thresholds and economic injury levels, (3) to estimate the effectiveness of control measures, and (4) to evaluate resistant varieties and lines of crop plants.

As is the case with population estimations, the methods used in damage evaluation differ, depending on whether the data are for research purposes or for practical management procedures. Mass screening of thousands of lines in a breeding program for resistance requires only a three-scale classification (e.g., heavily, moderately, or slightly damaged). However, a critical correlation between defoliation and yield loss may require measurements within classes of 5% defoliation or less.

A. Identification of the Type of Injury

Although correlation of a certain type of injury with the causal organism presents no difficulty in most cases, there may be instances in which symptoms are observed long before the causal organism is identified. When we find an ear of corn with a worm tunneling among the kernels there is little doubt about the cause–effect relationship. In the development of adventitious roots on soybeans stimulated by the feeding of the three-cornered alfalfa hopper, *Spissistilus festinus* (Say), the relationship is less obvious.

Damage by insects generally results from their feeding activities and, to a lesser extent, from certain types of oviposition processes. The types of damage inflicted by insects on their host plants are therefore as varied as are the kinds of feeding habits found among insects. Brues (1946) is still one of the best sources of information on the diversity of insect feeding habits.

Since a detailed discussion of every type of damage is beyond the limitations of this chapter, it will suffice for pest-management purposes to adopt the concept of direct and indirect pests proposed by Turnbull and

Chant (1961). *Direct pests* attack produce directly, immediately destroying a significant part of its value—relatively small populations may cause a significant economic loss. *Indirect pests* attack plant parts that may be physiologically related to yield but do not produce by themselves, and damage results only by intensive or extended infestation. The codling moth on apples, the corn earworm on sweet corn, or the boll weevil on cotton are examples of direct pests. The green cloverworm on soybeans and rootworms on corn are examples of indirect pests.

Sampling and measuring procedures must therefore be suited to either of these major categories of pests.

B. Measuring Damage by Direct Pests

Since the produce is the object of the sampling procedure, techniques to evaluate damage usually refer to absolute or relative numbers of damaged units, for example, ears of corn per 10 plants, apples per tree, bean pods per 10 ft of row, or cabbage heads per acre. Different levels of damage can be recognized, and this is sometimes necessary for the conversion of damage to actual value decrease. Thus Prasad (1963) classified cabbage heads as nonmarketable if feeding by the imported cabbageworm was more than two leaves deep or the heads were less than 2.5 in. in diameter. Feeding by caterpillars on the external leaves therefore was not enough to disqualify a cabbage head.

Damage to sorghum by the sorghum midge, *Contarinia sorghicola* (Coquillet), and by stinkbugs was investigated by Bowden (1965) in Ghana. The procedure used was as follows. Ten sprigs of spikelets were selected at random from heads in which flowering had ceased, along rows selected at random. All spikelets were removed, and 100 selected from the bulk by the successive halving method. These spikelets were dissected, and the contents recorded in four categories—sound, midge attacked, bug attacked and other. The percent of damaged spikelets in this scheme can then be converted to a percent yield loss.

LeRoux (1961) analyzed sampling procedures to evaluate apple damage by a complex of four pests for which the nature of the injury was clearly identified as: (1) irregular scarring of the fruit surface by the larvae in the immediate vicinity of the larval leaf shelter [the eye-spotted bud moth, *Spilonota ocellana* (D. & S.)]; (2) a pronounced and regular scarring of the fruit surface by the larvae or deep feeding within the young fruit near the larval leaf shelter [the fruit tree leaf roller, *Archips argyrospilus* (Walker)]; (3) light, semicircular egg-laying scars or deeper irregular feeding scars on the surface of the fruit by the adults or the larvae feeding within the apples [plum curculio, *Conotrachelus nenuphar* (Herbst)]; and (4) light or deep penetration into the fruit of the larvae with much brown frass [the codling moth, *Laspeyresia pomonella* (L.)].

Leaf clusters were the sampling units in LeRoux's experiments, and measurements were made to evaluate the mean number of leaf clusters per tree and the mean number of apples per leaf cluster. With this type of data available, it is possible to compute very accurate damage estimates. The analysis of the contribution to the total variance of trees between blocks, trees within blocks, and two levels and four quadrants within tree crowns permitted optimization of the sampling size for given levels of precision (see Section V).

C. Measuring Damage by Indirect Pests

Only two kinds of damage are discussed: (1) damage due to defoliation caused by insects with chewing mouthparts (e.g., grasshoppers, lepidopterous caterpillars, leaf beetles); and (2) damage caused by insects feeding on the root system.

1. Measurement of Defoliation

Experienced field entomologists are often capable of visually estimating with great accuracy the percent defoliation of a plant stand. This is a subjective estimation, however, and often estimations made by different persons in a field may differ very significantly.

Although for many practical purposes visual estimations are frequently used, classes of defoliation can seldom be set within less than 10% intervals. When more refined estimations are needed, and in the calibration of the visual estimation, other methods must be used.

The methods available are based on techniques designed to measure irregular areas. Planimeters and grids have been used by some researchers, but photometric devices have found more widespread use in recent years. Photoplanimeters measure the change in light transmission when a leaf is interposed between a light source and a photocell. Usually, changes in the photometric scale are linearly correlated with the area of the interposed leaf. A simple regression of the photometric readings on a series of known leaf areas allows conversion to units of area. Carman (1963) described a large-stage photoplanimeter for use with excised leaves (Fig. 9.14). More recently, instruments were developed that use advanced electronic technology and can be used either in the field

Fig. 9-14 Carman-type large-stage photoplanimeter with lower door open to show soybean leaf on the illuminated stage.

or in the laboratory with attached or excised leaves. One such instrument is the LI-COR area meter produced by Lambda Instruments (Lincoln, Nebraska) (Fig. 9.15).

To use these measuring devices in the assessment of defoliation, it is necessary to know the area of the whole leaves without feeding. One method of accomplishing this is to spray a block of plants to keep them insect-free while they grow at the same rate as the plants being defoliated. Although interplant leaf areas vary considerably, enough samples can be taken from whole and from defoliated plants to give a statistically meaningful result.

Another more tedious method to measure defoliation uses photocopies of the leaves. The copies are cut and weighed and the weights transformed into area measurements by using adequate conversion factors correlating the weights with known areas of the same paper. This method has been used to measure growth rates of soybean plants (Kogan, unpublished data).

2. *Measurements of Damage to the Root System*

Extensive studies on the effect of insect feeding on roots were carried out with the corn rootworm larvae complex feeding on corn (Chiang 1973). Among the techniques used to evaluate damage to root systems are (1) amount of damage to the entire root system; (2) percentage of nodes below the soil surface having injury; and (3) percentage of roots pruned on the first and second nodes below the soil surface. Indirect measurements use (1) the force needed to pull a plant from the soil, and (2) ratings of damage based on plant lodging (see Chiang 1973).

The vertical pull technique was developed by Ortman et al. (1968). It uses a grip which is attached to a cut corn plant and to a 1000-lb-capacity recording dynamometer (Fig. 9.16). The pulling weight is highly correlated with visual ratings of corn root damage, but soil factors can greatly influence the results of these measurements. The technique offers the great advantage of providing quantitative data free of bias by the observer.

D. Relationship Between Damage and Yield Reduction

Correlation of damage level with yield loss is usually simpler in direct pests. If we count the number of damaged apples in a barrel, we can easily calculate the percentage of damage and the consequent loss in yield. Grades of damage can be established to account for losses that

Fig. 9-15 LI-COR model LI-3000 portable area meter showing use on soybean leaves in the field (Courtesy of Lambda Instruments Corporation, Lincoln, Nebr.

Fig. 9-16 Corn-pull device used to estimate damage by corn rootworms (Adapted from Ortman et al. 1968.)

result in unmarketable fruit, or fruit of reduced quality but still marketable.

The need for sound crop loss evaluations in pest-management programs is generally recognized (Chiarappa et al. 1972). Chiarappa (1971) compiled extremely valuable information on crop loss methodology, and Judenko (1973) described some analytical methods for assessing yield losses caused by pests on cereal crops.

Experimental procedures in these evaluations usually consist of comparing yields of two sets of plants grown under nearly uniform conditions. One set is submitted to attacks by known insect populations, and the second set is kept free from attack. Different levels of attack are achieved by means of artificial infestation, screening of plants in field cages, and mechanical or chemical removal of pests. Most commonly, experimenters take advantage of natural infestations and use insecticides to screen out insect pests from plots to be kept undamaged.

Another approach consists of using plants grown under natural field conditions. Estimation of losses is made by regression analysis using yields of plants displaying different levels of injury.

One analytical procedure described by Judenko (1973) is exemplified by the attack of aphids on silage corn. The data used in this analysis and the computation of loss are shown in Table 9.2.

Table 9.2 Example of Estimation of Loss by Analytical Procedures on Silage Corn Attacked by Aphids[a]

Parameters Measured	Value
Total number of plants (T)	1200
Percentage of attacked plants (P)	25
Number of attacked plants (NAT)	300
Actual yield per unit area (ACT)	2100 kg
Mean yield per unattacked plant (a)	2 kg
Mean yield per attacked plant (b)	1 kg

Actual loss $= (a - b)\,NAT = 300$ kg

Expected yield in the absence of attack $=$ loss $+\ ACT = 2{,}400$ kg

$$\text{Percent economic loss} = \frac{\text{expected yield in the absence of attack}}{\text{yield loss}} \times 100 = 12.5\%$$

[a] Judenko 1973.

Several factors may complicate the assessment of losses. Judenko (1973) lists the following, among others: (1) various types of attack by one or by a complex of pests; for example, the European corn borer may cause damage from leaf feeding by the first-generation larvae, stalk tunneling, or direct ear injury; (2) unattacked plants neighboring damaged ones may produce a compensatory yield; (3) the pesticide used may have a direct effect on the yield in addition to the effect on the pest. These factors should be taken into account when designing experiments to assess crop losses.

The relationship between damage by indirect pests and yield loss is more difficult to assess. The effect of artificial defoliation on the reduction in yields of soybeans has been extensively investigated. Generally, experiments are carried out by hand picking one, two, or three leaflets to attain three levels of defoliation, and measuring the yields of plants under the various treatments. The effect of defoliation in soybeans varies considerably according to the stage of development of the plant when damage is inflicted. Prior to flowering most varieties can tolerate up to 30% leaf area reduction without any effect on yield. During the flowering period and during pod set and pod fill, plants are more sensitive to reduction in foliage area.

Artificial defoliation differs in many respects from actual insect feeding. Leaves are artificially removed in one operation, whereas insect feeding occurs over a period of days or weeks, during which time the plant is growing and compensating for some of the lost tissue. Furthermore, the leaf area removed by insect feeding is discontinuous and composed of many larger or smaller holes or edge indentations, whereas most defoliation studies remove whole leaflets or large leaf portions. As a preliminary piece of information, however, results of simulated damage have been very useful in pest-management programs for soybeans (see Chapter 10).

The relationship of root damage to yield reduction was established for the northern corn rootworm. Root damage was rated on a 1-to-6 scale, and it was observed that, for a normal yield of 125 bu per acre, there was a reduction of 5.8 bu per acre for every adjusted root damage rating unit (see Chiang 1973). Lodging of corn plants caused by rootworm feeding also resulted in yield losses when the ear weight of lodged plants was compared to that of standing plants in the same field.

In sugarcane it was observed that sugar yields decreased at a rate of about 1% for every 1% of internode bored by the sugarcane borer, *Diatraea saccharalis* (Fabricius). It was calculated that 1% of internode bored caused a weight loss that ranged from 0.38 to 0.54% according to the variety (Long and Hensley 1972).

V. DISPERSION AND THE SAMPLING PROGRAM

After choosing the technique(s) to be used for estimating population size and evaluating damage, it remains to decide how many samples to take. The answer to the question of how many is a subjective decision based on the following considerations:

1. How much manpower is available for taking and processing samples? What would this manpower be doing if not taking and processing samples?
2. How much accuracy is desired in the resulting data? What is the absolute lower limit of accuracy that is acceptable?

The sampling program chosen is necessarily a compromise between manpower expended and the accuracy of results. In some cases the decision is that insufficient manpower is available to meet even the minimum acceptable level of accuracy, in which case the whole idea of sampling must be scrapped or redesigned using a different sampling method. The relationship between accuracy and the number of samples can be established from statistical analysis.

There are basically two purposes for which sampling programs are used in insect pest management. In some cases we want to know if the population exceeds a given threshold; collecting data to answer this question is *decision-making* sampling. In other cases we must accurately estimate the population size regardless of what it is; this is *parameter-estimation* sampling. In the first case we can tolerate considerable sampling error if the sample mean is exceptionally far from the threshold; in the second case the sampling error must be equally small for all sample means. Decision-making sampling is used only in the operation of a pest-management system, while parameter-estimation sampling is used for both operating and designing such systems.

Both kinds of sampling require that we know something about the nature of the statistical distribution from which we are sampling. The statistical distribution is related to the spatial distribution and also to the sampling method. To understand this relationship it is necessary to understand the concept of randomness as used by statisticians.

In two dimensions a spatial pattern (i.e., spatial distribution) is random if every point on the surface is equally likely to be occupied by an individual. An alternate way of phrasing this definition is to say that knowing the location of one individual on the surface provides no information as to the location of any other individual.

Spatial patterns can deviate from randomness in either of two directions (Fig. 9.17). If the presence of an individual at one point increases the probability of another individual being nearby, then the spatial pattern will be clumped, whereas if it decreases the probability of another being nearby, the pattern will be more uniform than random. An extensive literature has been developed on the subject of spatial aggregation and its measurement. Our interest in this chapter, however, is limited to how spatial distribution affects the number of samples we must take.

A. The Poisson Distribution

If a population exhibits random spatial distribution and, if the act of taking one sample does not bias the numbers taken in any subsequent

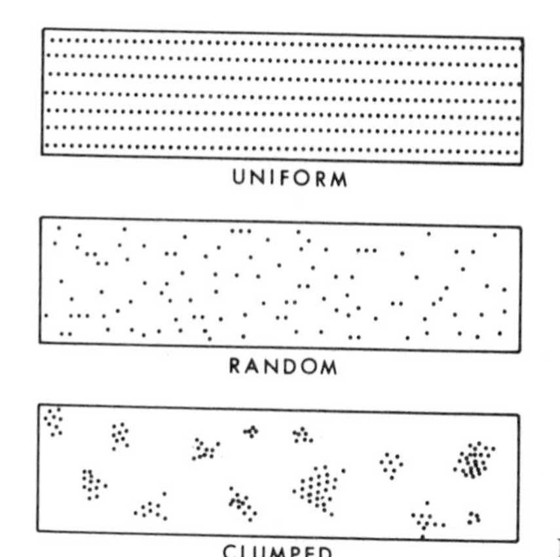

Fig. 9-17 Types of distribution.

samples, the number of individuals per sample unit will follow the Poisson statistical distribution. The probability that a unit will contain n individuals is

$$p_n = \frac{\lambda^n e^{-\lambda}}{n!} \tag{4}$$

where λ is the single parameter of the Poisson distribution. An important feature of this distribution is that λ is equal to both the mean and the variance. More details about the Poisson distribution and about how to calculate the mean and variance from sampling data can be found in any introductory statistics text; entomologists will find Wadley (1967), Bliss (1967), and Sokal and Rohlf (1969) relatively easy to read.

B. The Negative Binomial Distribution

If a population exhibits aggregated spatial distribution, and if samples are taken from randomly distributed points in space, the number of individuals per sample unit will follow one of the so-called contagious statistical distributions. One important feature of all contagious distributions is that the variance exceeds the mean. The most frequently used contagious distribution is the negative binomial, which is described by two parameters, the mean m and the exponent k. The expected frequency of zeros is

$$p_0 = \left(\frac{k}{m + k}\right)^k \tag{5}$$

and the probability that a sample will contain exactly n individuals $(n > 0)$ is

$$p^n = p_0 \frac{k(k+1)\cdots(k+n-1)}{n} \left(\frac{m}{m+k}\right)^n \qquad (6)$$

The parameter k can be estimated from the mean and variance of the sample data using

$$k = \frac{m^2}{s^2 - m} \qquad (7)$$

In the case that m is greater than 0.5, the above equation works well only for populations that are very highly aggregated. If m is large and the aggregation is only moderate, other more complex methods must be used to estimate k. Anscombe (1949) and Southwood (1966) discuss several alternate approaches.

C. Taylor's Power Law

Taylor (1961) discovered a most useful relationship between sample mean and sample variance. Regardless of what organism is considered or which sampling technique is used, the variance is related to the mean by this power law:

$$s^2 = am^b \qquad (8)$$

As few as three or four sets of data are sufficient to estimate the coefficient a and the exponent b, if the sets cover a wide range of mean densities. According to Taylor, the coefficient a varies with sampling technique and habitat, while the exponent b is constant for the species. The most convenient method for estimating a and b is via linear regression of log s^2 on log m.

D. Parameter Estimation

When sampling for parameter estimation, the usual criterion is that the estimate of mean density should be within 20% of the true value. More precisely, this criterion should read that we want to be 95% sure our estimate is within 20% of the true value. The required number of samples N can be estimated from a and b for any m using the appropriate value of the t distribution in the equation

$$N = 25t^2am^{b-2} \qquad (9)$$

where t must have degrees of freedom of $N - 1$ and a two-tailed probability level of 0.05. Equation (9) can be used for probability levels other than 95% by using other t values, and for accuracies other than 20% by replacing the coefficient 25 with $(100/c)^2$, where c is the accuracy as a percent of the mean.

Since the number of samples required depends on the mean number of insects per sample, it becomes necessary to know the mean a priori. Although this is obviously impossible, we often have an idea of what to expect. To be on the safe side, it is better to underestimate the mean density when equation (9) is used to determine N. If there is no basis for estimating m, the best approach is to take three or four samples and enter the mean of these in equation (9).

E. Sequential Sampling

In operation many pest-management programs do not demand exact estimates of population density, but rather require that we categorize the pest population, and possibly certain natural enemies, into density classes. Figure 9.18 presents a simplified hypothetical management system for a crop with a single pest and one important predator. This scheme is typical of the kind of recommendations that might be made for defoliators on soybeans prior to bloom. The only instance in which it recommends a heavy dose of insecticide is when pests far outnumber predators. At intermediate predator/prey ratios a corrective treatment is suggested, while at favorable ratios no treatment is required.

The successful operation of such a management system depends on one being able rapidly and efficiently to classify populations as high, medium, or low. For this the technique of sequential sampling is unsurpassed.

		PEST DENSITY		
		LOW	MEDIUM	HIGH
DENSITY OF PREDATOR	LOW	DO NOTHING	SPRAY WITH 4 OZ OF X	SPRAY WITH 16 OZ OF Y
	MEDIUM	DO NOTHING	DO NOTHING	SPRAY WITH 4 OZ OF X
	HIGH	DO NOTHING	DO NOTHING	DO NOTHING

Fig. 9-18 Chart of simplified pest-management program.

Sequential sampling plans have already been developed and are available in the technical literature for many important economic pests, particularly pests of vegetable crops and of forests (Fig. 14.1). To show how such a plan is developed and used, we will create one for green cloverworm on soybeans.

The concept of sequential sampling is as follows. Suppose we must decide whether the population density in a certain field is above or below the critical level of 20 larvae per foot of row. If the first few samples average 1 or less, then quite probably no more samples will be needed to conclude that the density is less than 20. If the average is 50 or more, we can probably safely conclude that the density exceeds 20. But if the first few samples average near 20, it will be necessary to take more samples before making the decision.

Waters (1955) gives the mathematical equations used to create a sequential sampling plan for any insect. Only three kinds of information are required:

1. The nature of the statistical distribution describing the variation among samples, including the numerical values of relevant constants (e.g., k if the distribution is negative binomial)
2. The values for α and β, the probabilities of calling a small population large and a large one small, respectively
3. The limits of each population class to be recognized, m_2 being the lower limit of the large class and m_1, being the upper limit of the small one.

The green cloverworm has been shown to follow the Poisson distribution (Pedigo et al. 1972), for which the decision line equations are

$$d_1 = bn - h_1 \qquad \text{(lower)} \tag{10}$$

$$d_2 = bn + h_2 \qquad \text{(upper)} \tag{11}$$

where

$$b = \frac{m_2 - m_1}{\ln m_2 - \ln m_1} \tag{12}$$

$$h_1 = \frac{\ln \dfrac{1 - \alpha}{\beta}}{\ln m_2 - \ln m_1} \tag{13}$$

and

$$h_2 = \frac{\ln \dfrac{1 - \beta}{\alpha}}{\ln m_2 - \ln m_1} \tag{14}$$

In this example we use $\alpha = \beta = 0.05$; that is, we want to be 95% certain of classifying the population into the proper class. The class limits are:

1. Low density – 4 larvae or less per foot of row
2. Medium density – 6 to 19 larvae per foot of row
3. High density–21 larvae or more per foot of row.

These conditions yield

$$d_1 = 4.93N - 7.26 \tag{15}$$

and

$$d_2 = 4.93N + 7.26 \tag{16}$$

for the pair of decision lines between the low- and medium-density classes, and

$$d_1 = 19.98N - 29.42 \tag{17}$$

and

$$d_2 = 19.98N + 29.42 \tag{18}$$

for the pair of lines between the medium- and high-density classes.

The most convenient form for using the above equations is a tabulation as given in Table 9.3. In practice the operation might go as follows:

1. Our first sample contains 6 larvae. We check Table 9.3 for $N = 1$ and see that, if our catch had equaled or exceeded 50, we could stop; otherwise we must continue sampling.

Table 9.3 Total Accumulated Number of Green Cloverworm Larvae That Must Be Taken in N Samples to Categorize a Population as Light, Medium, or Heavy

	Population Class		
N	Light	Medium	Heavy
1	–	–	50 or more
2	2 or less	–	70 or more
3	7 or less	23 to 30	90 or more
4	12 or less	27 to 50	110 or more
5	17 or less	32 to 70	130 or more
6	22 or less	37 to 90	150 or more
7	27 or less	42 to 110	170 or more
8	32 or less	47 to 130	190 or more
9	37 or less	52 to 150	210 or more
10	42 or less	57 to 170	230 or more

2. Our second sample contains no larvae, making the total catch 6 $N = 2$. Table 9.3 again indicates we must continue sampling.

3. Our third sample contains 1 larva, making the total catch 7 for $N = 3$. Table 9.3 indicates that a total catch of 7 or less for $N = 3$ means the population is probably light, and that we may stop sampling.

It is possible that sampling may continue indefinitely when open-ended sequential plans are used. The standard procedure is to indicate in the table a point at which sampling should stop even if the population has not been placed in one of the classes. Usually this is done simply by ending the table at the maximum value of N. In our example, if a total of 48 larvae had been taken after 10 samples, we would have stopped, and said that the population falls between light and medium.

VI. ECONOMIC THRESHOLDS

The concepts of economic threshold and economic injury levels are the backbone of any sound pest-management program (see Chapter 1 for definitions).

Although the importance of these criteria is universally recognized, entomologists have been slow in establishing economic injury levels for the major insect pests (Stern 1973). The reason is probably related to difficulties in making these determinations. In general, economic injury levels are not fixed in time or space for a given pest or a certain crop. It is not enough to establish one level for a pest on a given crop. Usually the level varies from region to region, from year to year with the fluctuations of the value of the crop and the cost of treatments, and with the stage of development of the plants. Certainly, economic variables are a primary consideration (Southwood and Norton 1973; see also Chapter 3).

The basic need to determine an economic injury level of a pest is to distinguish between its mere presence in a crop as opposed to the population density that will cause a loss in the quality or quantity of the product (Stern 1973).

The relationship between yield loss and population densities in some cases is linear. For example, the regression of the number of northern corn rootworms, *Diabrotica longicornis* (Say), on percent yield reduction was determined as:

$$y = 0.001 + 0.765x \tag{19}$$

where y is the percent loss and x is the number of larvae per plant (Chiang 1973). Other linear relationships were obtained by Prasad (1963), in determining the effect of feeding by the imported cabbageworm, *Pieris rapae* (L.), the cabbage looper, *Trichoplusia ni* (Hübner), the diamondback moth, *Plutella xylostella* (L.), and the cabbage aphid, *Brevicoryne brassicae* (L.). Significant reductions in yield were obtained with 2 cabbageworms, 1 cabbage looper, or 20 diamondback moth larvae per plant. An increase of 2 cabbageworms per plant caused a decrease of 43 to 142 g, according to the time infestation occurred after transplanting. A population increase of 10 aphids per plant caused an average decrease of 30 g per plant infected 4 weeks after transplanting. It is therefore evident that threshold levels vary with the species, as well as with the stage of plant development at which peak infestation occurs.

To compute economic thresholds, one needs a series of paired points which allow regression of yield reduction on population levels. Sampling of insect populations has been discussed before. Yield samples must be taken to detect some measurable loss in the quality and/or quantity of the product in plots with various levels of infestation. It is critical to take yield samples from the various plots within the same field and at the same time with all possible variables remaining constant, except for the technique used to establish the various levels of infestation (Stern 1973).

The establishment of these various levels of infestation is perhaps the most critical part of the experimental design in economic threshold determinations. Perhaps the most widely used technique is the caging of plants in the field and the entrapment within them of various population levels. In this case uninfested controls are needed to indicate the effect of the cages on the plants (Daugherty et al. 1964).

Cages may drastically change the microclimate around the plants and produce results that may be difficult to extrapolate to open-field conditions.

Techniques that use selective insecticides to produce uneven population levels were adopted by some investigators. In adopting such techniques the effect of the pesticides is assumed to be less disruptive than that of a cage.

Various levels of the *Heliothis* larval populations on cotton were obtained by Adkisson et al (1964), using endrin, methyl parathion, and a mixture of both. Table 9.4 shows the results obtained in the experiments performed in 1962. Results from this study permitted an estimation that 2000 to 2500 larvae per acre (1.5 to 2.0 larvae per foot row) were required to cause a significant yield loss to cotton.

Because of the dynamic nature of economic threshold evaluations and difficulties involved in empirical determinations, attempts are being made to obtain mathematical expressions that permit a ready readjust-

Table 9.4 Seasonal Averages of *Heliothis* Larval Population on Cotton under Three Kinds of Insecticide Treatments[a, b]

Record	Check	Endrin	Methyl Parathion	Methyl Parathion and Endrin
Larvae per acre	3636 *a*	977 *b*	3773 *a*	1658 *c*
Larvae per 10-ft row	2.8 *a*	0.6 *b*	2.9 *a*	1.3 *c*
Percent injured squares	16.9 *a*	7.8 *b*	15.0 *a*	3.3 *c*
Percent injured bolls	9.2 *a*	4.0 *b*	9.0 *a*	4.1 *b*
Yields	773 *a*	1708 *b*	1110 *c*	1683 *b*

[a] Adkisson et al. 1964.
[b] Means followed by the same letter did not differ at 0.05 level.

ment of thresholds as conditions change. An example of one such expression is given for soybeans in Chapter 10, Section VII.

VII. CONCLUDING REMARKS

The technology surrounding data acquistion is a large and growing field. The material in this chapter covers the methods most often used to acquire information for designing and operating insect pest-management systems. Southwood (1966) describes many variations on these methods, as well as several not mentioned here. Current papers in many scientific journals describe additional methods.

Without exaggerating too much, we might say that a new method is developed for every new sampling situation. Hence it is most important that a student understand why data is being gathered and the accuracy required of that data. Based on this understanding, he can design an appropriate sampling program.

REFERENCES

Adkisson, P. L., R. L. Hanna, and C. F. Bailey. 1964. Estimates of the numbers of *Heliothis* larvae per acre in cotton and their relation to the fruiting cycle and yield of the host. *J. Econ Entomol.* 57:657–663.

Anscombe, F. J. 1949. The statistical analysis of insect counts based on the negative binominal distribution. *Biometrics* 5:165–173.

Bliss, C. I. 1967. Statistics in biology. Vol. 1. McGraw-Hill, New York. 558 pp.

Bowden, J. 1965. Sorphum midge, *Contarinia sorghicola* (Coq), and other causes of grain sorphum loss in Ghana. *Bull. Entomol. Res.* **56**:169–189.

Brues, C. T. 1946. Insect dietary. Harvard University Press, Cambridge, Mass. 466 pp.

Carman, P. D. 1963. A large-stage photoelectric planimeter for leaves. *Appl. Opt.* **2**:1317–1321.

Chiang, H. C. 1973. Bionomics of the northern and western corn rootworm. *Annu. Rev. Entomol.* **18**:47–72.

Chiarappa, L., ed. 1971. Crop loss assessment methods: FAO manual on the evaluation and prevention of losses by pests, disease and weeds. Commonwealth Agricultural Bureau, Farnham Royal, Slough, England. 162 pp.

Chiarappa, L., H. C. Chiang, and R. F. Smith. 1972. Plant pests and diseases: assessment of crop losses. *Science* **176**:769–773.

Daugherty, D. M., M. H. Neustadt, C. W. Gehrke, L. E. Cavanah, L. F. Williams, and D. E. Green. 1964. An evaluation of damage to soybeans by brown and green stink bugs. *J. Econ. Entomol.* **57**:719–722.

Dietrick, E. J., E. I. Schlinger, and R. van den Bosch. 1959. A new method for sampling arthropods using a suction collecting machine and modified Berlese funnel separator. *J. Econ. Entomol.* **52**:1085–1091.

Geier P. W. 1960. Physiological age of codling moth females (*Cydia pomonella* L.) caught in bait and light traps. *Nature* **185**:709.

Gist, C. S., and D. A. Crossley, Jr. 1973. A method for quantifying pitfall trapping. *Environ. Entomol.* **2**:951–952.

Gosswald, K. 1935. Uber die Frasstätigkeit von Forstschädlingen unter Einfluss von Altersunterscheiden und der Einwirkung verschiedener Temperatur and Luftfeuchtigkeit and ihre praktische und physiologische Bedeutung. I. *Z. Agnew. Entomol.* **21**:183–201.

Green, G. W., and A. S. DeFreitas. 1955. Frass-drop studies of larvae of *Neodiprion americanus banksianae* Roh. and *N. lecontei* (Fitch) (Hymenoptera: Diprionidae). *Can. Entomol.* **87**:427–440.

Johnson, C. F. 1957. The distribution of insects in the air and the empirical relation of density to height. *J. Anim. Ecol.* **26**:479–494.

Judenko, E. 1973. Analytical method for assessing yield losses caused by pests on cereal crops with and without pesticides. *Trop. Pest Bull.* 2. 31 pp.

Kogan, M., W. G. Ruesink, and K. McDowell. 1974. Spatial and temporal distribution patterns of the bean leaf beetle, *Cerotoma trifurcata* (Forster), on soybeans in Illinois. *Environ. Entomol.* **3**:607–617.

LeRoux, E. J. 1961. Variation between samples of fruit, and of fruit damages mainly from insect pests, on apple in Quebec. *Can. Entomol.* **93**:680–694.

Lincoln, F. C. 1930. Calculating waterfowl abundance on the basis of banding returns. *USDA Circ.* **118**:1–4.

Long, W. A., and S. D. Hensley. 1972. Insect pests on sugar cane. *Annu. Rev. Entomol.* **17**:149–176.

Mail, G. A., and R. W. Salt. 1933. Temperature as a possible limiting factor in the northern spread of the Colorado potato beetle. *J. Econ. Entomol.* **26**: 1068–1075.

Menhinick, E. F. 1963. Estimation of insect population density in herbaceous vegetation with emphasis on removal sweeping. *Ecology* **44**:617–621.

Morris, R. F. 1964. The value of historical data in population research, with particular reference to *Hyphantria cunea* Drury. *Can. Entomol.* **96**:356–368.

Ortman, E. E., D. C. Peters, and P. J. Fitzgerald. 1968. Vertical-pull technique for evaluating tolerance of corn root systems to northern and western corn rootworms. *J. Econ. Entomol.* **61**:373–375.

Pedigo, L. P., G. L. Lentz, J. D. Stone, and D F. Cox. 1972. Green cloverworm populations in Iowa soybeans with special reference to sampling procedure. *J. Econ. Entomol.* **65**:414–421.

Pond, D. D. 1961. Frass studies of the armyworm, *Pseudaletia unipuncta. Ann. Entomol. Soc. Amer.* **54**:133–140.

Prasad, S. K. 1963. Quantitative estimation of damage to crucifers caused by cabbageworms, cabbage looper, diamondback moth and cabbage aphids. *Indian J. Entomol.* **25**:242–259.

Ruesink, W. G. 1972. The integration of adult survival and dispersal into a mathematical model for the abundance of the cereal leaf beetle, *Oulema melanopus* (L.). Michigan State Univ. Ph.D. Thesis, 80 pp.

Scheepers, C. C., and D. L. Gunn. 1958. Enumerating populations of adults of the red locust, *Nomadacris septemfasciata* (Serville), in its outbreak areas in east and central Africa. *Bull. Entomol. Res.* **49**:273–285.

Sokal, R. R., and F. J. Rohlf. 1969. Biometry. W. H. Freeman, San Francisco. 776 pp.

Southwood, T. R. E. 1966. Ecological methods with particular references to the study of insect populations. Methuen, London. 391 pp.

Southwood, T. R. E., and G. A. Norton. 1973. Economic aspect of pest management strategies and decisions. Pages 168–184 *in* Geier, P. W. et al. eds., Insects: Studies in population management. Vol. 1. Memoirs of the Ecological Society of Australia.

Stern, V. M. 1973. Economic thresholds. *Annu. Rev. Entomol.* **18**:259–280.

Taylor, L. R. 1961. Aggregation, variance and the mean. *Nature.* **189**:732–735.

Turnbull, A. L., and D. A. Chant. 1961. The practice and theory of biological control of insects in Canada. *Can. J. Zool.* **39**:697–753.

Wadley, F. M. 1967. Experimental statistics in entomology. Graduate School Press, USDA, Washington, D.C. 132 pp.

Waldbauer, G. P. 1964. Quantitative relationships between the number of fecal pellets, fecal weights and the weight of food eaten by tobacco hornworms,

Protoparce sexta (Johan.) (Lepidoptera:Sphingidae) *Entomol. Exp. Appl.* **7**:310–314.

Waldbauer, G. P., and M. Kogan. 1973. Sampling for bean leaf beetle eggs: Extraction from the soil and location in relation to soybean plants. *Environ. Entomol.* **2**:441–446.

Waloff, N., and R. E. Blackith. 1962. The growth and distribution of mounds of *Lasius flavus* (Fabricius) (Hym:Formicidae) in Silwood Park, Berkshire. *J. Anim. Ecol.* **31**:421–437.

Waters, W. E. 1955. Sequential sampling in forest insect surveys. *Forest Sci.* **1**:68–79.

Zippin, C. 1956. An evaluation of the removal method of estimating animal populations. *Biometrics* **12**:163–189.

10

ANALYSIS AND MODELING IN PEST MANAGEMENT

William G. Ruesink

In the preceding chapters you have read about the general principles of insect pest management and many of the tactics available for our use. Occasionally, we are so fortunate that use of a single tactic against a pest solves the problem to everyone's satisfaction. Examples can be found in the use of resistant varieties of plants, perhaps the best known being *Phylloxera* resistance in grapes (Chapter 4).

In most cases, however, a single action is not sufficient to satisfy all economic, ecological, and sociological considerations. The preferred strategy usually entails the use of two or more tactics against a single pest and, if several pests are of concern at one time, the number of considerations expands rapidly.

This chapter discusses the procedures involved in designing an effective pest-management system, and the need for a mathematical approach to the design process is emphasized. But first we look at the processes involved in the operation of pest-management systems, a look that will help us design better systems.

I. THE OPERATION OF PEST MANAGEMENT SYSTEMS

Pest management is a combination of processes which includes decision making, taking action against a pest, and obtaining the information to be

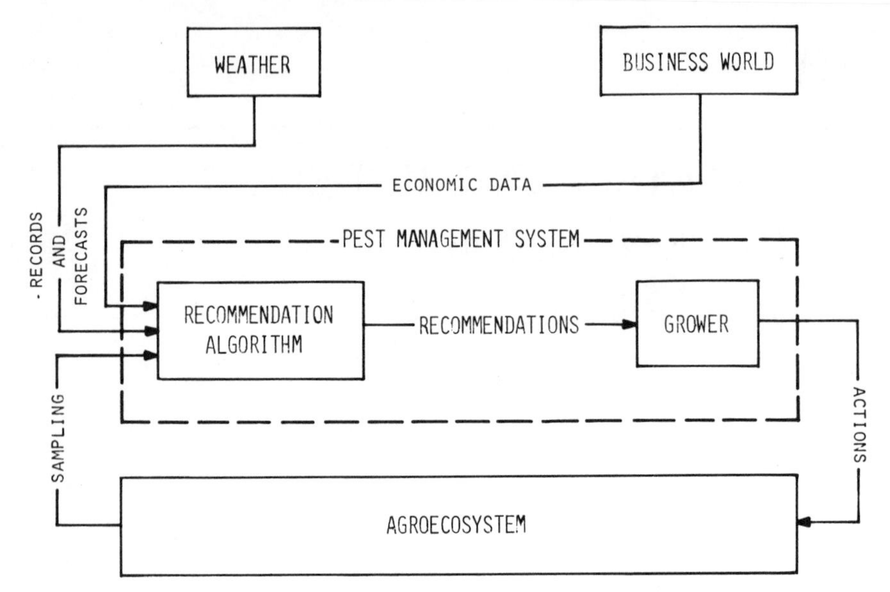

Fig. 10-1 Diagrammatic representation of a pest-management system in operation.

used in reaching these decisions. Figure 10.1 diagrammatically represents how pest management operates at the grower level. The actions taken by the grower are the result of personal decisions and are based on recommendations received from what we will call the *recommendation algorithm*. By definition an algorithm is a precisely defined sequence of rules stating how to produce specified output information from given input information in a finite number of steps (Knuth 1973). In the case of insect pest management, we want the output to consist of a suggested course of action, while the input consists of information received from several sources. From the agroecosystem must come estimates of pest and beneficial population densities, estimates of current state of crop growth and vigor, and possibly estimates of such stress factors as weed competition and soil moisture. From the weather component must come current weather data and perhaps forecasts. From the business world must come data on costs of potential control actions and estimates of the economic value of the crop.

This diagram applies to all pest-management systems, regardless of the form the algorithm takes. In the most primitive system the grower serves as his own algorithm; that is, he uses his intuition and experience to weigh the relative merits of alternate actions. The output is a recommendation to himself regarding what he should do. In a more advanced system, a trained individual (such as an extension specialist) replaces the grower as the algorithm, or more often, publishes a guide (such as an extension bulletin) which the grower can use. Guides are traditionally based on the intuition and experience of one or more professionals, hence partially substitute for having these people inspect the particular fields in question.

Recently, two new forms of algorithms have been developed. Both are based on the fact that mathematical and computer models are now available for many of our worst pest problems. One involves using the models to rewrite extension bulletins so that the grower can consider more options and use more information when making decisions. The second use involves the grower submitting information (from the agroecosystem, the business world, and the weather) to a computer, which uses the model to evaluate optimal and alternate strategies. Within minutes the grower receives recommendations specific to his situation, together with predicted cost/benefit analyses. The changing economics of agricultural production and computer systems suggest that a "computer on every farm" must be taken as a realistic possibility.

Two points must be made regarding Fig. 10.1. Time is not indicated anywhere in the diagram, rather time is considered to be in continuous flow through the entire system. In practice the particular algorithm being used tends to dictate the frequency of sampling the agroecosystem. Every day that a sample is taken, a recommendation results. The second point is that, when the algorithm produces a recommendation, the recommendation may be to "take no action." Hence the range of actions from which a grower must choose includes the option of doing nothing.

Recall that a sound pest-management system must include ecological and social considerations, neither of which is indicated in Fig. 10.1, for they have been internalized in the algorithm. The advantage of internalizing them is that the overall operation of the algorithm is greatly simplified, since no ecological or social data are required. The disadvantage is that, whenever ecological or social concerns change significantly, the algorithm must be either revised or replaced. Fortunately, changes tend to occur slowly in these areas, hence a given algorithm can be expected to be applicable for at least several years.

II. DESIGNING AN ALGORITHM: THE CONCEPT

We have already discussed what an algorithm does and some of the forms it may take. If you are in the business of using pest-management systems, rather than designing them, then perhaps you may wish to skip the rest of this chapter. However, many people find they can use a system better once they know something about how it was developed. If you expect to participate in the development of new systems or in the modification of existing ones, you should certainly read on. Considering the condition of many of the pest-control programs currently being used, most readers of this book will want to help change these programs.

Regardless of whether the algorithm is human intuition and experience or a computer simulation model, there is a set of information on which it is built. If this information is meager and/or inaccurate, the algorithm cannot operate effectively. If the information is precise and nearly complete, it has the potential of being highly accurate. This potential will probably not be realized, however, when the information is massive (i.e., the system complex) and the algorithm designed without the use of mathematical analysis and modeling.

First we look in considerable detail at the kinds of information that form the basis of any pest-management system. This information can be divided into three broad classes: (1) What is the objective of the management system? (2) What are the cause–effect relationships among the parts of the agroecosystem? (3) How do the various available tactics affect the parts of the agroecosystem?

Deciding on the objective of a pest-management system at first seems fairly simple. The species we call pests obviously cause problems for mankind. What we are after is a system that simply eliminates these problems. The only reason our objective becomes more complicated of course is that every system seems to have some undesirable environmental, economic, or social effects. Some of these are obvious: for example, hand picking of pests from large acreages of agricultural crops is economically absurd. Others are rather subtle: for example, the lethal concentration of certain organo chlorine insecticides (e.g., DDT) at higher trophic levels of a food chain was not known to occur until several years after these chemicals had been in widespread use.

Chapter 1 states that the objective of a sound system should be "the intelligent selection and use of pest-control actions that will ensure favorable economic, ecological, and sociological consequences." The difficulty arises in determining what is meant by favorable, and how to compare two potential strategies each of which includes some disagreeable features not directly comparable; for example, how should we compare a robin kill with increased soil erosion or with a loss in crop yield?

A related problem arises because the objectives of society (e.g., social well-being, a viable environment, an adequate food supply) differ from the objectives of the grower (e.g., personal profit, financial security). It is the responsibility of those involved in designing pest-management systems to ensure that both sets of objectives are represented in the same decision algorithm. The individual grower can be expected to place more importance on his own objectives than on the objectives of society in general and, if he must lose profit, to meet society's objectives, we can expect that these objectives will not be met.

A commonly used method of reconciling these differences is to legally prohibit or require a particular pest-management action. For example, there are laws prohibiting the use of a chemical as an insecticide until it is approved for such use by the federal government. In 1973 this power was used to help satisfy society's demand for a cleaner environment by removing DDT from the list of chemicals approved for agricultural use in the United States.

Some pest-management tactics operate efficiently only when everyone participates. Almost every cultural tactic directed against a pest with moderate dispersal capabilities falls in this category. Although cultural practices generally are not legal requirements, the destruction of tobacco stalks after harvest is a good example in which such a requirement would be useful. Late-instar larvae of the tobacco hornworm, *Manduca sexta* (Linneaus); tobacco budworm, *Heliothis virescens* (Fabricius); and corn earworm, *Heliothis zea* (Boddie), are all killed in significant numbers by this practice, so there are fewer moths to infest tobacco the next growing season (NAS 1969).

Many of these considerations are discussed in Chapters 1 through 3. The problem is mentioned here to emphasize the importance of having a specific objective, preferably written in some quantitative form. Moreover, it is clear that, if management actions are to be ordered or carried out by the grower, the objective must be heavily weighted toward maximizing grower profit. The trick is to ensure that the broader objective is also satisfied when grower profit is greatest.

The second broad class of information pertains to understanding the component parts of the agroecosystem and their interactions. At a minimum this includes the bionomics and life histories of the crop, the pests, and their natural enemies. Without knowing a pest's life history, for example, it is impossible to evaluate the merit of any tactic. An even better evaluation is possible if, in addition to bionomics, we understand the population dynamics of the various species involved. This understanding should include quantitative relationships indicating how such factors as temperature and food availability affect the fecundity, survival, and developmental rates of each life stage. All this information can result in confusion rather than enlightenment, unless we use some system to keep it organized. The best way to organize it is through the use of descriptive models: two popular forms are the Leslie matrix formulation and the computer simulation method. Another way that is often satisfactory is via life tables. These are discussed in more detail later in this chapter.

The third broad class of information pertains to understanding the individual tactics and how each affects the various component parts of the agroecosystem. The levels of understanding available here are analogous to those mentioned in the preceding paragraph. At least we must know the typical sequence of events following the application of the tactic. We must know, for example, if it operates by increasing pest mortality, by reducing pest fecundity, or by allowing the crop to tolerate the pest. An improved level of understanding involves knowing how these effect changes in time; for example, a resistant plant variety in the seedling stage may affect the pest in one way, and in a second way as the plant becomes more mature. Some field corn is known to exhibit antibiosis to European corn borer larvae up to about the flowering stage. After this time larval survival increases (Painter 1951). In addition, some of these same varieties have a good tolerance to stem breakage, a factor that has absolutely no impact on borer populations. These two complementary mechanisms of resistance have very different effects on the pest population, and the quantitative effect of the first factor has a definite variation with time. Again, an efficient way of handling the large quantity of information on tactics is through mathematical and/or computer models.

The word *model* has been used rather freely in this chapter without being defined. A good working definition is: a model is an imitation and representation of the real world. This definition can be elaborated on by discussing a model's uses and the typical forms it may take.

Experimental science progresses by manipulating portions of the real world to test hypotheses about its behavior. There are occasions, however, when this direct manipulation is not feasible for a variety of reasons. Perhaps the experiment costs far more than available resources permit. Perhaps it requires many years of experimentation, but the results are required soon. Perhaps the real system includes humans or other factors we do not wish to harm. These are all examples in which experimental science cannot proceed by directly operating on the real world.

Enter the model. Here is a representation of the real world on which the scientist may perform his experiments without the constraints of time, money, and disruption. Here is a substitute for the real world. The validity of the conclusions drawn by working with the model can never be better than the validity of the model, yet this does not require that a model necessarily have mathematical form. A perfectly good model for some needs may be an analogy. For example, an entomologist wanting to understand the factors affecting predatory attack used a blindfolded person tapping a desktop with her finger in search of circular paper disks (Holling 1959). However, if one hopes to utilize the power of electronic computers, he should endeavor to develop mathematical models.

III. CLASSES OF DESCRIPTIVE MODELS

The subject of descriptive models and their construction is much too broad and complex to cover in detail in this chapter. In fact, many books have already been written on the subject, some of which are cited at the end of this chapter. However, an introduction to the subject seems appropriate, so that the reader can appreciate the usefulness of models in pest management.

The following discussion is restricted to models for the dynamics of a single insect species. With some modification the remarks also apply to models of plant growth and models of management tactics.

A model may be discrete or continuous, it may be static or dynamic, and it may be deterministic or stochastic. In addition, it may consider temporal effects or spatial effects or both, and it may or may not distinguish among the various ages or life stages. In addition to these possibilities, the form of a model can vary considerably; for example, changes in population size may be modeled as a function of time alone; or as a function of birth, death, and developmental rates (all time-varying); or even as a function of these rates when the rates themselves are functions of weather, crop condition, parasitoid population, and so on (and *these* are time-varying).

Let us consider first the distinction between a static and a dynamic model. The dictionary definitions of these words give a reasonably precise feeling for the difference. *Static* is defined as "not moving or progressing," and *dynamic* as "relating to change." Generally, models are considered dynamic or static with respect to time, hence models that describe how events or populations change in time are called dynamic models.

Independent of whether a model is static or dynamic, it must also be either discrete or continuous. The distinction can be clearly demonstrated with the example shown in Fig. 10.2. The two plots represent the response (output) of a discrete (A), and a continuous (B), dynamic popu-

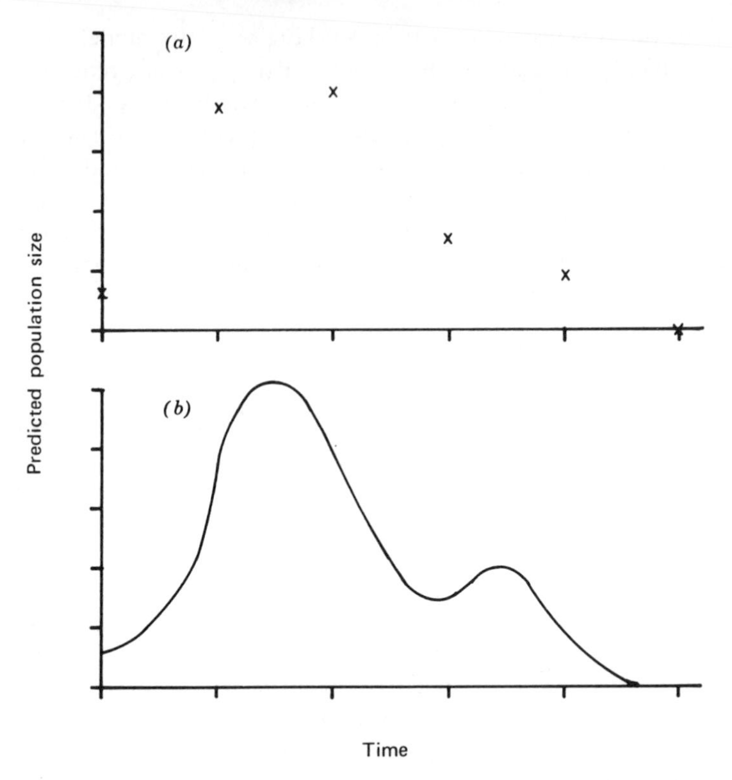

Fig. 10-2 Diagrammatic distinction between discrete (*a*) and continuous (*b*) descriptive models.

lation model. The continuous form produces results for every instant in time, whereas the discrete model says absolutely nothing about population size between the particular points plotted. Discrete models can be developed with any desired interval between points; if the interval is sufficiently small, there will be no lingering doubts about the behavior between points.

A deterministic model is one in which probability and uncertainty play no role. For any given set of conditions, such a model consistently produces identical results. Such models are justified whenever we are fairly confident of our understanding of cause–effect in the real world. A stochastic model includes one or more probabilistic statements; for example, we may say that, when conditions x and y occur simultaneously, a migratory flight will occur with probability 0.20.

Hence there are eight classes of population models defined by these three pairs of characteristics. In general, static models neglect too many important facts, and stochastic models are too cumbersome to manipulate. Consequently, most pest-management work is done with either dynamic-deterministic-discrete or dynamic-deterministic-continuous models. Whether a discrete or continuous form is chosen depends largely on the personal preferences of the persons writing the models and on how the models will probably be used. Mathematicians tend to prefer continuous forms, while biologists prefer discrete forms; continuous forms are amenable to mathematical manipulation, while discrete forms are easier to program for use on digital computers.

Continuous models are almost invariably written as a set of differential equations. If there is no distinction of age or maturity of individuals in the population and no consideration of spatial distribution, these will probably be simple differential equations. The best known simple continuous model is the Verhulst-Pearl logistic equation:

$$\frac{dx}{dt} = rx\left(1 - \frac{x}{k}\right) \tag{1}$$

where r is the intrinsic rate of natural increase, that is, the rate at which the population would grow if resources were unlimited and the individuals did not affect one another, and k is the carrying capacity, that is, the maximum attainable population size. More information on the derivation and use of this equation can be found in Pielou (1969).

Simple models such as the logistic equation have been used extensively in the development of ecological theory. There are those who warn against overreliance on models based on such grossly oversimplifying assumptions. "The use of unrealistic and inaccurate models even for expository purposes must, however, be approached with caution. If the inaccuracy of a model extends to large population processes such as overall population growth form, degree and persistence of fluctuations in abundance and age structure, presence of stable and unstable equilibria, or probability of random population extinction and recolonization, this will engender erroneous conclusions, even at the most general level. Without models which accurately represent natural populations it will be difficult to claim that we have achieved much generality in our concepts" (Streifer and Istock 1973).

The use of partial differential equations allows considerable improvement in accuracy while retaining the necessary level of generality. Barr et al. (1973) proposed as a starting point the forward Kolmogorov equation:

$$\frac{\partial x(z,t)}{\partial t} = \frac{1}{2}\frac{\partial^2[v(z,t)\ x(z,t)]}{\partial z^2} - \frac{\partial[r(z,t)\ x(z,t)]}{\partial z} - d(z,t)\ x(z,t) \tag{2}$$

where z is the measure of an individual's maturity, d is the mortality rate, r is the developmental rate, and v accounts for the dispersion of a cohort in time with respect to maturity. Streifer and Istock (1973) started with:

$$\frac{\partial x(t,z,m)}{\partial t} + \frac{\partial x(t,z,m)}{\partial z} + \frac{\partial [g(t,z,m)\, x(t,z,m)]}{\partial m} = -d(t,z,m)\, x(t,z,m) \tag{3}$$

where z and d are as above, m is the mass or other pertinent size variable, and g is the growth rate.

These two partial differential equations are good starting points for the mathematically inclined student who wishes to pursue population modeling. But most biologists find that sufficient accuracy can be obtained from discrete models, and that the degree of mathematical sophistication required to pursue discrete modeling is within their reach. The remainder of this chapter considers only discrete models.

Discrete models may appear in several forms. One rather familiar form is known as the Leslie matrix, in which the number of individuals present at time $t + 1$ can be computed from the number present at time t and the "projection matrix" $M(t)$ by the equation $\mathbf{x}(t+1) = M(t)\mathbf{x}(t)$. The elements of $\mathbf{x}(t)$ represent the various ages (or age classes if all classes are of equal duration), and $M(t)$ is of the form:

$$M(t) = \begin{bmatrix} F_1(t) & F_2(t) & \cdots & F_{n-1}(t) & F_n(t) \\ P_1(t) & 0 & \cdots & 0 & 0 \\ 0 & P_2(t) & \cdots & 0 & 0 \\ \cdots & \cdots & \cdots & \cdots & \cdots \\ 0 & 0 & \cdots & P_{n-1}(t) & 0 \end{bmatrix}$$

where $F_1(t)$ represents the age-specific fecundity at time t, and $P_1(t)$ represents the age-specific probability of survival to the next age group at time t.

Another common form for discrete models is a set of difference equations analogous to the differential equations used for continuous formulations. An example of a difference equation is: $x_2(t + 1) = P_2(t)x_2(t) + P_1(t)x_1(t)$. This equation says that the size of age class 2 at time $t + 1$ is some part of its size at t plus some part of what was in age class 1 at time t. Often a discrete model consists of a mixture of matrix and difference equations.

A third common approach for discrete models is to bypass mathematical formulation and write the model directly in FORTRAN or some special-purpose computer language such as CSMP, GASP, or GPSS. The major disadvantages in this approach are: (1) It is difficult to communicate the features of the model to anyone, since the programs tend to be difficult to read; and (2) considerable savings can often be effected in both programming effort and computer costs by performing some mathematical manipulations on the model prior to programming.

IV. HOW TO BUILD A DESCRIPTIVE DYNAMIC MODEL

In this section an efficient method for developing a complete descriptive model for any arbitrary system is presented (Coulman et al. 1972):

1. Determine what parts of the real world will be described by the model versus what will be considered environment.
2. Select the components (i.e., the subdivisions) of the system being modeled in a way that reflects the "primary functional characteristics of the modeling problem," that is, that will "capture the essence of the problem."
3. Each component is given a mathematical description which interrelates the inputs, outputs, and states of that component. In population dynamics a state is usually considered the population size (or density). And for the egg stage, for example, one input is oviposition rate, and usually a second is environmental temperature; one output is hatch rate and another might be "natural" death.
4. The last step is to couple the various components together via their inputs and outputs and then connect the system to the environment.

The process of developing a model via these steps is now illustrated using the cereal leaf beetle, *Oulema melanopus* (L.), (CLB) as an example.

Step 1: In modeling population dynamics the central concern is to understand, hence be able to predict, changes in population size as a function of time. Hence we restrict our system to include only the various life stages of the CLB and exclude everything else (e.g., host plant condition, weather, natural enemies). Figure 10.3*a* diagrammatically represents this situation.

Step 2: In selecting components there are two rather contradictory guidelines to consider. First, every group of individuals that can be recognized as having unique responses to potential environmental conditions should be modeled as a unique component, and second, the fewer components contained in a model, the easier it is to use. Figure 10.3*b* is a *system graph* for CLB population dynamics. Each circle represents a component, and the lines connecting components represent inputs and out-

(A)

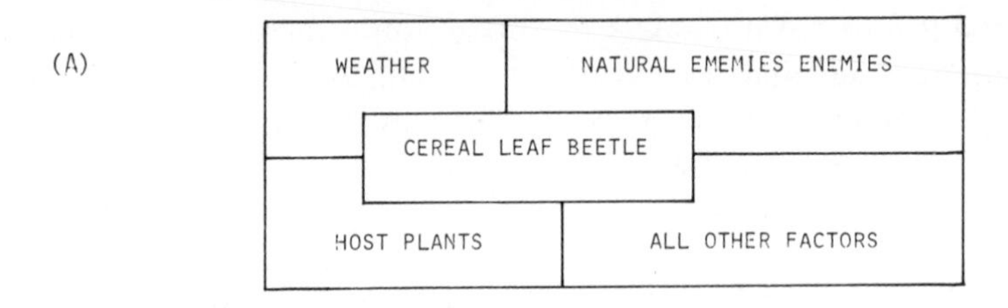

(B)

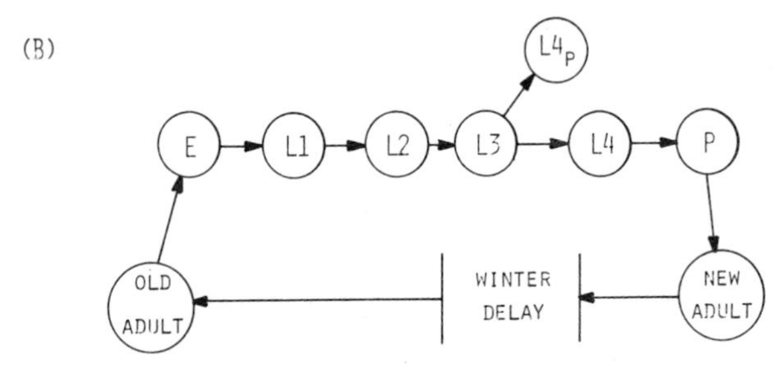

Fig. 10-3 (*A*) Separation of the world into the cereal leaf beetle system and its environment. (*B*) System graph indicating the interrelationship of the nine components of the CLB system (Adapted from Tummala and Ruesink, unpublished data.)

puts. In general, there is one component per life stage of the insect, except that parasitized larvae are distinguished from unparasitized ones. The main reason for doing this is that parasitized fourth-instar larvae mature into parasitoid pupae, whereas unparasitized ones become CLB pupae.

Step 3: This step involves all the biologically difficult parts. While steps 1, 2, and 4 are mostly conceptual or mathematical manipulations, in step 3 we must write our biological knowledge into mathematical equations. This is also the point where decisions must be made regarding how much detail to include in the model. Any gross assumptions or simplifications concerning either the biology or the mathematics must be explicitly written out. Furthermore, at this time we must determine which class of model we will develop.

In the CLB problem Tummala et al. (1975) chose a dynamic-deterministic-discrete form, using difference equations, while Barr et al. (1973) chose a dynamic-deterministic-continuous form. We look at the discrete formulation in detail.

A convenient approach to modeling the time lags caused by the developmental rates of the various life stages is to operate the model in discrete steps of physiological time. For the CLB physiological time was measured as degree-days above the base 48°F, and each step in the difference equations was set equal to 60 degree-days. (Degree-days per day = daily mean temperature minus base temperature.) With this step size, three steps are required to complete egg development, one step each for the five larval components, and seven steps for the pupal component, as shown in Fig. 10.4*a*. The equations interrelating the inputs, outputs, and states for the 17 subcomponents are given in Fig. 10.4*b–e*.

Step 4: The coupling of these subcomponents is achieved simply by equating the output of one to the input to the next, as indicated in Fig. 10.4*a*. The resulting set of difference equations (Fig. 10.5) is known as the model.

Having proceeded through the four steps, let us return to step 3 to complete some modeling glossed over in the paragraphs above, namely, writing explicit equations for, and/or assigning numerical values to, the

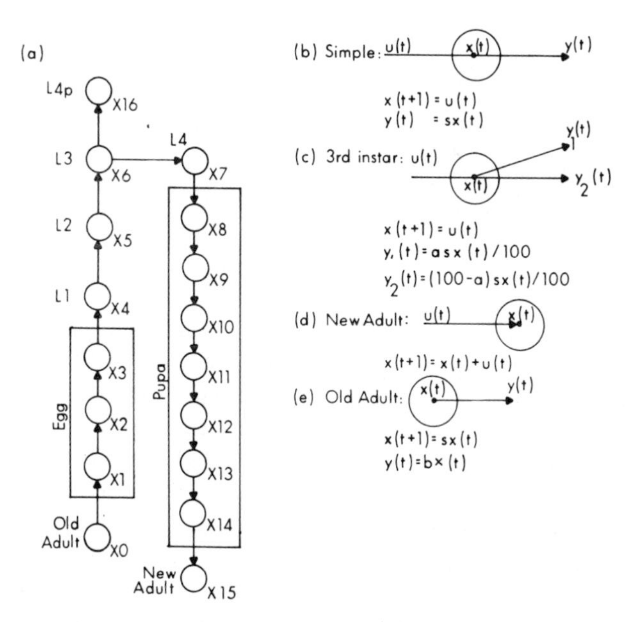

Fig. 10-4 (*a*) Revised system graph to demonstrate the technique of adding subcomponents to achieve proper time lags. (*b* to *e*) The four basic subcomponents types, where *y* represents output, *u* represents input, and *x* represents the population size. The parameters *s*, *a*, and b stand for age-specific survival, percent parasitism, and oviposition rate, respectively.

$$
\begin{aligned}
x_0(t+1) &= s_0 x_0(t) & s_0 &= .70\\
x_1(t+1) &= b x_0(t) & s_3 &= .90\\
x_2(t+1) &= x_1(t) & s_4 &= 1.00 \text{ if } x_0(0) < 1\\
x_3(t+1) &= x_2(t) & s_4 &= .99 - .46 \log x_0(0) \text{ if } x_0 \geq 1\\
x_4(t+1) &= s_3 x_3(t) & s_5 &= .70\\
x_5(t+1) &= s_4 x_4(t) & s_6 &= .55\\
x_6(t+1) &= s_5 x_5(t) & s_7 &= 1.00 \text{ if } x_0(0) < .06\\
x_7(t+1) &= (100-a) s_6 x_6(t)/100 & s_7 &= .66 - .28 \log x_0(0) \text{ if } x_0(0) \geq .06\\
x_8(t+1) &= s_7 x_7(t) & s_{14} &= .60\\
x_9(t+1) &= x_8(t) & a &= \text{determined externally}\\
x_{10}(t+1) &= x_9(t) & b &= 22.\\
x_{11}(t+1) &= x_{10}(t) & & \text{Initial conditions}\\
x_{12}(t+1) &= x_{11}(t) & x_0(0) &= \text{determined externally}\\
x_{13}(t+1) &= x_{12}(t) & x_i(0) &= 0 \text{ for } i \neq o\\
x_{14}(t+1) &= x_{13}(t) & &\\
x_{15}(t+1) &= s_{14} x_{14}(t) + x_{15}(t) & &\\
x_{16}(t+1) &= a\, s_6 x_6(t)/100 & &
\end{aligned}
$$

Fig. 10-5 The complete difference equation model for the population dynamics of the CLB. This model is of the dynamics-deterministic-discrete form and excludes spatial considerations.

parameters s_i, a and b. The right half of Fig. 10.5 contains these functions and parameters values as taken from Tummala and Ruesink (1975).

It must be emphasized that assigning constant values to this many of the parameters restricts the range of conditions to which the model applies. A preferred approach is to submodel the survival and oviposition rates as functions of the environment, which includes everything except the CLB. Were this done well, the model would apply to a great range of geographic, climatic, and environmental conditions. The difficulty in this latter approach is that in general entomologists do not possess an adequate understanding of the cause–effect relationship present in the ecosystem. If the mechanisms are not understood, they cannot be modeled; hence in many cases the best that can be done is to include a numerical constant which represents the average value of the available data.

A classic approach to obtaining parameter values for s_i, a, and b is via the construction and subsequent analysis of many life tables. A sometimes more efficient approach is to gather field data as if a life table were the goal, but then to analyze it using regression techniques or even direct fitting to the proposed model. An example of the regression approach can be found in Helgesen and Haynes (1972).

V. LIFE TABLES

Considering the important role that life tables have played in providing data on the operation of natural populations, it is appropriate that a section be dedicated to their development and analysis. It should be emphasized that constructing life tables is not an essential step in the model-building process. Rather, a life table is a convenient intermediate goal, one that is more easily realizable than the development of a complete descriptive population model. However, there are cases in which the immediate real need is for models; in these cases one should bypass the construction of life tables and proceed directly to an analysis of the quantitative causal relationship affecting survival, developmental, and oviposition rates.

The life table can assume several forms depending on the available data; for instance, there are *current life tables* and *cohort life tables,* the latter being the type commonly used in entomology. The construction of a cohort life table is described with the aid of an example drawn from work by Campbell (1969) on the gypsy moth (see Table 10.1).

The typical life table consists of five columns of information. The left column is labeled x, and it lists the age interval or age class under consideration. The next column, labeled lx, lists the number of individuals of an original cohort alive at the beginning of age class x. Next comes dxf which is the factor responsible for the observed mortality, and dx is the number of individuals of the cohort dying from cause dxf. The final column, headed $100qx$, is the proportion dying by cause dxf expressed as a percentage of lx.

The easiest way to construct a life table is to make frequent observations on a cohort, recording x and lx at each observation. With this type of data the life table can be constructed directly. The life table in Table 10.1 represents the average of several sets of observations.

In many cases in entomology, the insect we are dealing with has only one to three generations per year with essentially no overlap of generations. Under these circumstances it is convenient to consider each gen-

Table 10.1 Life Table Typical of Dense Gypsy Moth Populations in Glenville, N.Y.[a]

x, Age Interval	*lx*, Number Alive at Beginning of *x*	*dxf*, Factor Responsible for *dx*	*dx*, Number Dying during *x*	100*qx*, *dx* as percent of *lx*
Eggs	250	Parasites	50.0	20
		Other	37.5	15
		Total	87.5	35
Instars I–III	162.5	Dispersion, etc.	113.8	70
Instars IV–VI	48.7	Parasites	2.4	5
		Disease	29.2	60
		Other	12.2	25
		Total	43.8	90
Prepupae	4.9	Desiccation	0.5	10
Pupae	4.4	Parasites	1.1	25
		Disease	0.7	15
		Calosoma larvae	0.9	20
		Other	0.4	10
		Total	3.1	70
Adults	1.3	Sex (S:R = 30:70)	0.9	70
Adult ♀ ♀	0.4	—	—	—
Generation	—	—	249.6	99.84

[a] Campbell 1969.

eration a cohort, even though the individuals are not exactly uniformly aged. This approach results in one life table per generation, in which 100*qx* is the average mortality rate for the generation. The only difference in constructing a generation life table involves the difficulty in determining the number of individuals entering each life stage (or age class). Of the many techniques that have been used and proposed, one of the simplest is the area-under-the-curve technique of Southwood (1966). The results obtained compare favorably with those of more complex techniques.

The technique involves plotting population size (or density) versus time, and finding the area under the curve formed by connecting the points with straight-line segments (Fig. 10.6). This area has units of population times time; dividing this quantity by the developmental time of

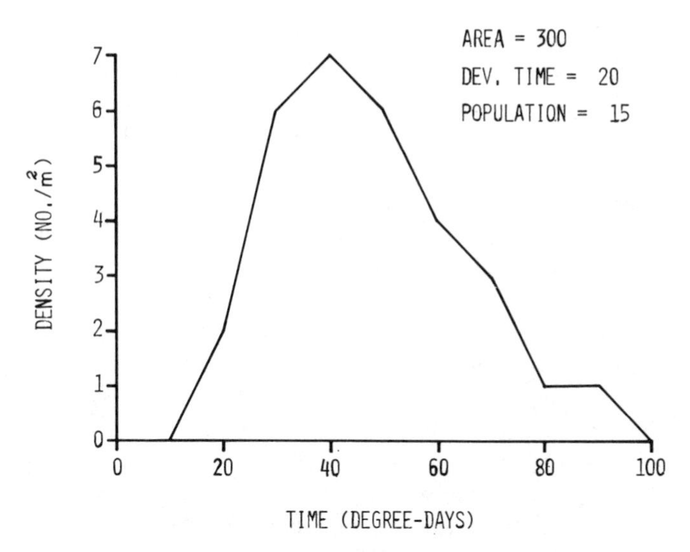

Fig. 10.6 Estimating total population passing through a life stage from density estimates using Southwood's (1966) area-under-the-curve technique.

the life stage under consideration results in a measure of the total number of individuals passing through this life stage in the generation. When temperatures are rather variable during the generation, it is best to use degree-day accumulations above the developmental threshold as the measure of time. Hence both the x axis and the developmental time constant have units of degree-days in Fig. 10.6.

This method measures the population passing through the median age of the life stage rather than the number entering the stage. If mortality is not great early in the stage, this is a reasonably good estimate of the number entering. But rather than trying to estimate the number entering each stage, a pragmatic alternative is to establish the age classes to overlap life stages. For example, instead of entering "first instar" in the x column, we could use "from midfirst to midsecond instar." Instead of considering mortality "during" an instar we could consider mortality "between" instars. And if the life table information is to be used in a difference equation descriptive model, it is more convenient to have mortalities defined between stages than during stages.

VI. USING MODELS TO BUILD AN ALGORITHM

In order for descriptive models to be used effectively in designing a pest-management system, we normally must have more than one or two of them. In fact, we must have models that describe enough of the agroecosystem so that we can evaluate the performance of our management system; our evaluation is based on the objective function written earlier. The objective is a complex statement which may include social, economic, and ecological considerations, in which case our descriptive models would also have to include these same considerations. In practice the objective function normally includes only biological and/or economic considerations, the other aspects being handled subjectively. But even in this case our descriptive models should include all important pest species, the significant natural enemies of each, a plant/crop model, and the impact of each control tactic on any feature of the plants or insects included in the models. In a typical agroecosystem there might be 4 pests of importance, a total of 10 important natural enemies, 1 plant, and 5 control tactics, for a total of 20 descriptive models. Because of the time involved in developing and validating descriptive models, most of the examples available of the use of models to design strategy are restricted to agroecosystems having a single key pest over a large geographic area or in which the management of one pest can be effected without upsetting the conventional programs used for the other pests. Examples are the alfalfa weevil in alfalfa (Chapter 12), the cereal leaf beetle in small grains, and mites in apple orchards (Chapter 13).

Once the descriptive models are at hand, several choices are available. The first one we discuss might be called the "weatherman" approach. An example of this can be found in the Alfalfa Action Pest Management Project begun at Purdue University in 1972. Descriptive models were used for the alfalfa crop and the alfalfa weevil in a forecasting program for alfalfa growers in Indiana. The system used past and predicted weather (including temperature, rainfall, etc.) as input for the models, and produced an alfalfa pest status forecast which was sent to county extension agents via a computer telephone hookup. The agents in turn notified the growers via conventional means of communication.

A second approach is to use models as a substitute for the real world and to perform the equivalent of field experiments on them. The results of these experiments can be field-verified and then written into a conventional extension bulletin for dissemination to the appropriate growers. This approach is analogous to the experimental approach commonly used in the past, only models now replace the field to a large extent. This has the advantage of permitting a large number of experiments in a relatively short time.

A third approach is to develop a cooperative program with someone trained in control theory, a branch of systems engineering. Mathematical and computer methods are available, which operate with the descriptive models and the objective function to produce a best possible control strategy. "Best" means that no other strategy deviates less from the stated objective. Of the available techniques one that is relatively well known is called *dynamic programming*; Shoemaker (1973) has applied this to a hypothetical problem in agricultural pest management. Several other techniques are finding some applications in entomology but, as pointed out by Caswell et al. (1972), there is no standard approach to optimization in ecological systems. The science is yet too young to have a well-marked road.

VII. BUILDING AND USING STATIC MODELS

Although the real world is full of time-varying phenomena, and although the only reasonable way to model time is with dynamic models, there are many cases in insect pest management in which more simple static models can be very useful. A good example is the determination of economic injury levels for soybean defoliators. The following model is based on work by Stone and Pedigo (1972).

To permit a relatively simple mathematical analysis, three simplifying assumptions were made:

1. The plant does not produce any new leaves during the time intensive defoliation is occurring.
2. The relationship between percent defoliation and percent yield reduction depends solely on the soybean stage of growth, and these relationships are assumed known.
3. If control is applied, all damage is avoided.

Allowing these assumptions, the theoretical economic injury level can be computed from the following set of data:

1. The leaf area present L_k per foot of row for beans in growth stage k
2. The amount of food F consumed by a larva from egg hatch to pupation
3. The expected bushels-per-acre yield with zero defoliation N
4. The dollars-per-bushel value of soybeans C_1
5. The dollars-per-acre cost of spraying C_2
6. The relationship between percent defoliation and percent yield loss (α_k and β_k from Fig. 10.7*a*).

The economic injury level in pest control is the level of infestation at which control costs exactly equal the benefits expected from exercising the control (Chapter 3). The benefits B in this case can be calculated from

$$B = C_1(NY_k/100) \tag{4}$$

where $NK_k/100$ represents the bushels saved. But from Fig. 10.7*a*,

$$Y_k = \alpha_k D + \beta_k D^2 \tag{5}$$

so

$$B = C_1 N(\alpha_k D + \beta_k D^2)/100 \tag{6}$$

In addition we know that, if population density (larvae per foot of row) at stage k is designated by P_k, percent defoliation is computable from the ratio of food consumed to total leaf area by:

$$D = 100FP_k/L_k \tag{7}$$

Substituting equation (7) into equation (6) gives

$$B = C_1 N \left(\frac{100\alpha_k FP_k}{L_k} + \frac{100^2 \beta_k F^2 P_k^2}{L_k^2} \right) \Big/ 100 \tag{8}$$

$$= \frac{C_1 NF}{L_k^2} (\alpha_k L_k P_k + 100\beta_k FP_k^2) \tag{9}$$

Equating total costs C_2 to total benefits B gives:

$$C_2 = \frac{C_1 NF}{L_k^2} (\alpha_k L_k P_k + 100\beta_k FP_k^2) \tag{10}$$

or

$$0 = 100\beta_k FP_k^2 + \alpha_k L_k P_k - \frac{C_2 L_k^2}{C_1 NF} \tag{11}$$

Solving for P_k gives

$$P_k = \frac{1}{2(100\beta_k F)} \left[-\alpha_k L_k \pm \sqrt{\alpha_k^2 L_k^2 - 4(100\beta_k F) \frac{-C_2 L_k^2}{C_1 NF}} \right] \tag{12}$$

which simplifies to

$$P_k = \frac{0.005L_k}{\beta_k F} \left(-\alpha_k + \sqrt{\alpha_k^2 + 400 \frac{C_2 \beta_k}{C_1 N}} \right) \tag{13}$$

The $\pm$ in front of the square root term in equation (12) was replaced with a $+$ because, when the $-$ possibility is considered, P_k becomes negative. A negative P_k is mathematically meaningful but biologically impossible.

Equation (13) is the static model for insect defoliation on soybeans. It applies to any species that exhibits a distinct population peak (to satisfy the assumption that the plant does not grow during defoliation) and for which the food consumption per individual is known.

For the green cloverworm, *Plathypena scabra* (Fabricius), in Iowa, the following values can be used for the parameters of the model: $F = 105$ cm^2 per larva, $C_1 = \$6$ per bushel, $C_2 = \$4$ per acre, $N = 35$ bu. per acre, $L_1 = 1180$ cm^2 per foot of row, $L_3 = 3540$ cm^2 per foot of row, $L_5 = 10{,}770$ cm^2 per foot of row, $L_7 = 14{,}970$ cm^2 per foot of row, and $L_9 = 14{,}430$ cm^2 per foot of row. Values for α_k and β_k are taken from Fig. 10.7*a*.

Green cloverworm outbreaks typically occur when the crop is setting and filling pods, which is stage 7. The economic injury level is calculated for this stage as:

$$P = \frac{0.005(14970)}{0.008(105)}\left[-0.032 + \sqrt{0.032^2 + 400(0.008)\,\frac{4}{6(35)}}\right] \tag{14}$$

$$= 89.1(-0.032 + \sqrt{0.062})$$

$$= 19.3 \text{ larvae per foot of row}$$

Figure 10.7*b* shows how this economic injury level varies with the stage of growth of the soybean plant.

VIII. CONCLUDING REMARKS

The subject of this chapter has been the building and use of models in pest management, with the primary emphasis on using models to help design a recommendation algorithm as defined by Fig. 10.1. In addition to those discussed here, models have many other uses in entomology. Specifically, descriptive models have two very important uses worth mentioning. First, the process of organizing existing knowledge on the population dynamics of a species into mathematical equations forces one to consider all aspects of the problem. If there are major gaps in the existing knowledge, the process of modeling reveals these gaps. Second, a completed model is an excellent teaching tool. It contains all relevant information about the dynamics of a population, and thus it can substitute in part for the many months or years of field experience usually necessary before a young scientist can understand the workings of a particular population.

In addition to developing and using models, an entomologist working in pest management must deal with two other kinds of mathematical

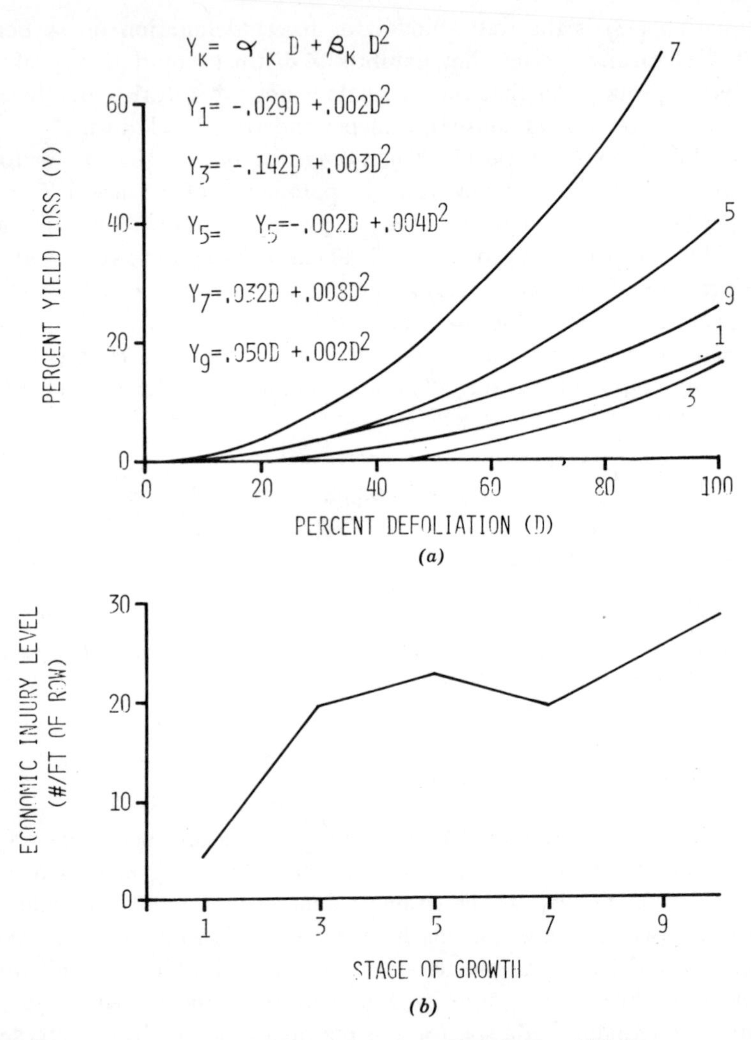

Fig. 10-7 (*A*) The second order regression of percent yield loss on percent defoliation based on mechanical defoliation experiments (Adapted Stone and Pedigo 1972.) (*B*) The economic injury levels predicted by equation (13) for green cloverworm for several stages of growth of the soybean plant.

analysis. Chapter 9 discussed sampling and measurement and, as pointed out, an adequate sampling program includes consideration of adequate sample size. The mathematics involved in weighing the trade-offs between accuracy and work load is generally not complex, but it does demand a certain level of competence in algebra and fundamental statistics. The second kind of analysis referred to here is the mathematical-statistical interpretation of field data. Some authors refer to this as empirical modeling. An example can be found in Helgesen and Haynes (1972), in which survival rates were computed from population density samples, and then regression analysis was used to relate the survival rates to population density. Their decision to use linear regression represents the simplest of several options available; a curvilinear regression would almost certainly be more compatible with population theory, but the mathematics would have been more complex.

The development of a sound pest-management strategy depends on our ability to understand all the significant and relevant interactions in the real world. Models serve as a crutch to aid our mental recall. The method of design is to manipulate our concept of the real world (our model) using potentially useful tactics in all possible combinations. We must then evaluate these combinations and choose the one that is best.

In general, the system dynamics are very important, and the pest-management system chosen as best must reflect those dynamics. The spray calendar once used for control of orchard pests ignores dynamics; it is a strategy that assumes this year's problems are identical to those of last year and the year before, hence the same sequence of tactics is optimal for all years. A better strategy would be a set of rules for deciding if and when each tactic should be used, depending on current pest, crop, weather, and economic conditions. Thus, when we speak of using models to develop an optimal strategy, we are looking for the set of rules (i.e., the algorithm) that will tell us what the proper action should be.

Systems science has only recently been applied to pest management, and we have much to look forward to in the future. There is no doubt that society and government will place increasing emphasis on this approach to problem solving (Haskins 1973), and this will add considerably to the demand on entomologists for innovative approaches to pest-management research and implementation. Persons with a broad range of expertise, and an even broader outlook on the role of entomology in society, are essential to the success of utilizing systems science in pest management. Society will tolerate inaction for a few years while we reorient ourselves to this task. But then we must produce. And our results must satisfy the demands of society, since "after this there is no place to go."[1] There is no better route to problem solving than the systems approach, and this will be the approach for many years to come.

REFERENCES

Barr, R. O., et al. 1973. Ecologically and economically compatible pest control. Pages 241–263. *In* Insects: Studies in Population Management. Memoirs 1, *Ecol. Soc. Australia,* Canberra.

Campbell, R. W. 1969. Studies on gypsy moth population dynamics. Pages 29–34 *in* Forest insect population dynamics. *USDA Forest Ser. Res. Paper* NE-125.

Caswell, H., H. E. Koenig, J. A. Resh, and Q. E. Ross. 1972. An introduction to systems science for ecologists. Pages 3–78 *in* B. C. Patten, ed., Systems analysis and simulation in ecology. Vol. II. Academic Press, New York.

Coulman, G. A., S. R. Reice, and R. L. Tummala. 1972. Population modeling: A systems approach. *Science* **171**:518–521.

Haskins, C. P. 1973. Science and social purpose. *Amer. Sci.* **61**:653–659.

Helgesen, R. G., and D. L. Haynes. 1972. Population dynamics of the cereal leaf beetle, *Oulema melanopus* (Coleoptera: Chrysomelidae): A model for age specific mortality. *Can. Entomol.* **104**:797–814.

Holling, C. S. 1959. Some characteristics of simple types of predation and parasitism. *Can. Entomol.* **91**:385–398.

Knuth, D. E. 1973. Computer science and mathematics. *Amer. Sci.* **61**:707–713.

National Academy of Science. 1969. Insect-pest management and control. National Academy of Science, Washington, D. C.

Painter, R. H. 1951. Insect resistance in crop plants. University of Kansas Press, Lawrence. 520 pp.

Pielou, E. C. 1969. An introduction to mathematical ecology. Wiley-Interscience, New York. 286 pp.

Shoemaker, C. 1973. Optimization of agricultural pest management. I: Biological and mathematical background. *Math. Biosci.* **16**:143–175.

Southwood, T. R. E. 1966. Ecological methods. Methuen, London. 391 pp.

Stone, J. D., and L. P. Pedigo. 1972. Development and economic-injury level of the green cloverworm on soybean in Iowa. *J. Econ. Entomol.* **65**:197–201.

Streifer, W., and C. A. Istock. 1963. A critical variable formulation of population dynamics. *Ecology* 54(2):391–398.

Tummala, R. L., W. G. Ruesink, and D. L. Haynes. 1975. A discrete component approach to the management of the cereal leaf beetle ecosystem. *Environ. Entomol.*

[1] Haynes, D. L. 1971. Pest management and the cereal leaf beetle. Paper presented at the meetings of the North Central Branch, Entomological Society of America, Chicago, March 1971.

EXAMPLES

11

COTTON INSECT PEST MANAGEMENT

H. T. Reynolds, P. L. Adkisson, and Ray F. Smith

INTRODUCTION

Production of cotton is ancient, dating back thousands of years. Cotton fabrics as old as 3000 BC have been found in the Indus Valley (Gulati and Turner 1928) and, in the New World, specimens dating to 2500 BC have been found in Peru (Bird and Mahler 1951). Over an even longer period of time, many species of insects, spider mites, and other pests have adapted to live and feed on cotton. Hargreaves (1948) listed 1326 species of insects on this crop worldwide. In the United States no less than 100 species of insects and mites are reported to attack cotton, although very few are so consistently serious as to be considered key pests. The crop is subject to attack from the time seeds are planted until harvest about 7 to 10 months later. All the plant parts may be attacked, but the most serious pests attack primarily the fruiting portions—squares (flower buds), blooms, and bolls, reducing both quantity and quality of the harvested lint and seed. Newsom and Brazzel (1968) calculated that more than 80% of the losses attributable to cotton pests were caused by species that attack the fruits.

Estimates of crop losses caused by insects are notoriously variable and poorly documented, but the National Cotton Council of America calculated, on the basis of estimates by the U.S. Department of Agriculture

(USDA) Crop Reporting Board, a loss of 1 bale of cotton lint for every eight produced—or the equivalent of one entire crop out of every nine production seasons—may be attributed to the arthropod pests of the crop. The average annual loss for the period 1951–1960 was calculated to be nearly $500 million. This is equivalent to a 19% loss of the potential crop (Table 11.1). This represents only potential yield in cotton not harvested and does not include the estimated $150 million farmers spend each year on pest-control costs. Clearly, such losses year after year are intolerable, and strongly suggest that prevailing cotton protection practices are less than ideal. Furthermore, as will be pointed out, the situation in some cases is such that the efficacy of cotton pest-control practices has deteriorated in recent years, making the development of better insect pest-management practices even more imperative.

Cotton is an important economic crop in world agriculture. Approximately 31 million ha are grown in the world each year. In some countries cotton provides a large part of all export earnings, and thus is par-

Table 11.1 Estimated Average Annual Losses Caused to Cotton in the United States by Insects and Spider Mites, 1951–1960[a, b]

	Loss from Potential Production		
Pest	Percent	Quantity (1000 bales)[c]	Value ($1000)[d]
Boll weevil	8.0	1,239	200,613
Bollworm and tobacco budworm	4.0	619	100,307
Plant bugs	3.4	527	85,261
Spider mites, pink bollworm, cotton aphid, cotton leafworm, thrips, cotton leaf perforator, cabbage looper, beet armyworm	3.6	558	90,276
Total	19.0	2,943	476,457

[a] Newsom and Brazzel 1968.

[b] Data from *USDA Agricultural Handbook* no. 291 (USDA 1965).

[c] Estimates based on full production with causes of loss estimated.

[d] Includes quality and quantity and assumes that market outlets are available for increased production with no change in average prices.

ticularly important to the economic development of many emerging nations. The crop plays a dominant role in the economy of many states in the United States cotton belt. It accounts for more than 50% of the income from agricultural crop production in more than half of the 19 states in which it is grown. Even in large agricultural states such as California and Texas, cotton is usually the most important cultivated crop.

Cotton is produced in a variety of environmental situations in the United States. Requisites common to all areas are a long growing season, fertile soil, warm temperatures, and adequate soil moisture. The oldest production area is the warm, humid Southeast, where soil moisture is dependent on rainfall. The arid southwestern states where irrigation provides moisture are the most recent in cotton production.

Cotton production in the United States ranges from about 11 to nearly 15 million acres each year. About 99% of the cotton produced represents several varieties of the species *Gossypium hirsutum,* an upland cotton with short- to medium-length, relatively coarse fibers. The remainder (1%) is the fine, long-fibered 'Pima' cotton, *G. barbadense.* The average yield of *G. hirsutum* in the United States is a little over one bale of lint per acre (a bale weighs approximately 500 lb), with some irrigated areas, for example, Arizona and California, averaging about two bales per acre. In a good production year, restricted areas, such as the Imperial Valley of California, have averaged as high as $3\frac{1}{2}$ bales per acre with some fields producing in excess of five bales. The seed ginned from the lint is also a valuable commodity. The extracted, refined oil is used in cooking and oleomargarine, for example, and the residual cottonseed cake is also valuable, often being used as high-protein cattle feed. For every pound of lint produced, there are about 2 lb of seed.

United States cotton growers, in contrast to producers of most other agricultural commodities, have organized and voluntarily tax themselves according to bales of lint produced. The nonprofit organization, Cotton, Inc., for example, receives $1 for every bale produced in the United States. These funds are used mainly to promote cotton and support research on cotton production and protection. Some states do likewise, often to support research on particular problems, such as California is doing at the present time to support broadly based research on the pink bollworm problem. It is clearly evident that this progressive outlook by the cotton industry has accelerated remarkably the "state of the art" of pest management, including the development of many new pest-management techniques.

I. THE COTTON ECOSYSTEM

The cotton ecosystem, despite its apparent simplicity, is a remarkably complex ecological unit with a multitude of interacting factors. The cottonfield is not a unit in isolation, but is part of an ecological system which includes associated agricultural fields of many types, woods, streams, weedy or uncultivated areas, and more. The major components of the cottonfield include the plants, weeds and other plants, the soil and its biota, overall conditioning of the physical and chemical environment, pest species with their natural mortality factors including disease and beneficial species, arthropod competitors for food and space, and overall conditioning of man including his management of the system (Smith and Reynolds 1972).

The basic foundation of insect pest management is based on a detailed knowledge of the crop ecosystem. This seems rather straightforward and simple, but in reality it is very complex and probably will never be understood completely. Furthermore, what is learned about the cotton ecosystem in the San Joaquin Valley of California, or the Salt River Valley of Arizona, or the Mississippi Delta may be indicative but not directly applicable to other areas. Each major production area differs to varying degrees, because different cotton varieties are grown under different management systems. Also, there may be distinct differences in the surrounding physical and crop environments in different production regions. In fact, neighboring fields in the same region may vary to some degree for the same general reasons.

Entomologists have progressed remarkably in their knowledge of the cotton agroecosystem and the interactions therein, even though much is still incompletely understood. For example, the relative value of beneficial insects is increasingly recognized, although much more quantification of their value is needed. This is particularly true with regard to the relation of total populations of the complex of beneficial species to pest populations.

A. The Cotton Plant

Only 4 cotton species of the 30 or more known are cultivated commercially. The two originally from the New World grown commercially are *Gossypium hirsutum* and *G. barbadense*. The two originally from the Old World are *G. arboreum* and *G. herbaceum*. The well-known long-staple cottons of the United Arab Republic, 'Tanguis' cotton of Peru, the 'Sea Island' cotton of the West Indies, and the 'Pima' cotton grown locally in the southwestern United States are derived from *G. barbadense*. The bulk of the world's cotton, including that grown in the United States, belongs to various strains selected from *G. hirsutum*.

Most cotton is grown as an annual crop, but a few areas of the world utilize the perennial nature of the plant to grow a second and, sometimes, a third crop from the same plants. These are cut off (ratooned) a few inches above the ground level at the end of each production season and allowed to regrow. This practice is known as growing *stub* or *ratoon* cotton (*Socas* in Latin America). In the United States this practice is discouraged, and in most areas is prohibited. In warm, arid regions stub cotton usually produces well and is favored by some growers where water for irrigation is scarce. Also, production costs may be less, since there is no need to prepare the field for planting. The ratooning of cotton is discouraged mainly because the plants permit increased pest population carryover, particularly the pink bollworm, and may provide a reservoir for a virus disease of cotton, "crumple-leaf," transmitted by the whitefly, *Bemisia tabaci* (Gennadius) (Dickson et al. 1954; Laird and Dickson 1959; Erwin and Meyer 1961).

Many strains of *G. hirsutum* have been developed which incorporate desirable characteristics such as quality and quantity of fiber produced, time of fruiting, and crop maturity. Most of these strains have been developed without consideration of possible varietal relationship to pest problems. Recently, however, more research is being conducted by teams including entomologists and other crop-production and -protection personnel on development of cotton strains resistant or tolerant to attack by important pest species.

Cotton plants follow a consistent and predictable pattern of growth. The pattern varies somewhat according to variety and environmental conditions under which the plant is grown, but may be defined for each situation (Delattre 1973). One such growth pattern is shown in Table 11.2.

The maximum boll-carrying capacity is genetically determined, but final yields are strongly influenced by the physiological capacity of the plant as modified by temperature, day length, soil fertility, moisture, and cultural practices, as well as by insects and other plant pests. Routine, periodic plant examination by the grower can provide a clear evaluation of the progress of fruiting and crop maturity.

Likewise, the importance of the growth and fruiting characteristics of the cotton plant to pest-management programs is becoming increasingly recognized. Eaton (1955), in characterizing the cotton plant, pointed out the presence of vegetative and reproductive fruiting branches, inde-

Table 11.2 Growth and Development Data for the Upland Cotton Variety 'Acala SJ-1' Grown in the San Joaquin Valley of California[a]

1. Plant emergence takes an average of 12 to 15 days.
2. The first true leaf (third leaf) appears 10 days after emergence.
3. The fourth leaf appears 2 to 4 days after first true leaf.
4. Time from emergence to square averages 35 to 55 days.
5. Time from square to white bloom averages 21 to 23 days.
6. Time from bloom to open boll averages 55 to 60 days for midseason fruit, but may range from 65 to 90 days for late-season bloom because of shorter days, lower temperatures, and other factors.
7. Squares formed 1 June to 10 July make the flowers of 25 June to 6 August, which contribute most to yield and quality.
8. Average number of days between flowers on same fruiting branch is 6 days; range is 4 to 8 days.
9. Average number of days between flowers on two successive branches is 3 days; range is 1 to 6.
10. Bolls reach full size 24 to 30 days after bloom.
11. Approximately 44% of blooms make bolls.
12. Approximate number of bolls per row foot to equal one bale per acre yield; 38-in. rows, average 8.0; 40-in. rows, 8.5.

[a] Smith and Falcon 1973.

terminate growth, and natural shedding of squares and small bolls. Flowering is progressive; shedding is very light in the early fruiting stage of the plant but increases with time. Boll retention likewise is excellent early in the season but declines as the season progresses. Over a season nearly two-thirds of the squares fail to produce bolls. This occurs in the absence of any pest species that may affect the fruiting pattern. Thus insect-produced square and small-boll shed may have little effect on yield if less than the natural shed of the plant (Eaton 1955; Newsom and Brazzel 1968; Pearson and Maxwell-Darling 1958). The ability of the plant to compensate and retain fruit to its physiological capacity is exemplified further by the fact that stands of 25,000 to 75,000 plants per acre produce about the same yield per acre under comparable growing conditions (Adkisson et al. 1964).

The characteristic average growth and maturity pattern of 'Acala SJ-1' cotton in the San Joaquin Valley of California is illustrated in Figs. 11.1 and 11.2 (Smith and Falcon 1973). Similar data are available for 'Delta Pine' cotton grown in the Imperial Valley and for cotton grown in most

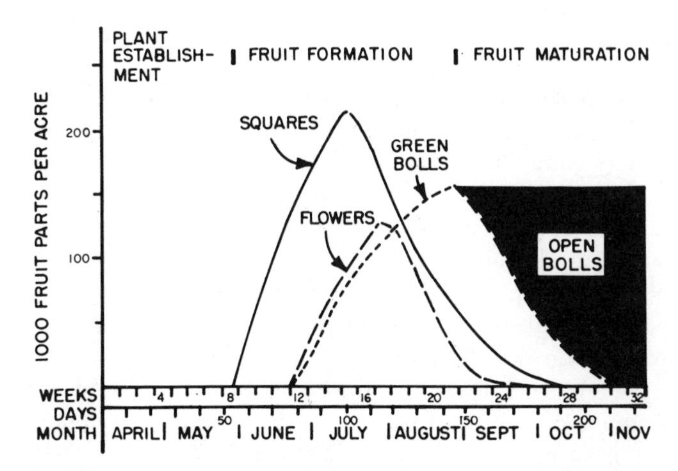

Fig. 11.1 Three main periods of plant growth and development and average fruiting profile of 'Acala SJ-1' in the San Joaquin Valley of California. (Smith and Falcon 1973.)

other states. Figure 11.2 graphically illustrates the great excess of fruiting forms produced by the plant that are not retained and do not make mature bolls. It is important to note that the early squares appearing from about June 1 to July 10 result in flowers about June 25 to August 6. These contribute the most to yield. Consequently, this is a critical period during which squares should be protected from excessive loss that may be inflicted by insect pests.

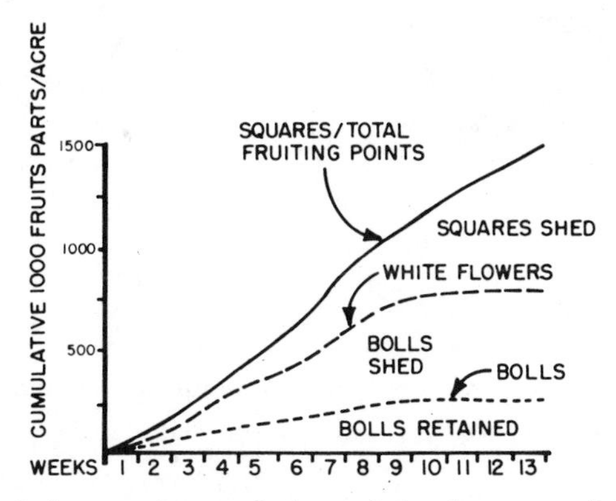

Fig. 11-2 Cumulative number and time of development of fruiting parts of 'Acala SJ-1 in the San Joaquin Valley. (Smith and Falcon 1973.)

B. Changes in Cotton Pests

Cotton pests are subject to many environmental pressures in addition to those resulting from the application of pesticides. Changes in the agro-ecosystem may eliminate a pest population or reduce it to insignificance. However, individuals in a surviving population of insects may become well adapted to newer conditions and assume greater importance as pests. Pest populations also may develop new characteristics which allow them to become better adapted to their environment or enable them to expand their geographical distribution, host range, or even host preference.

In the cotton-growing regions of Nicaragua, the cotton insect pest complex has changed markedly since the introduction of commercial cotton production in 1949 (Table 11.3). Insecticide usage increased from none at all or, at best, a few applications per season in the early 1950s to 10 or 12 at the end of the decade. Usage increased even more rapidly during the 1960s, and by 1966–1967 many fields were receiving in excess of 30 chemical insecticidal treatments per year. The cotton industry has passed through each of the stages of development discussed by Smith (1971a). The disaster phase occurred in the late 1960s, and the integrated

Table 11.3 Changes in Importance of Cotton Insect Pests in Nicaragua

	Order of Importance[a]		
Species	1956[b]	1965[c]	1970[d]
Alabama argillacea (Hübner)	2	5	4
Anthonomus grandis Boheman	1	3	1
Aphis gossypii Glover	4	M	M
Bemisia tabaci (Gennadius)	*	4	3
Creontiades femoralis Van Duzee	*	M	6
Heliothis spp.	3	1	2
Prodenia spp.	*	2	5
Sacadodes pyralis Dyar	5	*	*
Spodoptera spp.	*	7	8
Trichoplusia ni (Hübner)	*	6	7

[a] 1, most important; *, not important; and M, of minor importance.

[b] En el área de OIRSA, 1956 Lista de enfermedades de los cultivos principales existentes.

[c] Report to FAO by Ray F. Smith.

[d] Falcon 1971a.

control phase began almost immediately thereafter. This has continued into the 1970s (Chapter 1).

Throughout the first 20 years of cotton production in Nicaragua, the number of insect pests increased. It is interesting to note, however, that one, *Sacadodes pyralis* Dyar, virtually disappeared. This is an example of an insect species not being able to cope with the pressure created by the use of chemical insecticides.

Obviously, some insects may be released from natural control to become pests when the natural factors regulating their abundance have been minimized by abnormal occurrences. The cabbage looper, *Trichoplusia ni* (Hübner), offers an excellent example of this phenomenon in many areas of the Western world. In parts of the San Joaquin Valley, California, the cabbage looper is always present in and around cottonfields, but seldom becomes a pest. However, a catastrophic event, such as treatment of the field with a broad-spectrum chemical insecticide, may destroy the complex of predators that regulates looper abundance and the cabbage looper population then may explode.

Cotton insect pests also may change their behavior patterns or host range. In the Near East, for example, *Spodoptera littoralis* (Boisduval) apparently has changed its habits. The larvae once fed almost entirely on the leaves, but now the middle and late instars often attack bolls. In addition, this insect has lately become the worst pest of apples in Israel, although before 1950 it had never been recorded on that crop.

For many years the boll weevil was confined to the more humid regions of the southern United States where there were significant amounts of summer rainfall. It was assumed that the insect could not survive in the hot, arid regions of the southwestern United States. In the early 1950s, however, it gradually moved westward into some of these formerly unoccupied areas. It reached the lower end of El Paso Valley in Texas in 1961. During this same period the boll weevil crossed another barrier and moved above the Caprock in Texas. It now occurs over the eastern edge of the Texas high plains area (Bottrell et al. 1972).

C. Host-Plant Resistance

Until recently, little emphasis was placed on breeding commercial varieties of cotton for host-plant resistance to insects. This now has changed, and research has been accelerated in several laboratories across the cotton belt.

Host-plant resistance is not unknown in cotton. Varieties were bred successfully for resistance to the phloem-feeding species of *Empoasca* several years ago in Africa (Pearson and Maxwell-Darling 1958; Painter 1951). There were also reports of generally insignificant levels of resistance to other species—thrips, fleahoppers, cotton stainers, whiteflies, boll weevils, spider mites, and several lepidopterous pests. Nevertheless, the great potential for selection of plant resistance in the United States was largely ignored until very recently (Newsom and Brazzel 1968).

Small lysigenous glands containing the pigment gossypol are present in all parts of the cotton plant except the roots. This pigment is toxic to nonruminant animals and creates difficulties in processing cottonseed and cottonseed products. For these reasons efforts were made to develop varieties with virtually no gossypol content. Entomologists soon found low-gossypol cotton was highly susceptible to insect attack by several pests, some of which did not attack the normal, gossypol-containing varieties at all (Lukefahr and Martin 1966; Bottger et al. 1964; Murray et al. 1965). Gillham (1965), in fact, suggested that gossypol in cotton evolved as a protective measure in pest attack. Some entomologists, notably Lukefahr and Martin (1966) and co-workers, found that increased gossypol content in flowers and seeds makes cotton less susceptible to attack by several serious pests. These workers now have several advanced breeding lines of these types of cotton, which show great promise for resistance to *Heliothis* spp. and certain other lepidopterous pests.

A recent unpublished report listed most of the research progress in host-plant resistance at several laboratories. Much of it is quite preliminary. Characters known to confer some degree of resistance in cotton to major pests are listed in Table 11.4 by courtesy of F. G. Maxwell (personal communication). In addition, research in California (Leigh et al. 1968; Leigh and Hyer 1963; Tingey et al. 1973a and b) has identified a few varieties with some resistance or tolerance to lygus bug and spider mite attack.

Undoubtedly, continued progress can be anticipated in the development of resistant varieties. Close inspection of Table 11.4, however, indicates some of the complexities involved. The bracts on 'Frego' cotton, for example, do not close over the boll but are somewhat twisted and are perpendicular to the boll. This character affects the behavioral pattern of the boll weevil and causes reduced oviposition. But it has been noted in field trials that 'Frego' cotton is much more susceptible to attack by lygus bugs. Development of commercially acceptable resistant varieties is notoriously slow, but continued research effort may pay great dividends which far exceed research costs. Just this year (1974), a commercial nectarless variety was released in Mississippi, which is moderately resistant to *Lygus* spp., the cotton fleahopper, *Pseudatomoscelis seriatus* (Reuter), and *Heliothis* spp. (Schuster and Maxwell 1974).

D. Influence of Agronomic Practices on the Cotton Ecosystem

Any modification in prevailing practices in cotton production may create changes, sometimes subtle, in the plant and the environment. The attractiveness and suitability of the plant and environment to the pests may be altered and, in some instances, this may aggravate or ameliorate certain pest problems or even create new ones. Modifications in time of application and amounts of irrigation or fertilizer, use of different cotton varieties, plant and row spacing, cultivation practices, or time of planting and crop maturity may all have an impact on pest or beneficial species populations. Some modifications are designed especially to give cultural control of certain pests. In some cases these provide the first line of defense against pest insects, and utilization of other suppressive measures, for example, insecticides, is only supplementary.

Cotton growers in irrigated areas have long recognized that lush spots in a field are especially attractive to lygus bugs, *Heliothis* spp., and boll weevils, *Anthonomus grandis* Boheman. Similarly, overirrigation and overfertilization may produce lush plants which may not necessarily produce more cotton but often create more severe pest problems (Adkisson 1958; Leigh et al. 1969). Other changes in production practices also may have similar far-reaching effects.

Newsom and Brazzel (1968) reviewed many cultural practices useful in reducing cotton pest problems. They included an areawide uniform planting date, a cotton-free period, early crop maturity and harvest, destruction of infested bolls, destruction of alternate hosts, early and uniform stalk destruction, and the use of trap crops.

The present cultural program for control of the pink bollworm, *Pectinophora gossypiella* (Saunders), in Texas, provides an excellent example of a highly successful program. This program evolved from a series of studies which provided a very thorough understanding of the seasonal history of the species. The eggs of this serious pest are laid in protective sites on the fruiting forms of cotton, and the larvae, immediately after hatching, burrow into the squares or bolls. Because of this behavior, they are relatively safe from parasites and predators and are difficult to control with insecticides.

Early research by Ohlendorf (1926) showed that the pink bollworm diapauses as a last-instar larva and generally overwinters in the seed of

Table 11.4 Genetic Sources of Insect Resistance Found in Cotton and Currently Being Utilized in Breeding Programs (Texas, Mississippi, USDA, and Cooperating States of Louisiana and Missouri)[a]

Morphological or Chemical Characters Identical	Major Cotton Insects[b]					
	Boll Weevil	*Heliothis* Complex	Plant Bugs *Lygus lineolaris*, *Lygus hesperus*	Cotton Fleahopper	Spider Mites	Whiteflies
'Frego' (fg)	R (60–90% suppression)	N (insecticide coverage increased)	S	S	N	N
'Nectariless' (ne)	N	R (20–50% egg suppression)	R	R	N	N
'Smooth leaf' (glabrous)	N	R (60% egg suppression)	S	S	R–	N
High gossypol	N	R	N	R–	N	N
Pilose (pubescence)	R–	S	R?	R	N	S
'Okra leaf'	N (better insecticide coverage increased kill in squares)	N	N	N	N	R–

Red color	R (choice situation)	N	N	N	N	N
"X" factor (*G. hirsutum* wild races)	N	R	N	N	N	N
Oviposition suppression factor (OSF) (*G. hirsutum* wild races)	R (40%+ suppression of eggs)	N	N	N	N	N
Plant bug suppression factor ('Stoneville' wild races and other sources)	N	N	R	R	N	N
Gossypium barbadense ('Pima S-2')	N	R –	N	N	R	N
Earliness of maturity	R (escape)	R (escape)	N	N	N	R?

[a] Fowden Maxwell, Department of Entomology, Mississippi State University, personal communication.

[b] R, Resistant; N, no effect; S, susceptible.

bolls left in the field after harvest. This is the "weak link" that is vulnerable to attack. Later studies showed that the diapause is controlled by the photoperiod (Lukefahr et al. 1964; Adkisson et al. 1963), so that the first diapausing larvae of the year occur without fail during early September when day lengths are less than 13 hours (Adkisson 1964). The incidence of diapause then increases rapidly and attains a maximum in mid-October and early November. This response to photoperiod provides the key for control, since the onset of diapause may be predicted for any given location with great precision (Adkisson 1966).

The timing of cultural practices used in the production of cotton was modified to take advantage of this phenomenon in such a way as to obtain maximum reduction in the size of the overwintering pink bollworm population. The sequence of practices from the time of maturity of the crop in the fall to planting of the subsequent one is as follows. (1) Defoliate or desiccate the mature crop to cause all bolls to open at nearly the same time, expediting machine harvesting; (2) harvest the crop early, shred stalks, and plow under crop remnants immediately; (3) irrigate prior to planting in desert areas if water is available; (4) plant new crops during a designated planting period, which allows for maximum suicidal emergence of overwintering moths, that is, moths emerge and die before cotton fruit is available for oviposition (Adkisson and Gaines 1960).

Early-maturing varieties are very important to this program, since the development of most of the potential overwintering population of pink bollworms can be virtually prevented if the cotton can be defoliated or desiccated in late August or early September rather than in October or November (Adkisson 1962). This expedites the early and rapid harvest of the crop by mechanical harvesters. Mechanical strippers leave fewer larvae in the field than spindle pickers, and almost all larvae carried to the gin in seed cotton are killed by the ginning process (Robertson et al. 1959). The spindle picker may leave immature bolls in the field, which can harbor a source of infestation for the following year. However, if the stalks are shredded and plowed under, the larvae may experience a mortality of more than 90% (Wilkes et al. 1959; Adkisson et al. 1960; Noble 1969; Fenton and Owen 1953).

This combination of practices has reduced the pink bollworm in Texas to the status of a minor pest, and insecticides are seldom needed for its control. The value of the program is such that regulatory measures have been enacted by the Texas legislature to force compliance by cotton producers on an areawide basis.

The development of diapausing boll weevils also can be terminated or prevented by many of the measures used against the pink bollworm.

Early workers developed sufficient information to promote areawide destruction of cotton in early fall as the best means of weevil control in an era before highly effective insecticides had been discovered for control of the pest (Malley 1902; Hunter 1912). Since these early demonstrations, considerably more has been learned about the diapause of the boll weevil. Unlike the pink bollworm, the boll weevil diapauses as an adult and hibernates outside the cottonfield. Thus, in order to obtain maximum reduction in potential overwintering numbers, the prehibernating weevils must be killed before they leave the cottonfields for hibernation at nearby woody or bushy sites. This can be accomplished best by slightly modifying the pink bollworm control program to include the selective use of an organophosphorus insecticide during the harvest period. The insecticide is added to the defoliant or desiccant just before harvest. One or two additional insecticidal treatments may be required if the harvest season is prolonged.

This program has proved highly effective in reducing boll weevil numbers without adverse impact on the arthropod natural enemies of the bollworm or tobacco budworm, *Heliothis virescens* (Fabricius) (Brazzel 1961b; Brazzel et al. 1961b; Adkisson et al. 1966). In many years this combination of practices, applied on an areawide basis, has so reduced boll weevil numbers that they have not been able to attain economic proportions during the subsequent growing season. This program, because it conserves insect natural enemies, has also been very successful in averting economic outbreaks in cotton of the bollworm and tobacco budworm and serves as the foundation for the cotton insect control program in the boll weevil–infested areas of Texas.

In Nicaragua, in contrast, the emphasis must be on proper handling of crop stalks. Falcon and Daxl (1974) recommend that in Nicaragua the stalks be left undisturbed until the winter rains begin in May. As the stalks send out new leaves, boll weevil adults are attracted to them. The weevils are then collected by hand or killed with an insecticide. After a few days the stalks are cut and burned.

Unfortunately, the irrigated western states (Arizona and southern California) have been recently infested by the pink bollworm. Populations literally exploded, resulting in severe boll infestations. The cultural practices adopted successfully in Texas have been shown to be effective in these newly infested areas; however, western growers will not adopt them. In many years cotton grown in these hot desert areas will make a first crop and then set a smaller very profitable second crop. Early crop termination with chemicals and stalk destruction prohibits a second crop. As a consequence, growers in the Far West depend almost totally on insecticides for pink bollworm control at the present time.

In general, recommended agronomic practices that produce vigorous, physiologically strong cotton plants benefit insect pest management. Planting should take place when soil temperatures are sufficiently high to ensure prompt, uniform seedling emergence and growth. Planting should also be accomplished uniformly, so that harvest occurs when the season is relatively dry. Delayed harvest or nonuniformity of harvest in an area may have profound effects on pest populations, as it allows additional generations of pests to develop during the season. Significant benefits also may be gained from defoliation, rapid harvest, stalk shredding, and plowing under of crop debris if these practices can be accomplished uniformly and sufficiently early so as to prevent continued population increases of pests. Many of the benefits gained carry into the following season, since early harvest may also reduce numbers of pests that successfully overwinter.

As stated earlier, the cottonfield cannot be considered an ecosystem in isolation. New crops introduced into cotton production areas can complicate pest problems on cotton. For example, safflower in the San Joaquin Valley of California may develop an enormous population of lygus bugs. Lygus apparently has little impact on safflower seed production. When the safflower approaches maturity, it becomes unsuitable for the lygus bugs, and they emigrate—often in tremendous numbers—to cotton. The research of Mueller and Stern (1973, 1974) has demonstrated that it is far more economical and less disruptive to the insect natural enemies in the area to kill them with an insecticide in the safflower fields than in the cotton. Secondary problems often are incurred if cotton has to be treated with one or more insecticide applications for lygus control. These problems are circumvented when the pest is controlled on safflower before it leaves for the cottonfields.

When grown adjacent to cotton, certain crops may aggravate pest problems on the latter. Soybeans and certain forage crops can intensify *Heliothis* spp. and lygus bug problems on cotton. Seed alfalfa, sorghum, and possibly soybeans allow development of large populations of stinkbugs which migrate to cotton in late summer and cause severe damage to the bolls. In portions of Texas and Oklahoma, the fleahopper moves into cottonfields from several species of weeds. In parts of Arizona and southern California, when sugar beets are mature, *Empoasca solana* DeLong move from that crop to cotton where they may cause serious damage if sufficiently abundant. Other examples could be cited. Unfortunately, it is not always feasible to prevent pest movement from an outside area into cottonfields.

E. Ecosystem Diversity

Crop monocultures are often damaged more severely by pests than is the same crop located in an area with crop diversity. As a consequence, the assumption is frequently made that maximum diversity in an area is always desirable. Careful analysis, however, does not support this assumption, since in many situations certain pest problems are aggravated rather than reduced. In fact each situation must be carefully studied for its individual merits. In the present state of knowledge, it is difficult to generalize regarding potential pest problems and also regarding natural enemy reservoirs and possible effectiveness.

Recent studies in the San Joaquin Valley of California have demonstrated the value of two unusual types of ecosystem heterogeneity. Lygus bugs are considered a key pest of cotton in California. A major habitat of the bugs is provided by alfalfa fields which generally are interspersed throughout the cotton production area. When the alfalfa is cut, the lygus bugs leave in vast numbers, and frequently populations that far exceed economic thresholds quickly appear in the cottonfields. Since high populations of this insect do not cause economic losses in alfalfa fields, they should be kept there if possible. Stern et al. (1967) demonstrated that it was possible to do this by cutting the alfalfa in alternate strips rather than completely cutting the entire field as is normally done. The adult lygus bugs move to the uncut strips rather than fly to nearby cottonfields.

A second manipulation and type of heterogeneity was also demonstrated by Stern (1969). In this method strips of alfalfa about 20 ft wide are interplanted in cottonfields after every 128 rows, or about every 400 ft, across the field. Only one-half of the 20-ft strip is mowed at a time, leaving a 10-ft strip of succulent alfalfa which is much preferred by lygus bugs to the surrounding cotton (Fig. 11.3). It was demonstrated that lygus bugs concentrate in these strips, and the need for insecticides was virtually negated. Furthermore, beneficial parasitoids and predators were extremely abundant in these strips and, as they moved to and from the adjacent cotton, an added benefit resulted.

Both of the methods developed by Stern and colleagues have been proved effective by farmers. Although they appreciate their value, widespread adoption has not occurred to date. Each method requires special planning and operations. Irrigation procedures are more complicated also. It is simpler to treat with an insecticide, and it is expected that producers will continue to do this until the lygus become resistant to insecticides, or other problems force adoption.

Another aspect of heterogeneity was involved in a method Smith and Reynolds (1972) called the "dirty-field" technique. The principle under-

Fig. 11-3 Alfalfa strips planted through a cottonfield. The uncut portion of the strip traps and retains lygus bugs, keeping them out of the cotton.

lying this procedure is the maintenance of pest populations in cottonfields at low, uneconomic levels. The pest population then serves as a food source for insect natural enemies, keeping them in the cottonfields. Any technique that virtually eliminates a pest from a field, for example, an insecticide, is certain to reduce greatly populations of its specific natural enemies not only by killing them directly but also by eliminating the host used as a food source or for oviposition sites. Resurgences of pest species following treatment have become increasingly common on cotton (Newsom and Brazzel 1968). Theoretically, it appears possible that insecticides can be used to reduce the pest population to levels below economic thresholds without virtual elimination if the dosage applied is sufficiently selective to allow key beneficial species to survive at levels sufficiently high to provide regulation of the pest species.

II. THE ARTHROPOD COMPLEX

A large complex of insect and spider mite species is found in cotton ecosystems wherever the crop is grown. Of this large number, comparatively few are pests of economic significance. Many are beneficial species. A survey of predatory species of arthropods listed about 600 species inhabiting cotton in Arkansas (Whitcomb and Bell 1964). The 100 or more species reported to attack cotton in the United States are largely secondary and potential pests, only very few key pests are present in each production area. Newsom and Brazzel (1968) noted that less than two dozen species caused the annual production losses of nearly 20%. Of this loss, however, approximately 80% can be directly attributed to a few key pests of this crop. The percentage is probably higher, considering indirect losses inflicted by secondary pest resurgence brought on by use of insecticides to control the key pests.

A. Key Pests

Everywhere cotton is grown in the United States there are at least one or two key pests whose population must be managed or controlled if cotton is to be grown profitably. Obviously, in the case of such key pests, natural controls including insect natural enemies are inadequate to suppress populations below economic thresholds. It is precisely on such key pests that integrated pest management should be focused initially. It is their suppression without disruption that provides the key to pest-management programs in the cotton ecosystem. Present control technology for most key pests rests almost entirely on insecticides, repeated applications being made at frequent intervals. This creates serious upsets and imbalances in the arthropod complex, and secondary pest outbreaks usually follow. Also, treatments applied for control of key pests are the major contributors to environmental contamination problems originating in the cotton environment.

Key insect pests on cotton in the United States are comparatively few. The most devastating species is unquestionably the boll weevil, *Anthonomus grandis* Boheman. This pest invaded the United States from Mexico in 1892 and is now a serious pest from the Atlantic coast to the southern high plains of Texas (Fig. 11.4*A* and *B*). A second key pest, which also entered the country from Mexico, is the pink bollworm, *Pectinophora gossypiella* Saunders (Fig. 11.5*A* and *B*). This pest is normally controlled by cultural methods in most infested areas such as Texas but, in the arid, warm-winter conditions of Arizona and southern California, this species has been particularly destructive. The pink bollworm is not a pest east of the Mississippi River, nor is it serious generally north of Texas. The species, worldwide, is usually considered to be the

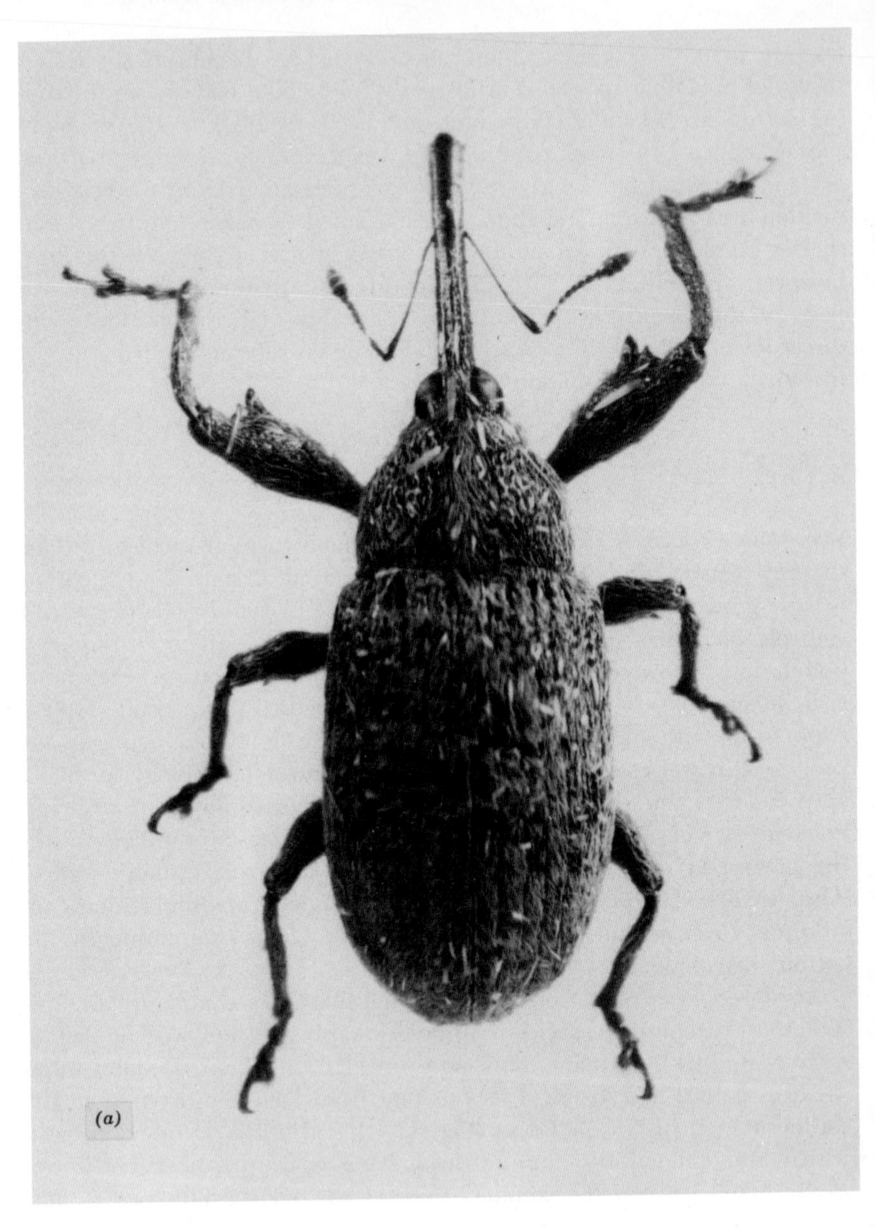

Fig. 11-4 (*a*) Adult boll weevil, *Anthonomus grandis* Boheman.

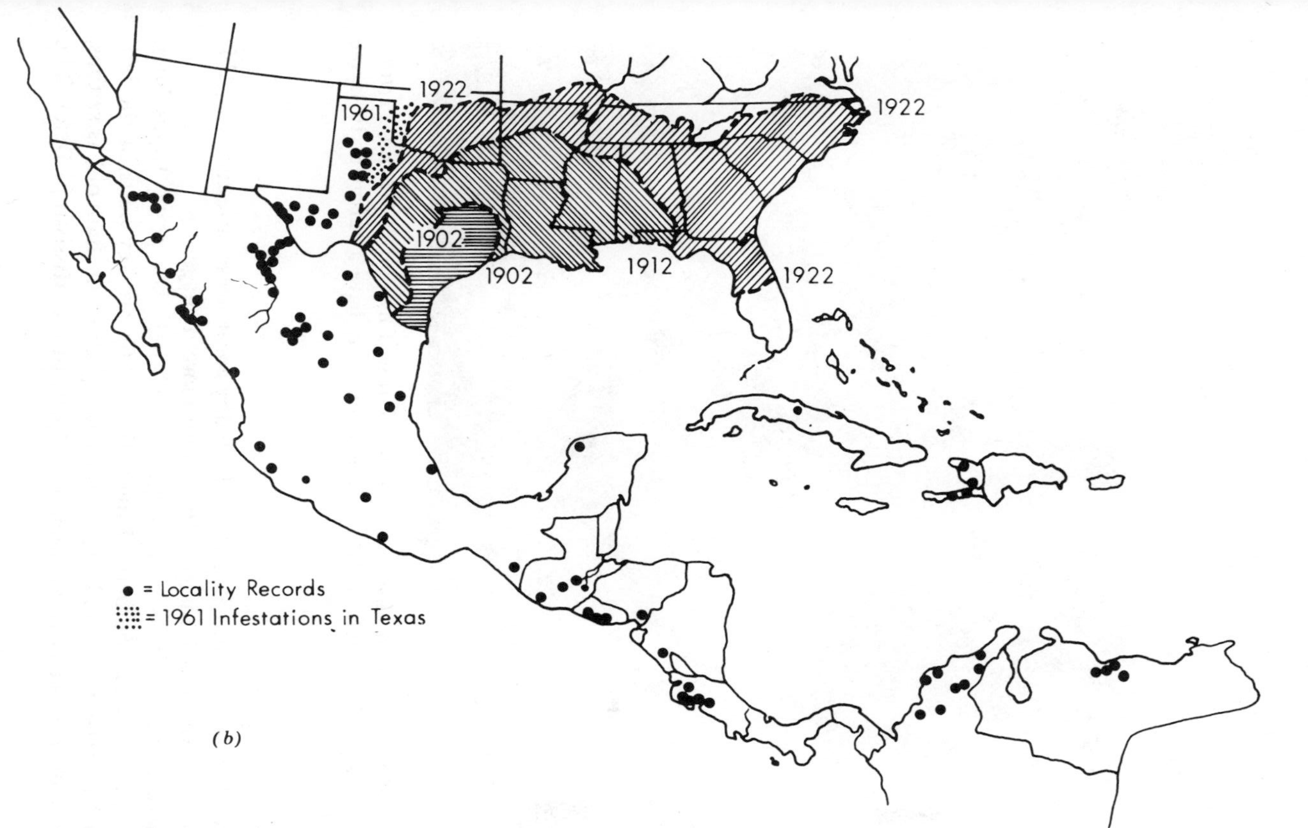

(b) **World distribution of *Anthonomus grandis* Boheman, 1961.**

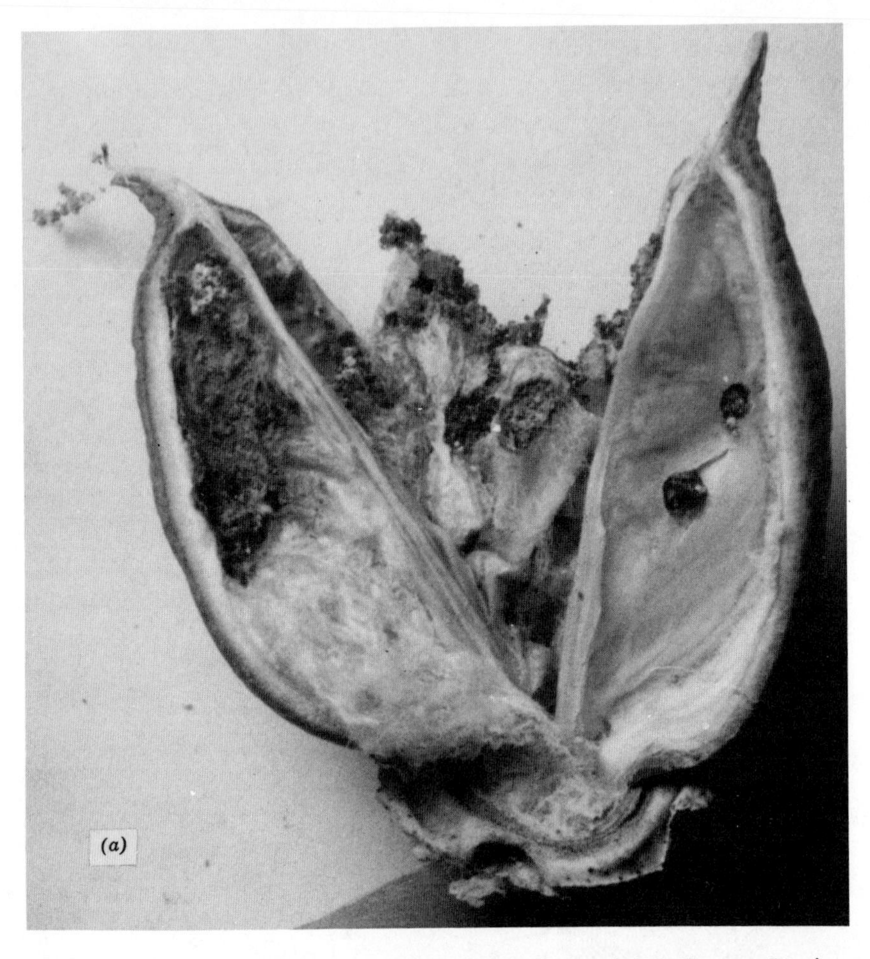

Fig. 11-5 Cotton boll almost completely destroyed by pink bollworm, *Pectinophora gossypiella* Saunders. Heavy infestations may have as many as 10 larvae per boll.

most destructive pest of cotton. In many local areas infested with this pest, however, other insects often cause more damage.

Several hemipterous plant bugs are definitely key pests. Lygus bugs, particularly *Lygus hesperus* Knight (Fig. 11.6), are key pests in the irrigated western states, and *Lygus lineolaris* (Palisot de Bauvois) is serious, although it is perhaps less than a key pest, in the Mississippi Delta. The

(*b*) Lint pulled out from immature boll to expose pink bollworm larvae in seed.

cotton fleahopper, *Pseudatomoscelis seriatus* (Reut.), can be considered a key pest in portions of Texas and Oklahoma.

The bollworm, *Heliothis zea* (Boddie), has long attracted attention as a pest of cotton. In recent years, however, the bollworm and tobacco budworm, *Heliothis virescens* (F.), have become particularly destructive, at times causing more damage than the boll weevil in the South. Many entomologists consider members of the bollworm complex to be less than key pests in the absence of ecosystem disruption which often follows insecticidal applications, particularly repeated applications made for such pests as boll weevils, lygus bugs, and fleahoppers.

Adkisson (1973) has listed the three major cotton-producing regions in the United States and the key pests in each. Each of the three has a different climate and diversity in the cotton ecosystem.

1. The irrigated deserts of the Far West where the key pests are the pink bollworm, lygus bugs, and the bollworm. The tobacco budworm

Fig. 11-6 Adult lygus bug, *Lygus hesperus* Knight. (Coutesy Thomas Leigh, University of California, Davis.)

recently has infested portions of this area, but its ultimate status as a pest is not known.

2. The semiarid regions of the southwestern United States where the boll weevil, fleahopper, bollworm, and tobacco budworm are the key pests.

3. The humid regions of the midsouthern and southeastern United States where the boll weevil, plant bug, bollworm, and tobacco budworm are the key pests.

There are many major pests of cotton in production areas of the world that do not occur in the United States. Although under tight quarantine, some of these likely would be key pests if they were to become established. Some of the principal pests in the cotton production areas of the world are shown in Table 11.5.

B. Occasional and Potential Pests

There is a large number of insect and spider mite species which at times are pests of cotton. In most cases population increase to damaging levels is sporadic, thus they cannot be considered key pests. Some of these species become problems almost exclusively when their insect natural enemies are killed by insecticidal treatments applied for control of key pests. As mentioned previously, changes in agronomic practices, to the use of different cotton varieties and to other production practices in the cotton environment, should be carefully evaluated for effect on insect populations. Care must be taken that changes do not allow great increases in numbers of occasional and potential pests.

C. Biological Control Utilizing Beneficial Insects

Although the cotton crop is most often grown on an annual basis, cotton agroecosystems contain a varied and complex entomophagous fauna (Tables 11.6 and 11.7). Indeed, it would not be possible to produce cotton crops economically, or at all, without the regulatory impact of parasitic and predaceous species on the pest complex. In many areas heavy use of chemical pesticides virtually eliminates these natural controls. As a result, there is often a tremendous resurgence of "primary target" pest species and the unleashing of secondary pests following the use of pesticides. There are many examples of such pesticide-induced events, including outbreaks of bollworm, cabbage looper; beet armyworm, *Spodoptera*

Table 11.5 World Distribution of Principal Cotton Insect Pests[a,b]

Order and Genus	Old World Species, Europe, Africa, Asia, and Australia	New World Species, North, Central, and South America
Coleoptera		
Alcidodes	Several species	—
Anthonomus	Present but not recorded on cotton	*grandis* Boheman
Eutinobothrus	—	Several species in South America
Podagrica	Several species	—
Sphenoptera	Several species, including *gossypii* Cotes	—
Syagrus	Several species	—
Lepidoptera		
Xanthodes	*graellsii* (Feisthamel)	—
Alabama	—	*argillacea* (Hübner)
Cosmophila	*flava* (Fabricius)	—
Diparopsis	*castanea* Hampson, *watersi* Roths.	—
Earias	Several species	—
Heliothis	*armigera* (Hübner)	*virescens* (Fabricius), *zea* (Boddie)
Sacadodes	—	*pyralis* Dyar

exigua (Hübner); cotton aphids *Aphis gossypii* Glover; spider mites, and many other species, in cotton in the United States, Mexico, and Central America.

Direct, harmful action on insect natural enemies is only one of the ways insecticides can disrupt natural control of pests. Disruption of food chains can also be important. For example, if aphids, spider mites, and thrips (the usual prey of abundant, omnivorous predators such as *Chrysopa* spp., *Nabis* spp., *Geocoris* spp., and *Orius* spp.), are eliminated from the cotton agroecosystem early in the season by chemical treatments which in themselves are not significantly harmful to the predators, the latter will either starve, emigrate, or cease to reproduce. Later, when strong-flying insects such as *Spodoptera* spp. or *Heliothis* spp. invade the treated fields, they are essentially free from predator attack and explosive outbreaks of these pests occur.

Table 11.5 (Continued)

Order and Genus	Old World Species, Europe, Africa, Asia, and Australia	New World Species, North, Centrla, and South America
Spodoptera	*exigua* (Hübner), *littoralis* (Boisduval)	*exigua*, *frugiperda* (Smith), *ornithogalli* (Guenée)
Sylepta	*derogata* (Fabricius)	—
Hemiptera		
Dysdercus	Several species	Several species
Austroasca	*terraereginae* Paoli	—
Empoasca	*facialis* (Jacobi), *devastans* Distant	*fabae* (Harris)
Helopeltis	*schoutedeni* Reuter	—
Horcias	—	*nobilellus*
Lygus	*vosseleri* (Poppius)	Several species in North America
Nezara	*viridula* (Linneaus)	*viridula*
Oxycarenus	Several species	—
Acarina		
Cecidophyes	*gossypii*	*gossypii*
Hemitarsonemus	*latus*	*latus* (Banks)

[a] Adapted from Pearson and Maxwell-Darling 1958, by Falcon and Smith 1973.

[b] Three important pests not listed are *Aphis gossypii* Glover, *Pectinophora gossypiella* (Saunders), and *Tetranychus telarius* (Linneaus), which are present in practically all cotton-growing areas of the world.

The importance of these predators was recognized by early entomologists who worked with this crop. Gorham (1847) was aware of parasitism of the cotton leafworm, and Riley (1873) listed many species of parasites and predators that still are important today. Before the invasion of the United States by the boll weevil, most of the insects (with the exception of the cotton leafworm) that attacked cotton generally occurred in such low numbers that they never became pests of any consequence. Occasionally, natural enemies failed to control one of these species, and an outbreak occurred. An exception was the cotton leafworm, an animal migrant from Central America, which frequently inflicted serious damage to cotton and whose control had to be achieved with arsenical insecticides.

When the boll weevil invaded the United States, an intensive search for parasites and predators of this pest began. More than 50 species of parasites and predators were discovered (Pierce 1908; Pierce et al. 1912).

Table 11.6 Parasites of Cotton Insect Pests in California[a]

Pest Species	Parasites
Beet armyworm, *Spodoptera exigua* (Hbn.)	*Apanteles laeviceps* Ashm. (Hymenoptera, Branconidae) *Apanteles marginiventris* (Cress.) (Hymenoptera, Braconidae) *Chelonus texanus* Cress. (Hymenoptera, Braconidae) *Meteorus vulgaris* (Cress.) (Hymenoptera, Braconidae) *Campoletis argentifrons* (Cress.) (Hymenoptera, Ichneumonidae) *Hyposoter exiguae* (Vier.) (Hymenoptera, Ichneumonidae) *Melanicheneumon rubicundus* (Cress.) (Hymenoptera, Ichneumonidae) *Therion californicum* (Cress.) (Hymenoptera, Ichneumonidae) *Trichogramma* spp. (Hymenoptera, Trichogrammatidae) *Lespesia archippivora* (Riley) (Diptera, Tachinidae) *Eucelatoria armigera* (Coq.) (Diptera, Tachinidae)
Cabbage looper, *Trichoplusia ni* (Hbn.)	*Apanteles glomeratus* (L.) (Hymenoptera, Braconidae) *Apanteles marginiventris* (Cress.) (Hymenoptera, Braconidae) *Microplitis brassicae* Mues. (Hymenoptera, Braconidae) *Hyposoter exiguae* (Vier.) (Hymenoptera, Ichneumonidae) *Copidosoma truncatellum* (Dalm.) (Hymenoptera, Encyrtidae) *Trichogramma* spp. (Hymenoptera, Trichogrammatidae) *Archytas californiae* Walk. (Diptera, Tachinidae) *Eucelatoria armigera* (Coq.) (Diptera, Tachinidae) *Voria ruralis* (Fall.) (Diptera, Tachinidae) *Winthemia quadripustulata* (Fabr.) (Diptera, Tachinidae) *Sarcophaga* spp. (Diptera, Sarcophagidae)
Cotton bollworm (corn earworm), *Heliothis zea* (Boddie)	*Apanteles marginiventris* (Cress.) (Hymenoptera, Braconidae) *Apanteles militaris* (Walsh) (Hymenoptera, Braconidae) *Chelonus texanus* (Cress.) (Hymenoptera, Braconidae) *Campoletis argentifrons* (Cress.) (Hymenoptera, Ichneumonidae) *Hyposoter annulipes* (Cress.) (Hymenoptera, Ichneumonidae) *Hyposoter exiguae* (Vier.) (Hymenoptera, Ichneumonidae) *Therion californicum* (Cress.) (Hymenoptera, Ichneumonidae) *Spilochalcis igneoides* (Kirby) (Hymenoptera, Chalcididae) *Trichogramma* spp. (Hymenoptera, Trichogrammatidae) *Prospaltella* spp. (Hymenoptera, Aphelinidae) *Eucelatoria armigera* (Coq.) (Diptera, Tachinidae) *Lespesia archippivora* (Riley) (Diptera, Tachinidae) *Gonia capitata* (DeGeer) (Diptera, Tachinidae) *Winthemia quadripustulata* (Fabr.) (Diptera, Tachinidae)

Table 11.6 (Continued)

Pest Species	Parasites
Cotton leaf perforator, *Bucculatrix thurberiella* (Busck)	*Apanteles bucculatricis* Mues. (Hymenoptera, Braconidae) *Cirrospilus* spp. (Hymenoptera, Eulophidae) *Closterocerus utahensis utahensis* Crawf. (Hymenoptera, Eulophidae)
Saltmarsh caterpillar, *Estigmene acrea* (Drury)	*Therion* spp. (Hymenoptera, Ichneumonidae) *Trichogramma* spp. (Hymenoptera, Trichogrammatidae) *Exorista larvarum* (L.) (Diptera, Tachinidae)
Western yellow-striped armyworm, *Spodoptera praefica* Grote	*Apanteles marginiventris* (Cress.) (Hymenoptera, Braconidae) *Chelonus texanus* Cress. (Hymenoptera, Braconidae) *Meteorus vulgaris* (Cress.) (Hymenoptera, Braconidae) *Campoletis intermedius* (Vier.) (Hymenoptera, Ichneumonidae) *Nepiera marginata* (Prov.) (Hymenoptera, Ichneumonidae) *Hyposoter exiguae* (Vier.) (Hymenoptera, Ichneumonidae) *Pterocormus difficilis* (Cress.) (Hymenoptera, Ichneumonidae) *Therion californicum* (Cress.) (Hymenoptera, Ichneumonidae) *Trachysphyrus tejonensis tejonensis* (Cress.) (Hymenoptera, Icheumonidae) *Archytas californiae* Walk. (Diptera, Tachinidae) *Eucelatoria armigera* (Coq.) (Diptera, Tachinidae) *Euphorocera claripennis* (Macq.) (Diptera, Tachinidae) *Aphiochaeta* spp. (Diptera, Phoridae)

[a] Adapted from van den Bosch and Hagen 1966.

However, it soon became apparent that the available natural enemies were not capable of controlling the boll weevil satisfactorily, and cotton growers turned to cultural methods and insecticides for crop protection.

Entomologists have not been able to quantify with precision the absolute effectiveness of arthropod parasites and predators in keeping cotton pest species below economic densities. That there are many effective insect natural enemies operating in cotton became immediately apparent with the widespread use of calcium arsenate. Arsenical treatments of cotton were rapidly followed by outbreaks of aphids and bollworms, two species that formerly were of little concern to growers (Folsom 1928; Dunnam and Clark 1941; Ewing and Ivy 1943). These outbreaks were found to result from the destruction of insect natural enemies in treated fields.

Table 11.7 Some Predaceous Arthropods of Cotton Insect Pests in California[a]

Order Neuroptera
- Family Chrysopidae
  - Green lacewings, *Chrysopa* spp.
  - Golden-eye lacewing, *Chrysopa oculata* (Say)
- Family Hemerobiidae
  - Brown lacewings, *Hemerobius* spp.

Order Hemiptera
- Family Anthocoridae
  - Minute pirate bug, *Orius tristicolor* (White)
- Family Lygaeidae
  - Big-eyed bugs, *Geocoris* spp.
- Family Nabidae
  - Damsel bugs, *Nabis* spp.
- Family Reduviidae
  - Assassin bugs, *Zelus* spp.
  - Spined soldier bugs, *Sinea* spp.

Order Coleoptera
- Family Malachiidae
  - Collops beetles, *Collops* spp.
- Family Coccinellidae
  - *Coccinella* spp.
  - *Hippodamia* spp.
  - *Olla* spp.
  - Spider mite-feeding lady beetle, *Stethorus picipes* Casey
- Family Anthicidae
  - Notoxus beetle, *Notoxus calcaratus* Horn

Order Diptera
- Family Syrphidae
  - Hover flies, *Syrphus* spp.

Order Thysanoptera
- Family Thripidae
  - Six-spotted thrips, *Scolothrips sexmaculatus*

Order Acarina
- Family Phytosciidae
  - Mites, *Phytoseiulus* spp.

[a] van den Bosch and Hagen 1966.

The value of natural biological control still was not fully appreciated by entomologists or cotton producers until after the introduction of synthetic organic insecticides. The widespread application of these chemicals elevated many insect species, namely, the bollworm, tobacco budworm, cotton leaf perforator, spider mite, cabbage looper, beet armyworm, and salt marsh caterpillar, to the status of major pests. In fact, the bollworm and tobacco budworm now probably inflict more damage to the crop each year than the boll weevil (Whitcomb 1970; Newsom 1970; Hagen et al. 1971; van den Bosch et al. 1971; Adkisson 1969).

More than 600 species of predators and parasites have been found in cotton where they do an effective job of keeping most insect pests below crop-damaging numbers (Whitcomb and Bell 1964). For example, where natural mortality factors are present, only a relatively small percentage of the eggs deposited by bollworm moths survive to become damaging late-instar larvae. Fletcher and Thomas (1943) found that predators destroyed 15 to 33% of the bollworm eggs, and from 13 to 60% of the first-instar larvae. In other studies, Bell and Whitcomb (1962) and Whitcomb and Bell (1964), indicated that as many as 41% of the bollworm eggs were destroyed by predators within 24 hours after deposition. In replicated study cages in which *Chrysopa* spp. larvae were released, Lingren et al. (1968) found that the egg population was reduced by 76% at 8 days, and 96% at 13 days.

Most other insect pests of cotton have important natural mortality agents. Cotton leaf perforator larvae, for example, are heavily attacked by predators, and there are several important parasites as well. About 70 to 80% of the larvae are killed by parasites in the leaf-mining stage, and almost an equal proportion of the remaining larvae is killed by other parasite species in the final stage.

Not all pests of cotton are regularly subjected to such satisfactory and desirable population regulation by beneficial insects. Parasitism and predation of the boll weevil and pink bollworm, for example, appear to be less than adequate. Predation on lygus bugs in the irrigated western United States is seldom sufficient by itself to keep populations below economic levels. This is mainly why the boll weevil, pink bollworm, and lygus bug are all considered key pests, whereas the potentially highly destructive bollworm is considered less of a threat in the absence of such ecological disturbances as insecticide applications. A single application of insecticide to the cotton often may be all that is required to reduce the effectiveness of the natural enemies to the point that a secondary pest outbreak is allowed to occur (Newsom and Brazzel 1968). Thus crop protection specialists presently are searching for methods that conserve natural enemies. This is being done mostly by minimizing the use of insecticides or timing applications so as to cause minimum disruption of the natural enemy complex.

Although a great number of predators and parasites is indigenous to cottonfields, there is a need for importation of additional biological control agents. This is particularly true because many of the major pests of cotton are of foreign origin. Relatively little effort has been made toward determining the effectiveness of biological control agents in these areas. A search for parasites and predators of the boll weevil in Central America and Mexico is particularly needed, since Bottrell (1974) has shown that a high level of biological control of the weevil may occur in areas where insecticides are seldom, if ever, applied to cotton. It is possible that many useful species may be available there. It is a well-known fact that countries having the greatest success with biological control are those that have imported the greatest number of parasite and predator species.

Augmentation of naturally occurring parasites and predators by programmed releases of laboratory-reared insects has provided satisfactory control of certain pest species, particularly *Heliothis* spp., in experimental plots (Ridgway 1969; Lingren 1969). However, techniques for large-scale rearing and release of insect natural enemies have not yet been developed to the point where augmentation of natural populations is practical for grower use.

One problem with the utilization of indigenous parasite and predator species is a general lack of understanding and appreciation of the details of their role in cotton agroecosystems. The time of their impact on the pest population is of considerable importance. In Nicaragua, for example, the beneficial insect fauna plays a major role in regulating pest insect abundance for 9 months of the year (January to September). Yet during the months of October, November, and December, when the cotton plants are at peak production, active stages of the beneficial insects are virtually absent. Their disappearance coincides with the period of greatest precipitation. Also, at this time, pest problems on cotton become most severe. Consequently, artificial control procedures, such as the application of chemical pesticides, have become the first line of defense. During the remainder of the year, arthropod natural enemies are important pest control agents, and artificial controls need be applied only to augment these natural controls.

Another problem that has prevented the proper utilization of parasites and predators is that the various cotton-growing regions of the world differ greatly as to the native predatory and parasitic species present and

Fig. 11-7 *Trichogramma* egg parasite on egg of the cabbage looper (Courtesy of E. J. Dietrick, Riverside, California.)

their relative importance. As one example, an egg parasite, *Trichogramma* spp., is abundant in desert cotton-producing areas of southern California and during the growing season may destroy an average of from 40 to 50% of bollworm and cabbage looper eggs (Fig. 11.7). In contrast, in the San Joaquin Valley farther north, eggs of these insects are rarely parasitized by *Trichogramma*.

1. Evaluation

Observation, experiments, and experience may eventually show that in any area only a few insect pests are economically important and that only a few of their natural enemies are influential. Nevertheless, as complete an inventory as possible should be made, since this is the basis for establishing the relative roles of the various species. To make such an inventory, the first step is intensive collection of the insect fauna in the cottonfields. All developmental stages of the pests present in an area should be collected and reared for parasite emergence in the laboratory. Population censuses and periodic sampling also provide knowledge on seasonal activity of cotton pests and their natural enemies. Once sufficient data has been obtained, definite relationships may be established between pest and natural enemy numbers. (For additional information see Falcon and Smith 1973; Whitcomb and Bell 1964.)

Both physical and biological properties of the environment determine the efficacy of natural enemies of cotton pests. Man is generally not in a position to manipulate the macroclimate, but he can influence the microclimate. This ability can lead to increased efficiency of the natural enemies.

In general, there are two ways of utilizing natural enemies of cotton pests in integrated control programs:

1. By means of a pest-management system that preserves and augments naturally existing populations of predators and parasites
2. By means of mass rearing of natural enemies for release as nuclei for further reproduction in the field at the expense of a given pest or pests or, what is more desirable, for release in large numbers as a population regulator.

2. *Preservation and Augmentation*

Utilization really amounts to preservation of the natural enemies known to attack those pests existing in a given cotton agroecosystem. Such preservation can be achieved in various ways:

1. Knowledge of the effects of different pesticides on the most important beneficial species, through experimentation, which permits the selection of pesticides not only according to their effects on the target pests concerned, but also according to their ability to leave alive as many beneficial forms as possible
2. Restriction of the use of broad-spectrum insecticides to cases of absolute need, controlled by entomologists responsible for technical assistance, backed by regulatory cultural practices, and adjusted to the ecology and needs of each locality
3. Development of alternative noninsecticidal methods of pest control that aid in the preservation of natural enemies.

3. *Mass Liberations*

The main objective of mass liberations of parasites or predators is to put into the cotton field beneficial organisms that will aid in suppressing populations of one or more pest species. This may be done by collecting parasites and predators from one crop on which they are abundant and transferring them to another crop on which they are not. A large suction machine has been employed in California; insect natural enemies are collected in alfalfa and sorted, and the desirable species are released into cottonfields.

Large numbers of parasites and predators also may be obtained through mass rearing. In the mass culture of entomophagous insects a major problem is the rearing medium. This may be partially solved by:

1. Production of the host on plant parts, such as cotton plants, melons, tubers, and the like, that are not normally infested in the field
2. Employment of a factitious host which can be reared and is acceptable to the parasite or predator
3. Use of artificial media for host or parasite.

The entomophagous insects most commonly released in cotton and other crops are *Trichogramma* spp. and *Chrysopa carnea* Stephens for control of *Heliothis* spp. These particular natural enemies are employed because they are the species most easily mass-cultured today in the laboratory. In the future there may be a greater choice of entomophagous species available to the crop-protection specialist. This will come only as techniques to culture more species economically are developed. Before mass releases of beneficial species can be recommended, cost analysis and careful studies on the effectiveness of this method should demonstrate feasibility.

D. Microbial Control

Epizootics of naturally occurring insect diseases frequently are observed. In fact, in many areas of the cotton belt, producers commonly depend on viral diseases of the cabbage looper for control of this pest. The cabbage looper virus is more efficient than most insecticides and provides control of the pest when numbers reach a sufficiently high density. Leaf feeding is readily evident, but population buildup is normally late enough in the season that cotton yields are not affected.

When naturally occurring mortality agents are assessed, it is readily seen that insect pathogens are important biotic entities which aid in regulating the abundance of many insect pest species. In some situations they may be all-important in keeping pest populations below damaging levels. The value of insect pathogens as population regulatory agents is most evident in epizootics which, under most situations, greatly deplete a host's numbers. Besides causing outright death, insect pathogens may interfere with insect development and reproduction and may lower insect resistance to attack by parasites, predators, and other pathogens. They also may influence the susceptibility of insects to control by chemical insecticides or other artificial methods (Falcon 1971b). Insect pathogens are not always identified or included in the analysis of naturally occurring mortality factors, most likely because of the difficulties of recognizing them because of their small size or the subtle effects they may exert on their hosts.

The importance of microbial control of cotton pests is very great. Of the arthropod pests on cotton, all, including the spider mite complex, are reportedly infected by one or more insect pathogens. For example, in the United States the lepidopterous species *Agrotis malefida* Guenée, *Alabama argillacea* (Hübner), *Chorizagrotis auxiliaris* (Grote), *Estigmene acrea* (Drury), *Feltia subterranea* (Fabricius), *Heliothis virescens* (Fabricius), *H. zea* (Boddie), *Pectinophora gossypiella* (Saunders), *Peridroma saucia* (Hübner), *Spodoptera exigua* (Hübner), *S. frugiperda* (J. E. Smith), *S. ornithogalli* (Guenée), *S. praefica* (Grote), and *Trichoplusia ni* (Hübner) have one or more reported virus diseases.

Insects are also attacked by several other microorganisms including bacteria, protozoa, fungi, rickettsiae, and nematodes. All these pathogens play an important role in the natural regulation of insect and mite numbers.

Of the pathogens known to be important disease organisms to the insect pests of cotton, *Bacillus thuringiensis* Berliner (BT) and the nuclear polyhedrosis viruses (NPV) of the bollworm, cabbage looper, and beet armyworm are of the most interest to researchers and commercial firms (Chapter 5). BT is registered for use on cotton and is available commercially. The NPV of *Heliothis* spp. has been labeled experimentally and widely tested on cotton. However, its continued use is dependent on protocols for labeling which are to be established by the U.S. Environmental Protection Agency (EPA).

Utilization

Despite the knowledge that most cotton insect pests are susceptible to one or more diseases, progress in evaluating and developing the use of insect pathogens has been slow. Much of the effort has been restricted to laboratory studies, and not enough work has been done to thoroughly evaluate the effectiveness of pathogens under field conditions. Because of their many inherent advantages, the use of pathogens in integrated control programs for cotton pests should be more widely investigated. Also, because many insect pathogens are inexpensive to produce, they have great potential as ideal pest-control agents for use in developing countries.

The basic approach to the use of insect pathogens for pest control involves (1) finding a suitable pathogen, (2) field collection or artificial mass culture of the pathogen in the laboratory, and (3) dissemination of the pathogen in an effective way so that the host is destroyed. This can be done either by the introduction and colonization of a pathogen as a permanent mortality factor in the host population, or by making repeated applications of a pathogen for temporary suppression of an insect pest. The latter approach has received the most attention in recent years.

Commercial preparations of BT and NPV can be applied as dusts or sprays in the same manner as chemical insecticides. The control attained has often been as good as that produced with insecticidal treatments, but not as consistent (Ignoffo 1970; Newsom and Brazzel 1968). Before insect pathogens will be used on a wide scale for pest control in cotton, they must (1) produce more reliable and consistent control of pest species; (2) be more economical to use; (3) be placed in formulations that have better shelf-life, are easier to apply, and are more persistent in the field than present ones; and (4) meet safety standards established by the EPA.

E. Chemical Control

1. *Use Patterns*

Most recent statistics indicate that of the approximately 500 million pounds of pesticides applied each year to United States cropland, 54%, or 270 million pounds, are insecticides. However, only about 5% of the total crop acreage is treated with insecticides, 47% of the total amount being applied to cotton (Eichers et al. 1970; Pimentel 1973). Also, it is important to note that the amount of insecticide used on cotton across the United States is disproportionate, almost half (46%) of the acreage receiving no treatment in most years. The largest percentage (79%) of the acres treated is in the Southeast and in the Delta states, and the smallest percentage (37%) is in the Texas southern plains and the northern Mississippi River Delta regions of Arkansas, Missouri, Tennessee, Kentucky, and Illinois (Pimentel 1973).

Available statistics show, in terms of total amounts of insecticide, that 75.5, 64.9 and 73.3 million pounds, respectively, were applied to United States cotton in 1964, 1966, and 1971. Total use declined 14% from 1964 to 1966, but acreage was down 32%, indicating a more intensive use during the latter year (Davis et al. 1970). The total amount applied increased from 1966 to 1971, but acreage of cotton also increased from 10.3 million in 1966 to 11.5 million in 1971.

Crop protection is but one of the many essential components of profitable crop production. It is true that the amounts of insecticide used on cotton and the costs involved in their purchase and application are quite high when compared with other field crops. For this reason many crop-protection specialists and environmentalists have an exaggerated view concerning pest control and pesticide usage on the crop. A better perspective might be gained by comparing the costs of insect control on the crop with the total costs of cotton production. Average percentage of total cost of production for the insecticides and fungicides used on cotton during the period 1966–1969 ranged from 3.7 to 4%. Much larger percentages were spent for power and equipment, ginning, land, and general overhead than for crop protection (Starbird and French 1972; Smith 1971b).

The average per-acre cost to cotton producers for insecticides in 1969 ranged from a high of $35.99 in southern California/southwest Arizona to a low of $0.53 in the Texas high plains. The high-cost areas are the boll weevil–infested portions of the cotton belt and the Arizona/southern California deserts where the pink bollworm (since 1966) has become a serious problem. Costs of control in the boll weevil area are greater in the high-rainfall areas of the Delta states and the Southeast than in the semiarid areas of Texas and Oklahoma (Starbird and French 1972), where the pest is less favored by the climate.

The drastic effect of a key pest (such as the boll weevil or pink bollworm that must be controlled by insecticides) on costs of production is well illustrated by comparing insecticidal costs in pink bollworm infested areas of California/Arizona or boll weevil–infested areas of the southeastern states with those in the Texas high plains where neither species is a pest. Cotton farmers in the high plains in most years produce a crop without the use of any insecticides, while California producers may spend as much as $35 per acre and those in the Southeast as much as $18 per acre.

The amounts of insecticides used on cotton also are closely tied to the productivity of the land. Areas having high-yield potential generally use much greater amounts of insecticide than low-yielding areas. This is the reason insecticide costs are much higher in the Mississippi River Delta states, for example, than in the dryland areas of Texas, although cotton in both areas may suffer heavy infestations of the boll weevil. The average per-acre production in the Delta states is more than twice that of the Texas drylands, and insecticide costs per acre are more than five times greater.

2. *Importance*

Chemical pesticides are useful and powerful tools for the management of pest populations (Chapter 7). Many are effective, dependable, economical, and adaptable for use in a wide variety of situations. Indeed, the use of chemical pesticides is the only known method for control of many of the world's most important pests of agriculture and public health. No other tool lends itself with such comparative ease to manipulation, and none can be brought to bear so quickly on outbreak populations.

It is recognized that pesticide chemicals have been and will continue to be an essential part of crop protection, but current practices in pesticide use have not always been sound, either in terms of food production or from the standpoint of human health and environmental quality.

The proper use of pesticide chemicals depends mostly on a continuing program of research and education. Each use of a pesticide should be judged on the basis of the potential positive values to be achieved as weighed against the possible negative values, such as those from residues on the harvested crop, occupational hazards to humans, hazards to pollinating and other beneficial insects, harmful effects on wildlife, and increases in environmental pollution.

3. *History*

The chemical control of the arthropod pests of cotton may be divided into four periods: (1) before the general use of insecticides on the crop, pre–Civil War times until 1923; (2) extensive use of calcium arsenate–nicotine sulfate dusts, 1924–1945; (3) the organochlorine insecticide era, 1945–1955; and (4) the organophosphorus insecticide era after the boll weevil became resistant to organochlorines, 1955 to present (Newsom and Brazzel 1968).

Before the invasion of the United States by the boll weevil in 1892, there was relatively little damage to cotton by insects or spider mites. There were no key pests, and only the bollworm, cotton leafworm, and the cotton aphid appeared infrequently in sufficient numbers to cause alarm to growers. Most cotton growers did not attempt to control these pests with insecticides, although a few used Paris green, London purple, or lead or calcium arsenate against the worms and nicotine sulfate against the aphids.

This situation changed when the boll weevil entered the United States and inflicted great losses to the crop. At this time an intensive search was begun for more effective insecticides and improved machinery for applying them. The first insecticide recommended for control of the boll weevil was an arsenical molasses spray containing Paris green, London purple, or lead arsenate (Townsend 1895; Malley 1902). However, the control produced by these concoctions was not satisfactory, because of poor formulations and inadequate machinery for applying the sprays.

The airplane introduced the era of chemical pest control to the cotton industry. The Ohio Experiment Station in 1921 first demonstrated the utility of aerial application by controlling a catalpa sphinx, *Ceratomia catalpae* (Boisduval), outbreak on catalpa trees. The success of this experiment suggested the possibility of using airplanes to control crop pest insects. The first airplane experiments on cotton were conducted at the USDA Federal Delta Laboratory by Coad and McNeil (1924) in 1922, who obtained effective control of the cotton leafworm with aerial applications of calcium arsenate dusts. In 1923, the two entomologists demonstrated that the boll weevil also could be effectively controlled with aerial applications of calcium arsenate dusts. The success of this method was quickly confirmed in Georgia and Texas during the period 1925–1927 (Post 1924; Thomas et al. 1929).

Dusting of cotton by airplane remained the principal method of applying insecticides to the crop until the early 1950s, when low-volume sprays were developed. Sprays have now replaced dusts to the point that cotton is seldom ever dusted (Fig. 11.8).

The cotton aphid, because of the destruction of its arthropod natural enemies, became a serious pest of cotton with the advent of the widespread use of calcium arsenate. Nicotine sulfate was added to the arsenical dust whenever aphids became a problem.

The bollworm also became much more serious during this period, and for the same reasons as the cotton aphids. However, calcium arsenate was fairly effective against very young larval bollworms and provided satisfactory control when applied to the crop on a regular basis at short intervals.

The importance of the cotton leafworm declined after the introduction of calcium arsenate dusts. This decline continued with the introduction of the new synthetic insecticides after World War II to the point that this species no longer is economically important in the United States.

Fig. 11-8 Airplane sprays are the most common method of insecticide treatment. Rates may range from 3 to 10 gal per acre, with 5 being about average.

The cotton fleahopper and lygus bugs became noticed as pests in the late 1920s, principally in the Southwest and the new irrigated areas of the West. Sulfur dusts offered the principal means for the control of these pests, as well as for occasional outbreaks of spider mites.

In summary, the modern era of intensive insecticidal treatment of cotton began in 1923 with the advent of airplane dusting of cotton. The principal insecticides used from this period until the introduction of organochlorine materials were dusts of calcium arsenate, nicotine sulfate, and sulfur. Most insecticidal use was confined to boll weevil–infested areas.

The era of very intensive insecticide treatment of cotton was ushered in by DDT, BHC, and toxaphene, which entered the scene at the end of World War II. These materials were followed by the introduction of other synthetic organic chlorinated hydrocarbon insecticides including aldrin, dieldrin, endrin, heptachlor, Strobane, and TDE. The new insecticides possessed two qualities of great importance: (1) high initial toxicity to the pest insects of cotton; and (2) sufficient persistence to control newly emerging insects or insects migrating from untreated to treated areas.

Organochlorine insecticides had a great impact on cotton production. For the first time cotton producers were able to gain highly effective control of all the arthropod pests of the crop. The impact of these insecticides was to stimulate an unprecedented demand by growers for almost complete control of pest insects. It then became profitable for producers to use fertilizer, irrigation, and long-growing, indeterminate varieties. Spectacular increases in yields were obtained at high profit levels for many years.

The apparent victory gained over the pest insects of cotton through the intensive use of organochlorine insecticides was not a lasting one. By the mid-1950s the boll weevil in Louisiana and Mississippi had developed resistance to most of these chemicals. The resistant strain of the pest spread rapidly across the southern and southwestern states, and all infested areas reported organochlorine resistant weevils by 1960 (Roussel and Clower 1955; Brazzel 1961a).

This problem was solved by a switch to organophosphorus and carbamate insecticides, mainly methyl parathion, azinphosmethyl, ethyl parathion, malathion, EPN, and carbaryl. Organophosphorus compounds were highly toxic to the boll weevil at relatively low rates, but these rates were not sufficient to control the bollworm and tobacco budworm. In order to gain control of all the major pest species, mixtures of chemicals were formulated containing DDT, toxaphene–DDT (and, to a lesser extent, endrin) for control of the bollworm and budworm; methyl parathion, azinphosmethyl, EPN or malathion were added for control of the boll weevil.

These mixtures were broad-spectrum formulations which also provided control of aphids, fleahoppers, plant bugs, spider mites, and leaf-feeding caterpillars. Of course, they also were highly toxic to insect parasites and predators. But during this period most cotton producers could not have cared less, as they were demanding insecticidal mixtures capable of "sterilizing" a field, that is, rendering it almost completely devoid of all insects. They had little concern for the insect parasites and predators they killed with these treatments (Adkisson 1969).

In the early 1960s, this situation shifted again. The bollworm and tobacco budworm developed a high level of resistance to organochlorine and carbamate insecticides (Brazzel 1963, 1964; Adkisson 1969; Harris et al. 1972). Now the pest-control situation in cotton suddenly reversed. The bollworm and tobacco budworm now were more important pests than the boll weevil. The problem of bollworm and tobacco bud-

Fig. 11-9. *Heliothis* larvae and damage to cotton boll. Bolls damaged by this pest are invariably invaded by boll rot organisms.

worm resistance was solved by increasing the dosage of methyl parathion from the 0.25 to 0.50-lb rate per acre per application used for boll weevil control to 1.0 to 2.0 lb per acre. Monocrotophos at 0.8 to 1.0 lb also was introduced, as were mixtures containing 2.0 lb of toxaphene, 1.0 lb of DDT, and 0.5 to 1.0 lb of methyl parathion. An immediate effect of increasing dosage rates was to increase costs of production. Yield remained high, but profits decreased (Adkisson 1969).

This situation prevailed until the late 1960s. Control was then directed mainly toward the bollworm and tobacco budworm (two pests formerly of secondary importance), while the boll weevil became less significant (Fig. 11.9).

In the late 1960s, there was another drastic change in the pest situation. The tobacco budworm in the lower Rio Grande Valley of Texas and northeastern Mexico became resistant to the organophosphorus insecticides. Many Valley producers treated fields 15 to 18 times with methyl parathion and still suffered great losses in yield. Others produced relatively high levels but made small profits because of the large costs incurred by intensive insecticidal treatment of the crop. Some saw their cotton destroyed regardless of treatment and, in fact, approximately 700,000 acres in northeastern Mexico were removed from cotton production because of the budworm (Atkisson 1969, 1972) (Chapter 7).

Organophosphorus-resistant tobacco budworms now occur in Texas, Louisiana, Arkansas, and other states to the east. The pest has developed such a high level of resistance that control has become very difficult with any insecticide presently registered for use on cotton.

Another drastic change in the pesticide usage pattern on cotton occurred when the EPA banned the use of DDT on cotton beginning with the 1973 season. DDT in combination with toxaphene provided satisfactory control of the boll weevil (methyl parathion was frequently added to this mixture at a low rate if weevils became extremely numerous), bollworm, cotton fleahopper, and plant bugs in cotton-producing states east of Texas. Because the toxaphene–DDT formulations had provided satisfactory control of a broad spectrum of pests, and resistance had been slow to develop in bollworm–tobacco budworm populations in the mid-South and Southeast, the cotton producers of these regions did not suffer the problems of the farmers in Texas and Mexico. The banning of DDT has forced southern cotton producers to shift to high rates of organophosphorus insecticides. These are usually applied in combination with toxaphene and, to a lesser extent, with endrin or chlordimeform for pest control. Thus the banning of DDT has had an unfortunate effect of increasing the selection pressure for development of organophosphorus-resistant pest strains. In due course the cotton producers of the South will suffer all the pest insect problems induced by organophosphorus insecticides that now occur mainly in Texas and Mexico. That is, chemical control will become more difficult, more costly, and less effective.

The evolution of insecticidal control of pest insects in cotton grown in the irrigated deserts of the western United States where the boll weevil does not occur has followed a similar course, except that the pest-control scheme has been dominated by lygus bugs, the bollworm and, more recently, the pink bollworm. In the California deserts lygus bugs have been the key pest for which insecticide treatment generally is started. Treatments also may be started for control of the bollworm. These treatments kill insect natural enemies, unleashing attacks of bollworms, cabbage loopers, cotton leaf perforators (Fig. 11.10*A* and *B*), salt marsh caterpillars, spider mites, and other pest species. Thus the desert regions may follow the same pattern as boll weevil–infested areas. Once insecticidal treatments are started, they must be continued because of target pest resurgence and secondary-pest outbreaks. This situation has led to the

Fig. 11-10 (*a*) Damage of the first three instars of the cotton leaf perforator. (*b*) Damage of the last two instars of the cotton leaf perforator in contrast to undamaged leaf.

intensive insecticidal treatment of cotton in desert areas (Smith and Reynolds 1972; van den Bosch et al. 1971).

Severe insecticide-induced problems of secondary pests led to the development of a simple system of integrated pest control in southern California in the late 1950s, which was centered on conserving the insect natural enemies of target and secondary-pest species. Broad-spectrum organochlorine insecticides were phased out. Selective dosages of certain organophosphorus insecticides were substituted, used under supervised conditions and only when pests caused crop damage. All these methods reduced the need for insecticides on the crop, and the system worked rather well. Beneficial insect populations again reached effective levels, and secondary pests virtually disappeared from the cotton environment.

This type of program in Arizona and southern California was ended by the pink bollworm invasion of the late 1960s. The crop came under intensive treatment with organophosphorus insecticides, and there is a potential for a great resurgence of secondary pests. Presently, the use of insecticides in these areas is as great as in the boll weevil–infested regions of the cotton belt (Smith 1971a).

4. *Selective Use*

Narrowly selective chemicals appear to offer an almost ideal means of pest control. Development of selectivity is extremely important to maintenance of pest-management programs. However, only a very few such chemicals have been discovered and developed for commercial use. All pesticides have some selectivity, but the range and degree among them is substantial. For many years much effort has been expended in seeking materials with relatively high toxicity to invertebrates and low toxicity to mammals. This is of course necessary for human safety, but differential toxicity to groups within the phylum Arthropoda must also be sought. The ultimate in specificity to enable prescription of a specific and unique chemical for each pest species is not imperative. However, effective materials are needed that are specific for *groups of pests,* such as aphids, plant bugs, locusts, caterpillars, weevils, and muscoid flies.

Fortunately, it is not always necessary to rely on the physiological selectivity of chemicals to obtain some of the specific effects required in integrated control and other pest-management systems. Ecological selectivity obtained by the discriminating use of even the most broad-spectrum insecticides can be employed in many cases for the development of effective, economical, and ecologically sound pest-control programs. The development of such programs is presently limited to some extent by a lack of knowledge of the ecology, biology, and behavior of pest–natural enemy–crop complexes.

Under integrated control systems the population dynamics of the pests or the pest abundance–crop damage relationship often is such that a high level of pest mortality is unnecessary. Instead of 95% mortality or higher, mortality of 75% or lower may be satisfactory and even desirable. Under these circumstances the low dosage of pesticide needed for the lower percentage mortality may permit the desired selective action between pest and beneficial forms.

The development of new highly specific pesticides will undoubtedly come very slowly, if at all, and they will be more costly than many of the pesticides available today. In the meantime the best use of the insecticides now available must be made through modification of dosages, formulations, timing of applications, methods of application, and other such techniques. It is often desirable to have a differential kill which leaves the balance in favor of the pest natural enemies. There is abundant evidence that this may be achieved for many pests with the present chemical tools.

5. *Impact*

Application of insecticides in agroecosystems is most commonly the factor that causes the most drastic reduction in arthropod natural enemies of pests in these systems. These effects have been demonstrated by resurgences of target pests and the change in status of formerly economically unimportant species into important pests. In Egypt, for example, the widespread use of organochlorines in cottonfields was followed by resurgence of the target insect, *Spodoptera littoralis* (Boisd.), and the unleashing of nontarget pests which formerly had been of little importance, such as *S. exigua* (Hübner), *Earias insulana* (Boisduval), aphids, and spider mites. Similar ecological disruptions and resultant pest outbreaks have occurred in cotton agroecosystems in many other parts of the world.

Pesticides have direct and indirect effects on natural enemies of cotton pests. Direct effects include death following contact with a toxic pesticide or feeding on pesticide-laden pollen, nectar, honeydew, or moribund prey in treated fields. Indirect effects occur with predators, such as *Geocoris* spp., *Nabis* spp., and *Orius* spp., which take moisture from the plant and as a result are adversely affected by feeding on the poisoned juices of plants treated with certain systemic insecticides (Reynolds 1971). Until more-selective pesticides are commercially available at reasonable expense, more judicious use of the pesticides now available should be made. This could be achieved, at least in part, through better knowledge of dosages, formulations, timing, and methods of application.

6. Resistance

Resistance to insecticides is an ability of selected insect populations to withstand exposure to dosages of pesticides that exceed that of a normally susceptible population—such ability being inherited by subsequent generations. All the most important arthropod pests of cotton in the United States have developed resistance to one or more pesticides. Resistance of cotton pests to insecticides has developed rapidly in recent years. Since 1947, when organic chemicals began to have wide usage on cotton, 23 species of insects and spider mites that attack the crop are known to have developed resistance, and several other species are strongly suspected of having developed resistance. One or more of these resistant species occur in localized areas in most cotton-producing states, from California to North Carolina. In most cases the pests are resistant to organochlorine insecticides, but four species of mites and the tobacco budworm are known to be resistant to organophosphorus compounds.

Table 11.8 Species of Arthropod Pests of Cotton That Have Developed Strains Resistant to Insecticides[a]

Organochlorine Compounds	Organophosphorus Compounds	Carbamates	Arsenicals
Boll weevil	Bollworm	Bollworm	None
Bollworm	Tobacco budworm	Tobacco budworm	
Tobacco budworm	Spider mites (*Tetranychus* spp.)		
Pink bollworm	Bandedwing whitefly		
Cotton fleahopper	Cabbage looper		
Lygus bugs	Cotton leaf perforator		
Cotton aphid			
Thrips (*Frankliniella* spp. and *Thrips* spp.)			
Cabbage looper			
Cotton leafworm			
Cotton leaf perforator			
Salt-marsh caterpillar			
Beet armyworm			
Southern garden leafhopper			
Consperse stinkbug			

[a] Modified from Newsom and Brazzel, 1968, and the 27th Annual Cotton Conference Report on Cotton Insect Research and Control, Dallas, Texas, 1974).

Resistance of most species continues to be restricted to relatively small areas, and no species is known to be resistant throughout the range of its occurrence. However, the boll weevil is known to be resistant in certain areas in 10 of the 11 states in which it occurs from Texas to North Carolina. A list of pests and the pesticides to which they have become resistant is shown in Table 11.8.

The development of resistance by the boll weevil in the mid to late 1950s provided the first impetus for cotton producers in a wide area to switch from organochlorine to organophosphorus and carbamate insecticides. The later development of resistance to organochlorines by the bollworm and tobacco budworm caused an even greater shift to the organophosphorus compounds. This, coupled with restrictions on the use of DDT and similar materials in California and Arizona because of drift hazards to other crops, had forced most cotton producers from Texas west to use organophosphorus compounds (in some areas these now are often combined with toxaphene) well before the EPA banned the use of DDT on cotton in 1973. This action, combined with the earlier development of resistant pest strains, has drastically reduced the use of organochlorines on cotton. Toxaphene, a chlorinated camphene, and endrin are the only exceptions, and considerable amounts still are applied to cotton in formulations containing methyl parathion. Table 11.9 and 11.10 illustrate the development of resistance in the bollworm and tobacco budworm.

Table 11.9 Development of Resistance by the Bollworm and Tobacco Budworm, as Expressed by Increased LD_{50} Values, from 1960 to 1965 to Certain Organochlorine and Carbamate Insecticides[a]

	LD_{50} (mg/g of larva)			
	Bollworm		Tobacco Budworm	
Compound	1960[b]	1965	1961[b]	1965
DDT	0.03	1000+	0.13	16.51
Endrin	0.01	0.13	0.06	12.94
Carbaryl	0.12	0.54	0.30	54.57
Strobane–DDT	0.05	1.04	0.73	11.12
Toxaphene–DDT	0.04	0.46	0.47	3.52

[a] Adkisson 1974.

[b] LD_{50} values for 1960 and 1961 were obtained from reports by Brazzel et al. (1961a) and Brazzel (1963).

Table 11.10 Comparative Resistance of Tobacco Budworms Collected from Various Locations to Methyl Parathion[a]

Location	LD_{50} (mg/g)		Increase in Resistance Over 1964, College Station, Texas
	Year	Larva	
College Station, Texas[b]	1964	0.01	Susceptible
College Station[b]	1968	0.04	4×
College Station[b]	1969	0.06	6×
College Station[b]	1970	0.14	14×
College Station[b]	1971	0.11	11×
College Station[b]	1972	0.51	51×
Weslaco, Texas[b]	1968	0.05	5×
Weslaco[b]	1969	0.11	11×
Weslaco[b]	1972	0.25	25×
Tampico, Mexico[c]	1969	0.46	46×
Tampico[c]	1970	0.61	61×
Monte, Mexico[c]	1970	2.01	201×
Giradot, Columbia[c]	1970	0.01	Susceptible
Lima, Peru[c]	1970	0.001	Susceptible

[a] Adkisson 1974.

[b] Data from Nemec and Adkisson 1973.

[c] Data from Wolfenbarger et al. 1973.

III. EVALUATION OF CONTROL NEEDS UTILIZING ECONOMIC THRESHOLDS

For the rational development of any pest-control program, it is essential to understand the relationship between pest-infestation levels and crop loss. From the broad view of human society, all crop losses are to be considered real losses; however, the costs of achieving the full crop potential may exceed the value of the potential benefit. Therefore it is necessary to determine *economic thresholds,* that is, the maximum pest population that can be tolerated at a particular time and place without a resultant economic crop loss.

The grower may consider only part of the reduction in yield or quality as a loss. His determination, made consciously or intuitively, on good or bad advice, is influenced by many factors such as the cost for crop protection, the cost of avoiding the potential loss, prevailing marketing conditions, and the ultimate use of the crop. To make an accurate judgment, the economic matrix must be understood and fully meshed with an understanding of the damage potential of the pest species.

Initially, economic injury thresholds can be determined tentatively on empirical evidence, that is, by deduction from experiences with the pest in the past. However, levels established in such a manner should be reviewed constantly and readjusted in accordance with changes in farming practices and with information obtained from continuing observations and specially designed experiments. Several methods for assessment of losses are possible. One approach is based on comparison of the yields from groups of plants treated identically in all respects, except that one is kept free of insect pests and the other is placed under attack by the specific pest being considered. Careful population monitoring is essential, and several years of research are normally necessary.

It is important to recognize that economic crop loss depends not only on the numbers of pests present, but on the plant reaction to the attack. In addition to considering pest abundance, age structure, and duration of attack, the stage of the plants attacked and especially the surplus amount of fruit being produced by the plants must not be overlooked.

Because of the variability in the effects of factors causing economic losses, long-term experiments are generally required to establish economic thresholds in relation to intensity of attack. Furthermore, long-term assessment of economic losses on untreated check plots also permits studies to determine levels of infestations and losses under natural conditions. The data thus obtained provide a useful point of reference to evaluate the efficacy of various control procedures tested at the same time in adjacent plots. For example, the actual yield and economic losses in plots treated with a chemical insecticide compared to an untreated check measure the effectiveness of the insecticide at a given level of infestation. For the results of these studies to be the most useful, proper experimental design and correct scientific procedure must be followed. Misleading results often have been obtained when check plots have been located in such a way that insecticidal drift from treated plots kills most of the insect natural enemies in the untreated plots, allowing them to be decimated by both primary and secondary pests.

Studies of the development of cotton plants and their interactions with pest populations under commercial field conditions have been aided by techniques adapted from those developed over many years by scientists of the Cotton Research Corporation (Munro and Farbrother 1969). Davidson (1973) developed a system whereby data concerning randomly sampled plants are read into a tape recorder: at each node, leaf size, leaf damage, fruiting point type, and damage are all summarized in four-code digits. These data are later transcribed to punched cards and processed by computer, which produces a numerical summary of the average plant. By means of a computer-controlled plotting device, a plant diagram is drawn, which shows the frequency distribution of fruiting points at each node and the extent to which shedding or pest damage has occurred. Such plant diagrams are useful for a rapid visual assessment of the situation, and for the extension worker, pest management specialist, or supervised control entomologist to discuss with the farmer. Ideally, pest-control decisions should be based on such a computer-aided comparison of the numerical data with patterns of growth and damage recorded under similar conditions in previous years, or a comparison with projected growth patterns derived from predictive models of cotton growth based on laboratory data (Hesketh et al. 1972). It is now possible to categorize the pattern of plant response to damage induced by some of the major cotton pests in California. Satisfactory progress is being made in simplifying these techniques to make the process practicable for use in commercial farming operations and as a method of obtaining greater feedback from the cotton plant component of the agroecosystem.

Also, large screen cages often have been used in which predetermined levels of certain pests are maintained, while in other cages no pests are allowed. Results are then correlated with similar levels of pest infestation and yield in large field trials.

A. Growth and Fruiting

As concerns cotton, one approach to establishing a reference for economic injury levels is to follow plant growth, fruit formation, and boll maturation systematically throughout the season for several seasons (e.g., 3 to 5 years). In this manner useful data can be accumulated about the carrying capacity of the plant and how it is influenced by soil type, soil fertility, moisture distribution, temperature, insects, and plant diseases. It will become evident what percentage of the fruiting forms constitutes surplus fruit production, what percentage must be protected, and at what time during the season. This information may then be related to what is known about the biodynamics and damage potential of each insect pest species as it appears, and the importance of naturally occurring biotic mortality factors in regulating the pest species.

Although in many areas of the world information on cotton plant growth and development has been collected over the years, little effort has been made to use these data to establish economic injury levels for pest insects. This approach has received increased interest in recent years and should be more widely employed in the future. The basic approach in utilizing a pest-control system based on plant growth requires that the field technician systematically examine the plants to determine their stage of development. This is done in measured areas which equal $\frac{1}{1000}$ acre, or other area of measurement. The readings obtained are multiplied by 1000 to translate them to an acre basis. In addition to the plant sample, the fieldman looks for pest insects and for the presence of naturally occurring biotic factors.

In this manner the field technician can determine the abundance of harmful and beneficial insects, and the square, flower, or boll count. Pest insect numbers, the amount of damage, the stage of development, and the degree of fruit formation and boll retention are then evaluated to provide the essential information for making decisions regarding pest control and overall crop management. Each field has to be viewed as a separate unit with its own unique limits, tolerances, and requirements.

B. Leaf Area

During several years of research, it has been noted that at certain growth periods cotton can suffer great reductions in leaf surface area without corresponding economic damage to the plant or crop. During the period of *plant establishment,* leaf production far exceeds plant growth needs. Leaf surface area may be reduced up to 50% during this period without significantly affecting ultimate yield and lint quality. Defoliation exceeding 50% may be tolerated during this period, the main detrimental effect being to delay the onset of fruiting.

During the *fruit-formation* period, leaf surface area becomes more important; apparently during this time only about 20% defoliation can be tolerated. During this period the plant also has the greatest uptake and utilization of water and nitrogen.

Once into the period of *fruit maturation,* leaf surface area needs greatly diminish, and defoliation up to 50% may again be tolerated. Not only may this large amount of defoliation be tolerated, it may be beneficial. When leaf surface area is reduced and the leaf canopy opened, air circulation and solar penetration are improved. This hastens boll maturity and inhibits boll rot.

C. Population Measurement and Prediction

No control measures should be undertaken against an insect pest unless it has been determined to be present and expected or known to be causing economic loss. From these basic criteria the importance of pest population measurement and prediction immediately becomes apparent. The fundamental approach to population measurement and prediction is to make regular samples of the developing crop and its surroundings in a systematic and standardized manner. The ultimate objectives are to measure qualitative and quantitative changes in the crop and associated flora and fauna, and to attempt to anticipate future events. Many techniques and procedures have been developed to achieve this objective. However, before relying on any of them, their accuracy and reliability must be determined.

To be most effective, integrated control of cotton pests should be carried out on an areawide basis, since the actions of one producer may affect the pest population in the field of another. This is especially true when insecticides are used. Thus a producer needs to have information concerning the relative abundance of a pest across the area, and the absolute abundance in his own fields.

D. Field Checking

There is no better method for the determination of pest insect levels than frequent inspection of individual fields. For some minor arthropod pests, such as thrips, aphids, and spider mites, casual observation by a practiced eye is sufficient to determine if and when chemical control is needed. However, for the major pests, including the boll weevil, bollworm, tobacco budworm, fleahopper, lygus bugs, and pink bollworm, direct counts of the pests themselves or estimates of the damage inflicted by them must be made on a per-unit basis. Insecticidal control should never begin until it has been determined that the pest population is of sufficient size to cause yield losses substantially in excess of treatment costs.

The determination of the pest population level at which to begin insecticidal treatment is not easy. The economic threshold for a given pest may vary with climate, crop condition, stage of fruiting, type of production practices used, numbers of beneficial arthropods and other pest species, cost of control, value of the crop, and many other variables known and unknown, some controllable and some not. Thus guidelines published in a list of recommendations concerning the numbers of a pest or amount of damage that constitute an economic infestation seldom are sufficient to cover all the situations producers may encounter during the season. The economic threshold is always changing. The threshold applicable today may not be good tomorrow, because insect populations and climatic and crop conditions constantly change. Because of the difficulties in establishing rigid economic thresholds for each pest, there is no substitute for the experienced pest-management specialist who has a trained eye for evaluating field conditions.

Specially trained personnel, often referred to as checkers or scouts should be used to make insect population counts in the fields. In addition to pest arthropods, beneficial insect populations are evaluated and full advantage is taken of those that are present. Furthermore, plant growth *and* fruit development, as well as damage, should be measured at regular intervals. In this manner unnecessary pesticide treatments are eliminated and control measures made more effective through better timing. The savings that accrue merely by eliminating one or more pesticide applications usually more than pay for the cost of employing field checkers for the season.

The grower may hire field checkers to make records and to submit them to him. He then determines the need for insecticides and applies them with his own equipment or arranges for custom applications. Another method of operation is for the grower to confer with a consulting entomologist who examines the fields and recommends the insecticides that should be applied or does the job himself. Although such arrangements are convenient, growers should be cautious in hiring or relying entirely on consulting entomologists employed by, or in any way affiliated with, pesticide marketing or application firms. Because the responsibility of these persons must be to the primary employer, their motivation may be to use more insecticide than is really needed for crop protection.

E. Use of Traps

Insect traps using fluorescent ultraviolet (black-light) lamps are useful in fixing the time of appearance and seasonal abundance of important insect pests (Chapter 9). Some of the more important cotton pests collected by light traps include Lepidoptera, primarily in the family Noctuidae—*Heliothis* spp., *Spodoptera* spp., *Trichoplusia* spp.—but also *Estigmene* spp. in the family Arctiidae. Commonly, the methods used to assess the need for control of these pests rely on detection of eggs or larvae in the field. But eggs and larvae appear only after there has been moth activity, and it is highly useful to have information on the timing of moth flights and the abundance of moths of these species. Such information alerts growers and entomologists to moth activity in the fields and possibly to the potential size of infestations. In this way field sampling can be intensified at the correct times, and control measures used with greater precision. While light traps are useful detection devices, they have not been effective for the control of insect pests.

Attractant traps using natural pheromones extracted from the bodies of insects or synthetic substances have come into use in recent years (Chapter 8). Traps containing a sex attractant extracted from the tips of abdomens of virgin female pink bollworm moths have been highly effective in trapping male moths. Caged virgin females have also proved effective in luring males into traps. A synthetic attractant, hexalure, is now available commercially and is being used instead of the natural lure in population detection and survey traps. Population sampling to determine the time to initiate insecticide treatments for pink bollworm control has been slow and laborious. It has been accomplished by collecting boll samples in the field, cracking them, and examining the interiors for larval infestation. Recent research by Toscano et al. (1974) has demonstrated the feasibility of timing applications by basing population prediction on the number of males attracted and caught in pheromone traps baited with hexalure.

A new pink bollworm population suppression technique is under investigation by Shorey et al. (1974). With this method, hexalure was continuously evaporated into the air of cottonfields throughout the cotton production season. The subsequent disruption of premating pheromonal communication between males and females of the pink bollworm resulted in a reduction in larval boll infestation comparable to that provided by commercial insecticide applications. It should be noted that scheduled insecticide applications are made at 5- to 7- day intervals in Arizona and southern California. If the promising technique of Shorey and co-workers proves feasible, the heavy reliance on insecticides may be unnecessary.

The natural sex pheromone of the pink bollworm recently has been identified and synthesized (Hummel et al. 1973). Named gossyplure, investigation has established it to be several 100-fold as attractive to male moths as the synthetic hexalure. It is anticipated that it will replace hexalure and likely will significantly improve the effectiveness of the technique described above for managing this pest.

Similarly, it appears that the male boll weevil pheromone will become a powerful tool in management of this insect. The pheromone acts as an aggregating signal for both sexes (Hardee et al. 1969). The synthetic preparation of this material, grandlure, is used for detecting low-level infestations, being particularly effective early in the season before the cotton begins to fruit. The weevils again strongly respond to grandlure later in the season when the weevil population starts to disperse. Bottrell (1972) noted a potential utility in regulating the dispersal track of low-density populations.

These findings and those of Lloyd et al. (1972) disclose implications as to how synthetic boll weevil pheromone possibly can be used in the development of novel, integrated pest-management strategies. In fact, some of these strategies have been tested on a small scale. These investigators showed that a strategy consisting of (1) sidedress treatments of a systemic insecticide (aldicarb) to a restricted portion (ca. 10%) of cottonfields and (2) sterile male boll weevils (the pheromone source) forestalled significantly the time in which damaging populations of native boll weevils developed in the fields. Native weevils that responded to pheromone released by sterile males were killed by the systemic insecticide. A system of this type appears to be especially feasible when integrated with other suppression techniques such as short-season cottons (Walker and Niles 1971) and practices of early stalk destruction to further reduce population density (Bottrell 1972).

Cross (1973) recently summarized much of the research progress since 1965. Noting the potential value of the pheromone in more intelligent use of insecticides for weevil control, he further felt that the pheromone in combination with other techniques, properly integrated, is so promising that there is hope for eradicating this pest from the United States. Indeed, preliminary programs are under way to this end. Some entomologists believe eradication is too optimistic over such a large area (approximately 10 million acres), but it does appear that, when the pheromone is used with other techniques, integrated pest-management programs are feasible with a concomitant reduction in insecticide load in the cotton environment (Entomological Society of America 1973).

IV. CONCLUSION

Research on cotton and the associated arthropod fauna has already progressed to the point where much of the knowledge necessary for successful pest-management programs is available. Unfortunately, this knowledge is seldom applied in the decision process at the farmer level. The success or failure of any system of pest management depends on systematic field sampling and on the experience and knowledge of the program manager in applying the data obtained. It is anticipated that cotton producers will employ increasing numbers of professional pest-management specialists for their expertise. Employed solely for their skill and experience in the application of pest-management strategies, these specialists will advise farmers in their decision-making process. Without such skilled advisers, pest-control measures may not be applied at the right time and in the right way, and programs would fail.

In integrated-control programs for cotton, the right time for certain of these measures, that is, stalk destruction, insecticidal control of prehibernating pests, or advice on crop-planting patterns, may be the year before the planting of the crop to be protected. The right way may mean the use of a selective dosage of an insecticide to conserve insect natural enemies, or the use of one that is not highly toxic (when such a choice of chemicals is available) to certain of the most beneficial species. The right way also could be the use of crop-management practices, for example, strip cutting of alfalfa to control lygus infestation in cotton rather than an insecticidal treatment. Also, in carrying out these measures, the specialist must be aware that the action that solves one problem may create another. For example, insecticidal control of the boll weevil or lygus bugs may induce an outbreak of bollworms. During the cropping season the program manager must also know how to make assessments of pest insect numbers and crop damage in order to make an accurate appraisal of the possible economic losses that may accrue to pest attack. This is necessary to determine the need for and timing of appropriate suppressive measures, be they the application of an insecticide, irrigation, the cutting of an alfalfa hay strip, or the termination of the crop with a defoliant.

Although many pages of scientific literature have been published that provide information on all these matters, pest control for the average producer still is more of an art than a science. This is the reason more research is needed, coupled with better educational programs for assisting cotton producers and pest-management advisers in the solution of problems in the field.

REFERENCES

Adkisson, P. L. 1958. The influence of fertilizer applications on populations of *Heliothis zea* (Boddie) and certain insect predators. *J. Econ. Entomol.* **51**:757–759.

Adkisson, P. L. 1962. Timing defoliants and desiccants to reduce overwintering populations of the pink bollworm. *J. Econ. Entomol.* **55**:949–951.

Adkission, P. L. 1964. Action of the photoperiod in controlling insect diapause. *Amer. Nat.* **XCVIII**(902):357–374.

Adkisson, P. L. 1966. Internal clocks and insect diapause. *Science* **154**:234–241.

Adkisson, P. L. 1969. How insects damage crops. Pages 155–164 *in* How crops grow—A century later. *Conn. Agr. Exp. Sta. Bull.* 708.

Adkisson, P. L. 1972. The integrated control of the insect pests of cotton. *Proc. Tall Timbers Conf. Ecol. Anim. Control Habitat Manage.* **4**:175–188.

Adkisson, P. L. 1973. The principles, strategies and tactics of pest population regulation and control in major crop ecosystems: The cotton system. Pages 274–282 *in* P. W. Geier, ed., studies in population management. Memoirs of the Ecological Society of Australia, Vol. 1.

Adkisson, P. L. 1974. Paper presented at FAO/IAEA research coordination meeting in ecology and behavior of the *Heliothis* complex as related to the sterile male technique. Monterrey, Mexico, April 1974.

Adkisson, P. L., R. A. Bell, and S. G. Wellso. 1963. Environmental factors controlling the induction of diapause in the pink bollworm. *J. Inst. Physiol.* **9**:299–310.

Adkisson, P. L., and J. C. Gaines. 1960. Pink bollworm control as related to the total cotton insect control program of Central Texas. *Tex. Agr. Exp. Sta. Misc. Publ.* 444. 7 pp.

Adkisson, P. L., R. L. Hanna and C. F. Baily. 1964. Estimate of the numbers of *Heliothis* larvae per acre in cotton and their relation to the fruiting cycle and yield of the host. *J. Econ. Entomol.* **56**(5):657–663.

Adkisson, P. L., D. R. Rummel, W. L. Sterling, and W. L. Owen, Jr. 1966. Diapause boll weevil control: A comparison of two methods. *Tex. Agr. Exp. Sta. Bull.* 1054. 11 pp.

Annual Cotton Report. 1972. Cotton Incorporated, Memphis, Tenn.

Bell, K. O., and W. H. Whitcomb. 1962. Efficiency of egg predators of the bollworm. *Arkansas Farm Res.* **11**:9.

Bird, J. B., and J. Mahler. 1951-1952. America's oldest fabrics. *Amer. Fabrics.* **20**:73–78.

Bottger, G. T., E. T. Sheehan, and M. J. Lukefahr. 1964. Relation of gossypol content of cotton plants to insect resistance. *J. Econ Entomol.* **57**(2):283–285.

Bottrell, D. G. 1972. New strategies for management of the boll weevil. *Proc. Fifth Ann. Tex. Conf. Insect, Plant Dis., Weed Brush Control, Tex. A & M Univ.*, pp. 67–72.

Bottrell, D. G. 1974. Biological control agents of the boll weevil. *Proc. Conf. Boll Weevil Res. Elimination Technol., Nat. Cott. Council, Memphis,* Tenn., *February 1974*. In press.

Bottrell, D. G., D. R. Rummel, and P. L. Adkisson. 1972. Spread of the boll weevil into the high plains of Texas. *Environ. Entomol.* **1:136–140.**

Brazzel, J. R. 1961a. Boll weevil resistance to insecticides in Texas in 1960. *Tex. Agr. Exp. Sta.* PR.2171. 4 pp.

Brazzel, J. R. 1961b. Destruction of diapause boll weevils as a means of boll weevil control. *Tex. Agr. Exp. Sta. Misc. Publ.* 511.

Brazzel, J. R. 1963. Resistance to DDT in *Heliothis virescens. J. Econ. Entomol.* **56**:571–574.

Brazzel, J. R. 1964. DDT resistance in *Heliothis zea. J. Econ. Entomol.* **57**:455–457.

Brazzel, J. R., H. Chambers, and P. J. Hammon. 1961a. A laboratory rearing method and dosage-mortality data on the bollworm, *Heliothis zea. J. Econ. Entomol.* **54**:949–952.

Brazzel, J. R., T. B. Davich, and L. D. Harris. 1961b. A new approach to boll weevil control. *J. Econ. Entomol.* **54**:723–730.

Coad, B. R., and G. L. McNeil. 1924. Dusting cotton from airplanes. *USDA Bull.* 1204.

Cross, W. H. 1973. Biology, control and eradication of the boll weevil. *Annu. Rev. Entomol.* **13**:17–46.

Davidson, A. 1973. Computerized plant growth analysis of the interactions of arthropod pests and other factors with the cotton plant. Ph.D. dissertation, Department of Entomological Science, University of California, Berkeley.

Davis, V. W., A. S. Fox, R. P. Jenkins, and P. A. Andrilenas. 1970. Economic consequences of restricting the use of organochlorine insecticides on cotton, corn, peanuts and tobacco *USDA Agr. Econ. Rep.* 178. 52 pp.

Delattre, R. 1973. Parasites et maladies en culture cotonnière. Manual phytosanitaire. Institut de recherches du coton et des textiles exotiques. Paris. 146 pp.

Dickson, R. C., M.McD. Johnson, and E. F. Laird. 1954. Leaf crumple, a virus disease of cotton. *Phytopathology* **44**(8):479–480.

Dunnam, E. W., and J. C. Clark. 1941. Cotton aphid multiplication following treatment with calcium arsenate. *J. Econ. Entomol.* **34**:587–588.

Eaton, F. M. 1955. Physiology of the cotton plant. *Annu. Rev. Plant Physiol.* **6**:299–328.

Eichers, T., P. Andrilenas, H. Blake, R. Jenkins, and A. Fox. 1970. Quantities of pesticides used by farmers in 1966. *U.S. Dept Agr. Econ. Res. Serv. Agr. Econ. Rep.* 179. 61 pp.

Erwin, D. C., and R. Meyer. 1961. Symptomatology of the leaf crumple disease in several species and varieties of *Gossypium* and variation of the causal virus. *Phytopathology* **51**(7):472–477.

Entomological Society of America. 1973. The pilot boll weevil eradication experiment. *Bull. Entomol. Soc. Amer.* **19**:218–221.

Ewing, K. P., and E. E. Ivy. 1943. Some factors influencing bollworm populations and damage. *J. Econ. Entomol.* **36**:602–606.

Falcon, L. A. 1971a. Progreso del control integrado en el algodón de Nicaragua. *Rev. Peru Entomol. Agr.* **14**(2):376–378.

Falcon, L. A. 1971b. Microbial control as a tool in integrated control programs. Pages 346–364 *in* C. B. Huffaker, ed., Biological control. Plenum Press, New York.

Falcon, L. A., and R. Daxl. 1974. Report to the government of Nicaragua on the integrated control of cotton pests (NIC/70/002/AGP) for the period June 1970 to June 1973. Food and Agricultural Organizations of the United Nations, Rome. 60 pp.

Falcon, L. A., and R. F. Smith. 1973. Guidelines for integrated control of cotton insect pests. Food and Agriculture Organization of the United Nations. AGPP:Misc/8. 92 pp.

Fenton, F. A., and W. L. Owen. 1953. The pink bollworm of cotton in Texas. *Tex. Agr. Exp. Sta. Misc. Publ.* 100.

Fletcher, R. K., and F. L. Thomas. 1943. Natural control of eggs and first instar larvae of *Heliothis armigera*. *J. Econ. Entomol.* **36**:557–560.

Folsom, J. W. 1928. Calcium arsenate as a cause of aphid infestation. *J. Econ. Entomol.* **21**: 174.

Gillham, F. E. M. 1965. The relationship between insect pests and cotton production in central Africa. Pages 405–422 *in* M. Taghi Farvar and J. P. Milton, eds., The careless technology—Ecology and international development. Natural History Press, Garden City, New York.

Gorham, D. P. 1947. The cotton worm. *DeBow's Commerc. Rev.* **3**:535.

Gulati, A. M., and A. J. Turner. 1928. A note on the early history of cotton. *Bull. no. 17, Tech. Ser. No. 12, Indian Central Cotton Committee.*

Hagen, K. S., R. van den Bosch, and D. L. Dahlsten. 1971. The importance of naturally occurring biological control in the western United States. Pages 253–293 *in* C. B. Huffaker, ed., *Biological control*. Plenum Press, New York.

Hardee, D. D., W. H. Cross, and E. B. Mitchell. 1969. Male boll weevils are more attractive than cotton plants to boll weevils. *J. Econ. Entomol.* **62**: 165–169.

Hargreaves, H. 1948. List of recorded cotton insects of the world. Commonwealth Institute of Entomology, London. 50 pp.

Harris, F. A., J. B. Graves, S. J. Nemec, S. B. Vinson, and D. A. Wolfenbarger. 1972. Insecticide resistance. Pages 17–27 *in* Distribution, abundance and control of *Heliothis* species in cotton and other host plants. *South. Coop. Ser.* 169.

Hesketh, J., D. N. Baker, and W. G. Duncan. 1972. Simulation of growth and yield in cotton. *Crop Sci.* **12**:395–398, 436–439.

Hummel, H. E., L. K. Gaston, H. H. Shorey, R. S. Kaae, K. J. Byrne, and R. M. Silverstein. 1973. Clarification of the chemical status of the pink bollworm sex pheromone. *Science* **181**:873–875.

Hunter, W. D. 1912. The control of the boll weevil. *USDA Farmer's Bull.* 500. 14 pp.

Ignoffo, C. M. 1970. Microbial insecticides: No-yes; now-when! *Proc. Tall Timbers Conf. Ecol. Anim. Control Habitat Manage.* **2**:41–58.

Laird, E. F., and R. C. Dickson. 1959. Insect transmission of the leaf crumple virus of cotton. *Phytopathology* 49(6):324–327.

Leigh, T. F., D. W. Grimes, H. Yamada, J. R. Stockton and D. Bassett. 1969. Arthropod abundance in cotton in relation to some cultural management variables. *Proc. Tall Timbers Conf. Ecol. Anim. Control Habitat Manage.* **1**:71–83.

Leigh, T. F., R. E. Hunter, and A. H. Hyer. 1968. Spider mite effects on yield and quality of four cotton varieties. *Calif. Agr.* **22**(10):4,5.

Leigh, T. F., and A. H. Hyer. 1963. Spider mite-resistant cotton. *Calif. Agr.* **17** (2):6,7.

Lingren, P. D. 1969. Approaches to the management of *Heliothis* spp. in cotton with *Trichogramma* spp. *Proc. Tall Timbers Conf. Ecol. Anim. Control Habitat Manage.* **1**:207–218.

Lingren, P. D., R. L. Ridgway, and S. L. Jones. 1968. Consumption by several common arthropod predators of eggs and larvae of two *Heliothis* species that attack cotton. *J. Econ. Entomol.* **61**(3):613–618.

Lloyd, E. P., W. P. Scott, K. K. Shaunak, F. C. Tingle, and T. B. Davich. 1972. A modified trapping system for suppressing low density populations of overwintering boll weevils. *J. Econ. Entomol.* **65**:1144–1147.

Lukefahr, M. J., and D. F. Martin. 1966. Cotton-plant pigments as a source of resistance to the bollworm and tobacco budworm. *J. Econ. Entomol.* **59**: 176–179.

Lukefahr, M. J., L. W. Noble and D. F. Martin. 1964. Factors inducing diapause in the pink bollworm. *USDA Tech. Bull.* 1304.

Malley, F. W. 1902. Report of the boll weevil. Agricultural and Mechanical College of Texas, College Station, Texas. 70 pp.

Mueller, A. J., and V. M. Stern. 1973. *Lygus* flight and dispersal behavior. *Environ. Entomol.* **2**(3):361–364.

Mueller, A. J., and V. M. Stern. 1974. Timing of pesticide treatments on safflower to prevent *Lygus* from dispersing to cotton *J. Econ. Entomol.* **67**(1): 77–80.

Munro, J. M., and H. G. Farbrother. 1969. Composite plant diagrams in cotton. *Cotton Grow. Rev.* **46**:261–282.

Murray, J. C., L. M. Verhalen, and D. E. Bryan. 1965. Observations on the feeding preference of the striped blister beetle, *Epicauta vittata* (Fabricius), to glanded and glandless cottons. *Crop. Sci.* **5**:189.

Nemec, S. J., and P. L. Adkisson. 1973. Organophosphate insecticidal resistance levels in tobacco budworm and bollworm populations in Texas. Pages 18–25 *in* Investigations of chemicals for control of cotton insects in Texas. *Tex. Agr. Exp. Sta., Dep. Entomol., Tech. Rep.* 73-20.

Newsom, L D. 1970. The end of an era and future prospects for insect control. *Proc. Tall Timbers Conf. Ecol. Anim. Control Habitat Manage.* **2**:117–136.

Newsom, L. D., and J. R. Brazzel. 1968. Pests and their control. Pages 367–405 *in* F. C. Elliot, M. Hoover, and W. K. Porter, eds., Advances in production and utilization of quality cotton: Principles and practices. Iowa State University Press, Ames.

Noble, L. W. 1969. Fifty years of research on the pink bollworm in the United States. *USDA Agr. Handb.* 357. 62 pp.

Ohlendorf, W. 1926. Studies of the pink bollworm in Mexico. *USDA Bull.* 1374.

Painter, R. H. 1951. Insect resistance in crop plants. Macmillan, New York.

Pearson, E. O., and R. C. Maxwell-Darling. 1958. The insect pests of cotton in tropical Africa. Empire Cotton Growing Corporation and Commonwealth Institute of Entomology, London. Eastern Press, London. 355 pp.

Pierce, W. D. 1908. Studies of parasites of the cotton boll weevil. *USDA Bur. Entomol. Bull.* 73.

Pierce, W. D., R. A. Cushman, and C. E. Hood. 1912. The insect enemies of the cotton boll weevil. *USDA Bur. Entomol. Bull.* 100.

Pimentel, D. 1973. Extent of pesticide use, food supply, and pollution. *J. N.Y. Entomol. Soc.* **LXXXI**:13–33.

Post, G. B. 1924. Boll weevil control by airplane. *Georgia State Coll. Agr. Bull.* 301.

Reynolds, H. T. 1971. Recent developments with systemic insecticides for insect control on cotton. *Summ. Proc. West. Cotton Prod. Conf.,* pp. 18–20.

Ridgway, R. L. 1969. Control of the bollworm and tobacco budworm through conservation and augmentation of predaceous insects. *Proc. Tall Timbers Conf. Ecol. Anim. Control Habitat Manage.* **1**:127–144.

Riley, C. V. 1873. Fifth annual report of the noxious, beneficial and other insects of the state of Missouri, made to the State Board of Agriculture, pursuant to an appropriation for this purpose from the legislature of the state. *Eighth Annu. Rep. State Board Agr. 1872,* pp. 160–168.

Robertson, O. T., V. L. Stedronsky and D. H. Currie. 1959. Kill of pink bollworms in the cotton gin and the oil mill. *USDA Prod. Res. Rep.,* 26. 22 pp.

Roussel, J. S., and D. F. Clower. 1955. Resistance to the chlorinated hydrocarbon insecticides in the boll weevil (*Anthonomus grandis* Boh.). *La. Agr. Exp. Sta. Circ.* 41.

Schuster, M. F., and F. G. Maxwell. 1974. The impact of nectariless cotton on plant bugs, bollworms, and beneficial insects. *Proc. 1974 Beltwide Cotton Res. Conf. (Nat. Cotton Counc., Dallas, Tex., January 1974):* 86–87.

Shorey, H. H., R. S. Kaae, and Lyle K. Gaston. 1974. Sex pheromones of Lepidoptera. Development of a method for pheromonal control of *Pectinophora gossypiella* in cotton. *J. Econ. Entomol.* **67**:347–350.

Smith, Ray F. 1971a. Fases en el desarrollo del control integrado. *Bol. Soc. Entomol. Peru* **6**:54–56.

Smith, Ray F. 1971b. Economic aspects of pest control. *Proc. Tall Timbers Conf. Ecol. Anim. Control Habitat Manag.* **3**:53–83.

Smith, R. F., and L. A. Falcon. 1973. Insect control for cotton in California. *Cott. Grow. Rev.* **50**:15–27.

Smith, R. F., and H. T. Reynolds. 1972. Effects of manipulation of cotton agro-ecosystems on insect pest populations. Pages 373–406 *in* M. T. Farvar and J. P. Milton, eds., The careless technology—Ecology and international development. Natural History Press, Garden City, New York.

Starbird, I. R., and B. L. French. 1972. Costs of producing upland cotton in the United States, 1969. *USDA Econ. Rev. Serv. Agri. Econ. Rep.* 229. 47 pp.

Stern, V. M. 1969. Interplanting alfalfa in cotton to control lygus bugs and other insect pests. *Proc. Tall Timbers Conf. Ecol. Anim. Control Habitat Manage.* **1**:55–69.

Stern, V. M., R. van den Bosch, T. F. Leigh, O. D. McCutcheon, W. R. Sallee, C. E. Houston and M. J. Garber. 1967. Lygus control by strip-cutting alfalfa. *Univ. Calif. Agri. Ext. Serv.* AXT-241. 13 pp.

Thomas, F. L., W. L. Owen, J. C. Gaines and Franklin Sherman III. 1929. Boll weevil control by airplane dusting. *Tex Agr. Exp. Sta. Bull.* 394.

Tingey, W. M., T. F. Leigh, and A. H. Hyer. 1973a. Three methods of screening cotton for ovipositional nonpreference by lygus bugs. *J. Econ. Entomol.* **66**(6):1312–1314.

Tingey, W. M., T. F. Leigh, and A. H. Hyer. 1973b. Lygus bug resistant cotton. *Calif. Agr.* **27**(11):8, 9.

Toscano, N. C., A. J. Mueller, V. Sevacherian, R. K. Sharma, T. Niilus, and H. T. Reynolds 1974. Insecticide applications based on Hexalure® trap catches versus automatic schedule treatments for pink bollwarm moth control. *J. Econ. Entomol.* **67**:522–524.

Townsend, C. H. T. 1895. Report on the Mexican cotton boll weevil in Texas (*Anthonomus grandis* Boh.). *Insect Life* **7**:295–309.

van den Bosch, R. and K. S. Hagen. 1966. Predaceous and parasitic arthropods in California cotton fields. *Calif. Agr. Exp. Sta. Bull.* 820. 32 pp.

van den Bosch, R., T. F. Leigh, L. A. Falcon, V. M. Stern, D. Gonzales and K. S. Hagen. 1971. The developing program of integrated control of cotton pests in California. Pages 377–394 *in* C. Huffaker, ed. Biological control. Plenum Press, New York.

Walker, J. K., Jr., and J. A. Niles. 1971. Population dynamics of the boll weevil and modified cotton types: Implications for pest management. *Tex. Agr. Exp. Sta. Bull.* 1109. 14 pp.

Whitcomb, W H. 1970. History of integrated control as practiced in the cotton fields of the south central United States. *Proc. Tall Timbers Conf. Ecol. Anim. Control Habitat Manage.* **2**:147–155.

Whitcomb, W. H., and K. Bell. 1964. Predaceous insects, spiders, and mites of Arkansas cotton fields *Arkansas Agr. Exp. Sta. Bull.* 690.

Wilkes, L. H., P. L. Adkisson, and B. J. Cockran. 1959. Stalk shredder tests for pink bollworm control. *Tex. Agr. Exp. Sta.* PR-2095. 2 pp.

Wolfenbarger, D. A., M. J. Lukefahr, and H. M. Graham. 1973. LD_{50} values of methyl parathion and endrin to tobacco budworms and bollworms collected in the Americas and hypothesis on the spread of resistance in these lepidopterans to these insecticides. *J. Econ. Entomol.* **66**:211–266.

12

FORAGE CROPS INSECT PEST MANAGEMENT

Edward J. Armbrust and George G. Gyrisco

WHERE THE BUFFALO ROAM

Down through countless ages our grasslands and pastures accumulated nitrogen. Lightning in the sky fixed small amounts of nitrogen oxides which fell with the rain. Soil bacteria further enriched the humus with nitrogen.

Native legumes took nitrogen from the air and fixed it in the soil by a remarkable plant mechanism which still defies our full understanding. So—our grasslands and pastures flourished. Then, as the white man's herds displaced the buffalo, our prairies began to decline. Every ton of grass-fed beef took 50 lb of nitrogen from the range. Wind and water erosion removed additional fertility. This is one reason why intelligent management is now essential to improve our grassland and legume vegetation which in some ways supports the health and welfare of every one of us (Phillips Petroleum Company 1958). Insect pests play an important role in the decline of forage crops, and they must be managed properly if forages are to be grown successfully.

In the broad sense all feed consumed by livestock can be classified as forage. In the more restricted, commonly used sense, however, we think of forage as the roughages, mostly hay and pasture. Cereal grains and other concentrates are excluded. The principal forages are largely grasses, and legumes such as alfalfa, clover, and vetches. In this chapter we deal primarily with alfalfa, because it is one of the world's most valuable cultivated forages and it is an ideal food for nearly all classes of livestock. It is higher in feeding value than all other commonly grown hay crops, and it produces about twice as much digestible protein as clover and about four times as much as grass and clover hay mixture or corn silage.

Besides alfalfa's capacity to produce heavy yields of high-quality feed, its ability to improve soil is widely recognized. Alfalfa contributes nitrogen and organic matter to the soil, increases water filtration rates, and improves soil structure to reduce erosion. Anyone who has walked through a field of alfalfa after a rainfall soon realizes that a considerable portion of the water clings to the plants and must later pass down the plant to the soil or be lost by evaporation. The desirable attributes of alfalfa as a forage plant, its soil-improving ability, and its adaptation to a wide diversity of soil and climatic conditions have led to its use as a forage crop that far exceeds the use of any other single legume or grass species. In 1971, over 27 million acres of alfalfa or alfalfa hay mixtures were harvested.

The alfalfa ecosystem is unique among field-crop systems in that it represents a relatively long-lasting, well-established perennial system which exists nationally over a wide variety of climatic, geographical, and edaphic conditions (Fig. 12.1). This creates many subsystems, and the interactions of these with other specific agroecosystems or natural systems are equally varied. For example, in the South and Southwest, alfalfa–cotton systems are common, while in some western states alfalfa–sugar beet systems are more common. In the Midwest there are alfalfa–corn–soybean complexes, and in the East alfalfa–orchard or alfalfa–pasture systems can be found.

A field of alfalfa supports a wide variety of insects (Gyrisco 1958; Osborn 1939; Pimentel and Wheeler 1973). These include insects destructive to alfalfa or other crops, pollinating insects, species that inhabit the alfalfa because of the lush growth but have very little effect on the crop, and many predators and parasites of insect pests of alfalfa or other neighboring crops. Osborn (1939), in *Meadow and Pasture Insects,* states that McAtee found in a meadow an average population of 13,650,710 animal objects per acre of soil in March, even though he neglected the vertebrate population and omitted many of the most populous kinds of insects and many hibernating insects which become a large part of the fauna later in the season. Because of the perennial growth habits of alfalfa and other forage crops, many pests, predators, and parasites overwinter in these crops before dispersing to neighboring crops.

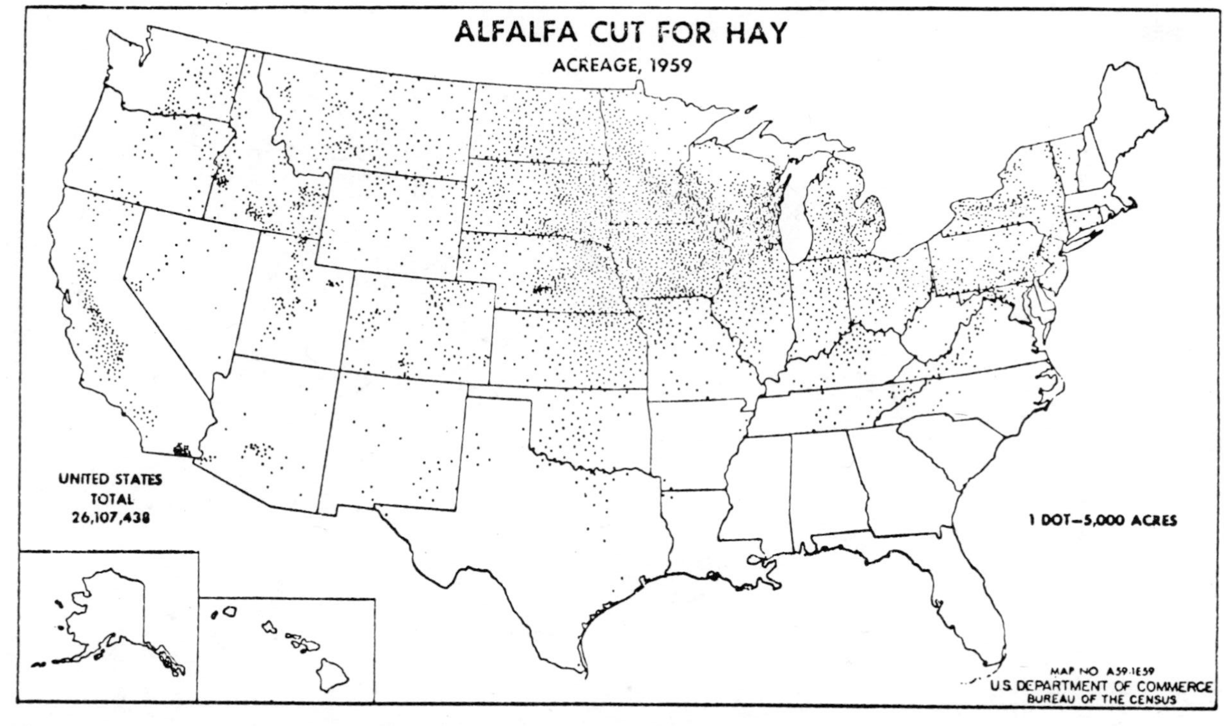

Fig. 12-1 Distribution of alfalfa hay production in the United States.

Several hundreds of species of insects are known to attack forage crops. If we select only species of insects that are generally of economic importance every year on red clover, for example, in the Northeast, the number of important species can be limited to about 20. If they are studied in relation to the plant and its parts (Fig. 12.2), we find that the plant is completely attacked from its root nodules to its seeds. This is but one of the many examples of an insect complex on a forage crop.

The annual cost of insect damage to alfalfa forage and seed production amounts to millions of dollars. Insects also impair forage quality and reduce the persistence and longevity of the crop. Of the destructive

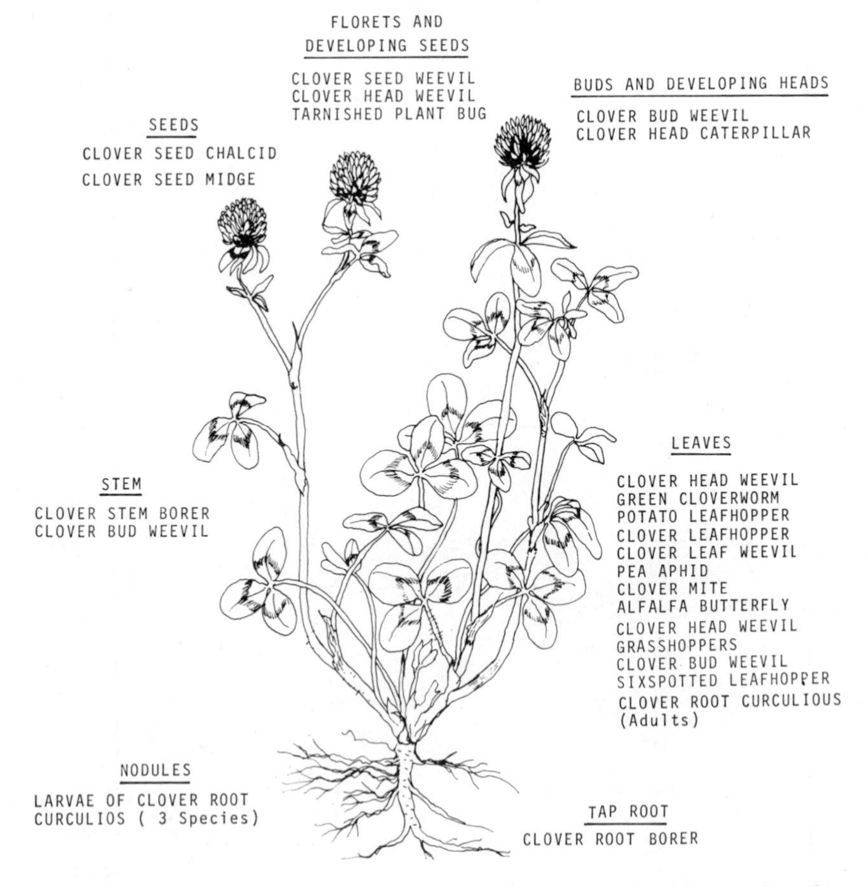

Fig. 12-2 Important insect pests of red clover in relationship to the plant and its parts.

alfalfa insect pests, all are unquestionably harmful to a certain degree, but certain species such as the alfalfa weevil, *Hypera postica* (Gyllenhal); potato leafhopper, *Empoasca fabae* (Harris); and army cutworms, *Euxoa auxiliaris* (Grote); are especially serious pests because of their widespread occurrence, generally high levels of infestation, and the severe damage they can cause when populations are not controlled. Some insect pest problems are local or regional in nature, while others, such as the alfalfa weevil, are essentially national.

Because the alfalfa weevil is the most important single insect pest of alfalfa in the United States, we limit our discussion to an insect pest-management system for the weevil. Other insect pests, parasites, and predators are considered in the total insect pest complex on alfalfa whenever possible. Occasionally, management practices for principal insect pests may cause secondary insect pest species to become real problems. For example, it is felt by many researchers that recent outbreaks of the meadow spittlebug in some eastern and midwestern states may be attributed to intensive spraying of alfalfa with insecticides for alfalfa weevil control. In New York a new leaf miner problem may be developing for the same reason.

In developing any type of insect pest-management system, care must be taken to account for all direct and indirect consequences that might arise out of one or more of the many management practices available to the grower. Many alfalfa insect residents are potential threats to alfalfa production, but do not always attain pest status. Naturally occurring biological control is largely responsible for this. Everything that is done to the crop, such as irrigation, harvesting, and pest control, ideally should be done so as to have a minimum negative impact on natural enemies. Harvesting alfalfa and chemical controls should be of particular concern with regard to their adverse effect on natural enemies. Removal of all the alfalfa plant material except for the stubble represents a drastic change in the insects' environment. The insects must adapt, migrate, or die. As a comparison, for human beings it would be like removing all homes, plants, and food from a county or state, for example, or perhaps an even larger area.

Alfalfa is a crop that can tolerate small amounts of damage. Because of this, it is an ideal crop for use in developing a pest-management system, in that the insect pests do not need to be completely eliminated. This is particularly important when considering the use of parasites as a control measure. A short-range complete kill of the insect pest could indirectly result in loss of a particular parasite, because its food source would no longer be present and the parasite would starve to death.

However, crops sold on the fresh market or processed for human consumption often are legally and aesthetically unacceptable if they contain insect parts or show evidence of insect damage. For example, slight feeding on the surface of an apple can make it unmarketable, but feeding from low numbers of insects on alfalfa may not cause much loss from the standpoint of market value or on-the-farm usage.

I. ALFALFA WEEVIL BIOLOGY

In developing an insect pest-management program, it is very important to fully understand the life history and biology of the organisms to be managed. An outline of the general biology of the alfalfa weevil and its most common parasite will facilitate further discussion of a specific insect pest-management program for alfalfa.

The alfalfa weevil was first discovered in the United States in 1904 (Titus 1907, 1910) near Salt Lake City, Utah, where it was probably introduced from Europe. For nearly 50 years the weevil remained confined to 12 western states. In 1952, however, it was discovered in Maryland (Bissell 1952; Poos and Bissell 1953), and from there it spread rapidly throughout the eastern, southern, and midwestern states. Presently, it is distributed throughout all the contiguous 48 states.

Most researchers feel that these 1904 and 1952 discoveries represent populations of two distinct but morphologically indistinct strains, namely, the eastern and western strains (Blickenstaff 1965). A diagrammatic representation of the biology of the eastern strain is shown in Fig. 12.3, and the life stages are shown in Fig. 12.4. Essentially, the basic biology of the two strains is identical, except for minor differences which are pointed out where important (Armbrust et al. 1970; Blickenstaff 1969; Davis 1967; Koehler and Gyrisco 1963; Pienkowski et al. 1969). The biology and control practices of these two strains are extremely dependent on climatic and geographic factors. Because of the diversity of alfalfa production and management and biological differences of this pest in relation to crop development, a management program cannot be identical nationwide.

The adult is approximately $\frac{1}{4}$ in. long and has a distinct snout. Newly emerged adults are brown and have a definite darker line extending centrally down their back. New adults emerge in the spring, feed for a short time, and then fly from the alfalfa to wooded areas, fencerows, and other protected areas where they enter aestivation or a resting period (Poinar and Gyrisco 1962; Titus 1910). The eastern strain of the weevil returns

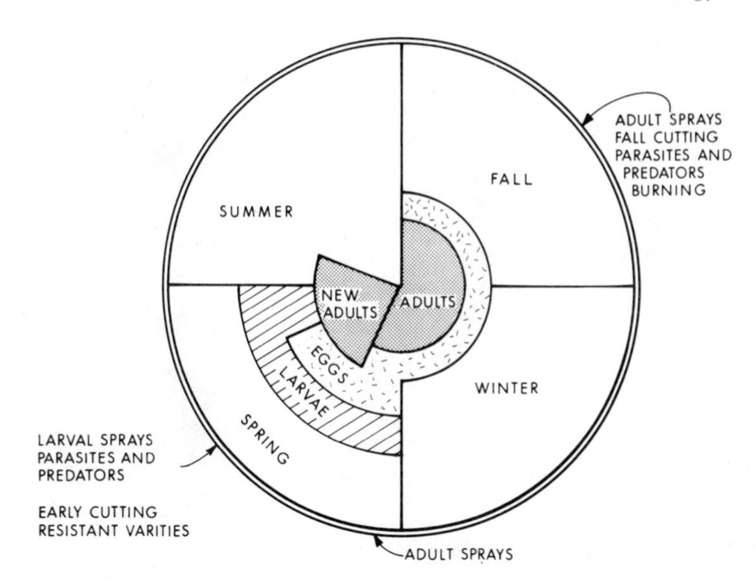

Fig. 12-3 Diagrammatic and seasonal representation of the biology of the eastern alfalfa weevil strain. Control methods are listed with reference to season.

to the alfalfa in late summer or early fall (Prokopy and Gyrisco 1963), while the western weevil usually does not return until spring. Researchers in the East have also noticed spring flight activity which no doubt involves more adults returning or movement within fields or between fields (Poinar and Gyrisco 1962).

The adults that return to the alfalfa in the fall continue their sexual development and, if temperatures permit, oviposit during the fall and throughout the winter in some southern regions (Armbrust et al. 1966; Niemczyk and Flessel 1970). In some areas egg laying is terminated or slowed down, depending on the winter temperatures of different geographic locations. Therefore in the eastern weevil one notices a negative gradient in the amount of fall or winter egg laying from south to north. Generally, these eggs overwinter and hatch in the spring and egg laying resumes. Because of this biological difference in the ratio of fall and winter eggs to spring eggs and crop development in the spring, the alfalfa weevil is much more difficult to control in the southern part of its range than in the more northern or western regions. The alfalfa in the South does not usually attain much growth before fall- and winter-laid eggs hatch and the tiny first-instar larvae begin to feed on the growing tips where they are protected from insecticide sprays and some predators.

Fig. 12-4 Life stages of alfalfa weevil, *Hypera postica* (Gyllenhal), and larval damage to alfalfa. (Left) Top to bottom: cocoons, larva, damage. (Right) Top to bottom: eggs, adults.

In areas where fall or winter egg laying does not exist or is at least minimal, the adult weevil begins its major egg-laying period in the spring about the time the alfalfa plant starts to grow; thus, by the time the eggs hatch, the crop may have attained considerable growth. The presence of 100 larvae per square foot of alfalfa 10 in. tall is certainly not as disastrous as the same number on alfalfa half as tall. These are important points that need to be considered in developing a management program. Also, it is important to remember that the alfalfa weevil is a winter-active insect; it does not go into hibernation or become inactive for the entire winter as do many insects, but rather the adults are active during periods of warming weather throughout the winter. In many areas it is during these warm periods that a large number of eggs are deposited.

The eggs are oval-shaped, approximately $\frac{1}{32}$ in. long, and bright yellow in color when first laid. As incubation progresses, the eggs turn a darker yellow, then brown, and just prior to hatching become black in appearance since the larval head can be seen through the egg shell (Essig and Michelbacher 1933). The eggs are laid in clusters of about eight to nine per cluster within the stems of growing alfalfa, dead stems, or stubble. A female is capable of ovipositing 600 to 800 eggs, but some researchers have reported that individual females have laid over 1500 eggs. This represents a high reproductive potential, thus a low adult density can account for a devastating larval population.

The newly emerged larva is about $\frac{1}{32}$ in. long and yellowish green with a black head. As the larvae grow, they move out of the growing plant tips and their color changes to a light green and then a darker green. When fully grown, they are about $\frac{3}{8}$ in. long and have a wide dorsal white stripe down the back and a faint stripe on either side. The eggs hatch and the larvae appear over an extended period of time, depending on temperature and, most important, on egg-laying patterns during the fall, winter, and spring. When oviposition extends from fall to spring, the spring larval feeding period can likewise be extended over a longer period of time than in those areas where eggs are laid only in the spring.

The influence of spring weather conditions on larval development is also an extremely important factor which determines the destructiveness of the weevil. A cool, prolonged spring results in extended spring egg laying, but larval growth is hindered more than plant growth, so that the crop becomes mature before weevil injury is severe. In warmer seasons or geographic areas, larval populations may develop faster than plant growth, thus causing severe damage to the alfalfa. Again, it becomes apparent that climatic conditions play a very important role in managing this pest.

II. CONTROL TOOLS

With this brief biological background in mind, we can now consider some of the control measures available to reduce alfalfa weevil populations. We present these as separate topics and in relation to weevil biology. As we discuss these control measures, the importance of the preceding biological information becomes apparent. A considerable amount of laboratory and field data dealing with chemical, biological, and cultural methods of weevil control is available (see Selected Readings). In order to manage this pest, these data need to be carefully interpreted with respect to their implications for integrated control, and then utilized in a management program in the field. With forage crops—unlike cash crops such as vegetables, fruits, and some ornamentals—we are dealing with a commodity that nets a small margin of profit per acre. The grower benefits from some of the lower-cost control methods such as crop rotation; variation in time of planting and harvesting; destruction of crop residues, trash, weeds, and volunteer plants; cultivating; use of resistant plant varieties; and biological methods including those involving predators and parasites.

A. Chemicals

The proper use of insecticides has produced tremendous increases in quantity and quality of pasture, silage, and hay, which have more than paid for their application (Fig. 12.5). Some of the problems one must consider before including insecticides in a pest-management system on forage crops are similar to those encountered on other crops, but a listing of these considerations points up the fact that a recommendation to use insecticides of any kind is not a simple matter. One must consider:

1. Cost and availability of insecticides
2. Availability and efficiency of application equipment
3. Formulation of the toxicant
4. Effect on beneficial insects, as well as livestock, man, and other warm-blooded animals
5. Effect of insecticides on the palatability of the forage

Fig. 12-5 Untreated strip of alfalfa between first-crop alfalfa treated with an insecticide. Larvae have begun to skeletonize the leaves and alfalfa looks frosted. This is typical of heavy weevil damage.

6. Long-term effects of the insecticide in rotation with other crops and the possible absorption of the toxicant by food crops or the impairment of growth, flavor, or odor of those food crops
7. Effects of the pesticide on the flavor and odor of, and as possible residue in, the milk of dairy cows and/or the meat of animals slaughtered for human consumption.

We cannot consider only the impact of an insecticide on the pest; we must also consider its impact on predators, parasites, and other insect pest species whenever possible. Insecticides should be used as a tool to manage an alfalfa weevil population from the standpoint of population reduction in order to allow some of the other control tools to be used more effectively. They should be employed carefully for control of a population that has reached damaging levels to avoid harmful residues on the alfalfa hay or in the soil, and to preserve the predators, parasites, and pollinators that inhabit or visit the alfalfa fields. From the human health aspect, considering the end use of alfalfa, it is extremely important that this crop be kept free of impurities, including insecticide residues. Dairy products and meat that come from animals dependent on alfalfa feed are

Fig. 12-6 Alfalfa defoliated by alfalfa weevil larvae. Note the severe damage to the growing plant tip.

basic foods for the population in general and are a particularly important food source for growing youth. Many forms of wildlife utilize the alfalfa ecosystem, and their presence must be considered.

It is common practice for alfalfa growers to use insecticides as larval sprays in the spring on first-crop alfalfa, or as stubble sprays after the first harvest. These treatments control the pest adequately but may have adverse effects on active predators and parasites also present. Besides being detrimental to these natural enemies, the timing of sprays in the spring is very critical, in that within a matter of a few days weevil larvae can practically defoliate an entire crop (Fig. 12.6). In the spring growers are busy planting other crops and cannot be bothered with continually checking their alfalfa fields, nor can they always stop planting operations to spray alfalfa. From the standpoint of management, spring larval sprays leave much to be desired. Insecticide sprays of this type can be considered action-type sprays, because they are applied when pest population levels develop to sufficient size to cause economic damage. The insect pest population is not actually being managed, but rather is being destroyed at a point in time when it is beyond management or at least when it will be almost out of control if nothing is done.

In order to manage alfalfa insect pests effectively with chemical sprays, one needs to consider carefully the biology of each individual pest by itself and also in relation to other pests. When you know that a problem will surely develop, it is not good management practice merely to wait until the problem is out of hand. To reduce or eliminate any problem, whether it is entomological, social, racial, or political, we must first fully understand the background and development of the problem. With insect pest problems, this involves the study of biology.

Specifically, with the alfalfa weevil, knowledge that in some areas the adults lay a considerable number of eggs in the fall leads one to consider fall insecticide sprays aimed at reducing adult populations and, in turn, fall and winter oviposition (Wilson and Armbrust 1970). A program of this type could be termed preventive, in that the object is to prevent the problem from developing to a point where drastic measures are required. This is considered a better management practice. Sprays of this type are less detrimental to predators and parasites. In areas where fall or winter oviposition does not occur, preventive sprays might be used in the early spring after adult movement into the fields and just before spring oviposition (Niemczyk and Flessel 1969). The important point is that, in order to manage the adult weevil effectively with preventive chemical sprays, certain biological data are needed—in this instance, information on the time and magnitude of the adults' return to the alfalfa, and their oviposition habits.

Careless intensive spraying can have drastic adverse effects on the alfalfa ecosystem, thus creating new or intensive insect problems. Certain control tools such as changing cultural practices, introduction of natural enemies of certain pests, and breeding better plant varieties, have resulted in a real reduction in the amount of insecticides applied on alfalfa. Further development and increased use of alternate methods of pest control through a well-established pest-management system will undoubtedly further reduce the amount of insecticides used in forage-crop production. Many of the major alfalfa pests have well-established natural enemies which can become more efficient through proper management.

B. Cultural Practices

The alfalfa weevil was a destructive pest long before the development of modern insecticides, and during those early days growers relied greatly on cultural practices and botanical insecticides (Fig. 12.7) for control.

Fig. 12-7 Early spraying (Reeves et al. 1920. Courtesy of United States Department of Agriculture.)

Many of the early methods were quite drastic and of little practical value. One of the first methods employed in Utah was a wire or brush drag which was used to crush or smother the weevils. Reeves et al. (1916) suggest that weevils can be killed in late winter or early spring by irrigating the fields with very muddy floodwater, thus burying the insects under a deposit of fine mud. Some of these early cultural practices were sound in principle, but impractical in use. However, the value of a vigorous stand, early plant maturity, or early cutting and clean culture is as great today as it was then.

Today some growers cut the alfalfa late in the fall, thereby making the field unattractive to migrating adults, and those adults that do establish themselves in the field find few satisfactory oviposition sites. Modern-day technology has developed specialized pieces of equipment, which burn propane gas or other fuels, which are used to flame the alfalfa in the fall or very early spring, thus burning up the plant material containing alfalfa weevil eggs (Fig. 12.8). In some parts of the country, it is practical to time the cutting of the first crop so that a majority of weevil larvae are removed or killed in the process. Since this practice does not destroy adults, some precautions may be needed on the second-crop

Fig. 12-8 Experimental propane gas flamer used to burn alfalfa plant material during late fall through winter to destroy alfalfa weevil eggs.

alfalfa because oviposition may continue as the plant resumes growth. Again we emphasize that the use of these cultural and mechanical approaches to insect control requires detailed knowledge of the insect pest's biology for any particular geographic area.

C. Biological Control

The classic concept of biological control is the importation of parasites to control introduced pests. Parasites of the alfalfa weevil were first introduced into the United States from Europe during 1911 (Chamberlin 1924, 1925). By 1919 parasites were well established in certain areas in the western United States. *Bathyplectes curculionis* (Thomson), an internal wasp parasite of the larvae (Fig. 12.9), is perhaps the most widely distributed introduced weevil parasite. In many areas it has become an important biological factor in alfalfa weevil control, while in other areas it has not been as promising. An understanding of parasite biology, the mortality factors influencing fluctuations in pest and parasite populations, and the impact of the parasite on the pest is essential if successful

Fig. 12-9 Adult and cocoon of *Bathyplectes curculionis,* an internal larval parasite of the alfalfa weevil.

population-management systems are to be developed for the alfalfa ecosystem. We must know as much about the parasite as we do about the pest.

Other parasites have been imported into this country, and several native parasites have been recovered from the alfalfa weevil. We suggest Brunson and Coles (1968) for additional information.

D. Plant Resistance

Notable advances have been made in developing alfalfa varieties resistant to insect pests. For the spotted alfalfa aphid, it has paid off 500-fold. Plant resistance provides a built-in insurance against insect attack at a very low cost to the grower and to the public, and at the same time reduces problems that may be associated with the use of insecticides, including damage to beneficial insects and the accumulation of hazardous residues in soil or on plant materials.

Developing alfalfa varieties resistant to the alfalfa weevil has proved to be one of the most difficult problems undertaken in alfalfa breeding. First, none of the hundreds of alfalfa varieties and introductions evaluated in U.S. Department of Agriculture (USDA) or state programs was found to have sufficient resistance to give adequate control. Second, handling plants and insects on the large scale required for this type of research was laborious and costly. The USDA laboratory at Beltsville, Md., alone evaluated about 2 million plants in a search for weevil resistance. 'Team' alfalfa, a multiple-pest-resistant variety with moderate resistance to the alfalfa weevil, was released in 1969 from the Beltsville program, in cooperation with the agricultural experiment stations of Maryland, North Carolina, and Virginia (Barnes et al. 1970). Agronomically, this variety is not suited for all geographic areas; its resistance to weevil attack has not always been sufficient, nor is it resistant to bacterial wilt.

III. DEVELOPING A PEST-MANAGEMENT SYSTEM FOR ALFALFA

We have briefly discussed some of the control tools available to alfalfa growers. There are indeed others, and imagination and research will undoubtedly develop additional ones and improve present practices. We cannot expect to control the alfalfa weevil or other alfalfa insect pests with a single-factor approach. Parasites will not eliminate the weevil, nor will the use of resistant or tolerant alfalfa varieties alone do the job. The presently recommended insecticides are either too residual with reference to pesticide tolerances, or not persistent enough for good weevil control. Many insecticides are too toxic for humans, predators, or parasites and, in many cases, little is known about their effects on the environment. In order to manage alfalfa insect pests properly, growers must use an integration of many cultural, chemical, and biological factors. Figure 12.10 shows a diagrammatic representation of how the use of insecticides and parasites can be effectively combined for an integrated alfalfa weevil management system. The insecticide is being used to slow down the larval population at a time when parasites are not active in the field. This allows the parasites to act as an effective control tool at a later date when the insecticide is no longer effective. By slowing down the initial larval population, the parasite can keep the population below the economic threshold. This type of system can also employ fall or very early spring sprays for adult control, thus reducing the initial larval population through reduced egg laying. Parasites may not be effective

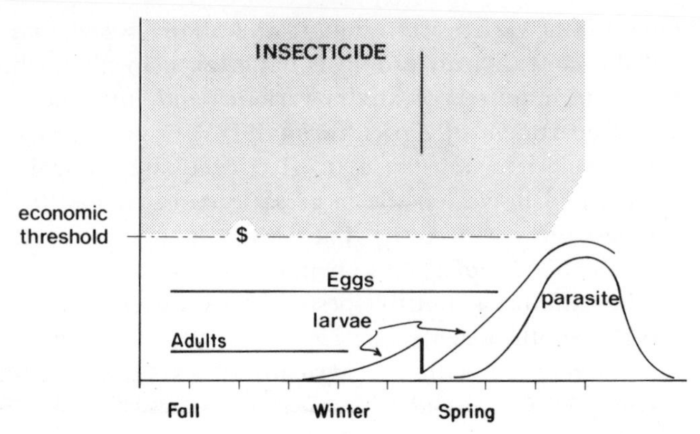

Fig. 12-10 Diagrammatic representation of insecticides and parasites for integrated alfalfa weevil control.

each year, but the potential exists and this must be considered and used in developing the management system.

In many respects alfalfa growers have been practicing pest management for years, but at a very rudimentary level. Researchers, extension personnel, and growers have not considered this pest management, and rightly so, because often the various control approaches have not been viewed with respect to each other. All the variables of the alfalfa ecosystem need to be combined with the available control approaches to create the best possible management strategy for maximum alfalfa production. This is a complex problem. Many entomologists and other specialists throughout the country have combined their research efforts on a national basis to solve this problem. The objectives of their program[1] are:

1. Synthesis and collection of all published and unpublished scientific information on the alfalfa agroecosystem
2. Collection of additional pertinent biological and ecological information required to formulate a sound insect pest-management program
3. Development of a general mathematical model for the alfalfa agroecosystem
4. Field verification of the model
5. Field implementation of the model at the grower level
6. Provision of the basis for evaluating and optimizing crop production and insect pest-management strategy for society.

[1] National Science Foundation Grant, Alfalfa Subproject, "The Principles, Strategies and Tactics of Pest Population Regulation and Control in Major Crop Ecosystems."

The analysis and utilization of published and unpublished data on the biology, ecology, and control of alfalfa insect pests (especially the alfalfa weevil) present immediate possibilities for developing models that permit analysis of integrated control systems. These data need to be integrated into a system that uses each factor in a complementary fashion. Researchers are using modeling as a tool to provide a comprehensive view of the complex alfalfa ecosystem. They have attempted to isolate and formulate into a mathematical model those components of the system that are the most important as they affect the system's behavior. These efforts have developed models to simulate plant growth dynamics, alfalfa weevil population dynamics, parasite and predator population dynamics, and economic decision-making processes. These individual models are being tied together into a comprehensive generalized strategy model (Fig. 12.11). Like any research project of practical importance, these research efforts need to be verified in the field and tested at the grower level.

The alfalfa plant is perhaps the most important single factor to be considered for, after all, it is the basic reason for developing an insect pest-management program. Miles et al. (1973) and Bula et al. (1974) have developed a simulation model of alfalfa growth response to environmental conditions (Fig. 12.12). This type of plant model can be used with alfalfa insect pest, parasite, and predator models to implement an alfalfa management program actively.

Purdue University researchers and extension personnel, cooperating with other states and the USDA as part of a pilot program for alfalfa pest management, are working with grower-cooperators in northern and southern Indiana to acquire an intensive sample of the major alfalfa-growing regions of that state. Their efforts have been to quantify insect development and plant growth in order to issue management advisories on a weekly basis. This service is somewhat different from normal extension activity in that (1) sampling was more concentrated and frequent, and (2) recommendations were tailored to the situation and made available immediately on teletypewriters linked on-line to a computer. Purdue's approach, like that of other research programs, has combined the efforts of research and extension personnel jointly and simultaneously with continuous feedback in both directions to plan, develop, construct, and implement comprehensive management concepts. With this type of team effort, reliable and acceptable results can be achieved. Figure 12.13 summarizes the components required for the design and implementation of Purdue's approach to a pest-management system for alfalfa.

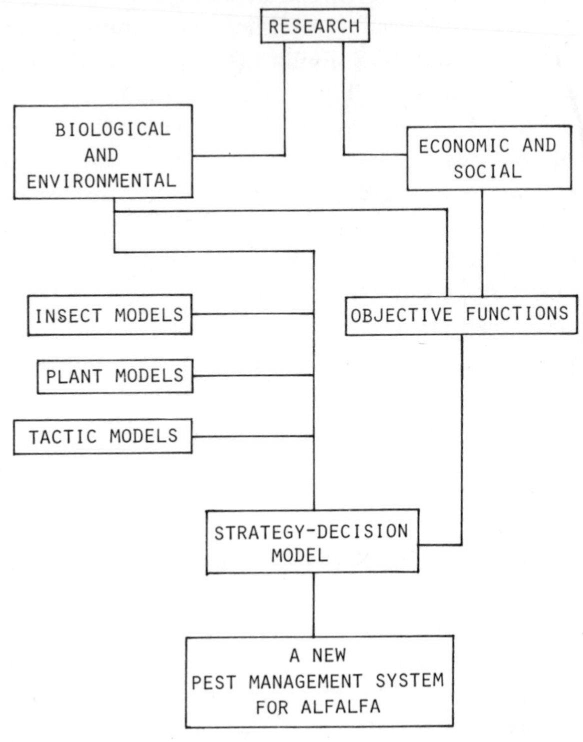

Fig. 12-11 Research areas that contribute to models for an alfalfa insect pest-management system.

Using modeling and computer tactics to describe the dynamics of the alfalfa ecosystem and to integrate control approaches has stimulated basic and applied research dealing with the various components. Many times researchers have found that the necessary data were lacking or insufficient. For example, most life stages of the alfalfa weevil are extremely difficult to sample in terms of absolute numbers. In developing a pest-management system, we need to be able to efficiently monitor and survey field populations of the pest. For extension personnel or the grower, the insect net is a practical survey tool, in that its use is not time-consuming and the insects can generally be counted in the field. The insect net has the disadvantage that it measures only relative numbers and not absolute numbers. Counts taken with an insect net can be affected by weather, alfalfa conditions, and by the person using the net.

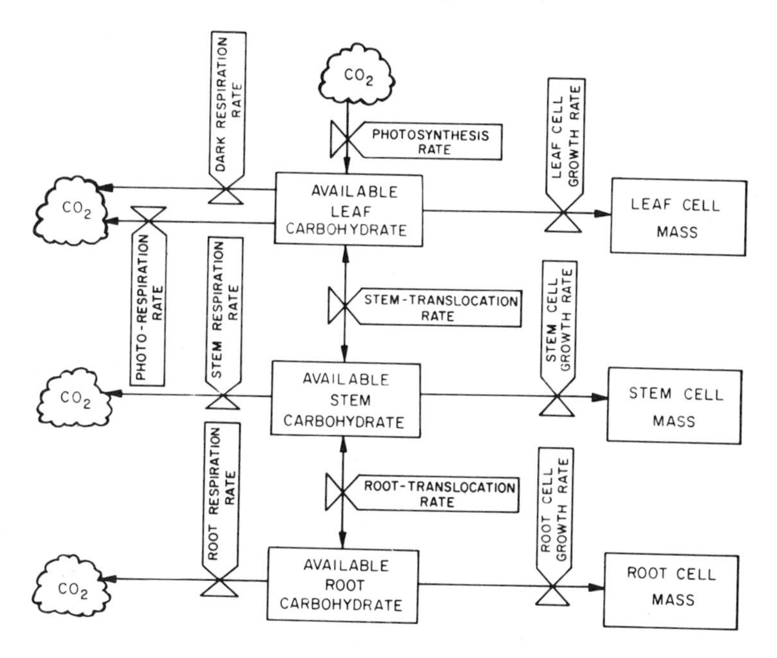

Fig. 12-12 Diagrammatic representation of alfalfa plant growth model. (Purdue University.)

Researchers are presently investigating the possibilities of quantifying sweep net counts to absolute values. In other words, the effectiveness of the sweep net is being tested so that counts may be related to absolute densities under any set of variables.

If we assume that some highly effective control agent that wipes out the weevil with no detrimental effect to the ecosystem will not be found, we are faced with the realization that the alfalfa weevil or any other alfalfa insect pest must be controlled and not destroyed. In other words, we need to learn how to live with the alfalfa weevil. This is especially true when we consider some of the biological methods of control. We must be able to live with a certain number of alfalfa weevils in order to perpetuate the existence of biological control agents. A decade ago research emphasis was placed on how many weevils we could kill. Now we must ask ourselves how many we can live with. A consideration of this question leads one into the area of levels of economic injury. We must address ourselves to the question, How much damage can we tolerate

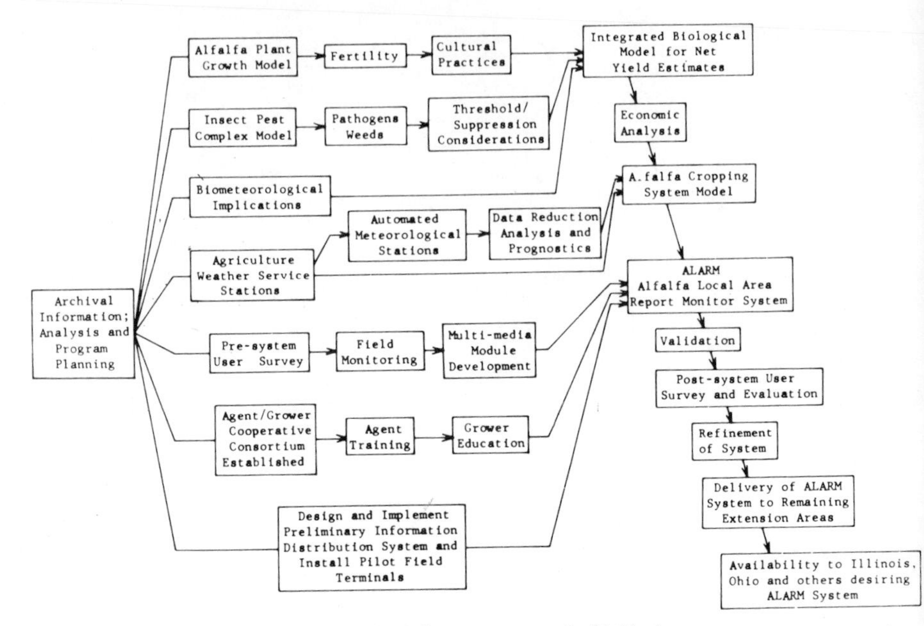

Fig. 12-13 Components required for one type of alfalfa insect pest-management system (Purdue University.)

before an individual control approach or combination of approaches is used? In considering this question we must remember that, with alfalfa, yield per acre based on weight cannot be the only consideration, since the most valuable portion of the alfalfa plant is the leaf material. The weight of the leaf material is far less than that of the stem, but unfortunately both weevil larvae and adults concentrate their feeding on the leaves. We must therefore base yield considerations on quality as well as quantity. We must also consider geographic location, since weevil development in relationship to crop growth varies considerably and is dependent on geographic location.

Each alfalfa region of the United States has its unique insect problems which require special pest-management innovations. However, we feel that, in the aggregate, through team and cooperative efforts, these problems can be integrated into a nationwide alfalfa pest-management scheme which can bring about efficient, ecologically tenable alfalfa insect pest control. Each of the control tools needs to be used in the light of known biological and ecological data for the best possible management of insect pests with maximum benefit to the grower, as well as to society in general.

REFERENCES

Armbrust, E. J., R. J. Prokopy, W. R. Cothran, and G. G. Gyrisco. 1966. Fall and spring oviposition of the alfalfa weevil, *Hypera postica,* and the proper timing of insecticide applications. *J. Econ. Entomol.* **59**:384–387.

Armbrust, E. J., C. E. White, and S. J. Roberts. 1970. Mating preference of eastern and western United States strains of the alfalfa weevil. *J. Econ. Entomol.* **63**:674–675.

Barnes, D. K., C. H. Hanson, R. H. Ratcliffe, T. H. Busbice, J. A. Schillinger, G. R. Buss, W. V. Campbell, R. W. Hemlken, and C. C. Blickenstaff. 1970. The development and performance of Team alfalfa—A multiple pest resistant alfalfa with moderate resistance to the alfalfa weevil. *Crop. Res. ARS* 34-15. 41 pp.

Bissell, T. L. 1952. *U.S. Bur. Entomol. Plant Quar. Coop. Econ. Insect Rep.* **2**:4.

Blickenstaff, C. C. 1965. Partial insterility of eastern and western U.S. strains of the alfalfa weevil. *Ann. Entomol. Soc. Amer.* **58**:523–526.

Blickenstaff, C. C. 1969. Mating competition between eastern and western strains of the alfalfa weevil, *Hypera postica. Ann. Entomol. Soc. Amer.* **62**: 956–958.

Bruson, M. H., and L. W. Coles. 1968. The introduction, release, and recovery of the alfalfa weevil in eastern United States. *U.S. Dep. Agr. Prod. Res. Rep.* 101. 12 pp.

Bula, R J., G. E. Miles, D. A. Holt, M. M. Schreiber, and R. M. Peart. 1974. Physiology and computer simulation of alfalfa growth. *Purdue Exp. Sta. Res. Bull.* R.B. 907.

Chamberlin, T. R. 1924. Introduction of parasites of the alfalfa weevil into the United States. *USDA Dep. Circ.* 301. 9 pp.

Chamberlin, T. R. 1925. A new parasite of the alfalfa weevil in Europe. *J. Econ. Entomol.* **18**:597–602.

Davis, D. W. 1967. How different are the eastern and western forms of the alfalfa weevil? *Utah Acad. Sci.* **44**:252–257.

Essig, E. O., and A. E. Michelbacher. 1933. The alfalfa weevil. *Calif. Agr. Exp. Sta. Bull.* 567. 99 pp.

Gyrisco, G. G. 1958. Forage insects and their control. *Annu. Rev. Entomol.* **3**: 421–448.

Koehler, C. S., and G. G. Gyrisco. 1963. Studies on the feeding behavior of alfalfa weevil adults from eastern and western United States. *J. Econ. Entomol.* **56**:489–492.

Miles, G. E., R. J. Bula, D. A. Holt, M. M. Schreiber, and R. M. Peart. 1973. Simulation of alfalfa growth. *Amer. Soc. Agr. Eng. (ASAE) Paper* 73-4547.

Niemczyk, H. D., and J. K. Flessel. 1969. Development and testing of a preventative program for control of the alfalfa weevil in Ohio. *J. Econ. Entomol.* **62**:1197–1202.

Niemczyk, H. D., and J. K. Flessel. 1970. Population dynamics of alfalfa weevil eggs in Ohio. *J. Econ. Entomol.* **63**:242–247.

Osborn, H. 1939. Meadow and pasture insects. Educators Press. Columbus, Ohio. 288 pp.

Phillips Petroleum Company. 1958. Pasture and range plants. Native legumes. 25 pp.

Pienkowski, R. L., F. Hsieh, and G. L. LeCato, III. 1969. Sexual dimorphism and morphometric differences in the eastern, western, and Egyptian alfalfa weevils. *Ann. Entomol. Soc. Amer.* **62**:1268–1269.

Pimentel, D., and A. G. Wheeler, Jr. 1973. Species and diversity of arthropods in the alfalfa community. *Environ. Entomol.* **2**:659–668.

Poinar, G. O., Jr., and G. G. Gyrisco. 1962. Flight habits of the alfalfa weevil in New York. *J. Econ. Entomol.* **56**:241.

Poos, F. W., and T. L. Bissell. 1953. The alfalfa weevil in Maryland. *J. Econ. Entomol.* **46**:178–179.

Prokopy, R. J., and G. G. Gyrisco. 1963. A fall flight of the alfalfa weevil in New York. *J. Econ. Entomol.* **56**:241.

Reeves, G. I., P. B. Miles, T. R. Chamberlin, S. J. Snow, and L. J. Bower. 1916. The alfalfa weevil and methods of controlling it. *U.S. Dept. Agr. Farmers Bull.* 741. 16 pp.

Reeves, G. I., T. R. Chamberlin, and K. M. Pack. 1920. Spraying for the alfalfa weevil. *U.S. Dept. Agr. Farmers Bull.* 1185. 20 pp.

Titus, E. G. 1907. A new pest on the alfalfa. *Deseret Farmer* **3**:7.

Titius, E. G. 1910. The alfalfa leaf weevil. *Utah Agr. Exp. Sta. Bull.* 110. 72 pp.

Wilson, M. C., and E. J. Armbrust. 1970. Approach to integrated control of the alfalfa weevil. *J. Econ. Entomol.* **63**:554–557.

SELECTED READINGS

The research papers dealing with the alfalfa weevil alone number over 700. Most of these have been compiled into a published bibliography with two supplements as listed below.

Cothran, W. R. 1966. A bibliography of the alfalfa weevil, *Hypera postica* (Gyllenhal). *Bull. Entomol. Soc. Amer.* **12**:151–160.

Cothran, W. R. 1968. A bibliography of the alfalfa weevil, *Hypera postica* (Gyllenhal), Suppl. I. *Bull. Entomol. Soc. Amer.* **14**:285–288.

Cothran, W. R. 1972. A bibliography of the alfalfa weevil, *Hypera postica* (Gyllenhal), and the Egyptian alfalfa weevil, *Hypera brunniepennis* (Boheman), Suppl. II. *Bull. Entomol. Soc. Amer.* **18**:102–108.

We have listed below several other sources for general information on forage crop insect pests and crop production. As a further source, the annual proceedings of the National Alfalfa Improvement Conference and the Certified Alfalfa Seed Council are suggested as excellent information sources.

Gyrisco, G. G. 1958. Forage insects and their control. *Ann. Rev. Entomol.* **3**:421–448.

Hanson, C. 1972. Alfalfa science and technology. American Society of Agronomy. 812 pp.

Hughes, H. D., M. E. Heath, D. S. Metcalf, eds. 1966. Forages. Iowa State College Press, Cedar Falls. 755 pp.

13

TREE FRUIT PEST MANAGEMENT

B. A. Croft

INTRODUCTION

Tree fruits referred to in this chapter include deciduous types such as apples, peaches, pears, plums, cherries, and apricots. Compared to cotton, discussed in Chapter 11, deciduous tree fruit orchards are more permanent ecosystems and provide greater habitat continuity for a variety of phytophagous and entomophagous arthropods. Apples, for example, have more than 500 insects feeding on them throughout the world (Slingerland and Crosby 1930). In a survey of arthropods occurring in Wisconsin apple orchards, more than 760 species were reported by Oatman et al. (1964). Although many were incidental migrants or natural enemy species, over 100 were phytophagous on apple trees or fruits. Of these, only 43 were of economic importance, and less than 10 were serious pests.

If undisturbed by pesticides or management practices that diminish diversity (e.g., intense groundcover control), deciduous tree fruit ecosystems remain relatively stable, and many potential arthropod pests are maintained below tolerable levels by complexes of diseases, predators, and parasitoids. However, there are plant diseases like apple scab, *Venturia inaequalis* (Cke), introduced arthropods like the codling moth, *Laspeyresia pomonella* (L.), or native insects similar to the plum curculio,

Conotrachelus nenuphar (Herbst), which are without effective natural enemies, and key pests of this type are capable of destroying a large portion of a fruit crop. Their importance was dramatically demonstrated by Glass and Lienk (1971) in a survey of the insect and mite damage that occurred in a New York apple orchard after insecticides and miticides (not fungicides) were discontinued for 10 years. They observed no appreciable control of the codling moth; plum curculio; red-banded leaf roller, *Argyrotaenia velutinana* (Walker); and apple maggot, *Rhagoletis pomonella* (Walsh); and for the next 8 years an entire apple crop was rendered consistently unmarketable by these four species.

Because deciduous fruits are high-value crops and have considerable consumer appeal, they must be essentially free of pest damage. The low tolerance to and high potential for damage from certain species requires that preventive pesticide applications be used in most deciduous fruit-growing areas of the world. When pesticides are intensively applied, problems of arthropod resistance (Hough 1963; Fisher 1960), induced secondary-pest outbreaks (Huffaker et al. 1969), and greater chemical use patterns often develop. To minimize these side effects, research during the past decade has focused on replacement of pesticides with alternative methods of control where possible. For certain pests considerable progress has been achieved, but for many others pesticides are still the only effective and economical control measures available.

In this chapter a case history of a deciduous fruit pest complex and management system is presented. Annual key pests, periodic key pests, induced secondary-pest complexes, natural control features, and factors affecting the population dynamics of each group are discussed. Successes and problems in developing and implementing nonchemical strategies or in integrating pesticides and biological control agents are reviewed and related to similar pest-management programs developing in other major deciduous fruit-growing regions.

I. AN APPLE ORCHARD ECOSYSTEM

The principal disease and arthropod pest elements of apple orchards of midwestern and eastern North America (hereafter referred to as "this region") are presented as examples of pests in a deciduous fruit ecosystem. In describing the features of this region, it is stressed that ecologically equivalent species and relationships often occur in different fruit-growing areas and on other deciduous crops, even though species composition may be relatively specific to each. Because of these relationships, the methods and philosophy of management applied to one system often are applicable to others. More detailed descriptions of worldwide deciduous fruit ecosystems and pest-management programs can be reviewed in Madsen and Morgan (1970), Kirby (1973), Wildbolz and Meier (1973), and Hoyt and Burts (1974).

In the apple orchards of this region, there are principally two annual key diseases, apple scab and powdery mildew; and four annual key insect pests, codling moth, red-banded leaf roller, plum curculio, and apple maggot (Table 13.1). All six directly damage apple fruits, and it is neces-

Table 13.1 Principal Disease and Arthropod Pests Occurring in Apple Orchards of Midwestern and Eastern North America

Annual key diseases

- *Venturia inaequalis*, apple scab
- *Podosphaera leucotricha*, powdery mildew

Annual key insects

- *Laspeyresia pomonella*, codling moth
- *Argyrotaenia velutinana*, red-banded leaf roller
- *Conotrachelus nenuphar*, plum curculio
- *Rhagoletis pomonella*, apple maggot

Potential key insects

- Large complex of species; see state or Canadian provincial fruit bulletins for individual area species complexes

Secondary induced arthropods

- Scales
  - *Quadrapidiotus perniciosus*, San Jose scale
  - *Lepidosaphes ulmi*, oystershell scale
- Aphids
  - *Aphis pomi*, apple aphid
  - *Anuraphis roseus*, rosy apple aphid
  - *Eriosoma lanigerum*, woolly apple aphid
- Leafhoppers
  - *Typhlocyba pomaria*, white apple leafhopper
  - *Empoasca fabae*, potato leafhopper
- Mites
  - *Panonychus ulmi*, European red mite
  - *Tetranychus urticae*, two-spotted spider mite
  - *Aculus schlechtendali*, apple rust mite

sary to ensure that the combined infection or infestation level of all pests does not exceed 1% of the fruit at harvest (economic injury level). The detailed biology of each of these species can be reviewed in publications by Anderson (1956), apple scab and powdery mildew; Putman (1963) and Chapman and Lienk (1971), codling moth; Oatman and Jenkins (1962) and Chapman and Lienk (1971), red-banded leaf roller; Armstrong (1958), plum curculio; and Dean and Chapman (1973), apple maggot.

II. ANNUAL KEY DISEASE PESTS

A. Apple Scab

This pest, *Venturia inaequalis* (Cke), (Fig. 13.1a) is a severe apple pest in the wet-humid fruit-growing areas of the world. Disease symptoms occur primarily as dark spots on leaves and fruits. Infected fruits show uneven growth and often are deformed. Disease inoculum is transmitted by ascospores released from old apple leaves when conditions become wet in the spring and infections rapidly develop, provided moisture and temperature conditions are favorable. Secondary infections from conidia occur repeatedly after bloom to the end of the season. During most seasons apple scab is a severe problem and, unless fungicides are applied, complete destruction of leaves and damage to almost 100% of the fruit occurs annually in the study region.

B. Powdery Mildew

Podosphaera leucotricha (E. & E.) Salm. (Fig. 13.1*b*) is favored by drier climatic conditions as compared to apple scab, and generally is more prone to occur on susceptible varieties (e.g., 'Jonathan'). This mildew fungus is disseminated as conidia which germinate when temperature and humidity are favorable and produce fungal plants which infect leaves, flowers, shoots, and fruits. Leaf lesions first appear as whitish, feltlike patches of mycelia. In time, destruction of tree foliage can reduce fruit size and affect tree vigor, while infections on fruit cause russeting. In this region powdery mildew is less important than apple scab, but fungicides are usually applied on an annual basis for control.

III. ANNUAL KEY INSECT PESTS

A. Codling Moth

This tortricid moth, *Laspeyresia pomonella* (L.) (Fig. 13.1c), is a key pest in most apple-growing areas of the world, with the exception of Japan

Fig. 13-1*A–F* Damage from two annual key diseases and the adult stage of four annual key insect pests. (*A*) apple scab. (*B*) Powdery mildew. (*C*) codling moth. (*D*) Red-banded leaf roller. (*E*) Plum curculio. (*F*) Apple maggot.

and parts of Asia. It attacks a variety of cultivated hosts including apple, pear, apricot, and a greater number of wild rosaceous fruits. The principal injury is caused by larvae which tunnel into the fruit to feed on the seeds. If left unchecked in this region and in most fruit-growing areas, more than 25% of the fruit will be "wormy" at harvest. In many areas large reservoirs of moths are produced outside commercial orchards, and invasion by migrating adults is always a problem. In most of this region, the codling moth has two or more generations, but is single-brooded in extreme northern areas (e.g., Nova Scotia).

B. Red-Banded Leaf Roller

This tortricid moth, *Argyrotaenia velutinana* (Walker) (Fig. 13.1*d*), was of minor importance in fruit orchards of midwestern and eastern North America prior to the introduction of DDT. It has since become a major pest. Its change in status has been attributed to chemical destruction of natural enemies, or to changes in its general response to apples and their culture. For this reason there is some question whether it is a key or a secondary pest. Larvae feed on apple foliage and on the surface of apple fruits, and may damage up to 50% of them seasonally. The red-banded leaf roller has two to four generations per season and develops on a wider variety of wild plants than does the codling moth. Outbreaks of this pest in apples are sporadic, and may be related to the dynamics of populations associated with wild hosts in close proximity to orchards.

C. Plum Curculio

This weevil, *Conotrachelus nenuphar* (Herbst) (Fig. 13.1*e*), is principally a pest of fruits in the midwestern, northeastern, and southeastern United States, and larvae can infest apples to an extent comparable to the codling moth and red-banded leaf roller. In spring adult females injure apples by making crescent-shaped egg punctures on the surface of fruit. During the late summer newly emerged adults feed on apple fruit, making numerous holes in the skin. The plum curculio mostly overwinters outside orchards in protected woodlots, debris, rock piles, and bushy fencerows. Only one generation is produced each year in most areas of the study region.

D. Apple Maggot

The distribution of the apple maggot, *Rhagoletis pomonella* (Walsh) (Fig. 13.1*f*), is mostly confined to the study region where the orchardist considers it the most feared apple pest. This arises from difficulties in recognizing maggot infestations in the early stages of development. In Glass and Lienk's (1971) survey of pest levels after insecticide treatments were discontinued, this fly was the most serious pest after the first year, and annually infested an average of 87% of the fruit for the next 8 years. Similar damage levels are not uncommon for many orchards in the study region. This species often disperses from unkept apples or wild hosts in the vicinity of an orchard; it is considered a late-season pest, as the adult emerges when the fruit is relatively large. Only one generation is produced per season.

IV. POTENTIAL KEY PESTS

On a more sporadic and less destructive scale, there are a variety of additional direct fruit pests not classified in the previous category. They include such lepidopterous species as the eye-spotted bud moth, *Spilonota ocellana* (D. & S.); fruit tree leaf roller, *Archips argyrospilus* Wlk.; oriental fruit moth, *Grapholitha molesta* (Busck): lesser appleworm, *Grapholitha prunivora* (Walsh) (Fig. 13.2*a*); green fruitworm, *Lithophane antennata* (Wlk.) (Fig. 13.2*b*); oblique-banded leaf roller, *Choristoneura rosaceana* (Harr.); and a large number of nonlepidopterous species (see individual state or Canadian provincial extension bulletins to determine specific complexes). Although many of these species can become serious pests, natural enemies or insecticides applied for control of key insect pests render them only occasional problems. They can dramatically influence pest-management decisions from time to time, but hereafter they are referred to only in specific examples. The biology of these species and others of this type can be reviewed in Metcalf et al. (1962), or in state and Canadian provincial fruit-control bulletins. The remaining discussion mostly deals with the two key diseases, the four key insect pests previously reported, and the four secondary pest complexes discussed hereafter.

V. SECONDARY INDUCED PESTS

These arthropods are complexes of species and do not feed directly on apple fruits but on leaves or woody portions of apple trees. In most cases their economic injury level is considerably higher than that of direct fruit pests, and this feature increases possibilities for their natural control by diseases, predators, and parasitoids. To varying degrees they are induced to pest status by insecticide treatments applied for key insect control. Their biologies are treated in Metcalf et al. (1962) and Oatman (1960).

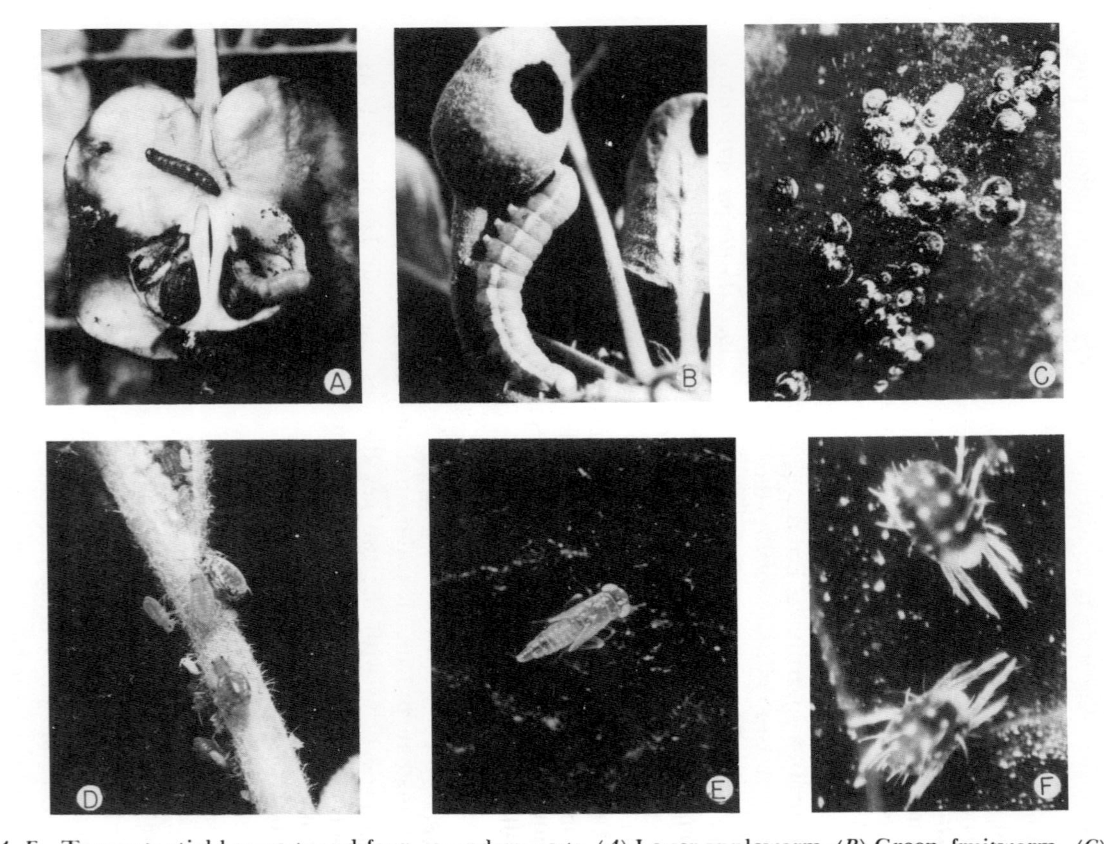

Fig. 13-2*A–F* Two potential key pests and four secondary pests. (*A*) Lesser appleworm. (*B*) Green fruitworm. (*C*) San Jose scale. (*D*) Rosy apple aphid. (*E*) White apple leafhopper. (*F*) European red mite.

A. Scales, Aphids, and Leafhoppers

These insects have comparable life histories and cause similar damage to leaf and woody portions of apple trees. The most abundant scales, the San Jose, *Quadrapidiotus perniciosus* (Comst.) (Fig. 13.2*c*), and the oystershell, *Lepidosaphes ulmi* (L.), suck plant fluids from apple trees. At high densities they almost entirely cover bark areas and occur on the surfaces of apples. The most destructive aphid pests, the apple aphid, *Aphis pomi* DeG.; the rosy apple aphid, *Anuraphis roseus* Baker (Fig. 13.2*d*); and the woolly apple aphid, *Eriosoma lanigerum* (Hausmann); attack rapidly growing shoots and can deform fruits and root systems; certain species may transmit apple diseases (Plurad et al. 1965). Leafhoppers of the genera *Typhlocyba, Empoasca,* and *Erythroneura* cause foliage to develop whitish stippled areas where dense populations have fed, and the resulting injury lowers the normal photosynthetic production of the plant. The most important species are the white apple leafhopper, *Typhlocyba pomaria* McA. (Fig. 13.2*e*), and the potato leafhopper, *Empoasca fabae* Harris. Both aphids and leafhoppers produce honeydew secretions which are substrates for development of sooty molds. These growths may occur on apple fruits and render them unmarketable. If dense populations of any of these three pest groups feed on apple trees for protracted periods of time, fruit, foliage, and general plant growth may be reduced. In subsequent years tree vigor may decline, and heavily infested trees may eventually die.

In unsprayed orchards effective complexes of predators and parasitoids provide appreciable biological control of scales, aphids, and leafhoppers. The most effective natural enemy of the San Jose scale is the aphelinid parasite *Prospaltella perniciosi* Tower. A mite predator, *Hemisarcoptes malus* (Shimer), and the chalcid parasite *Aphytis mytilaspidis* (LeBaron) are major factors regulating the density of the oystershell scale in Quebec orchards (Smarasinghe and LeRoux 1966).

Parasitism of the apple aphid and rosy apple aphid is relatively low, but predatory syrphids, lacewing flies, anthacorid and mirid bugs, cecidomyids, and coccinellids provide substantial control. The woolly apple aphid has fewer effective predators, but the parasitoid *Aphelinus mali* (Hald) is relatively common throughout this region and provides appreciable control in apple orchards on a worldwide basis. Probably to a lesser extent than scales and leafhoppers, apple aphids are under a certain degree of biological control in orchards where no insecticides or selective insecticides are used. Although not entirely induced to pest status by insecticide treatments, there is ample evidence that the severity of aphid problems is closely correlated with natural enemy destruction by pesticides.

Several hemipteran and arachnid predators attack leafhoppers, and the white apple leafhopper is parasitized in the late nymphal and adult stages by the dryinid *Aphelopus typhlocybae* Muesebeck and the mymarid egg parasite *Anagrus armatus* Ashmead (Armstrong 1935). Natural enemy complexes provide adequate biological control of leafhoppers in unsprayed apple blocks but, in treated orchards, parasite emergence coincides with the period of greatest insecticide application for control of several key insect pests (e.g., red-banded leaf roller and plum curculio). When sprays are applied, these natural enemies are greatly reduced and leafhoppers, which are resistant to many chemicals, subsequently attain outbreak proportions.

B. Phytophagous Mites

The European red mite, *Panonychus ulmi* (Koch) (Fig. 13.2*f*), two-spotted spider mite, *Tetranychus urticae* (Koch), and apple rust mite, *Aculus schlechtendali* (Nalepa), are the most destructive mites occurring in this region. Each of these mites damages apple foliage by removing the cellular content from apple leaves until they gradually give the leaf a fine yellow-green (spider mites) or silvery (rust mites) appearance. Heavy infestations may cause bronzing of leaves and some defoliation. In time mite feeding causes poor sizing and color quality of fruit, excessive fruit drop during the season, reduced blossom and fruit set in subsequent years, and a general decline in tree vigor.

Predatory mites, spiders, and several predaceous insects control phytophagous mites in orchards where pesticides are not sprayed. Phytophagous mites have developed resistance to many broad-spectrum chemicals, while most compounds remain toxic to their natural enemies. When insecticides are applied for key pest control, phytophagous mites are released from regulation by natural enemies, populations increase to outbreak levels, and miticides are applied several times during a season.

VI. SEASONAL OCCURRENCE OF ORCHARD PESTS

In addition to the principal diseases and arthropods described, there are other pests of lesser importance, and the composition of this group varies from area to area in the study region. In Table 13.2 the major pests and some of the less important species occurring in New York apple orchards are listed. Also, the relationship between the plant growth stage (or spray application date) and the seasonal occurrence of each species is indicated. Although constructed to show periods when treatments for chemical control may be necessary, this table also indicates when each pest is likely to be present in the orchard. During a growing season, there is no period when one or more destructive pests are not likely to occur and, during certain periods (e.g., petal fall, third cover), several may be simultaneously present.

VII. APPLE PEST-MANAGEMENT METHODS

A. Chemical Control

The development of organosynthetic pesticides provided growers with simple and effective measures for controlling fruit pests, and for the past $2\frac{1}{2}$ decades these toxicants have been used almost exclusively in most fruit-growing areas of the world (see Chapter 7). The reader is referred to Barnes (1959) and Madsen and Morgan (1970) for a detailed discussion of these programs. Whereas pesticides are useful and effective, they are not an unmixed blessing, and in some cases their intensive and unilateral use causes deleterious side effects on pests, natural enemies, and crop-associated ecosystems. DDT was the first synthetic compound to be widely used, but residue problems and resistance to this chemical prompted a subsequent shift to several organophosphorus or carbamate chemicals. As did DDT, most of the newer compounds contribute to secondary-pest outbreaks, and specific treatments for mites, aphids, scales, and leafhoppers are often required. In extreme cases certain secondary pests (e.g., mites, white apple leafhopper, also have acquired resistance to a variety of pesticides and have become almost impossible to control by chemical means.

Integrated chemical and biological control programs (reviewed in detail in Section VIII.) developed during the past decade provide an alternative to unilateral chemical pest control by using selective placement, timing, or physiologically selective pesticides for key pest problems, and biological control of certain secondary orchard pests (Hoyt 1969; Swift 1970; Asquith 1971; Rock 1972; Croft and Thompson 1975; Meyer 1974). When practiced, these programs provide for significant reductions in the amount of pesticides applied and costs for control (Hoyt and Caltagirone 1972; Asquith 1972). This is also the case with minimal application tech-

Table 13.2 Seasonal Occurrence of Apple Pests in New York Apple Orchards[a]

	Diseases								Lepidoptera											
									Tortricidae											
Plant Growth Stages and Typical Spray Application Dates	Apple Scab	Powdery Mildew	Black Rot	Apple Rusts	Fire Blight	Brooks Fruit Spot	Sooty Blotch	Fly Speck	Codling Moth	RBLR[b]	OBLR[c]	FTLR[d]	OFM[e]	LAW[f]	ESMB[g]	3LLR[h]	Green Fruitworms	Canker Worms	Apple Bark Borer	Tentiform Leaf Miner
Dormant, 4/12															X					
Silver tip, 4/20	X																			
Green tip, 4/23	X			X						X										
½ in. green, 4/27	X	X	X	X						X		X			X		X			
Tight cluster, 4/29	X	X	X	X								X					X			
Pink, 5/5	X	X	X	X	X							X					X	X		X
Blossom	X	X	X	X	X															
Petal fall, 5/27	X	X	X	X						X		X					X			
First cover, 6/9	X	X	X	X		X	X		X	X	X			X		X			X	X
Second cover, 6/24	X	X	X			X	X		X		X			X		X			X	X
Third cover, 7/7	X	X				X	X	X	X	X			X	X					X	X
Fourth cover, 7/22	X					X	X	X	X	X	X			X		X			X	
Fifth cover, 8/5	X		X			X	X	X		X	X		X		X	X				
Sixth cover, 8/18	X		X										X		X					

[a] Adapted from data prepared by K. Trammel, New York State Agricultural Experiment Station, Geneva, New York. X indicates control measures may be necessary.
[b] Red-banded leaf roller.
[c] Oblique-banded leaf roller.
[d] Fruit tree leaf roller.
[e] Oriental fruit moth.
[f] Lesser appleworm.
[g] Eye-spotted bud moth.
[h] Three-lined leaf roller.

Table 13.2 (Continued)

Plant Growth Stages and Typical Spray Application Dates	Cole-optera	Dip-tera	Homoptera							Hem-iptera	Hymen-optera	Mites					
	Plum Curculio	Apple Maggot	Rosy Apple Aphid	Apple Aphid	Wooly Apple Aphid	Oyster-shell Scale	San Jose Scale	Lecanium Scale	WA Leaf-hopper[i]	Tar-nished Plant Bug	Euro-pean Apple Sawfly	Euro-pean Red	Two-spotted	Apple Rust Mite	Four-spotted	Mc-Daniel	Clover
Dormant, 4/12			X					X									
Silver tip, 4/20																	
Green tip, 4/23																	
½ in. green, 4/27			X	X			X			X		X					X
Tight cluster, 4/29										X		X					X
Pink, 5/5			X	X						X		X					
Blossom																	
Petal fall, 5/27	X					X	X	X	X	X	X			X			
First cover, 6/9	X					X	X	X	X			X		X			
Second cover, 6/24	X	X		X					X			X		X			
Third cover, 7/7		X		X	X							X	X	X	X	X	X
Fourth cover, 7/22		X	X	X	X				X			X	X	X	X	X	
Fifth cover, 8/5		X		X	X				X			X	X	X	X	X	
Sixth cover, 8/18		X		X	X				X			X	X	X	X	X	

[i] White apple leafhopper.

niques such as ultralow-volume spraying (Howitt et al. 1966), alternate middle-row applications (Lewis and Hickey 1967), improved monitoring-reduced application methods (Batiste et al. 1973; Hagley 1973), and supervised control in which intensive pest surveys are used as the basis for applying pesticides (Thompson et al. 1973; Eves and Keenan 1973).

Although each of these newer methods adds refinement to chemical control programs, none of them totally eliminates the need for pesticides. At best, 12 to 20 judicious applications are still essential in each case, and these newer approaches are possible only because effective pesticides, particular formulations, and certain selective compounds have been developed. One should not conclude from this statement that improved selective chemical methods and nonchemical methods have not been developed for specific fruit pests. On the contrary, many have been demonstrated, and these are reviewed in the remaining portions of this chapter. Instead, this generalization reflects the difficulties encountered in implementing an all-inclusive nonchemical control system for an entire pest complex, and the slow rate at which biological control measures can be integrated into control systems in which pesticides are regularly applied.

B. Apple Diseases

In this study area apple diseases are controlled primarily with fungicide sprays applied periodically from bud break to harvest. As a result of improvements in fungicidal compounds, only 10 to 15 sprays are applied annually, compared with the 20 to 25 sprays required before 1945. Most spray treatments are for apple scab, with supplemental treatments for fire blight, apple rusts, or powdery mildew.

One method of timing treatments for apple scab is by stage of flower development, since a correlation exists between tree development and fungus development. A second timing method is to spray every 7 to 10 days early in the season and every 10 to 14 days in summer. In some fruit-producing regions sprays are made more frequently, but only one-half of the orchard is covered per application (Lewis and Hickey 1967). The main disadvantage of these scheduling methods is that in dry years, when there is less infection, the number of sprays applied is excessive as compared with the number applied in years when spray timing is based on rainfall and infection periods.

Techniques for timing orchard disease control treatments according to the developmental stage of the disease pathogens involved are being developed, and in some instances are being used on a regional or area basis.

These approaches are not new, but effective methods for following pathogen development generally have not been available in the past. Recently, spore traps, such as a rotorod sampler, have been used to collect apple scab ascospores (Jones and Sutton 1974), and a selective culture medium has been used for detecting fire blight bacteria (Miller and Schroth 1972). Monitoring of temperature and leaf wetness also provides data for use in predicting infection periods based on the Mills system (Mills and LaPlante 1951). Once it is determined that infection will occur, eradicative chemicals are applied. Still better predictions and timing appear possible through the development of disease-prediction models based on current meteorological data (Jones and Sutton 1974).

C. Codling Moth

In areas where two or more generations of the codling moth are produced each season, effective natural control seldom occurs, even though this pest is attacked by a variety of predators and parasitoids (Putman 1963). In certain circumstances entomophagous diseases, including several fungi (Russ 1964) and a naturally occurring granulosis virus (Falcon et al. 1968), are relatively effective as control agents, but as yet no effective pathogen has been developed and sold for commercial field use.

In Nova Scotia, where pioneering work in pest management began over 2 decades ago, the codling moth produces only one generation each season, and effective biological control often is provided by woodpeckers which feed on overwintering larvae, and by *Trichogramma* parasitoids, an anystid mite, and several mirid predators which attack the egg and summer larval stages (Pickett et al. 1958; MacPhee and MacLellan 1971). The effect of not using broad-spectrum chemicals for codling moth control has a marked influence on associated pest systems. The European red mite is regulated by a complex of predatory mites and insects. The oystershell scale is controlled by a predaceous mite and a parasite. On occasion other pests become problems (e.g., apple maggot) but, in these instances, relatively selective pesticides provide excellent control and do not disturb the biological regulating features of the system. As indicated by this success, where the codling moth is the only major insect pest and selective chemical or nonchemical measures are used, many secondary pests return to a natural control condition in which they are regulated by their natural enemies. Where the codling moth is multivoltine, this same type of approach is less effective (Oatman 1966) and is not an acceptable control program.

A basic requirement of any pest-management program is an effective survey method for estimating the density or damage potential for key pest species. For the codling moth accurate sampling is difficult, because of the normally low densities occurring in commercial orchards and the moth's cryptic behavior. If reliable estimates of either overwintering larval, pupal, emerging adult, or subsequently oviposited egg populations could be measured before damage occurred, the need and timing for sprays could be more precisely determined and the number of spray applications reduced appreciably.

Various lures including bait or light traps, caging systems for overwintering larvae, or heat-unit developmental models have been used to sample or predict emergence of adult populations with some success. More recently, the use of a synthetic sex attractant or pheromone (Roelofs and Comeau 1971) in a sticky trap proved to be a useful sampling tool for this pest (Chapters 8 and 9). In western (Batiste et al. 1973) and midwestern North America (Hagley 1973), pheromone traps have been used to detect moth emergence and to time pesticide applications with minimal effort and considerable accuracy (Fig. 7.2). In these programs initial male moth catch in the pheromone trap is used as a starting point for adult emergence, and by putting out caged mated females and observing egg hatch (Batiste et al. 1973), or by predicting hatch from developmental models (Hagley 1973), an estimate of when to spray can be timed more precisely.

On an experimental basis in Michigan (Riedl and Crofts 1974) and Nova Scotia (MacLellan 1974), pheromone traps also have been used to estimate expected seasonal infestation levels of the codling moth. To date, predictions have been based on correlating the number of moths caught with the damage levels at the end of the season. Research to relate trap catches to the dynamics of moth activity and density is currently underway. If these relationships can be determined, the pheromone trap may serve as a specific and reliable predictive device.

Another possible use of pheromones is for direct codling moth suppression by attracting sufficient males to sticky traps to suppress or prevent mating, or by flooding the environment with sex pheromones which confuse males in their orientation to virgin females (Chapter 8). At the present time both strategies are in the experimental stages of development (Trammel 1972).

Effective suppression of the codling moth by release of sterile males (Chapter 8) was initially demonstrated in British Columbia in abandoned and commercial apple orchards during 1966–1968 (Proverbs et al. 1966, 1969). Recent tests have been conducted in an area of 2200 ha (including 200 ha of apples), and this control measure continues to appear effective. British Columbia researchers estimated the cost for genetic control of the codling moth to be approximately $139 per acre. This is almost twice the amount spent for chemical control. They suggest that sterile-male release programs for this pest may be technically feasible in certain isolated fruit-growing areas, but may not be an economically acceptable practice (Hoyt and Burts 1974).

Integrated use of sterile-male releases, minimal insecticide applications, and sanitation by removing or spraying noncommercial trees has provided successful management of the codling moth in the isolated Wenas Valley of central Washington State (Butt et al. 1972; Butt et al. 1973). During 1970, a 97% reduction in the overwintering codling moth larval population was achieved by spraying. In 1971, this density was further reduced by treating a small number of noncommercial trees and releasing sterile males in commercial plantings. These experiments demonstrated that the sterile-male technique for codling moth control is likely to be most effective in areas where moth populations are somewhat localized and alternate source trees are limited. Western fruit-growing areas of North America appear to be suitable, but it is doubtful that these methods could be successfully implemented in midwestern or eastern North America or other fruit-growing areas where wild fruits and alternate hosts are abundant.

In summary, no nonchemical control method has been developed that provides as effective and economical control of multivoltine codling moth populations as do insecticides. However, most chemicals sprayed for control of this pest are toxic to natural enemies of orchard pests, and their use limits integration of biological control measures for several other fruit pests (see Section V). It has been suggested that combined utilization of several nonchemical methods (biological control, sterile-male technique, pheromones, etc.) might provide acceptable codling moth control. While these integrated programs would be costly to implement, their use may be necessary if this pest develops resistance to currently used insecticides (Hoyt and Burts 1974).

D. Red-Banded Leaf Roller

When preventive insecticidal sprays are applied for codling moth, plum curculio, and apple maggot, control of this pest is seldom a problem. If suitable selective chemical or nonchemical methods were developed for these three species, red-banded leaf roller populations would be somewhat suppressed by a complex of hymenopterous parasitoids (Oatman and Jenkins 1962; Glass 1963). Isolation of an active sex attractant or pheromone for the red-banded leaf roller (Roelofs and Feng 1968) provides for improved monitoring of this tortricid. Moth suppression by placing pheromone-baited sticky traps in orchards also is a possible control strategy (Glass et al. 1970). In the laboratory and field, granulosis virus for control of this pest has been tested with considerable success (Glass 1963). *Bacillus thuringiensis* Berliner is effective (Oatman 1965), and *Bergoldia clistorhabdion* Wasser & Steinhaus can reduce populations to extremely low levels in this region (Oatman and Jenkins 1962). Although several effective disease agents are available, none has been widely used for commercial control of the red-banded leaf roller. This probably is related to the availability of individual insecticides which provide excellent control of this pest and the three other direct key pests, while the aforementioned diseases are less broad-spectrum and only effective on one or two of the four species.

E. Plum Curculio

This weevil is most frequently attacked by natural enemies in the larval stages, and the parasitoid *Aliolus curculionis* (Fitch) is the most important biological control agent in this region. In Ontario this species and a complex of less important parasitoids parasitized from 7 to 27% of weevil larvae during a 5-year study period (Armstrong 1958). In other parts of the region, no significant natural control has ever been reported. A preliminary investigation of sterilization of male plum curculios has been conducted by releasing irradiated weevils into laboratory cages containing normal populations of both sexes. The number of larvae produced was reduced by 90%, and it was recommended that field trials to test the effectiveness of this method be implemented (Jacklin et al. 1970).

In general, efforts to develop alternative nonchemical measures for the plum curculio have been limited, as compared to research efforts expended on the codling moth and red-banded leaf roller. Lack of study is probably due to these factors: (1) The plum curculio's distribution as a pest is limited to central and eastern areas of North America; (2) adult weevils are difficult to sample and study because of their high mobility and tendency to be distributed in areas other than the orchard; and (3) a certain degree of selectivity can be obtained by applying insecticides for plum curculio control to areas where the natural enemies of other fruit pests are not appreciably affected (e.g., soil, fencerows, woodpiles, border trees).

F. Apple Maggot

Because of its larval habits, this fruit fly is somewhat protected within the apple fruit and is relatively immune to biological control agents (Monteith 1972; Rivard 1968; Dean and Chapman 1973). As a survey tool color-attractive sticky traps (Maxwell 1969) or spheres (Prokopy 1968) have been used to estimate fly emergence and density levels. However, when color and bait attractants are combined, a more reliable fly estimate is obtained (Howitt and Connor 1964). Recently, a chemical that inhibits oviposition of the apple maggot has been reported (Prokopy 1972). It is possible that distributing this compound at the proper time may provide partial or complete control of this pest.

G. Potential Key Pests

Pests in this group often are either controlled by natural enemies or pesticides applied for key pest control. For example, the fruit tree leaf roller often is kept under natural control in the study area by egg desiccation, mortality of dispersing larvae, or reduction of larvae and pupae by about 25 species of parasitoids (Paradis 1961). Similarly, the eyespotted bud moth, *Spilonota ocellana* (D. & S.), is attacked by a variety of parasitoids, predators, and diseases which often provide effective natural control (Legner and Oatman 1963; Jaques and Stultz 1966).

If pesticides applied for annual direct key pest control are the main reason insects of this group seldom achieve pest status, these species should become greater pest problems as chemicals are withdrawn. However, if natural enemies are effective but have been somewhat suppressed by chemicals, these pests should be further controlled by predators, parasitoids, and diseases as individual pesticides are withdrawn. The response of individual species in this group to reduced chemical usage probably can be evaluated only by implementing selective chemical or nonchemical programs and observing their influence. As was found in Nova Scotia, it is likely that several pests of this type can be completely controlled by biological control agents when no chemicals or selective chemicals are used, while others will require monitoring and periodic suppression with selective insecticides.

H. Scales, Aphids, and Leafhoppers

Applications of broad-spectrum insecticides for key insect control greatly limit the use of most nonchemical control measures for this pest group.

An exception to this generalization is the host-plant resistance developed to the woolly apple aphid in certain commercial apple variety rootstock, for example, 'Northern Spy' (Knight et al. 1962). European workers have shown that systemic insecticides can be translocated into plants and taken up by plant-feeding insects without affecting the natural enemies of aphids and scales appreciably (Chaboussou 1961, Mathys 1966). Integrated control of aphids and scales can also be attained by applying oil-insecticide sprays during the early-season dormant period before natural enemies are active (Madsen and Morgan 1970). When direct insecticidal treatments are needed for key insect, scale, or aphid control, proper use of these compounds conserves the most important natural enemies of these pests, and partial biological control can be achieved. The toxicity of a wide range of contact toxicants to the principal parasitoids and predators of these pest groups has been evaluated (Chaboussou 1961; Bonnemaison 1962; Benassy et al. 1964; Billiotti et al. 1960; Mathys and Guignard 1967).

I. Phytophagous Mites

Compared to the tree fruit pest life systems previously discussed, more progress in developing effective nonchemical control methods has been achieved for phytophagous mites than for any other group of pests (Huffaker et al. 1969). This success undoubtedly is due to the variety of effective natural enemies attacking these pests on a worldwide scale (McMurtry et al. 1970). Also, several natural enemies of this group have developed tolerance or resistance to toxic insecticides (Croft 1972). In the study area several effective integrated control programs have been developed in which biological control and chemical agents are compatibly used together. In New Jersey (Swift 1970), North Carolina (Rock 1972), Illinois (Meyer 1974), and Michigan (Croft 1974), the predatory mite *Amblyseius fallacis* (Garman) provides for effective integrated control of plant-feeding mites. In Pennsylvania (Asquith 1971), emphasis is placed on the biological control potential of a coccinellid beetle, *Stethorus punctum* (Le Conte), and in Canada conservation of entire complexes of natural enemies is stressed (Pickett et al. 1958; Sanford and Herbert 1970; Parent 1967). In the remainder of this chapter, an example of an integrated control program wherein the predatory mite *A. fallacis* is the key natural enemy is discussed.

VIII. AN INTEGRATED MITE-CONTROL PROGRAM

An outline proposed by Haynes et al. (1973) is followed (Fig. 13.3) in describing the components of this management system. Basic elements include (1) a description of the real insect population or specific predator–prey system including pests, natural enemies, alternate prey, and other

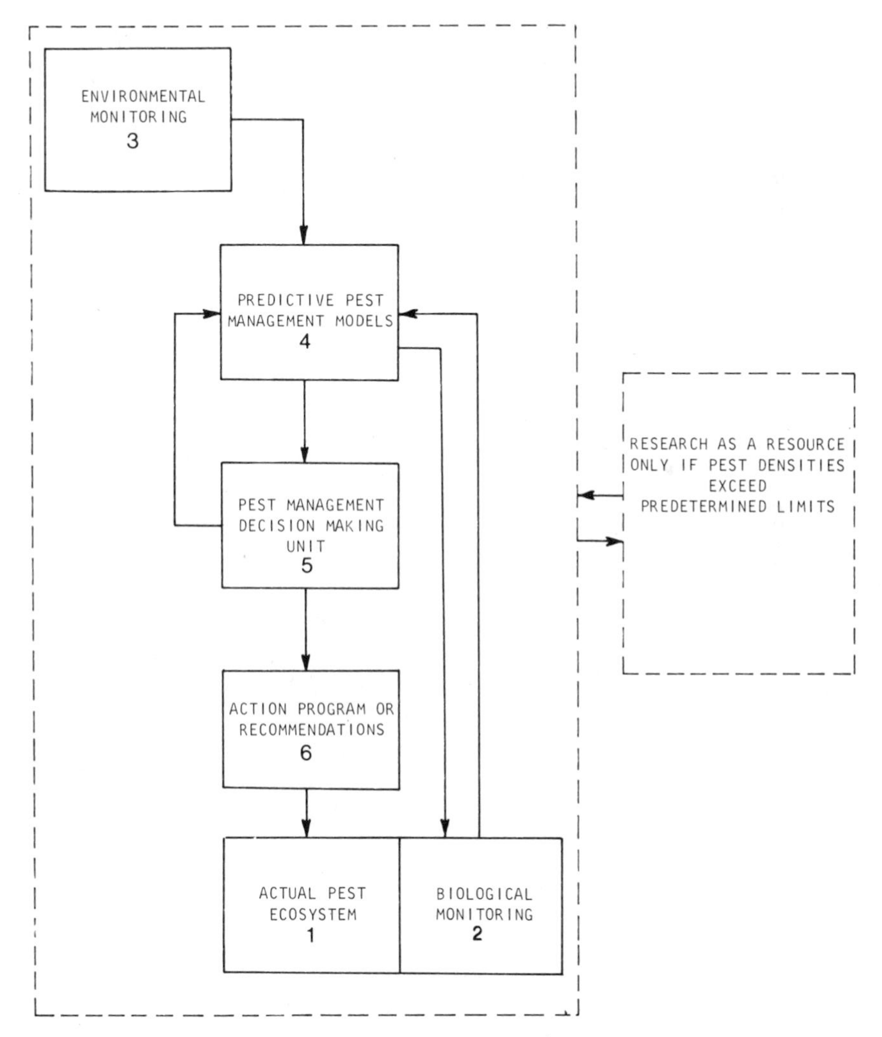

Fig. 13-3 Components of an on-line pest-managment system. Adapted from Haynes et al. 1973.)

associated species; (2) a biological monitoring component or survey method to measure the densities of each species in time; (3) an environmental monitoring system to supply "real-time" or on-line climatic inputs to each species life system; (4) species component and management models which describe the effects of various variable inputs (biological, climatic), and control manipulations on real arthropod populations; (5) an extension service group which interrogates and translates management model outputs into recommendations to growers; and (6) an implementation action program carried out by the grower which directly effects changes in the real insect population. After item 6 is executed, the management (feedback) loop is again coupled by biological (2) and environmental monitoring (3) to determine if the real insect population is responding to the implementation action program (6) as the management model (4) has predicted. Pest-management systems of the future will probably need all of the six features cited. More than any other fruit pest system discussed, phytophagous mite research programs are the most advanced in the development of these six elements.

A. The Real Life System

In Chapter 1 the basic features of the phytophagous-predatory mite system were illustrated (Fig. 1.4). In this section the biology of four species, the European red mite, the two-spotted spider mite, the apple rust mite, and the predator *Amblyseius fallacis* are briefly reviewed.

The European red mite is strictly an arboreal fruit (apple, plum, peach) pest which overwinters within trees in the egg stage. In spring immature red mites hatch and move to foliage where six to eight generations per season may develop. In late summer adult female mites lay overwintering eggs on small twigs and bark, and the seasonal cycle is completed. In spring two-spotted spider mites principally occur on the groundcover beneath the apple tree after emerging as adult females from overwintering sites in debris or under bark near the tree base. During mid to late summer, these mites may enter fruit trees and cause damage, but in late season they again return to the tree base to overwinter as adult females. Apple rust mites overwinter as adults under apple bud scales. In spring they migrate to foliage where they increase and may cause damage. Adult female *A. fallacis* overwinter in debris beneath the tree or on the tree trunk near the base. In spring they feed on two-spotted spider mites and other phytophagous mite species in the groundcover. From mid-June to the end of the growing season, they migrate to fruit trees and prey on developing European red mite, two-spotted spider mite, or apple rust mite populations.

During a growing season, *Amblyseius fallacis* and pest mite populations develop or can be managed in a variety of ways. The type of management applied or the degree of successful biological control by *A. fallacis* is determined by the number of predators in relation to the density of pest mites present in the orchard (predator/prey ratio). For example, Fig. 13.4*a* shows field-collected sample data in which an effective ratio of *A. fallacis* to spider mites (0.40:2.0 per leaf) is present in early season and predators provide complete biological control of spider mites before the latter exceed 5 per leaf. Figure 13.4*b* shows a less effective ratio of predators to prey (0.25:8.0 per leaf), and in this case pest mites reach a density of almost 20 per leaf before control is attained

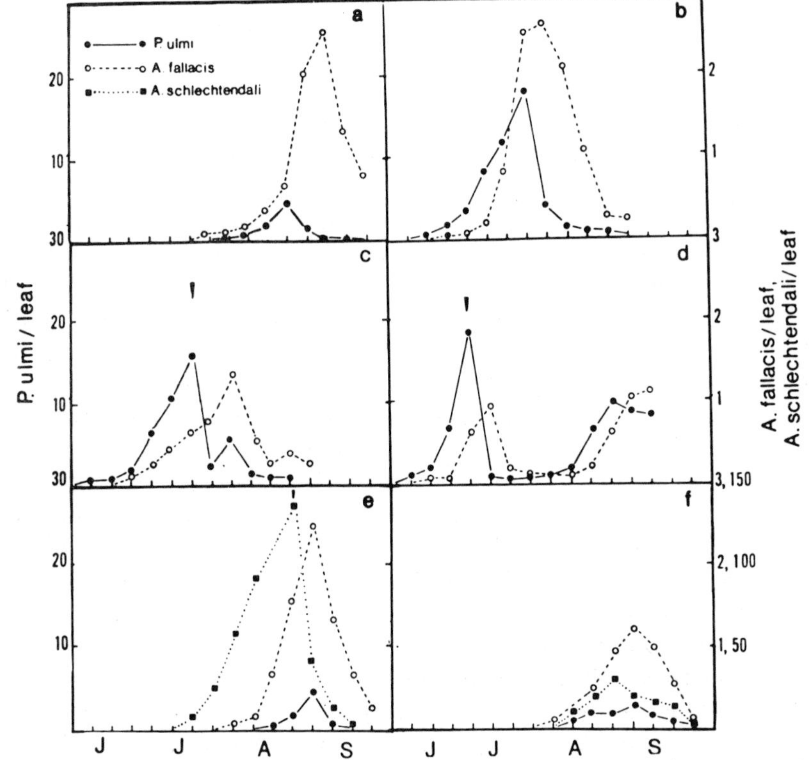

Fig. 13-4 Population dynamics of *Amblyseius fallacis* and plant-feeding mites in Michigan apple orchards.

(economic threshold is 15 mites per leaf). Figure 13.4*c* illustrates how a selective miticide can be used to reduce but not eliminate pest mites and thus change the predator/pest ratio from ineffective to effective. Figure 13.4*d* shows the effect of completely eliminating pest mites by using a full-strength miticide, the subsequent starvation of predators, and a second outbreak of pests (note the absence of late-season control or regulation by predators as shown in Fig. 13.4*c*). Figure 13.4*e* to *f* demonstrates how apple rust mites may serve as alternate prey for predators and influence control of spider mites by *A. fallacis.* In Fig. 13.4*e* apple rust mites enter apple trees before spider mite populations develop; *A. fallacis* develops on rust mites and becomes well distributed throughout the tree; a selective chemical which controls rust mites, aphids, and leafhoppers is applied as spider mites begin to increase; rust mites are chemically controlled and predators, which are unaffected by the spray, switch prey and quickly destroy spider mite populations. In Fig. 13.4*f* a similar condition is shown in which rust mites are present only at moderate levels; predators are able to enter trees before spider mites, but provide complete biological control of both plant-feeding pests without sprays for rust mites.

The seasonal density relationships presented in Fig. 13.4*a* to *f* show the potential of *Amblyseius fallacis* to control pest mites in the absence of toxic pesticides. Normally, in orchards of the region from 5 to 20 pesticide applications are applied each season for key disease and insect pest control. An interesting feature of this predator, and one that allows considerable flexibility in integrating chemicals and predators, is that *A. fallacis* has acquired resistance to a number of organophosphorus (Motoyama et al. 1970; Croft and Nelson 1972) and carbamate (Croft and Meyer 1973) insecticides. In addition, other pesticides are physiologically selective (e.g., most fungicides, insecticides like endosulfan, and miticides like Omite and Plictran), while others which normally are toxic to predators can be used with proper timing or placement. For example, selectivity can be obtained by spraying the toxic insecticides diazinon or demeton for scale and aphid control early in the growing season before mite predators are present in fruit trees. Research to measure the levels of resistance present in predator populations in commercial orchards and the toxicity of most pesticides used for key disease and insect control (Croft and Nelson 1972) has provided for the development of pesticide use recommendations (Croft and Thompson 1974). These provide growers with information on how to use a variety of alternative pesticides for key pest control during any period of the growing season. By selecting and timing chemical applications properly, the phytophagous-predatory mite life system is not directly influenced by pesticides, and *A. fallacis* can provide effective biological control of mites.

B. Biological Monitoring

As shown in Fig. 13.4*a* to *c*, the management alternative of whether to expect *Amblyseius fallacis* to provide complete biological control before spider mites (European red and twospotted) exceed 15 active mites per leaf or 150 apple rust mites per leaf (economic threshold levels), or to spray a selective acaricide, reduce the prey, and establish a more favorable predator/prey ratio, depends on the assessment of each of the mite population densities before pest mites reach damaging levels. To do this, one needs a sampling scheme that accurately measures the densities of pests and predators in the management unit (in this case a 10-acre apple block). Decisions on when to sample, how often to sample, size of sample, and so on, must be made before one can accurately estimate the predator-pest mite densities and reliably predict interaction results.

For this predator–prey system, a biological monitoring plan that optimizes sampling precision and cost for sample processing has been developed. Basic sampling features include (1) varietal preference selections; (2) proper sampling of orchard, tree, and leaf units; (3) minimal sample size determinations for leaves and trees evaluated sequentially at sample intervals of 25 leaves (5 trees and 5 leaves per tree); and (4) obtaining accurate life-stage distribution estimates as starting conditions for management model simulations (see Section VIII. C). With a mobile sampling laboratory and limited computing facilities, samples of pests and predators are taken and counted, biological control possibilities estimated, management decisions recommended, and subsequent sample times, sizes, and costs forecast. After the implementation action program (6 in Fig. 13.3) is made, a feedback sample estimate is made to verify that effective pest control has been achieved.

C. Management Models

In Fig. 13.5 an empirically derived management index based on a 5-year study of biological control of spider mites by *Amblyseius fallacis* is shown. As spider mites increase from 1 to 15 per leaf, their control by predators, the population level they attain, and the damage they do all depend on how favorable the ratio of predator to pest is in the orchard.

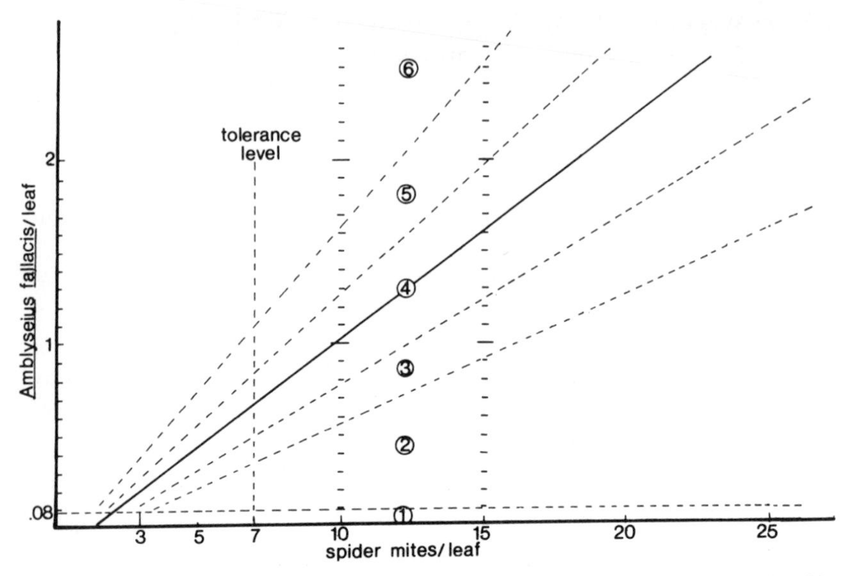

Fig. 13-5 A decision-making index for estimating biological control of spider mites by *Amblyseius fallacis.*

When sample estimates of prey density per leaf are plotted on one axis and predator numbers per leaf on the other, the likelihood of biological control occurring can be forecast (Fig. 13.5 and Table 13.3). For example, in Fig. 13.5 it is not necessary to make a mite control decision until spider mites exceed 7 per leaf. If prey mites number more than 7 per leaf and only a few *A. fallacis* (less than 0.08 per leaf) are present, a full-strength miticide spray is recommended (see region 1 in Fig. 13.5 and Table 13.3). If the *A. fallacis*/spider mite ratio falls in regions 2 or 3, an application of a selective miticide spray is necessary to establish a more favorable predator/prey relationship. The chemicals and rates recommended for these regions will reduce but not eliminate spider mites (complete elimination of spider mites would cause predators to starve to death or leave the apple tree), and are nontoxic or slightly toxic to *A. fallacis* (compare Figs. 13.4*c* and *d*). If a plot falls in region 2, the miticide Plictran is recommended for substantial but not complete spider mite control; in region 3 either Omite or a lower rate of Plictran is used. After either of these miticides is applied, pests are reduced and predators increase and regulate spider mites at low population densities for the remainder of the season. If a predator/prey ratio falls in region 4, the possibilities for complete biological control of spider mites are 50-50,

Table 13.3 Spray Recommendations for Predator/Prey Ratio Regions in Figure 13.3

Region	Suggested Recommendation	Probability for Biological Control
1	As bronzing appears, spray recommended miticide at full rate	Very low
2	If bronzing appears, spray Plictran[a] 50 WP at 2 oz/100 gal or 8 oz per acre	Equal to or less than 10%
3	If bronzing appears, spray Omite[b] 30 WP at $1\frac{1}{4}$ lb/100 gal or 5 lb per acre or Plictran 50 WP at $1\frac{1}{2}$ oz/100 gal or 6 oz per acre	Greater than 10% but less than 50%
4	Wait 1 week; biological control should occur soon; if not, spray Omite at $1\frac{1}{4}$ lb or Plictran $1\frac{1}{2}$ oz/100 gal	Approximately 50%
5	Same as 4	Greater than 50% but less than 90%
6	Wait 1 week; biological control is almost certain	Greater than 90%

[a] Trademark name for tricyclohexylhydroxytin.
[b] Trademark name for 2-(*p-tert*-butylphenoxy) cyclohexyl 2-propynyl sulfite.

and the grower is advised to wait and see if pests decline (Table 13.3). Plots falling in regions 5 or 6 indicate a high likelihood of complete biological control.

Although useful in making mite-management decisions, the empirical index in Fig. 13.5 lacks flexibility and is not applicable to all environmental conditions. For example, developmental time for the European red mite and the predator *A. fallacis* are differentially affected by temperature. At low temperatures the rate of development of the predator becomes slower than that of the prey, and thus the predator's effectiveness is reduced if cool environmental conditions persist for a long time period. To correct for this effect, temperature and other variable factors have been incorporated into a dynamic population model for biological control of spider mites by *A. fallacis* (Fig. 13.6). The following are its basic elements: (1) a within-season, multigeneration population development for both predator and prey as life-stage components [e.g., egg (E), larvae (L)] is simulated as a function of environmental climatic factors and an initial predator and pest life-stage distribution; (2) density-dependent stage mortalities of both European red mite and *A. fallacis* other than predation or cannibalism by *A. fallacis* are assumed to occur between molts and are subtracted from the populations at short time intervals; (3) coupling of the subtractive effects of *A. fallacis* predation on the prey population as a function of a predator preference consumption matrix which in turn is affected by prey and predator densities; and (4) a season-to-season overwintering component for predator and prey, which includes predator dispersal and groundcover interaction phases for *A. fallacis* and two-spotted spider mites.

D. Environmental Monitoring

The population dynamic model described in Fig. 13.6 is sensitive to inputs from various climatic features including temperature, humidity, rainfall, and others. Before "real-time" decisions can be made based on current conditions in the field, an on-line environmental monitoring

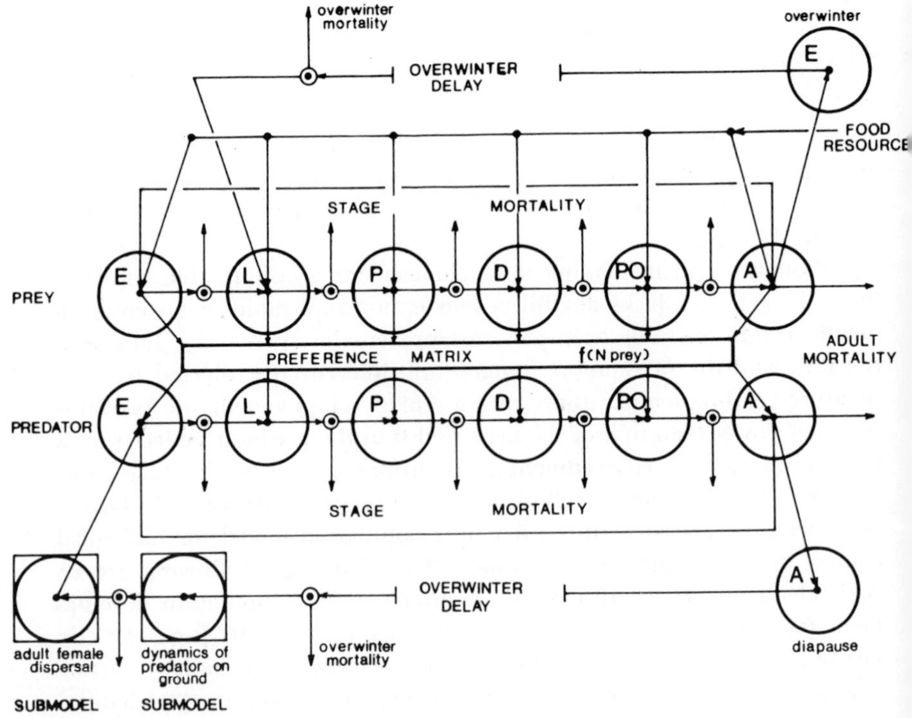

Fig. 13-6 Population dynamics model for estimating biological control of *Panonychus ulmi* by *Amblyseius fallacis*.

network to feed in changing climatic conditions must be established. At present, Aviation Weather Network (AWN) information from four regional airports in the Michigan fruit belt provides readings of current environmental conditions at 2-hour intervals. Through correlations relating historical weather trends, data for major airport weather stations can be extended to about 50 substations in Michigan's fruit belt. Research studies are underway to determine at what resolution level (leaf, tree, orchard level), macro- or microclimatic weather data must be obtained to estimate accurately the effect environmental features have on the real insect population (see Haynes et al. 1973 for further details on an environmental monitoring network).

E. Extension Service

As previously mentioned, it is the role of extension service personnel to interrogate management models for proper pest-management decisions and interpret them for grower uses. As shown in Table 13.4, a weekly printout using an empirical index (Fig. 13.5) is made from a series of predator/prey ratios found during the summer in a grower's orchard. Data of this type are used weekly by extension personnel in making spider mite–control recommendations. Table 13.3 shows that a miticidal spray at one-half full dosage was applied on 12 July and *A. fallacis* controlled the European red mite thereafter. By using the population model (Fig. 13.6) and entering the predator/prey ratio 0.16:12.3 sampled on 10 July (Table 13.4) and including the proper climatic inputs, a simulation run of the expected dynamics of predator and prey was forecast for the extension agent. Without spraying, prey levels would have increased to about 85 per leaf, and serious damage to the foilage would have occurred.

F. Action Program or Recommendation

By using the outputs of these models, extension personnel, insect scouts, or growers are ready to go into the orchard, inspect it to see if conditions reflect those projected by the model, and make a recommendation to spray or not to spray. Subsequently, the spray is or is not applied; at a later period a sample is again taken and the results are compared with those forecast by the model. At this point the feedback is complete; either a new decision is made, or control has been achieved for the season. Whereas growers normally spray from two to three miticide appli-

Table 13.4 Computer Printout of Weekly Samples and Recommendations Using an Empirical Index for Plant-Feeding Mite Management (see Fig. 13.3 and Table 13.3)

Sample Date	European Red Mite	Two-Spotted Mite	*Amblyseius fallacis*	Stigmaeidae	Rust Mites	Suggested Recommendation	Probability for Biological Control
6/27	2.5	0.0	0.16	0.00	0	Do not spray; destructive mite density too low	$0.50 < P < 0.90$
7/3	8.6	0.0	0.04	0.00	P	As bronzing appears, spray recommended miticide at full rate	Very low
7/10	12.3	0.0	0.16	0.00	0	If bronzing appears, spray Plictran 50 WP-8 oz. per acre	$P < 0.10$
7/17	5.2	0.0	0.20	0.00	0	Do not spray; destructive mite density too low	$0.10 < P < 0.50$
7/24	2.0	0.0	1.11	0.00	0–20	Do not spray; destructive mite density too low	$P > 0.90$
8/1	0.9	0.0	0.53	0.00	0	Do not spray; destructive mite density too low	$P > 0.90$
8/7	0.0	0.0	0.20	0.00	0	Do not spray; destructive mite density too low	$P > 0.90$
8/14	0.1	0.0	0.32	0.00	0–20	Do not spray; destructive mite density too low	$P > 0.90$

cations per season, they averaged less than one per season when these tools were available (Croft 1974).

IX. RESISTANT PREDATORY MITES

The greatest potential for integrating insecticides for key insect control and natural enemies for biological control of mites is in areas where predators of spider mites have acquired resistance to certain insecticides. As noted previously, in several central and eastern areas of the United States, insecticide-resistant *Amblyseius fallacis* predators provide substantial biological control of plant-feeding mites where integrated mite-control programs are practiced. In Washington State apple orchards, growers monitor population levels of the insecticide-resistant predator *Typhlodromus occidentalis* Nesbitt and its spider mite prey, *Tetranychus mcdanieli* McGregor. Before the program was developed, the problem of controlling the *T. mcdanieli* spider mite was a familiar one. The pest had developed resistance to one acaricide after another, and chemical control was often not satisfactory. Entomologists subsequently developed a selective spray program which allowed predators with previously acquired resistance to persist and control pest mites in the presence of insecticide sprays applied for key insect pest control. The combination of selective chemicals and resistant predators provided a less costly, more permanent control of pest mites than did pesticides alone (Hoyt and Caltagirone 1972). Other similar integrated programs involving apples in British Columbia (Downing and Aarand 1968) and Utah (Davis 1970), pears in Oregon (Westigard 1971), and peaches (Hoyt and Caltagirone 1972) in California utilize effective biological control of pest mites by insecticide-resistant strains of *T. occidentalis.*

X. SUMMARY AND CONCLUSIONS

The status of pest management in the case study region in many ways reflects the development and progress achieved in deciduous fruit areas worldwide. Because of the low tolerance for damage to a commodity highly rated for its consumer appeal, and the inability of natural enemies to keep certain key fruit pest populations below damaging levels, pesticide-dominated programs are and probably will continue to be the mainstay of deciduous fruit pest control for some time. At present, few fully tested alternatives are available that are as effective and economically competitive, and this is especially true for any complex of key fruit pests.

If significant nonchemical control is achieved in the near future, it probably will be in fruit-growing areas where key fruit pests are few (e.g. western North America, Nova Scotia), and is less likely to be in areas where large pest complexes are present (e.g., the case study region).

Greater success in achieving effective nonchemical control of deciduous fruit pests has been attained with certain indirect pests. As exemplified by the pest-management programs developed for plant-feeding mites, effective nonchemical or selective chemical programs can be developed and result in substantial reductions in the total amount of pesticide applied. Although the mite-management programs are a significant advance, it should be noted that these programs are not ends in themselves, and that they do not solve the greater problem of controlling key fruit pests. In contrast, if direct key pests could be controlled by means other than chemicals, it is likely that most indirect secondary pests could be readily managed by biological or selective chemical measures.

Because of a limitation of space, many additional advances in the areas of deciduous fruit pest biology, sampling, toxicology, physiology, behavior, ecology, and the development of other control measures could not be discussed in this chapter, but these can be reviewed in Barnes (1959), Madsen and Morgan (1970), Kirby (1973), Wildbolz and Meier (1973), and Hoyt and Burts (1974). Most have contributed in some measure to the general advancement of deciduous fruit pest management and certainly to the more judicious application of pesticides. As with pest management in all crops, the process of developing an effective system is always an unfinished task needing refinement, further improvement, and periodic updating in response to pest adaptations and changing technology. This is especially true for deciduous fruit systems in which we have just begun to move from programs in which pesticides are such a dominant control element to those in which nonchemical measures are more common. If and as this occurs, there will be a critical need for more extensive research on all methods of control of direct key pests of deciduous fruits, and even more importantly to integrate these into management systems that consider cost-benefits and environmental protection.

REFERENCES

Anderson, H W. 1956. Diseases of fruit crops. McGraw-Hill, New York. 501 pp.

Armstrong, T. 1935. Two parasites of the white apple leafhopper. *Annu. Rep. Entomol. Soc. Ont.* **66**:16–31.

Armstrong, T. 1958. Life history and ecology of the plum curculio, *Conotrachelus nenuphar* (Herbst) (Coleoptera: Curculionidae), in the Niagara Peninsula, Ontario. *Can. Entomol.* **90**:8–17.

Asquith, D. 1971. The Pennsylvania integrated control program for apple pests—1971. *Penn. Fruit News* **50**:43–47.

Asquith, D. 1972. Economics of integrated pest management. *Penn. Fruit News* **51**:27–31.

Barnes, M. M. 1959. Deciduous fruit insects and their control. *Annu. Rev. Entomol.* **4**:343–362.

Batiste, W. C., A. Berlowitz, W. H. Olson, J. E. DeTax, and J. L. Joos. 1973. Codling moth: Estimating time of first egg hatch in the field—a supplement to sex attractant traps in integrated control. *Environ. Entomol.* **2**:387–391.

Benassy, C., H. Bianchi, and H. Milaire. 1964. Experimentation en matiere de lutte integree dans deux vergers francais. *Entomophaga* **9**:271–280.

Billiotti, E., C. Benassy, H. Bianchi, and H. Milaire. 1960. Premiers essais experimentaux d'acclimatation en France de *Prospatella perniciosi* Tower, parasite specifique importe de *Quadraspidiotus perniciosus* Comst. Compt. *Rend. Hebd. Seance Acad. Agr. France.* **46**:707–711.

Bonnemaison, L. 1962. Toxicite de divers insecticides de contact ou endotherapiques vis-á-vis predateurs et parasites des Pucerons. *Phytiat. Phytopharm.* **11**:67–84.

Butt, B. A., J. H. Howell, H. R. Moffit, and A. E. Clift. 1972. Suppression of populations of codling moths by integrated control (sanitation and insecticides) in preparation for sterile-moth release. *J. Econ. Entomol.* **65**:411–414.

Butt, B. A., L. D. White, H. R. Moffit, D. O. Hathaway, and L. G. Schoenleber. 1973. The integration of sanitation, insecticides and sterile moth releases for the suppression of populations of the codling moth in the Wenas Valley of Washington. *Environ. Entomol.* **2**:208–212.

Chaboussou, F. 1961. Action de divers insecticides et notamment de certains produits endotherapiques vis-á-vis d' *Aphelinus mali* Hald. evaluant a l'interior du puceron lanigere du pommier, *Eriosoma lanigerum* Hausm. *Rev. Pathol. Veg. Entomol. Agr. France.* **40**:17–29.

Chapman, P. J., and S. E. Lienk. 1971. Tortricid fauna of apple in New York. *Spec. Pub. N.Y. Agr. Exp. Sta.* 122 pp.

Croft, B. A. 1972. Resistant natural enemies in pest management systems. *Span* **15**:10–22.

Croft, B. A. 1974. Unpublished data.

Croft, B. A., and R. H. Meyer. 1973. Carbamate and organophosphorus resistance patterns in populations of *Amblyseius fallacis*. *Environ. Entomol.* **2**:691–695.

Croft, B. A., and E. E. Nelson. 1972. Toxicity of apple orchard pesticides to Michigan populations of *Amblyseius fallacis*. *Environ. Entomol.* **1**:576–579.

Croft, B. A., and W. W. Thompson. 1975. Integrated control of apple mites. *Ext. Bull. Mich. St. Univ., Ext. Publ.* In press.

Davis, D. W. 1970. Integrated control of apple pests in Utah. *Utah Sci.* **31**:43–48.

Dean, R. W., and P. J. Chapman. 1973. Bionomics of the apple maggot in eastern New York. *Search Agr.* **3**:1–63.

Downing, R. S., and J. D. Arrand. 1968. Integrated control of orchard mites in British Columbia. *B. C. Dep. Agr. Entomol. Branch Publ.* 7 pp.

Eves, J. D., and L C. Keenan. 1972. Washington deciduous tree fruits pest management project. *1st Annu. Rep. Wash. St. Univ. Ext. Publ.* 20 pp.

Falcon, L. A., W. R. Kane, and R. S. Bethell. 1968. Preliminary evaluation of a granulosis virus for control of the codling moth. *J. Econ. Entomol.* **61**: 1208–1213.

Fisher, R. W. 1960. Note on resistance to DDT in the codling moth, *Carpocapsa pomonella* (L.) in Ontario. *Can. J. Plant Sci.* **40**:580–582.

Glass, E. H. 1963. Parasitism of the red-banded leafroller, *Argyrotaenia velutinana*. *Ann. Entomol. Soc. Amer.* **56**:564.

Glass, E. H., and S. E. Lienk. 1971. Apple insects and mite populations developing after discontinuance of insecticides: 10 year record. *J. Econ. Entomol.* **64**:23–26.

Glass, E. H., W. L. Roelofs, H. Arn, and A. Comeau. 1970. Sex pheromone trapping red-banded leafroller moths and development of a long-lasting polyethylene wick. *J. Econ. Entomol.* **63**:370–373.

Hagley, E. A. C. 1973. Timing sprays for codling moth (Lepidoptera: Olethreutidae) control on apple. *Can. Entomol.* **105**:1085–1089.

Haynes, D. L., R. K. Brandenburg, and D. P. Fisher. 1973. Environmental monitoring network for pest maganement systems. *Environ. Entomol.* **2**: 889–899.

Hough, W. S. 1963. Resistance to insecticides by codling moth and red-banded leafroller. *V. Agr. Exp. Sta. Tech. Bull.* 166. 32 pp.

Howitt, A. J., and L. J. Connor. 1964. The response of *Rhagoletis pomonella* (Walsh) adults and other insects to trap boards baited with protein hydrolysate. *Proc. Entomol. Soc. Ont.* **95**:134–136.

Howitt, A. J., E. J. Klos, P. Corbett, and A. Pshea. 1966. Aerial and ground ultra low volume applications in the control of diseases and pests attacking deciduous fruits. *Mich. Agr. Exp. Sta. Quart. Bull.* **49**:90–102.

Hoyt, S. C. 1969. Integrated chemical control of insects and biological control of mites on apple in Washington. *J. Econ. Entomol.* **62**:74–86.

Hoyt, S. C., and E. C. Burts. 1974. Integrated control of fruit pests. *Annu. Rev. Entomol.* **19**:231–252.

Hoyt, S. C., and L. E. Caltagirone. The developing programs of integrated control of pests of apples in Washington and peaches in California. Pages 395–421 *in* C. B. Huffaker, ed., Biological control. Plenum Press, New York.

Huffaker, C. B., M. van de Vrie, and J. A. McMurtry. 1969. The ecology of tetranychid mites and their natural control. *Annu. Rev. Entomol.* **14**:125–174.

Jacklin, S. W., E. G. Richardson, and C. E. Yonce. 1970. Substerilizing doses of gamma irradiation to produce population suppression in plum curculio. *J. Econ. Entomol.* **63**:1053–1057.

Jaques, R. P., and H. T. Stultz. 1966. The influence of a virus disease and parasites on *Spilonota ocellana* in apple orchards. *Can. Entomol.* **90**:1035–1045.

Jones, A. L., and T. B. Sutton. 1974. Unpublished data.

Kirby, A. M. A. 1973. Progress in the control of orchard pests by integrated methods. *Commonw. Bur. Hort. Plant Crops. Horst.* Abstr. **43**:1–16, 57–65.

Knight, R. I., J. B. Briggs, A. M. Massee, and H. M. Tydemann. 1962. The inheritance of resistance to woolly aphid, *Eriosoma lanigerum* (Hasmn.) in the apple. *J. Hort. Sci.* **37**:207–218.

Legner, E. F., and E. R. Oatman. 1963. Natural biotic control factors of the eyespotted bud moth, *Spilonota ocellana* on apple in Wisconsin. *J. Econ. Entomol.* **56**:730–732.

Lewis, F. H., and K. D. Hickey. 1967. Methods of using large airblast sprays on apples. *Penn. Fruit News* **46**:47–53.

MacLellan, C. R. 1974. Unpublished data.

MacPhee, A. W., and C. R. MacLellan. 1971. Ecology of apple orchard fauna and integrated pest control in Nova Scotia. *Proc. Tall Timbers Conf.* **3**: 197–208.

Madsen, H. F., and C. V. G. Morgan. 1970. Pome fruit pests and their control. *Annu. Rev. Entomol.* **15**:295–320.

Mathys, G. 1966. Possibilities de lutte contre le pour de San Jose par la methode biologique et integree. Pages 53–64 *in* Proceedings of the FAO symposium on integrated pest control, Rome, October 1965. Vol. 3. Food Agricultural Organization of the United Nations, Rome.

Mathys, G., and E. Guignard. 1967. Quelques aspects de la lutte biologique contre le pori de San Jose (*Quadraspidiotus perniciosus* (Comst.) a' l'aide de l'aphelinide *Prospaltella perniciosi* Tow. *Entomophaga* **12**:223–234.

Maxwell, C. W. 1969. Observations on bird tanglefoot as a direct method of apple maggot control. *J. Econ. Entomol.* **62**:495–496.

McMurtry, J. A., C. B. Huffaker, and M. van de Vrie. 1970. Ecology of tetranychid mites and their natural enemies: A review. I. Tetranychid enemies: Their biological characters and the impact of spray practices. *Hilgardia* **40**:331–390.

Metcalf, C. L., W. P. Flint, and R. L. Metcalf. 1962. Destructive and useful insects. McGraw Hill, New York. 1087 pp.

Meyer, R. H. 1974. Management of phytophagous and predatory mites in Illinois orchards. *Environ. Entomol.* **3**(2):333–340.

Miller, T. D., and M. N. Schroth. 1972. Monitoring the epiphytic populations of *Erwinia amylovora* on pear with a selective medium. *Phytopathology* **62**:1175–1182.

Mills, W. D., and A. A. LaPlante. 1951. Diseases and insects in the orchard. *Cornell Ext. Bull.* **711**:21–27.

Montieth, L. G. 1972. Status of predators of the adult apple maggot, *Rhagoletis pomonella,* in Ontario. *Can. Entomol.* **104**:257–262.

Motoyama, N., G. C. Rock, and W. C. Dauterman. 1970. Organophosphorus resistance in an apple orchard population of *Typhlodromus* (*Amblyseius*) *fallacis. J. Econ. Entomol.* **63**:1439–1442.

Oatman, E. R. 1960. Wisconsin apple insects. *Wisc. Agr. Exp. Sta. Tech. Bull.* 548. 29 pp.

Oatman, E. R. 1965. The effect of *Bacillus thuringiensis* Berliner on some lepidopterous larval pests, apple aphid and predators and on phytophagous and predaceous mites on young apple trees. *J. Econ. Entomol.* **58**:1144–1147.

Oatman, E. R. 1966. Studies on integrated control of apple pests. *J. Econ. Entomol.* **59**:368–373.

Oatman, E. R., and L. Jenkins. 1962. The biology of the red-banded leaf roller, *Argyrotaenia velutinana* (Wlkr.), in Missouri with notes on its natural control. *Univ. Mo. Res. Bull.* 789. 14 pp.

Oatman, E. R., E. F. Legner, and R. F. Brooks. 1964. An ecological study of arthropod populations on apple in northeastern Wisconsin: Insect species present. *J. Econ. Entomol.* **57**:978–983.

Paradis, R. O. 1961. Essai d'analyse des facteurs de mortalite chez *Archips argyrospilus* (Walk.). *Ann. Entomol. Soc. Que.* **6**:59–69.

Parent, B. 1967. Population studies of phytophagous mites and predators on apple in southwestern Quebec. *Can. Entomol.* **99**:771–778.

Pickett, A. D., W. L. Putman, and E. S. LeRoux. 1958. Progress in harmonizing biological and chemical control of orchard pests in eastern Canada. *Proc. 10th Int. Cong. Entomol. (Montreal 1956)* **3**:169–174.

Plurad, S. B., R. N. Goodman, and W. R. Enns. 1965. Persistence of *Erwinia amylovora* in the apple aphid *Aphis pomi* De Geer, a probable vector. *Nature* **205**:206.

Prokopy, R. J. 1968. Sticky spheres for estimating apple maggot adult abundance. *J. Econ. Entomol.* **61**:1082–1085.

Prokopy, R. J. 1972. Evidence for a marking pheromone deterring repeated oviposition in apple maggot flies. *Environ. Entomol.* **1**:326–332.

Proverbs, M. D., J. R. Newton, and D M. Logan. 1966. Orchard assessment of the sterile male technique for control of the codling moth, *Carpocapsa pomonella* (L.) *Can. Entomol.* **98**:90–95.

Proverbs, M. D., J. R. Newton, and D. M. Logan. 1969. Codling moth control by releases of radiation-sterilized moths in a commercial apple orchard. *J. Econ. Entomol.* **62**:1331–1334.

Putman, W. L. 1963. The codling moth, *Carpocapsa (Cydia) pomonella* (L.): A review of its bionomics, ecology and control on apple with special reference to Ontario. *Proc. Entomol. Soc. Ont.* **93**:22–60.

Riedl, H., and B. A. Croft. 1974. A study of pheromone trap catches in relation to codling moth damage. *Can. Entomol.* **106**:525–537.

Rivard, I. 1968. Synopsis et bibliographic annotte sur la mouche de la pomme *Rhagoletis pomonella* (Walsh). *Mem. Entomol. Soc. Que.* **2**:1–158.

Rock, G. C. 1972. Integrated control of mites. *Agr. Exp. Serv., Fruit Insect Note,* A-1, *N. C. State Univ. Publ.* 18 pp.

Roelofs, W., and A. Comeau. 1971. Sex attractants in Lepidoptera. *Proc. 2nd Int. Cong. Pestic. Chem. IUPAC, Tel Aviv, Israel.* **3**:91–114.

Roelofs, W. L., and K. Feng. 1968. Sex pheromone specificity tests on the tortricidae, an introductory report. *Ann. Entomol. Soc. Amer.* **61**:312–316.

Russ, K. 1974. Über ein bemekenswertes Auftreten von *Beauveria bassiana* (Bols.). Vuill. an *Carpocapsa pomonella* (L.). *Pflanzenschutzberichte* **31**: 105–108.

Sanford, K. H., and H. J. Herbert. 1970. The influence of spray programs on the fauna of apple orchards in Nova Scotia. XX. Trends after altering levels of phytophagous mites or predators. *Can. Entomol.* **102**:592–601.

Slingerland, M. V., and C. R. Crosby. 1930. Manual of fruit insects. *Macmillan,* New York. 503 pp.

Smarasinghe, S., and E. J. LeRoux. 1966. The biology and dynamics of the oystershell scale, *Lepidosaphes ulmi* (L.) on apple in Quebec. *Ann. Entomol. Soc. Que.* **11**:206–292.

Swift, F. C. 1970. Predation of *Typhlodromus* (A.) *fallacis* on the European red mite as measured by the insecticide check method. *J. Econ. Entomol.* **63**: 1617–1618.

Thompson, W. W., C. F. Stephens, L. G. Olsen, and J. F. Cole. 1973. Michigan apple pest management annual report—1972. *Mich. State Univ. Ext. Publ.* 48 pp.

Trammel, K. 1972. The integrated approach to apple pest management and what we are doing in New York. *Proc. N.Y. State Hort. Soc.* **117**:37–49.

Westigard, P. H. 1971. Integrated control of spider mites on pear *J. Econ. Entomol.* **64**:496–501.

Wildbolz, T., and W. Meier. 1973. Integrated control: Critical assessment of case histories in affluent economies. *Ecol. Soc. Aust. Mem.* **1**:221–231.

14

FOREST INSECT PEST MANAGEMENT

Ronald W. Stark

I. THE FOREST "PEST" PROBLEM

Forest ecosystems are by nature much more diverse than agroecosystems. The latter are man-made and largely man-controlled. Our forests were naturally created, and while they have been drastically altered (and frequently eliminated) by man, the vast bulk of them still retain their complex nature. Today there are all stages of forests, from agricultural types to complete wilderness, and there are different usages for each.

Forest nurseries with a 1- to 2-year planting cycle represent the closest comparison of forest and agriculture. Christmas tree plantations with a 3- to 10-year cropping cycle have the uniform spacing and age distribution, single-crop species, and containment attributes of field crops, but are more complex in that they are exposed over a longer period of time to a greater variety of injurious insects. Commercial monocultures for pulpwood or small-dimension lumber, while still maintaining the three features mentioned above, greatly increase the exposure time to injurious agents of all kinds. This is altered in some management systems by introducing different species of trees, thus increasing the complexity of the plantation environment, hopefully to confound the insect foragers. The other extreme is the natural forest which lives in harmony with its forest insects and other inhabitants of its ecosystem. These also vary in the kinds of species present and their proportions, the degree of age variation, and the uses to which they are put.

Such natural forests are generally highly resistant to forest pests. They have evolved over millions of years with insects, diseases, mammals, and a multitude of plants to a more-or-less harmonious balance. This is not to say that insects do not feed on or kill trees in natural forests, that diseases do not weaken and kill trees, or that beavers or porcupines cease their utilization of trees. It is only when their activities interfere with man's desires, economic or aesthetic, that they become undesirable or unacceptable. The problems of forest pest management thus differ from agricultural pest management, largely in time, space, and usage, with attributes peculiar to each.

All plants are exposed to a spectrum of insects from the seed stage to maturity or the age of harvesting. However, the length of time taken to grow a tree to commercial size may be anywhere from 40 to 100 years. During this period there are insect herbivores peculiar to each growth period. There are many insects that prefer to feed on young seedlings. If the tree survives this period to the sapling stage, there is a different complex of insects to face; this complex may be replaced by a different set at the pole stage, and when the tree reaches maturity there is yet another group of insects which feed on it.

The chances of tree damage occurring are magnified by the fact that over this long period of time the trees are exposed to a multitude of environmental factors which may render them more susceptible to insect attack or less resistant to it. Unless controlled, forests tend to be overstocked, that is, in nature, there are usually more plants per acre than can survive. This leads to competition for water and nutritive elements, and less vigorous trees are more liable to attack than vigorous ones. This in itself is not necessarily a problem, since these trees eventually die anyway, but their presence can contribute to increases in insect populations to a level where they attack even vigorous trees.

Other environmental hazards that contribute to population increases or outbreaks of forest insects are forest pathogens. There are many of these, some attacking the needles, some the branches, some the stem, and some the root system. Over time, these can weaken the tree, rendering it susceptible to attack by insects normally contained at low levels in healthy stands. Almost any natural disturbance over which man has little or no control, such as fire, flood, drought, windstorms, or lightning, can create conditions in the forest that permit insect populations to increase to damaging levels. Over the long time span of forest growth, the probability of such events occurring increases.

Another major difference between agroecosystems and forest ecosystems is in space. While there are vast acreages under agricultural production, these are largely in relatively small units under the strict control of specific sociological segments, that is, family farms to corporate farms. Also, while there may be many thousands of contiguous acres, they are or can be broken up by variation in crops planted, or are naturally discontinuous. Forest regions tend to be more discrete, and in any particular area more or less continuous and uniform over vast areas. However, ownership and supervision varies from private small family holdings, to large corporate holdings, to vast areas supervised by state and federal agencies. Thus a forest may encompass hundreds or even thousands of square miles, more or less uniform with respect to the crop present, but without the controlled supervision present in agroecosystems. A significant problem stemming from this is accessibility, which is vital to pest management. Agricultural systems have accessibility built in, and commonly each plant can be reached easily with modern machinery. Conversely, there are large tracts of forest lands that contain no roads or trails. Observation must be maintained by remote means such as aerial or satellite photography, and treatment in such areas is virtually impossible.

Allied with the space problem somewhat, but unique to forest ecosystems as opposed to agroecosystems, is the question of usage. The agroecosystem is man-made with specific objectives, and pest-management objectives are clear-cut. The principal objective is maximizing productivity of a commercial crop. Forests, however, serve a multitude of uses, not all of them commercial. The multiple-use aspects of forests and the mixed ownership of forests by private and public agencies places a variety of constraints on forest pest management not faced in agricultural pest management. These have increased in magnitude in recent years and are profoundly affected by the interests of the people involved. Few people know or care about the pest-management techniques used on a corporate farm in southern Idaho or Texas, but the nation identifies with the forests of Yellowstone or Yosemite National Park.

The legal constraints on pest management are common to both agricultural and forest ecosystems, varying only in the degree of application. For example, the direct statutory constraints of the Federal Insecticide, Fungicide and Rodenticide Act (FIFRA) (passed in 1947 and amended in 1959 and 1964) and its most recent successor, the Federal Environmental Pesticide Control Act (21 October, 1972), administered by the U.S. Environmental Protection Agency, place severe constraints on both practice and research in both ecosystems. As defined in this act, pesticides include insecticides, nematocides, fungicides (including wood preservatives), disinfectants, herbicides, rodenticides, animal poisons and repellents, plant regulators, plant defoliants, and plant desiccants, and has specific restrictions on interstate transport and usages of these materials. FIFRA and the National Environmental Policy Act of 1969 (Public Law 91-190) and the Environmental Quality Improvement Act of 1970 (Public Law 91-224) reflect an increasing national concern over the effects of pesticides in the ecosystem. This has its advantages, particularly for the development of ecologically sound pest management, but also its disadvantages in restricting or inhibiting extensive tests with natural viruses, bacterial diseases, juvenile hormones, and sex attractants and so on.

The user constraints in the forest ecosystem may be examined by stratifying users into interest and attitudinal groups to which the forest manager (private or public) must respond (Campbell 1973). To further simplify it, consider a particular pest, the Douglas fir tussock moth, *Hemerocampa pseudotsugata* McDunnough.

1. *Suburbanites.* The one or two fir trees on his lot are the most valuable trees in the neighborhood. The urticating hairs from the larvae may cause conjunctivitis or allergic reactions in his family.
2. *Users and managers of public and private recreation areas.* The degree of shade in a campground affects the use pattern of that campground. The presence of unfamiliar, messy, and loathsome (to some) caterpillars drives people away, in addition to causing the potential health problems mentioned above. A fishing stream covered with dead carcasses of larvae affects that activity.
3. *Users and managers of natural areas.* Reactions may vary from those mentioned in item 2 to the other extreme, awe and pleasure in witnessing a natural phenomenon.
4. *Users and managers of commercial forests.* Reactions here are the closest to those displayed by agricultural pest managers. Any insect feeding on a crop is the "enemy." Defoliation reduces the growth of the tree, resulting in longer rotation and an actual dollar loss; trees may be killed, resulting in changes in the cutting plan at increased cost or outright loss of the tree; regeneration may suffer; and the weakening of the tree may invite other destructive insect activity.

If all these interest groups are contiguous, as they are in many parts of Oregon, Washington, and Idaho, their input to pest-management decisions is often at odds and difficult to resolve. The complexity is further magnified by the fact that ownership of the forests is divided between public agencies (state and federal) and the private sector, which have different policy guidelines. Private ownership is divided among recreational, investment, and commercial interests, and commercial ownership objectives are comparable to those in agricultural crop production. Public ownership objectives vary from commercial use to retention of wilderness areas. Between these two extremes are varying categories of parks of all types, natural areas, and watershed areas. Policy and political considerations vary from state to state and between the large land-owning federal agencies such as the U.S. Forest Service, the Bureau of Land Management, the Bureau of Indian Affairs, the Army, the Air Force, and even the Navy.

We have attempted to outline the major problems of pest management in the forest ecosystem that are unique or different from those in the agroecosystem. Briefly, the problems differ in space because of the vast contiguous forested areas, in time because of the life cycle of the forest, and in what we call "usage" because of the multitude of uses to which a forest may be put. Not included in the above discussion is the affinity that the bulk of the public feel for their forests, related to its multiplicity of uses. The forests are used by many to "re-create" their spirits, even vicariously through television or movies. Intangible as it may be, there exists a spiritual relationship between the general public and forests. This relationship adds to the problems of pest management in that, before reaching decisions and implementing strategies, any pest-management system must include consideration of it to a greater degree than is required in the agroecosystem.

II. THE EMERGENCE OF PEST MANAGEMENT

Efforts to suppress populations of insects causing damage to our forest resources have largely avoided the treadmill of pesticide dependence suffered by the agricultural industry. There have been and still are efforts to apply the simplistic and ineffective tactics of pest control hitherto used in agriculture, namely, the use of DDT and other insecticides against the spruce budworm, *Choristoneura fumiferana* (Clemens); the gypsy moth, *Porthetria dispar* (L.); the lodgepole needle miner, *Coleotechnites milleri* (Busck); and most recently the Douglas fir tussock moth, *Hemerocampa pseudotsugata* McDunnough. The fact that agricultural entomologists are trying to get off the treadmill does not deter many of the proponents of this "direct control" philosophy. Fortunately, in forest entomology these victims of past procedural philosophy are becoming outnumbered, because the very complexity of managing a forest resource has forced the integration of many disciplines, and significant developments in these have merged to focus on ecologically based pest *management*.

The emergence of pest management is most timely, since many of our forest insect problems have been increasing in magnitude, in part because of our attempts to follow the agricultural control example, but more importantly because of our inexpert exploitation of the forests and abuses of them. The exhaustion of virgin stands has created the necessity for more perceptive management of forests to meet growing demands for the production of fiber, as well as competitive demands on forest resources such as recreation, mining, and watershed storage. A potential future demand on forest productivity of incalculable magnitude is as an energy source. As the supply of our fossil fuels, coal and oil, declines, there will be increasing pressures on the renewable source of energy—the forest.

Forest pest management has emerged as a discipline in itself based on two productive areas of scientific investigation (population dynamics and forest ecology), the intelligent application of economic theory, the development of a myriad of tactics for the suppression of insect populations, the phenomenal capabilities of the computer and the development of techniques for handling vast amounts of data, the development of simulation modeling, and the application of systems analysis to the whole.

The study of population dynamics of an insect provides the key to the regulation or management of populations of that insect. The aim of population dynamics is to determine the factors that cause populations of insects to rise and fall. To do this it is necessary to study populations of insects over a reasonably long period of time and under various conditions. It is essential to know the reproductive capacity of the insect and those factors that result in reductions of the population from the beginning of any particular life cycle. For the purposes of pest-management, population dynamics provides (1) measurement and evaluation of population size and distribution, (2) clues to suppression tactics, and (3) some predictive ability.

1. A quantitative sampling method is necessary to measure population size. However, it is not enough just to know how many insects are there; this must be related to their feeding activity and damage.

An example is that derived for the lodgepole needle miner, *Coleotechnites milleri* (Busck). The application of this for forest pest-management decision making is obvious (Fig. 14.1).

2. The study of causes of population fluctuations can provide clues to suppression tactics, as well as guides to avoidance of disruptive tactics.

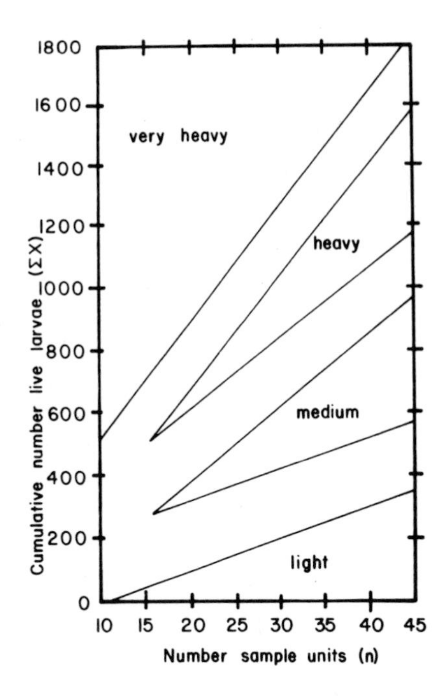

Fig. 14-1 A schematic example of a sequential sampling system. The cumulative sum of live larvae is plotted against the number of repetitive samples. From other studies the number of larvae are related to the degree of defoliation, and "defoliation classes" are established: very heavy, heavy, medium, and light. When the plotted cumulative sum falls into one of the classes, the level of damage is known, which aids in deciding whether action is necessary or not. (Adapted from Stevens and Stark 1962).

Results of such studies are presented in various ways, the two most common being life tables (Table 14.1) and survivorship curves (Fig. 14.2), which are merely graphical presentations of the results from life tables.

From the hypothetical example given, we can see that insect predators, parasites, and woodpeckers are responsible for the bulk of the mortality. Detailed studies of these may lead to methods to increase their numbers and effectiveness (or to avoid decreasing them).

3. A series of life-tables studies of successive generations of an insect provide some predictive ability. For example, if generation after generation there is a relatively constant percentage mortality, we can predict what the population will be at the end of the generation when we measure the number present at the beginning. Also, if we apply a suppression tactic and can measure the mortality effect it imposes, we can again predict the end result on the population. If the tactic is disruptive of existing mortality factors, such as an insecticide that does not discriminate between targets, we can again predict what the effect on the surviving target insect will be by virtue of elimination of the natural controls. Even more important, given the length of time a forest is exposed to damage, is the potential long-range predictive capability sought in population dynamics. For example, we know that the Douglas fir tussock moth

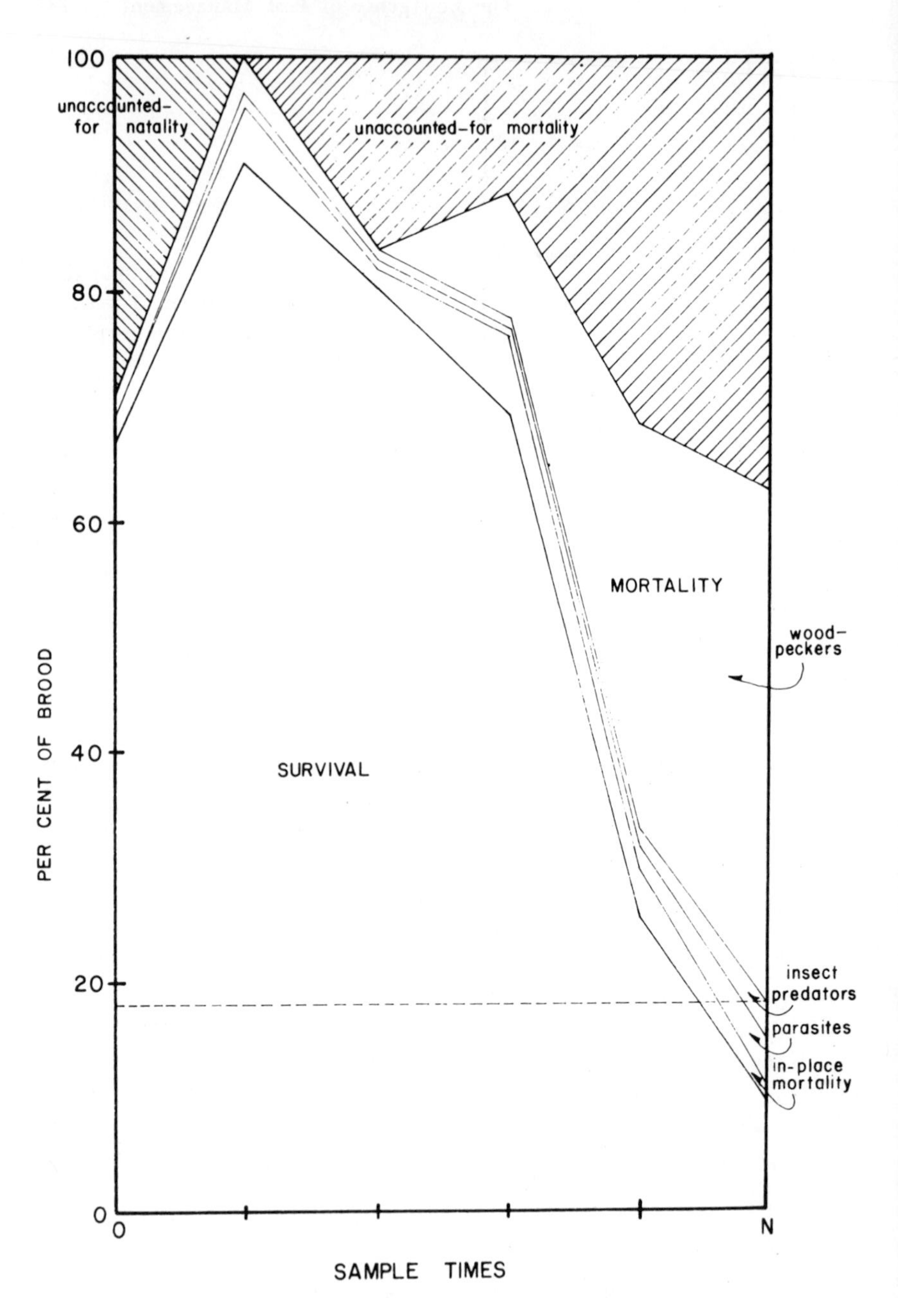

Fig. 14-2 A schematic survivorship curve. The percent surviving and the mortality inflicted by various mortality factors are plotted against sample time. The increase at the beginning occurs because the first sample was taken while the insect (western pine beetle) was still attacking, hence while mortality was occurring new progeny were being added. (Adapted from Stark and Dahlsten (1970.)

Table 14.1 Hypothetical Life Table for the Western Pine Beetle[a]

Sample Interval or Stage, x	Number Alive at Start of Sampling Interval, lx	Factor Responsible for Mortality, dxF	Number Dying during Sampling Interval, dx	Percentage Mortality, $100rx$
Egg	1000	Predators	100	10
		Desiccation	100	10
	−200	Total	200	20
Larval	800	Predators	200	25
		Parasites	100	12.5
		Woodpeckers	300	37.5
		Desiccation	50	6.25
		Unknown	50	6.25
	−700	Total	700	87.5
Emerging adults	100 −50	Predators	50	50
Successful emergence	50	—	Generation mortality 950	95
Sex ratio 1:1	25; 25			

[a] Simplified and hypothetical but based on actual observations (Stark and Dahlsten 1970).

increases explosively in numbers every 7 to 10 years, persists in an outbreak phase for 3 to 4 years, and then collapses, usually as a result of a natural virus disease. What we do not know yet is where and why the outbreaks occur. Further, we know that a prolonged drought in western coniferous forests is almost invariably followed by an increase in some bark beetle populations. Coupling meteorological records with population studies permits us to anticipate such problems.

Forest ecology has followed somewhat the same pattern as insect ecology and has developed its own variation of population dynamics—stand dynamics. Beginning with the existing forest, or a newly regenerated one, foresters are able to estimate for a given area the probable nature of that forest at maturity. Such predictions have been developed primarily for commercial use and have not included insect dynamics as a specific component. However, for insect pest management this is essential, since the determination of what should be done when an infestation occurs must be based on a realistic estimate of what will occur if nothing is done. More is said about this in Section III.

Of particular significance to pest management are ecological studies showing the interdependency of the various organisms that use the tree as a host—many of which we call pests. Organisms such as mistletoe, root diseases, rusts, and myriads of insects interact to place stress on their host, the tree. Physical and biotic factors of the environment mold the tree to a degree, and the ultimate appearance of trees and whole stands provides an indicator that may be used in pest management. Combining the symptoms that separate a tree or forest under stress from healthy vigorous trees or forests results in a rating system for risk or hazard to insect outbreaks. Our knowledge of this is very limited but, as in insect population dynamics, the study of stand dynamics provides keys to effective stand management.

Another contribution to emerging pest management is the intelligent application of economic theory. To date, in forest insect control the response to widespread insect damage has been largely emotional. Pest control has become procedural, and careful consideration has not been given to the economic benefits gained by treatment weighed against all the costs (Chapters 1 and 3). We have stated that forests are used for a multiplicity of purposes. Each use has a different set of values, and the effect on these should be included in the decision to expend control dollars, particularly since the bulk of control expenditures in the forest is from the public treasury.

For example, in any one national forest there may be five uses on which an insect outbreak may have an impact: timber production, domestic grazing, recreation, wildlife, and watershed. A tree-killing insect such as a bark beetle can possibly have a negative effect on timber production and domestic grazing, but the effect might be positive on wildlife by improving habitat, hence hunting, and on watershed by reducing the amount of water utilized. The impact on recreation might be positive or negative; the outbreak might present a natural attraction such as the Ghost Forests of Tioga, or reduce campground usage because of loss of shade. The point is that, for intelligent pest management, the pros and cons of the impacts on the multiple-use environment of the forest must be weighed before treatment is prescribed. Of course, in the case of a private, commercial forest, the impacts are more easily defined.

The development of alternative tactics for suppression of pests has been covered in other chapters. We only add that forests provide possibly a greater scope of cultural practices than do agroecosystems—at least intensively managed ones. The forest manager has time to manipulate stocking, species composition, and age structure, and to correct obvious problems.

Computer technology is making a significant contribution to modern forest pest management. Because of their phenomenal speed and storage capacity, the newest computers can describe the extremely complex models of natural processes (Chapter 10). Biology and mathematics have created a new type of mathematics hitherto largely shunned because of the extreme variability of biological processes. A host of ancillary disciplines has emerged, which permit us to mimic and then simulate biological processes—such as the fate of a forest stand over a 100-year period—with only limited excursions to the forest to resolve questions.

Systems analysis is not a terribly profound science. Any well-run business is an example of a good systems analytical approach. It involves mathematical, statistical, and mechanical techniques, and procedures for analyzing how a system works. The steps to be followed (and these are described by example in Section III) are:

1. *Definition of the system.* The example to be presented below has been defined as the mountain pine beetle–lodgepole pine ecosystem.

2. *Analysis of the system.* This analysis includes determining the components of the system and how they interact to perpetuate the system. At this point practicality must dictate the level of investigation. Our objective is not to determine the ecological principles of life. A systems analyst examines in minute detail existing information before embarking on the search for new information to supply missing factors in the well-defined equation.

3. *Modeling the ecosystem.* Having defined and analyzed the components of the pest ecosystem, the analyst attempts to model it, using all the variables thought to be important. The objective is to mimic the real system as accurately as possible so as to be able to tinker with it and, from the point of view of pest management, improve it by reducing pest populations (Chapter 10).

4. *Simulate the ecosystem.* With a model that resembles the real thing, the entomologists, foresters, mathematicians, and analysts then begin to change the ecosystem—on the computer. Their questions might all be prefaced with "what if," for example, "What if we thinned this stand to 200 basal feet per acre? Would that reduce the probability of attack by the mountain pine beetle?" Based on realism, the simulators are able to test theories on the computer that would be difficult or impossible to test in the forest.

5. *Testing the model.* No simulation or game is complete or acceptable until it has been tested in the field. After the pest suppression strategies are decided on and their potential effect on the pest and on the ecosystem estimated, it is time to face reality.

6. *Refining the model.* Testing the model inevitably points out deficiencies in the theory which require additional data for its support or refutation, or to direct investigation of an important but unrecognized causal factor; so the process begins anew; as a system should be, it is a multiloop feedback system, constantly being adjusted and refined.

III. A PEST MANAGEMENT APPROACH

Tangible recognition of the emergence of the new pest-management concept described here is the endorsement and support of a multidisciplinary, multiinstitutional project on pest management by the National Science Foundation, the U.S. Environmental Protection Agency, and the USDA (Agricultural Research Service and U.S. Forest Service). Begun in 1972, this mammoth undertaking addresses itself to six major national crop ecosystems—alfalfa, citrus fruits, cotton, pines, pome and stone fruits, and soybeans—and involves up to 200 scientists from 19 universities and state and federal agencies. The pines crop ecosystem subproject includes three major bark beetle species in North America, the mountain pine beetle, *Dendroctonus ponderosae* Hopkins; the southern pine beetle, *Dendroctonus frontalis* Zimmerman; and the western pine beetle, *Dendroctonus brevicomis* LeConte, and their principal host(s). In this subproject there are 25 scientists from the Universities of California and Idaho, Texas A & M, Washington State, and Virginia Polytechnic Institute, and 23 scientists from experiment stations and regional offices of the U.S. Forest Service. The mountain pine beetle–lodgepole pine ecosystem research objectives are presented here as an example—the objectives of all pine beetle studies are the same; there are only variations in procedures. It is believed that the methodolgy and technology developed from this project will be transferable to other forest pest problems.

A. Objectives of Pines Pest Management

"The overall objective is to develop a pest-management system which will achieve population regulation of the pests to levels compatible with tolerable socio-economic threshold values using methods which operate in concert with the ecological working of the ecosystem" (Huffaker 1973). Procedural objectives are:

1. To develop models to evaluate the impact of bark beetles throughout their range, considering *all* values implicit in the resource
2. To develop models that permit determination of local and regional trends in beetle activity
3. To develop strategy systems which, if implemented, will utilize integrated, ecologically harmonious population-regulation methods
4. To develop a pest management system which, utilizing the evaluations of item 1, and the determinations and predictions of 2, will permit rational decisions to be made in applying the results of 3.

The generalized mountain pine beetle–lodgepole pine pest management structure is presented in Fig. 14.3 and discussed in sequence.

B. Pest Population, Data and Modeling

The tasks implied in the flow chart are to describe the dynamics of mountain pine beetle populations (Berryman 1974), to construct mathematical models to predict population trends and resultant tree mortality, and to predict the effects of simulated management strategies. The dynamics of the population are determined by repetitive sampling of all factors affecting the population. Predictive models utilize the sampling data as well as population correlates, for example, phenological and climatic indicators, and may not in themselves be biologically meaningful. However, simulation models should be biologically meaningful in terms of the beetle–host ecosystem and should operate continuously over many population cycles. Simulation models are used for testing various management strategies, and predictive models can be used for validating the performance of the more complex simulation model.

The most optimistic expectation from modeling of the pest population will be to provide the other two modeling groups, host-tree and stand and socioeconomic impact, with a prediction of population events a generation (of the beetle) or more in advance (Fig. 14.3).

C. Host Tree and Stand Data and Modeling

The purpose of these studies is to provide growth projections for a lodgepole pine stand which take into account the prognostications for the insect population group (Campbell 1973).

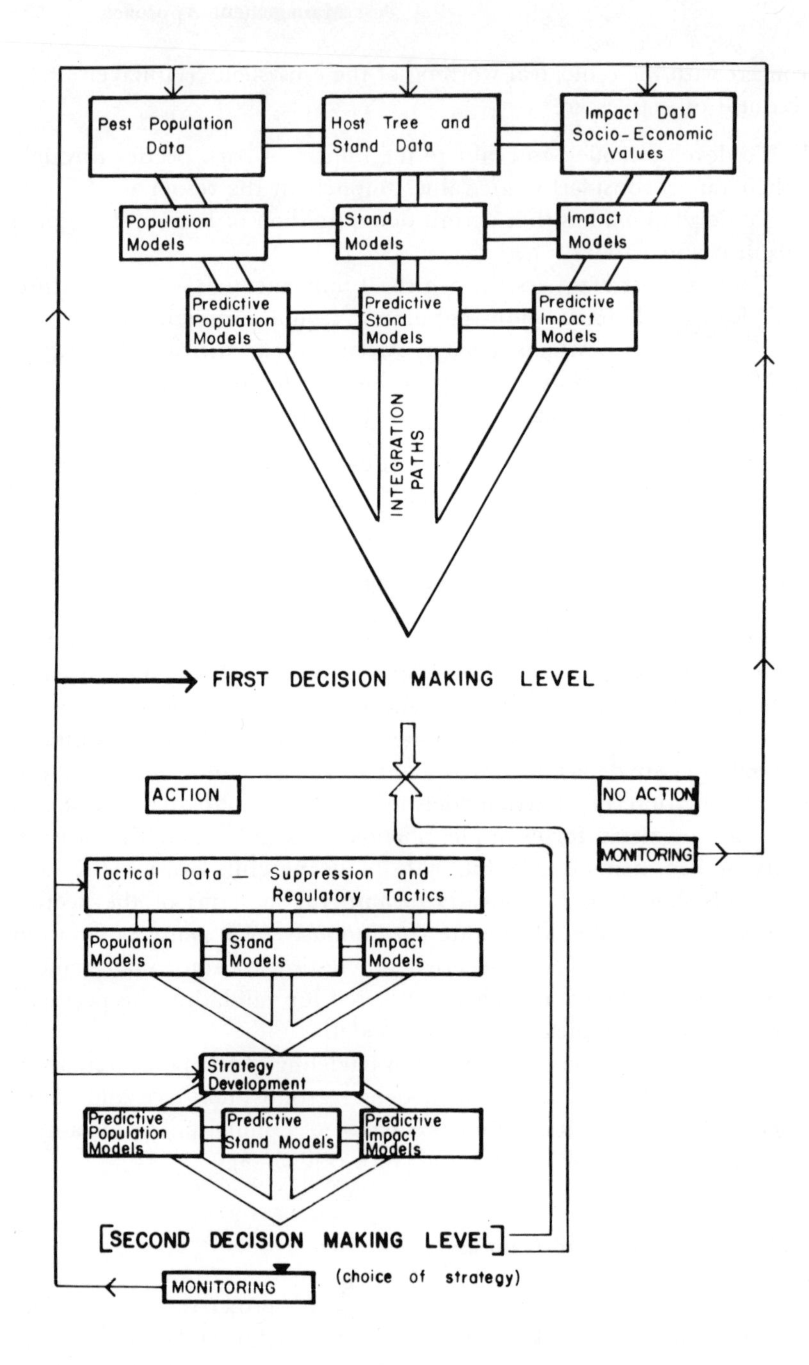

Fig. 14-3 A systems structure for forest pest management.

If it is assumed that we know the use(s) of the forest, the next step is to describe the stand conditions (Stage 1974) and classify them in similar groups which will respond to feasible, similar treatments. Then simulation models are designed for each stand classification, which present sequences of yields under different management alternatives (Stage and Alley 1972; Myers 1968). These yield predictions provide input for the forest manager to select the optimum mix of management activities. The technique used presently for this is Timber RAM (Stage 1973), which provides the manager a mix of prescriptions based on the stand development predictions. For pest-management purposes the predictive populaton models can be incorporated into this process beginning at the point where stand prognoses are undertaken. The various events that may occur in the development of a stand and the changes wrought by various management alternatives create a complex, exponential series of possibilities, but again these can be computerized (Myers 1968).

The success of this process depends on continuous monitoring of the pest situation and accurate population prediction in stands having different characteristics, accurate stand inventories in various situations, and feasible strategies for managing the stand. Hopefully, the insect and stand modelers have devised predictive models that describe what will happen to the pest population and what will happen to any particular stand, given the pest situation. At this focal point the results of the impact group are most meaningful.

D. Impact Data and Models

The impact group must evaluate the impact of the insect damage predicted by the population dynamics and stand dynamics group in terms of the potential or real uses of the forest stand (Huffaker 1973; Navon 1971). Most public forest land and much private land is planned for multiple use, and values, not necessarily monetary, must be assigned to them in order for the person(s) making pest-management decisions to act.

The damage done by the mountain pine beetle is precisely measurable in terms of killed trees (Fig. 14.4). Secondary effects of killed trees are used to determine economic impact on five major uses of the forest, and aesthetic impact on recreation (Fig. 14.5). Aesthetic impact may include purely political as well as emotional considerations. It should not be assumed that these impacts are all negative. If, for example, an area is restricted by law from timber harvests, there is no economic timber loss from beetle-killed trees.

Fig. 14-4 Typical mountain pine beetle damage in lodgepole pine stands. Grand Teton National Park, 1971. (Photo courtesy of U.S. Forest Service, Intermountain Region, Division of Timber Management, Ogden, Utah.)

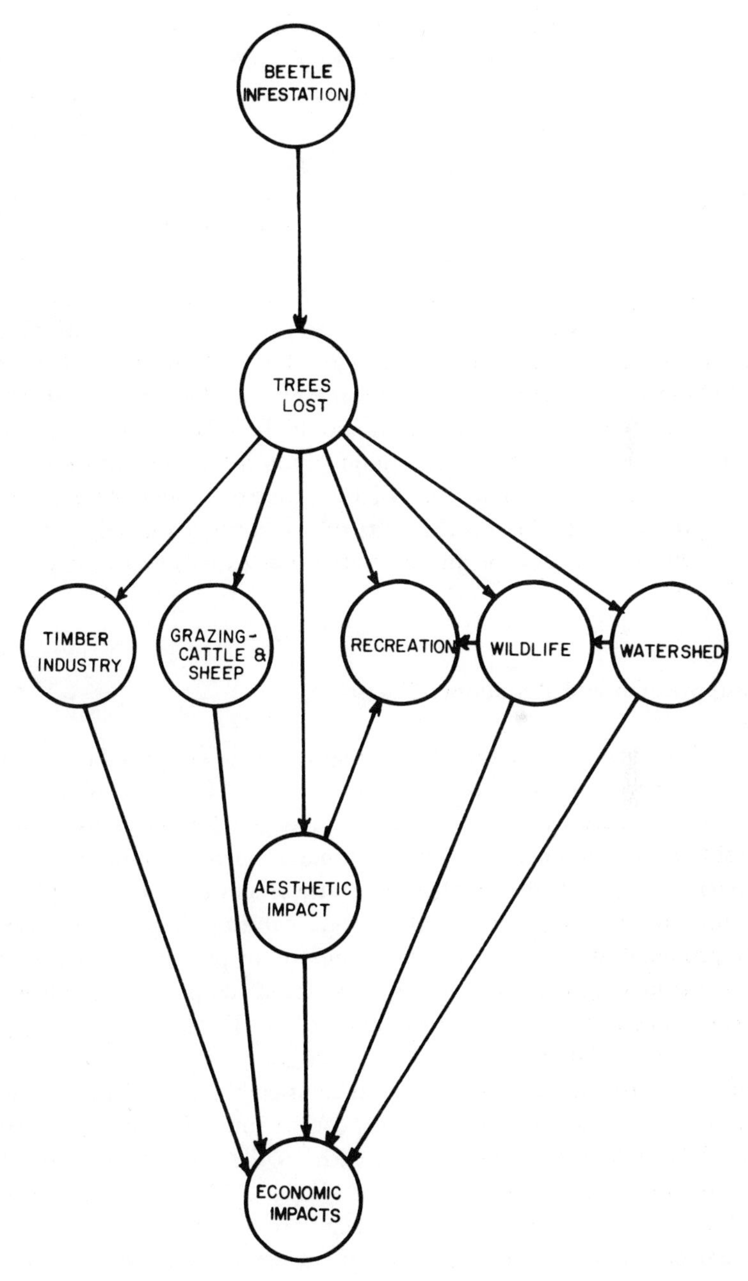

Fig. 14-5 Conceptual model of economic impact. This could apply to any tree-killing insect and, with adaptation, to sublethal defoliators and others.

The kinds of data used to determine economic and aesthetic impact on uses other than timber (including any wood fiber product) involve such factors as availability of forage and costs in removing problem trees (grazing), public reaction to beetle damage as determined by questionnaires (recreation), animal censuses and reactions and activity level of hunters and fishermen (wildlife), and changes in watershed storage capability and/or measurements of run-off (watershed).

As implied by the integrating paths in Fig. 14.3, none of these investigations operates alone, much of the data accumulated in each is of use to the others, the models interact, and the predictive models are almost totally integrative. Their products provide the material for the first decision-making level, that is, whether in any given insect infestation situation regulatory or suppressive action should be taken. Figure 14.3 does not clearly show that this first decision-making level must also know the costs of the second decision, that is, the strategy chosen, before *implementing* the first decision. If the costs of the strategy outweigh the estimated savings, no action or an alternative less expensive strategy might be selected.

E. Suppression and Regulatory Tactics

The various tactics for reducing and regulating insect populations have been thoroughly covered elsewhere in this text. However, as described above, the forest ecosystem, at least in managed forests, is particularly amenable to the use of silvicultural and management techniques, and for intelligent integrated pest management these must be fully exploited. Thinning, rotation age cycle, and site adaptability all show promise in "insectproofing" stands attacked by the mountain pine beetle. These and other forest-management practices, such as stand composition, are feasible and can sometimes be done profitably or at little or no cost from recovery of material.

The task of the pest manager, then, is to assemble strategies for various situations. The effect of these strategies can be simulated through predictive models to determine the effect of each on the insect population, the resultant forest stand, and the probable impact. These can be stratified according to degree of suppression or regulation and cost, and they can be weighed against the consequences of no action and relative gain of each and a strategy chosen.

CONCLUDING REMARKS

Simplistically, insect pest management consists of a decision or set of decisions to deal with an insect we have defined as a pest. The decision-making process assimilates information from various disciplines and deals with this information in various ways in order to reach decision(s).

The beginning of the process is the insect in the crop system. We must know the size of the population and the physical damage it inflicts in relation to the size of the crop it inhabits. We must interpret this damage in value terms, economic, social, political, or aesthetic. We must develop measures to predict what will occur, given action or no action, and we must develop tactics and strategies for regulation and suppression of forest insect pests that are ecologically compatible with the multiple-use ethic of the forests.

Pest management is not for the solitary scientist or practitioner, particularly in forest pest management. It requires the skills of many disciplines melded by a "science management system."

REFERENCES

Berryman, A. A. 1974. Management of mountain pine beetle populations in lodgepole pine ecosystems: A cooperative, interdisciplinary research and development project. Symposium on Management of Lodgepole Pine Ecosystems, Washington State University, October 9-11, 1973, pp. 629–650.

Campbell, R. W. 1973. The conceptual organization of research and development necessary for future pest management. Pages 23–28 *in* R. W. Stark and A. R. Gittins, eds., Pest management for the 21st century. Natural Resource Series no. 2. Idaho Research Foundation, Moscow, Idaho.

Huffaker, C. B. 1973. Integrated pest management. The principles, strategies and tactics of pest population regulation and control in major crop ecosystems. Integrated summaries. (NSF GB 34178) Vol. 1, pp. 108–131.

Myers, C. A. 1968. Simulating the management of even-aged timber stands. *USDA Forest Serv. Res. Paper* RM-42. 32 pp.

Navon, D. I. 1971. Timber RAM. . . .A long range planning method for commercial timber stands under multiple-use management. *USDA Forest Serv. Res. Paper* PSW-70. 22 pp.

Stage, A. R. 1973. Prognosis model for stand development. *USDA Forest Serv. Res. Paper* INT-137. 32 pp.

Stage, A. R. 1974. Stand prognosis in the presence of pests: Developing the expectations. Symposium on Management of Lodgepole Pine Ecosystems, Washington State University, October 9–11, 1973, pp. 233–245.

Stage, A. R., and J. R. Alley. 1972. An inventory design using stand examinations for planning and programming timber management. *USDA Forest Serv. Res. Paper* INT-126. 17 pp.

Stark, R. W., and D. L. Dahlsten, eds. 1970. Studies on the population dynamics of the western pine beetle, *Dendroctonus brevicomis* LeConte (Coleoptera: Scolytidae). University of California, Division of Agricultural Science. 174 pp.

Stevens, R. E., and R. W. Stark. 1962. Sequential sampling for *Evagara milleri*. *J. Econ. Entomol.* 55:493.

SELECTED READINGS

Bulla, L. A., Jr. 1973. Regulation of insect populations by microorganisms. *Ann. N. Y. Acad. Sci.* **217**:1–243.

Geier, P. W., L. R. Clark, D. J. Anderson, H. A. Nix, eds. 1973. Insects: Studies in population management. *Ecol. Soc. Austr. Mem.* 1. 295 pp.

Northeastern Forest Experiment Station. 1969. Forest insect population dynamics. Proceedings of the forest insect population dynamics workshop, West Haven, Conn., 23–27 January 1967. *USDA Forest Serv. Res. Paper* NE-125.

Northeastern Forest Experiment Station. 1971. Toward integrated control. Proceedings of the third annual northeastern forest insect work conference. *USDA Forest Serv. Res. Paper* NE-194. 129 pp.

Stark, R. W., and A. R. Gittins, eds. 1973. Pest management for the 21st century. Natural Resource Series no. 2. Idaho Research Foundation, Moscow, Idaho. 102 pp.

15

PEST-MANAGEMENT STRATEGIES FOR THE CONTROL OF INSECTS AFFECTING MAN AND DOMESTIC ANIMALS

Robert L. Metcalf

Insect pests of man and higher animals fall readily into two categories: (1) creeping, crawling, buzzing, stinging, and biting forms which, although they may cause a great deal of annoyance, at worst are stingers or bloodsuckers. Familiar examples include the American cockroach, *Periplaneta americana* (L.); various species of ants, bees, and wasps; horseflies and deerflies (Tabanidae); the bedbug, *Cimex lectularius* L.; and the pubic louse, *Phthirus pubis* (L.). Category (2) includes insect pests that serve as vectors of specific human and animal diseases. Their depredations encompass a much wider scope, often resulting in persistent illness and sometimes death. Examples of pests in this latter category include the *Anopheles* mosquito vectors of malaria; the human body louse, *Pediculus humanus humanus* L., vector of epidemic typhus; the plague flea, *Xenopsylla cheopis* (Rothschild), vector of plague; the tsetse flies, *Glossina* spp., vectors of African trypanosomiasis; the Triatominae bugs, vectors of American trypanosomiasis; and the *Simulium* blackfly, vector of onchocerciasis. Certain insects may be both general pests and disease vectors depending on areas of infestation, for example, the housefly,

Musca domestica L., which is a vector of trachoma virus and bacterial dysenteries in many parts of the world, and the German cockroach, *Blattella germanica* (L.), which is a nuisance everywhere but is also strongly incriminated in the spread of viral hepatitis.

A distinction is drawn sharply between the two groups in this chapter, because for vector-borne diseases pest management has the requisites of (1) urgency in dealing with epidemic or highly endemic situations, (2) flexibility for deployment under a wide variety of ecological and cultural situations in many countries, (3) efficiency in achieving a very high degree of control in order to suppress or eradicate the disease, and (4) simplicity and economy commensurate with the resources of developing nations.

Economic injury levels and economic thresholds are usually very much closer to the general equilibrium position for insect vectors of disease than for pest species that are merely annoying (Chapter 1, Section III. E). This is clearly demonstrated by public response to warnings about encephalitides transmitted by *Culex tarsalis* Coquillett and other vectors. The presence of Venezuelan equine encephalitis in horses in the United States in 1971 resulted in aerial, ultralow-volume spraying of over 30 million acres with malathion at 6 oz per acre, although there is little reason to suppose that mosquito densities were higher on the average in that year. Managers of mosquito abatement districts are familiar with the public clamor for spray programs that follows announcement of a few cases of mosquito-borne diseases, often by the same groups that have protested the use of spray programs for pest mosquito control.

Pest-management programs for insect vectors of human diseases are perhaps more intricate than those customarily devised for insect pests of agriculture, and are usually deployed on a state, national, continental, or even global basis under the aegis of local or national health services or the World Health Organization. However, because of the ecological complexities of the relationships between pathogen, insect vector, and human host or vertebrate reservoir, the opportunities for multifaceted attack are usually much greater. Thus the use of prophylactic antimalarial drugs such as chloroquine, directed at the causative *Plasmodium,* can play an important role in malaria control, and the eradication of reservoirs of wild host animals has been employed in control programs aimed at African trypanosomiasis (sleeping sickness), plague, and American trypanosomiasis (Chagas' disease).

I. VECTOR ECOLOGY

The spectacular successes resulting from the employment of DDT and other modern insecticides against the arthropod vectors of human and animal diseases, particularly in the control of malaria and typhus, have obscured the necessity for understanding vector ecology and the especially complex interrelations between vector, pathogen, human victim, and possible animal reservoir. WHO (1972) has described the consequences: "Recently the use of chemicals has encountered two complications, namely, the development of resistance by the vectors and the disturbing accumulation of these chemicals in the environment. Thus it has become necessary to develop alternative methods of control—genetic, biological, and environmental. This revision of strategy requires a knowledge of the basic mechanisms that govern the abundance of vectors. The importance of vector ecology in all control programmes against vector-borne disease is thus clearly demonstrated."

Pest-management strategies promise to provide the most useful and generally acceptable long-term solutions to the control of vector-borne diseases. Understanding vector ecology and the intricate interrelationships of the vector with the pathogen and its human host and animal reservoirs of the disease is a considerably more complex undertaking than understanding the ecology of a typical insect pest of agricultural crops.

A. Vector–Disease Systems

Chapter 10 explored the systems analysis and modeling of pest-management systems as applied to the suppression of agricultural pests. It is evident that these systems are relatively complex, yet can be described in dynamic models, that is, limitations and representations of the real world, which can be used to (1) ensure a logical analysis of the components and their interactions, (2) allow quantitative testing of biological assumptions, and (3) allow simulation of the pest-management system and prediction of the effects of natural changes and planned interventions of the various parameters.

Such models of the dynamics of a vector population and its interactions with the pathogen and vertebrate host are generally more complex and have larger numbers of variables than models of pest–crop systems (WHO 1972). Moreover, the objectives of a sound pest-management system, to "ensure favorable economic, ecological, and sociological consequences" (Chapter 1), become more difficult to determine quantitatively as we attempt to weigh the effects of human disease on a community, national, or global basis. Nevertheless, the consequences of unsatisfactory strategies are likely to be far more calamitous, for example, the epidemic of malaria in Ceylon in 1968–1969, which involved more than 2.5 million cases resulting from premature termination of DDT residual spraying and lack of ancillary control measures (WHO 1971). Therefore systems analysis and modeling seem to be critical factors for planning pest-management strategies in the control of vector-borne diseases.

Modeling a vector–disease system involves:

1. Determination of the part of the real world to be described by the model versus natural environmental factors which operate outside the model
2. Selection of the components or submodels to reflect the primary functional characteristics of the system (WHO 1972): (*a*) life table of the immature stages of the vector, (*b*) infection of the vector, (*c*) extrinsic cycle of the pathogen, and (*d*) infection of the host
3. Description in quantitative terms (mathematical) of each component to interrelate the inputs, outputs, and population stages
4. Coupling the various components together via inputs and outputs and connecting the system to the environment. A flow chart for such a generalized vector–disease system is presented in Fig. 15.1 (WHO 1972).

Chapter 10 describes how to build a descriptive dynamic model applicable to any pest-management system. This technique has been applied to the dynamics of malaria by Macdonald et al. (1968). Their model is particularly suited to study of the pest-disease management approach incorporating suppression of the mosquito vector by source reduction, naturalistic control, larviciding, and adulticiding, together with chemotherapy of the infected human population.

The critical components of the model and the mathematical relationships involved are: m = anopheline density in relation to man, $a \doteq$ average biting rate in man per mosquito per day, b = proportion of anophelines that are infective, p = probability of anopheline survival through one whole day, n = days for completion of extrinsic cycle of parasite, h = proportion of population infected in 1 day (inoculation rate), x = proportion of people showing parasitemia, L = limiting value of proportion of people infected when equilibrium is reached, r = proportion of affected people who have received one infective inoculum only, who revert to uninfected state in 1 day, t= time in days, and z_0 = number of infections in community from single primary nonimmune case. Then the basic rate of infection is:

$$z_0 = \frac{ma^2bp^n}{r(-\log_e p)} \tag{1}$$

and the net rate is

$$z = \frac{z_0(-\log_e p)}{ax - (\log_e p)} \tag{2}$$

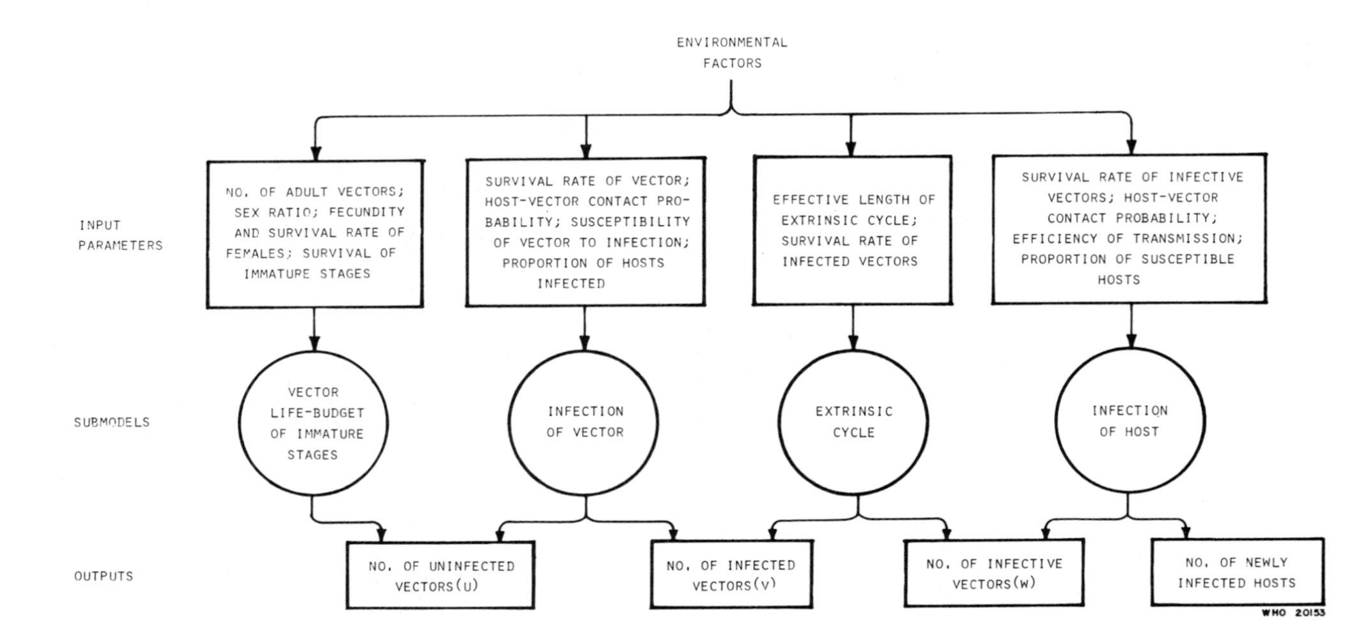

Fig. 15-1 An information flow chart for a vector-disease system. (From World Health Organization. Vector Ecology Tech. Rept. Ser. No. 501, 1972.)

The inoculation rate is

$$h = \frac{z_0 r x(-\log_e p)}{ax - \log_e p)} \tag{3}$$

$$h = \frac{z_0 r x_{t-i}(-\log_e p)}{ax_{t-i} - \log_e p)} \tag{4}$$

where i = coefficient of infectivity, about 30 for *Plasmodium falciparum* (Welch) and 18 for *P. vivax* (Grassi and Welch).
When $h < r$,

$$\frac{dx}{dt} = h(1 - x_t) - (r - h)\, x_t \tag{5}$$

When $h > r$, from equation (4),

$$\frac{dx}{dt} = h(1 - x_t) \tag{6}$$

The limiting value of equation (5) is

$$L_x = (-\log_e p)/a(z_0 - 1) \tag{7}$$

and the limiting value of equation (6) = 1.0.
The stability index is:

$$a/(-\log_e p) \tag{8}$$

and this represents the mean number of bites on a human that a typical mosquito vector makes during its lifetime. This determines stability, because it is a density-dependent mechanism.

Using models such as the one above in computer simulation of vector–disease systems assists in the development of optimal control strategies. Thus Macdonald et al. (1968) showed by simulation that mass drug therapy could play an important role together with insecticidal interventions in the control and eradication of malaria. Conway (1970) points out that optimizing control strategies in regard to malaria control involves determining the "minimum cost strategy which reduces the vector population to a level at which the incidence of malaria is zero." Conway devised a computerized basic anopheline population model which incorporates submodels for (1) development, (2) survival, (3) migration, and (4) fecundity. Each of the submodels is in turn broken down to further subsidiary models [e.g., for fecundity: (*a*) copulation, (*b*) egg maturation, and (*c*) oviposition)]. The output of each of the subsidiary models is relatively distinct, but linked as the input to another subsidiary.

Four control techniques were simulated on the model:

1. Insecticides to control adult anophelines
2. Insecticides to control larvae and pupae
3. Sterile-male releases to control adult fertility
4. Source reduction of breeding sites.

Output from the simulations provided predictions about the effects of combinations of the control techniques on adult anopheline populations. This type of model can easily be linked with the Macdonald epidemiological model to provide information on the effect of simulated control actions on both vector populations and parasite infections in a human population.

B. Life Tables of Immature Stages

The life-table concept has been shown to be of value (Chapters 1, 7, and 10) in delineating the density-dependent and density-independent factors that cause changes in insect pest densities. Life-table analysis of vector populations identifies mortality factors which may be manipulated to reduce pest populations in pest-management programs and may suggest improvements in the integrated use of insecticides. Life-table analysis can also produce parameters, such as the net reproductive rate, that are of basic importance in programs of genetic control (Chapter 8). Remarkably, considering their potential importance, there are few thorough life-table analyses of insect vectors. Southwood et al. (1972) in an extensive survey of the "life budget" of *Aedes aegypti* (L.) breeding in water jars in Bangkok, produced the life-table information shown in Table 15.1. Under these conditions the variation in number of adults emerging was largely the result of mortality factors rather than natality factors, and it appeared that the critical factor was nutrition of the larvae. Clearly, any change in the larval habitat substantially increasing larval survival can increase disease transmission by the adults.

1. *Larval Habitats*

Detailed knowledge of the larval ecology of arthropod vectors, especially mosquitoes, was a basic component of control techniques a generation ago (TVA 1947), was often ignored during the age of insecticides, and is being revived today under source reduction approaches to mosquito pest management (WHO 1973). Understanding larval ecology is needed to (1) map breeding areas, (2) plan control measures, and (3) evaluate the

Table 15.1 Life Table for *Aedes aegypti* Breeding in Water Jars in Bangkok[a]

Growth period, *x*	Mean number alive at beginning of *x*	Mortality factor, *dxF*	Mean number dying, *dx*	Percent mortality, 100*gx*
Eggs	100	Physiological	85.1	85
Instars I–II	14.9	Starvation	5.0	33
Instar III	9.9	Starvation	5.5	56
Instar IV	4.4	Starvation	2.3	52
Pupa	2.1	—	—	—
Generation	—	—	97.9	97.9

[a] Southwood et al. 1972.

impact of water resource development on the epidemiology of vector-borne diseases (Hess et al. 1970). Ecological classifications of natural factors and plant types associated with mosquito production have proved invaluable in reservoir management for naturalistic control (Hess and Hall 1945). Species preference for sites of larval development is wide-ranging, even in a closely knit genus such as *Anopheles* (Table 15.2).

Table 15.2 Relation of Aquatic Plant Types to Production of *Anopheles quadrimaculatus* Larvae[a]

Type	Example	Mean Number of Larvae per square foot
Flexuous	*Echinochloa*, wild millet	7.9 ± 1.8
Submerged	*Najas*, naiad	6.7 ± 2.0
Carpet	*Eragrostis*, teal grass	6.5 ± 2.7
Floating mat	*Alternantheria*, alligator weed	6.2 ± 1.9
Erect, leafy	*Aster*, aster	4.2 ± 0.9
Coppice	—	3.9 ± 1.4
Pleuston	*Lemna*, duckweed	3.2 ± 0.6
Woods	*Salix*, willow	1.9 ± 0.6
Erect, naked	*Eleocharis*, spike rush	1.6 ± 0.5
Floating leaf	*Nelumbo*, lotus	1.1 ± 0.7

[a] Hess and Hall 1945.

Source reduction or destruction may be the only feasible way to control malaria vectors that are primarily exophagous, for example, *A. bellator* Dyar and Knab and *A. cruzi* Dyar and Knab, which breed in epiphitic bromeliads in forest trees and which bite humans working or living on the margins of forested areas.

2. *Larval Development*

Rates of larval development are strongly influenced by environmental factors such as temperature, humidity, and light. Temperature is particularly critical, and for *Anopheles quadrimaculatus* Say, Huffaker (1944) found the following relationship between environmental temperature and developmental time from eclosion to emergence:

12.1° C, 1572 hours	22.0° C, 319 hours	29.6° C, 190 hours
15.1° C, 768 hours	24.7° C, 245 hours	32.3° C, 178 hours
19.2° C, 501 hours	27.2° C, 210 hours	34.6° C, 203 hours

C. Infection of the Vector

The susceptibility of a potential vector to infection by a pathogen is poorly understood. For the transmission of some viruses, this process may be purely mechanical, as with a contaminated hypodermic needle. Vector susceptibility to infection may involve genetic variations in the pathogen, that is, virulence; and with encephalitis viruses both endemic and epidemic strains are known which infect mosquito species differently. It is probable that genetic variations in the susceptibility of vector species exist for most pathogens, as has been demonstrated with *Aedes aegypti* to *Brugia* and *Wuchereria* filariae (Macdonald and Ramachandram 1965). These genetic controls must affect vital physiological factors involved in nutrition, development, and reproduction of the pathogen in the body of the host. This relatively unstudied area seems to be fruitful for study and exploitation in disease control.

D. Extrinsic Cycle

Although some vectors may act as purely mechanical transmitters of pathogens, for example, the housefly and various bacteria, and the German cockroach and hepatitis virus; many pathogens have intimate physiological involvement with their vectors and spend an obligatory period of development, the extrinsic cycle, in the vector, during which they develop to infective form. Examples are the malaria *Plasmodia* (Fig. 15.2),

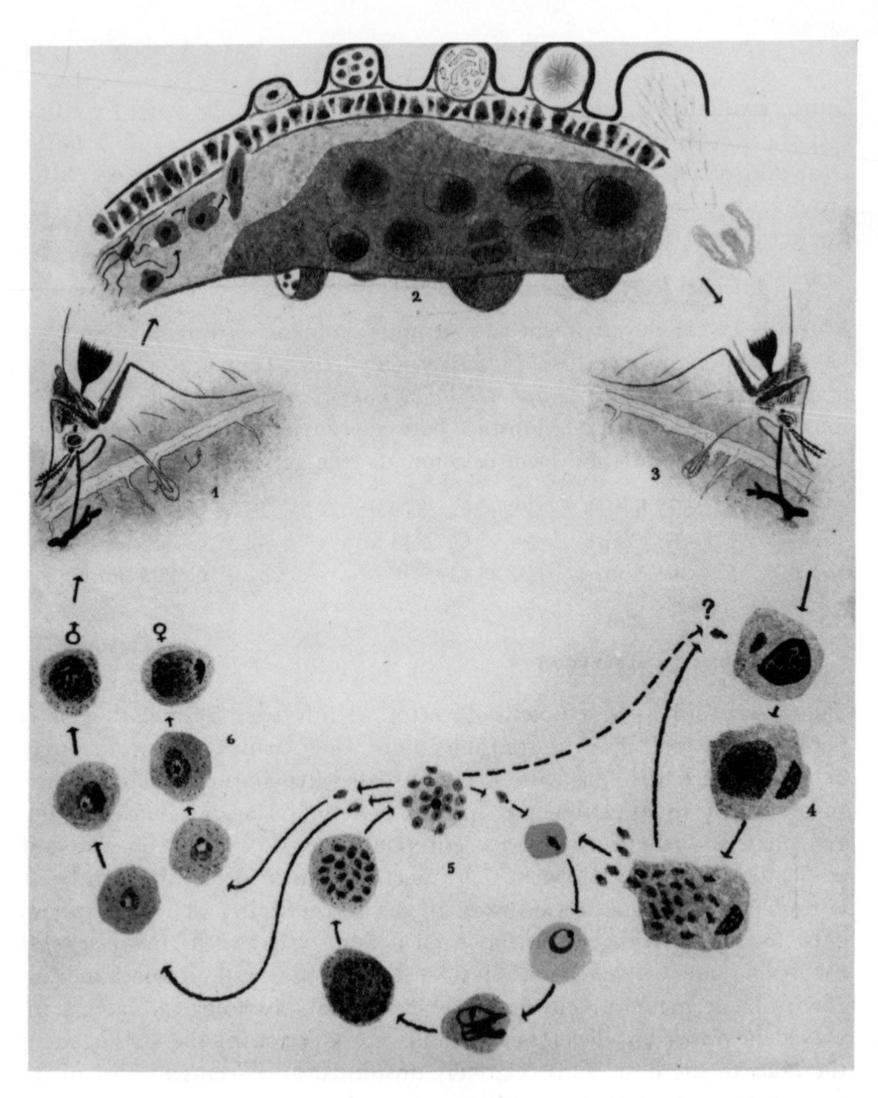

Fig. 15-2 Life cycle of the malaria *Plasmodium*. (1) A female *Anopheles* mosquito ingests blood containing sexual forms of the *Plasmodium* from a sick person. (2) Oocysts produced by the sexual *Plasmodium* develop in the stomach of the mosquito. (3) The oocyst bursts, liberating sporozoites which migrate through the blood to the salivary glands of the mosquito. The healthy person bitten by the mosquito becomes infected. (4) The *Plasmodium* first develops in the liver of the infected person. (5) Subsequently, the parasites develop and multiply in an asexual cycle in the red blood cells of the infected human. This phase concludes with an attack of malaria, with fever, anemia and other symptoms. (6) Gametocytes develop in the blood of the infected person, leading to a sexual cycle. (From World Health Organization.)

which develop from gametocytes to sporozoites in the *Anopheles* mosquito over a 10- to 14-day period, the filarian *Onchocerca volvulus* (Leuckart) which requires about 1 week to grow from microfilariae through three larval stages in the *Simulium* blackfly before becoming infective, and *Trypanosoma* of African sleeping sickness which requires 3 weeks to mature in the tsetse fly vectors, *Glossina* spp. The length of the extrinsic cycle is directly affected by environmental temperatures.

E. Infection of the Host

This area represents the crucial interaction between vector, pathogen, and host involved in disease transmission. The ecological patterns may be almost those of random chance, as in the attachment of young "seed ticks" of *Boophilus annulatus* (Say), which climb vegetation and remain quiescent waiting for contact with an animal to which they transfer; or they may involve the lengthy flight of an adult female mosquito, which drawn by perception of heat, carbon dioxide, and lactic acid, seems unerringly to find its way into a human habitation. For the vector to be effective in inoculation of the hosts, so essential to the continued survival of the pathogen, a certain level of infectivity must exist in the vector population; and for pathogens with elaborate extrinsic disease cycles such as malaria, filariasis, and trypanosomiasis, the pathogen must exist in the vector in the proper physiological state, for example, sporozoite or filarian larva, to invade and colonize the host. Within this enormous range of ecological complexity, there must exist specific points of vulnerability or "weak links" which should be exploitable in terms of disease control. The resting habits of adult mosquitoes or tsetse flies engorged with a host blood meal are good examples. Many others will doubtless be revealed from intensive ecological study.

1. *Vector Population*

Critical estimates of the size of the vector population and its age composition are essential to adequate modeling and systems analysis. Techniques for the determination of populations are discussed in Chapter 9. These comprise absolute estimates such as marking-release-recapture, usually with fluorescent dyes, using the formula (WHO 1972):

$$P = \frac{an}{r}$$

where P = total population, a = number of marked individuals released, n = total number trapped, and r = marked individuals recaptured.

Relative estimates such as the number of larvae or pupae per unit area, the number of adults per house or room, the biting rate, or the adult emergence rate have been widely used in lieu of absolute estimates.

Age structure of the population is important to systems analysis, and has been variously estimated by changes in the morphology of the female reproductive system of the mosquito following the gonotrophic cycle, using follicular relics, or observing the daily growth rings of the cuticle. The aim should be to determine the daily survival rate of a stable population.

2. *Dispersal of Vectors from Their Breeding Sites*

This is of major importance in the epidemiology of some vector-borne diseases. Triatominae bugs as vectors of Chagas' disease may be largely domestic, confining themselves to human dwellings, as does *Triatoma infestans* (Klug) in an anthropotic or peridomicilliary cycle; or wild species, invading dwellings or attacking humans in the field in a zooanthropotic or intradomicilliary cycle. Lice commonly spend their entire lives on their host. Cockroaches and houseflies range only as far as is necessary to find food. Man and his transportation systems have greatly extended the dispersal of all insect vectors.

Flight range is an important part of dispersal for flies and mosquitoes. *Simulium* species may vary in flight range from short distances characteristic of *S. neavei,* which is confined to individual river basins, to *S. damnosum,* which may fly up to 100 km and can easily disperse from one river system to another. These blackflies tend to disperse further in wet than in dry seasons (WHO 1973c). Such varying dispersal patterns complicate vector control procedures. A useful index of flight dispersal is the distance from the larval habitat where standard percentages of marked populations are recaptured, for example, the FD_{50} or FD_{90}.

3. *Host-Vector Contact*

The behavioral patterns that determine the contact between vector and vertebrate host during the search for and indulgence in a blood meal are perhaps the most critical factors in the epidemiology of vector-borne diseases (Hamon et al. 1970). Their quantitative assessment is essential for systems analysis and modeling of pest–disease management programs. *Host preference* ranges from exclusive feeding on a single species of host, as is typical of the bloodsucking lice, Anoplura, for example, *Pediculus humanus*; to generalized zoophilic attack of groups of large and small vertebrates as in many Culicidae, Simuliidae, and Tabanidae; to indiscriminate feeding of Triatominae bugs on a large variety of hosts. *Culex pipiens* has races that seem to feed exclusively on birds, and other races, morphologically indistinguishable, which feed generally on mammals. *Culex tarsalis* Coquillett, an important vector of encephalitis, feeds on a wide range of birds and mammals. Host preferences can be determined by comparing biting rates in traps baited with alternative hosts, or by analyzing blood meals by the precipitin technique. Careful allowance must be made for the relative availability of various hosts in calculating the forage ratio or percentage of blood meals on a specific host/percentage of this host in the available population of hosts, where ratios < 1 = host avoidance and > 1 = host preference.

Most important malaria vectors are anthropophilic and prefer to feed on man. However, such predominantly zoophilic species as *Anopheles culicifascies* Giles in India and *A. aquasalis* Curry in South America have become important vectors where there is a scarcity of cattle, for example, because of their replacement by tractors in the Guianas (Hamon et al. 1970). Such behavioral traits of mosquitoes may be critical to the success or failure of residual house spraying for malaria control. Behavioral resistance or increased physiological irritability to DDT residues has been suggested to exist in such vectors as *A. pseudopunctipennis* Theobald in Mexico, *A. albimanus* Wiedemann in Central America, and *A. sundaicus* (Rodenwaldt) in Java, Indonesia. These species are endophagous, but are readily repelled by contact with DDT-treated surfaces and leave sprayed houses before acquiring a lethal dose.

Diurnal and nocturnal feeding preferences of the vector are also of importance in relation to sites of feeding and accessibility of the host. The frequency of feeding is determined by the *biting cycle* or interval between blood meals, which is related to the gonotrophic cycle, but is also influenced by temperature and humidity and may be completely interrupted under unfavorable climatic conditions unless, as in *Anopheles atroparvus,* overwintering and disease transmission continue in houses during winter.

These interrelationships are simplified among nonflying vectors such as lice, fleas, mites, and ticks for which vector density is generally correlated with host density, for example, the prevalence of lice in prison camps or barracks. Host movement patterns and densities, for example, nomadism, may determine vector contacts and infection sites.

4. *Resting Habits of Vectors*

These may be of considerable importance in pest-disease management. The success of residual house spraying for malaria control depends on the resting preference of engorged female *Anopheles* for dark corners of walls and ceilings. The tsetse flies, *Glossina* spp., typically rest, after engorgement, on tree trunks or branches near the ground, and *G. swynnertoni* has been eradicated over large areas in Africa by selective treatment of branches of the proper size and height with dieldrin or endosulfan (Chapter 7, Section III. A. 4).

5. *Vectorial Capacity*

This is defined as the product of all the intervening factors that produce an infection in the vector and enable it to transmit the infection to a host (WHO 1972). Thus it is an operational means of assessing the relative importance of a specific vector in relation to a particular disease. Vectorial capacity represents the interaction of (1) the physiological and biochemical factors that determine the susceptibility of an arthropod vector to become infective with the pathogen, and (2) the ecological factors such as population density and longevity, dispersal and flight range, and host preference and feeding patterns that determine probability of successful contact with the host.

Inital efforts to replace the present and increasingly unsatisfactory reliance on exclusively insecticidal control of insect vectors involve integrated control by combinations of chemical and biological methods. Hopefully, this will become a transition phase leading to a systems approach to pest-management technology with analysis of cost/benefits to public health and environmental quality, heavy dependence on source reduction of the vector, and ecologically planned integrated control (NAS 1973).

II. PEST MANAGEMENT IN THE CONTROL OF VECTORS OF HUMAN DISEASES

For a time after the development of DDT, lindane, and other modern synthetic organic insecticides, it appeared as if the exclusive use of these alone could conquer all vector-borne diseases. Insecticides still remain as the single most important control intervention (see Chapter 7). The mass delousing of the inhabitants of Naples in 1943 by application of 10% DDT dusting powder to the clothing (Fig. 15.3), to control the human louse, *Pediculus humanus humanus* L., the vector of epidemic typhus, *Rickettsia prowazekii* da Roche-Lima, successfully averted a wartime epidemic of the disease. Human delousing with insecticidal dusting powders is a good example of the proper use of insecticides applied as

Fig. 15-3 Use of DDT dusting powder for typhus control in Afghanistan (From World Health Organization.)

precisely as possible to the critical site of pest attack (Fig. 15.3). Use of DDT dusting powder for louse control became the standard postwar remedy for louse and typhus control, and successfully disinfested over 2 million persons in Japan and Korea in the winter of 1945–1946. However, during the winter of 1950–1951, human lice in Korea were found to be almost totally resistant to DDT, and this resistance became worldwide wherever DDT had been widely applied (Brown and Pal 1971). In critical areas 1% lindane dust replaced DDT for control, and subsequently louse resistance developed to lindane and to pyrethrins as well. Presently, 1% malathion dust is the weapon of choice, but resistance has developed in Africa, and it appears that 5% carbaryl dust may be a useful alternative (WHO 1973). The problem is complicated by the very high safety standards required of insecticides to be applied directly to the human body (WHO 1973a).

This use of louse powders is at present the only practical strategy for control of human lice and of epidemic typhus. It conforms to the basic aims of pest management in that the insecticide is applied precisely to the site of host–vector contact and with a minimum of disturbance of the surrounding environment. The incorporation of this use of louse powders, as an emergency measure to deal with epidemic typhus, into a general program of improving human sanitation and hygiene is the essence of the pest-management approach.

A. Mosquitoes

Mosquitoes must be accorded a preeminent position among the natural enemies of man, both from their vexatious attacks on the human person and because of their extraordinary roles as vectors of important human diseases. Despite the zealous employment of DDT and other modern insecticides, the biting of about 80 species of *Anopheles* mosquitoes still transmits malarial *Plasmodia* to at least 100 million new human victims each year. The common house mosquito, *Culex pipiens quinquefasciatus* Say (*Culex fatigans*), is the principal vector of filariasis, caused by *Wuchereria bancrofti* (Cobbold) and *Brugia malayi* (Brug), which afflicts more than 250 million persons (Bruce-Chwatt 1971). The *Aedes aegypti* mosquito transmits the deadly yellow fever virus and dengue fever virus to man in various tropical regions, and other species of mosquitoes are the vectors of more than 80 other viruses attacking man, including Venezuelan and St. Louis encephalitis in the Americas, Japanese encephalitis in Southeast Asia, o'nyong-nyong fever in Africa, and Chikungunga fever in Asia and Africa (NAS 1973).

1. *Current Status of Control*

"Mosquito control today is in a state of crisis. For the past 30 years, mankind has been almost completely dependent on synthetic organic insecticides. Today, the very properties that made these chemicals so useful, long residual action and toxicity for a wide spectrum of organisms, have brought about serious environmental problems. Moreover, mosquito resistance to chemical pesticides has caused the failure of many vector control campaigns . . . in areas of California, encephalitis transmitting mosquitoes are resistant to virtually all larvicides. . . ." (NAS 1973). Even the highly successful malaria eradication program of WHO, which has eliminated malaria from 36 countries with a total population of 710 million and has large-scale programs in another 27 countries (WHO 1971), has serious difficulties due to the developing resistance of *Anopheles* mosquitoes to insecticides. At the most recent tally (Brown 1971), 15 species of malaria vectors are resistant to DDT, and 36 to dieldrin and lindane, the three insecticides on which the residual house-spraying technique for malaria eradication was based. These resistant malaria vectors have disrupted malaria eradication efforts in Central America, the Middle East, Indonesia, and more recently in India and Pakistan (AID 1973). "Physiological resistance has become one of the major threats to the success of global malaria eradication" (Bruce-Chwatt 1970), and it now seems clear that exclusive reliance on regular applications of insecticides will not solve the problem of mosquito attacks on humans, either at the community- or countrywide level.

2. *Source Reduction*

This is a basic technique of insect pest management particularly applicable to mosquito control. It involves the deliberate modification of the aquatic environment in which mosquito larvae develop, to render it unsuitable for mosquito production. As a component of a mosquito pest-management system, source-reduction efforts range from total removal of standing water by drainage, filling, grading, ditching, or diking; through water level manipulations, and fluctuations and changes in water quality to manipulation of the "intersection line" where plants provide shelter for mosquito larvae, by mowing, clearing, and changing the plant species composition (WHO 1973c).

Historically, source-reduction projects have had major public health import through mosquito control: in the construction of the Panama Canal in 1904–1907, in the Pontine Marsh reclamation project in Italy in the 1930s, in the reservoir preparation and water management programs of the Tennessee Valley Authority in 1935–1945 (Fig. 15.4), and in

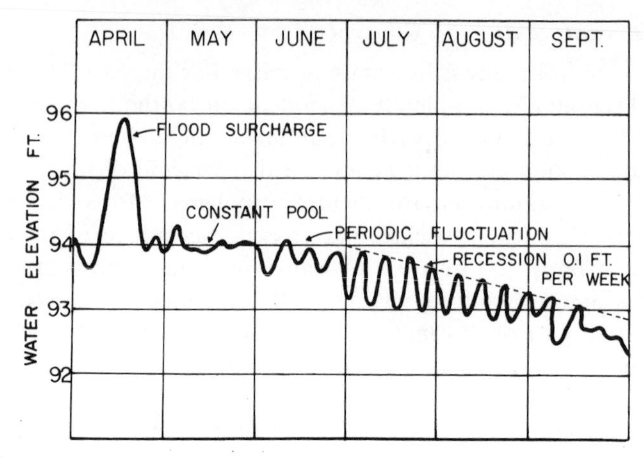

Fig. 15-4 Water level management to control mosquito breeding in Tennessee Valley Authority Reservoirs. (After TVA 1947.)

the irrigation drainage systems of the Nile flood plain. Source reduction of domestic water containers was a major factor in the elimination of *Aedes aegypti* from Brazil in the 1930s.

Source-reduction techniques are of almost endless variety, and their employment alone and together with water conservation and land reclamation programs is limited only by the imagination and technical competence of the planner and by the facilities at his disposal. Detailed discussions of source-reduction methods have been presented in TVA (1947) and WHO (1973). The techniques involve a detailed knowledge of mosquito ecology both in larval and adult habitats, careful planning, and engineering surveys and design of the measures to be implemented. Source-reduction operations almost always involve land and water use planning and the support of communities and agencies whose cooperation is essential for the construction and operation of the project.

3. *Insecticides*

These are important components of pest-management programs for mosquito control. However, problems of mosquito resistance and environmental quality that have arisen from the routine use of organic insecticides suggest that changes must be made in the insecticide approach.

Larviciding is an appropriate adjunct to source reduction. It offers the advantages of killing mosquitoes in the innocuous larval stage before the adults develop and disperse, and of providing a quick and relatively inexpensive way to prevent an incipient attack by the adults. The disadvantages are the temporary nature of the control and the inevitable contamination of the aquatic environment and damage to nontarget organisms. The decision to use larviciding as a part of a pest-management program depends on the nature and accessibility of the breeding sites, the ecology of mosquito larva and adult, and its role as a pest nuisance or a disease vector. There are obvious differences in the problems caused by the breeding of *Aedes nigromaculis* (Ludlow) in irrigated alfalfa pastures, *Anopheles quadrimaculatus* in large lakes and reservoirs, and *Culex fatigans* in household containers, discarded vessels, and small urban pools.

Useful larvicides must be rapidly biodegradable, of low toxicity to fish, crustacea, and molluscs, and of relatively low hazard to humans (WHO 1973a). Insecticides especially recommended for mosquito larviciding (WHO 1973b) are outlined in Table 15.3. Their properties are discussed more fully in Chapter 7. Larviciding should be utilized for mosquito control only as part of a coordinated pest-management program and only where other methods of control are impracticable and where degradation of environmental quality will be negligible. Consequently, larviciding should be carried out at as low a dosage and over as limited an area as possible. Spot treatments of heavy larval infestations are often very practicable. Detailed directions for larviciding operations are given in WHO (1973b).

Adulticiding, the insecticidal control of adult mosquitoes, can be a pest-management component at any point after emergence of the adult. Ultralow-volume aerial spraying of 98% malathion at 4 to 6 oz per acre has been used to kill adult *Aedes aegypti* and *Culex* spp. mosquitoes in cities and rural areas and is a powerful emergency weapon in controlling epidemic outbreaks of yellow fever, dengue fever, and encephalitis (WHO 1971). Humans in habitations provide an attractive lure for the adult female mosquito, and the daily use of space sprays of pyrethrins, synthetic pyrethroids, or dichlorvos from flitguns or aerosols can not only provide an acceptable life-style in areas infested with mosquitoes but also can interrupt the transmission of malaria and other mosquito-borne diseases. This measure is most effective when combined with adequate house screening and mosquitoproofing.

Residual spraying of the interiors of dwellings with DDT at 1 to 2 g/m^2 of surface is the standard technique for the malaria control program of WHO (Fig. 15.5). This employment of DDT is an appropriate pest-management technique which uses human attraction to lure the adult female into a "trap" (the house) where all interior resting places

Table 15.3 Larvicides for Mosquito Pest Management[a]

Insecticide	Formulation	Dosage (Active Ingredient)[b]	Use
Fuel oil or kerosene	With 0.5% spreader (Triton X-100)	40–80 l/ha	General for resistant larvae
Mosquito oil (e.g., Flit MLO)	—	9–19 l/ha	General for resistant larvae
Copper aceto-arsenite[c] (Paris green)	5–50% D	1 kg/ha	Stomach poison for anopheline larvae
Pyrethrins	0.07% in 66% kero-sene–water emulsion	46.5–465 l/ha	Very safe larvicide
Temephos (Abate) (OMS-786)	50% EC, 50% D, 1% Gr	55–110 g/ha	Safe larvicide, slow release rate
Chlorpyrifos (Dursban) (OMS–971)	24–48% EC	55–110 g/ha	Effective in polluted water, fairly persistent, may be toxic to wildlife
Chlorpyrifos methyl (OMS–1155)	48% EC	55–100 g/ha	Safe larvicide
Fenthion (OMS–2)	46–84.5% EC, 1% Gr	55–110 g/ha	Persistent in polluted water, may be toxic to wildlife
Malathion (OMS–1)	57–95% EC, 3–5% D, 1% Gr	400–550 g/ha	Safe, general-purpose larvicide

[a] WHO 1973b.

[b] 1 kg/ha = approximately 1 lb per acre.

[c] Questionable use because of persistent environmental pollution.

are lethal to the blood-engorged female. When carried out rigorously, efficient malaria control results because the infected female *Anopheles* is killed during the 10- to 14-day interval between gametocyte formation and sporozoite development (Fig. 15.2), and the transmission of malaria is interrupted at a "weak link." The use of the highly persistent DDT in houses seems to offer little risk to human health and minimal effects on the quality of the environment (WHO 1971).

Mosquito resistance to DDT has become widespread, and several other residual insecticides have been found effective, including lindane and the biodegradable malathion, fenitrothion, and propoxur (Table 15.4) (WHO 1973a). Their use in ecologically selected spot treatments, for

Fig. 15-5 Residual spraying of DDT for malaria control in India. (From World Health Organization.)

example, dark corners of houses, applied to preferentially attractive surfaces of cardboard or cheesecloth, could further improve the pest-management image of residual spraying.

4. *Naturalistic Control*

This is defined by WHO (1973b) as the destruction of mosquitoes or prevention of their development by any natural force, animate or inanimate. This broad term includes both *natural control,* the sum total of regulating effects from climate and the physical character of the environment and from the natural presence of parasitoids, predators, and diseases; and *biological control,* the introduction and encouragement of

Table 15.4 Adulticides for Mosquito Pest Management[a]

Insecticide	Formulation	Dosage	Use
DDT (OMS-16)	WP	1–2 g/m^2	Residual spray
Lindane (OMS-17)	WP	0.5 g/m^2	Residual spray
Dieldrin (OMS-18)[b]	WP	0.5 g/m^2	Residual spray
Propoxur (OMS-33)	WP	2 g/m^2	Residual spray
Fenitrothion (OMS-43)	WP	2 g/m^2	Residual spray
Malathion (OMS-1)	WP	2 g/m^2	Residual spray
	98%	90–180 g/ha	Ultralow-volume spray
Dichlorvos (OMS-14)	20% in deodorized oil-resin strip		Aircraft disinsectization, household space fumigation
Pyrethrins (synergized)	25% extract, 8% piperonyl butoxide, 8% in deodorized oil		Space spray
Resmethrin (OMS-1206)			Space spray

[a] WHO 1973b.

[b] Dieldrin has caused severe health problems in spraymen (Hayes 1959), and this together with the large number of resistant anopheline vectors has discouraged its use.

natural enemies (Metcalf et al. 1962). The cumulative effects of these natural forces on mosquito survival are drastic, as indicated by the life table for *Aedes aegypti* (Table 15.1). The major natural forces contributing to mosquito suppression are climatic—wind and wave action which may strand eggs or larvae to be destroyed by desiccation, or carry adults to unfavorable locations; evaporation and run-off which may remove water from breeding sites, or carry larvae to unfavorable locations; unfavorable temperature and humidity for adult survival; predatory aquatic insects, fish and birds; and parasitic diseases caused by bacteria, protozoa, fungi, and nematodes. These naturalistic control factors should become major components of pest-management programs. However, because of the rather special ecological niches of immature mosquitoes, natural enemies have not usually been utilized in a major way for integrated control (NAS 1973), and progress has been hindered by exclusive reliance on chemical control.

a. *Mosquito Fish.* These fish play a major role in regulating larval populations in fish pools, ponds, rice paddies, and marshes. The mosquito fish, *Gambusia affinis* Baird and Girard, which has been distributed throughout the warmer parts of the world, feeds voraciously on mosquito larvae and has been used in control programs since the early 1920s. In ponds and small bodies of water this fish has effected excellent control of anopheline and culicine larvae and, in warm climates, *Gambusia* populations often become self-perpetuating. It is said that six *Gambusia* can completely control mosquitoes in a 5- to 10-m^2 pool (WHO 1973b). However, healthy *Gambusia* are not compatible with employment of regular dosages of organochlorine larvicides, and their use has been neglected during the past 20 years. *Gambusia* is a small (5 cm) very hardy fish, and has been dropped into bodies of water by aircraft without injury. In the absence of mosquitoes it subsists on other organisms. *Gambusia* is ovoviviparous and reproduces very rapidly, a single female producing 200 to 300 young per season.

A major drawback in the use of *Gambusia* is the lack of an efficient mass culture method. There is also need for substantial study of the ecology of this species in relation to other aquatic organisms, including other species of fish used as food. Studies need to be made of the safety of various mosquito larvicides to *Gambusia*. This use of *Gambusia* in naturalistic control can be improved by stabilizing water levels, manipulating aquatic vegetation, shoreline improvement, and development of a compatible approach to chemical larviciding. The use of mosquito fish has the advantages of low cost, self-perpetuation, compatibility with water quality, and encouragement of growth of game fish. Operational methods with *Gambusia* are detailed in WHO (1973b).

Other species of fish that may play a similar role as predators of mosquito larvae include the guppy *Poecilia*; the salt marsh killifish, *Fundulus heteroclitus* (L.); the three-spined stickleback, *Gasterosteus aculeatus* (L.); the cyprinodonts *Aphanias dispar* and *Epiplatys senegalensis*; the carp, *Panchax panchax*; and annual fish able to survive semi-desiccation as eggs buried in moist soil, *Nothobranchius guentheri* and *Cynolebias bellottii*. The use of most of these mosquito fish is still in the experimental stage (WHO 1973b) and, before trials are undertaken, careful scientific inquiry should be made into the possible adverse effects of importing any exotic species into a new habitat.

b. *Invertebrate Predators.* These include the predaceous diving beetle larvae (Dytiscidae), the hellgrammite, *Corydalus cornutus* (L.), and other neuropterous larvae (Sialidae); and aquatic true bugs (Corixidae, Nepidae, Notonectidae, and Belastomatidae); and dragonflies (Odonata). Perhaps the most promising invertebrate predator for naturalistic control is the predaceous mosquito larva of the genus *Toxorhynchites (Megarhinus)*. The most promising is said to be *T. brevipalpis* of East Africa, which consumes about 250 mosquito larvae during its development (NAS 1973). Female *Toxorhynchites* are active searchers for mosquito-breeding habitats almost inaccessible to man, and this species could become a useful adjunct to mosquito pest management, especially in urban areas (Trpis 1973). Techniques for mass rearing have been developed. Disadvantages include the cannibalistic habits of *Toxorhynchites* and its susceptibility to all chemical larvicides. Much more study on this predator is needed.

c. *Diseases.* Mosquito larvae are the hosts of a variety of disease organisms. Those that show promise for the naturalistic control of mosquito populations include parasitic nematodes, protozoans, fungi, and pathogenic bacteria and viruses (Chapman et al. 1970; Roberts 1970; NAS 1973). The parasitic nematode *Reesimermis nielseni* (Mermithidae) naturally infects 22 species of mosquitoes and has infected another 33 species in the laboratory. The nematode lives in the mosquito larva's body, consumes vital tissues, and emerges to a free-living adulthood. The nematodes can be mass-cultured, and preparasitic larvae, when sprayed on breeding areas at $1000/m^2$, have controlled some species of *Anopheles*.

The parasitic Microsporidia of the genus *Nosema* may have potential use in mosquito control. *Nosema algerae* heavily parasitizes *Anopheles albimanus* larvae and has prevented the establishment of laboratory colonies. The spores of *Nosema* can be stored and applied by spraying, and at 1000 spores per milliliter have produced 70 to 80% infection in field trials. Other species of *Anopheles* are not as susceptible to *N. algerae*. Difficulties include lack of mass production and spore storage, and uncertainties as to pathogenicity in other organisms.

Many species of the parasitic fungi *Coelomomyces* infect only mosquito larvae and may produce up to 90% mortality of larval populations, destroying the fat body. The sporangia of *Coelomomyces* have been successfully used in field experiments. Major difficulties include lack of mass-production techniques, and little knowledge of species specificity and possible pathogenic effects on other organisms.

The crystalline spores of *Bacillus thuringiensis* Berliner have been successfully developed as a microbial insecticide, BT, and are used in control of lepidopterous larvae that are agricultural pests (Chapters 6 and 7). The strains commercially available have little or no toxicity to mosquito larvae, but *B. sphaericus* (BA-068) isolated from *Culex tarsalis* larvae may have a potential for mosquito control. At a dosage of 9×10^5 spores per milliliter, BA-068 has shown promise in the control of the pasture mosquito, *Aedes nigromaculis*.

Clearly, most of the biological control agents for mosquito control are in the early developmental stages. The mosquito fish *Gambusia* has been used for practical control for many years, and this approach needs re-emphasis. The other biological control methods need the development of practical, inexpensive mass-rearing techniques, field demonstrations of effectiveness against many species of mosquitoes, and proof of safety to nontarget organisms of the environment, including man. Their practical use lies in the future and they could become important components of pest-management programs.

B. Housefly

Initial efforts to control the housefly, *Musca domestica* L., with DDT residual sprays were so sensational that there was talk of eradicating this insect (Chapter 1). The widespread use of DDT in southern Italy in 1946 reduced infant mortality from gastroenteric disease to 7.7 per 1000 births, down from 31.3 the previous year. DDT resistance in the housefly vector, which became nearly worldwide within a few years after widespread application, increased infant mortality to 18.0 per 1000 in southern Italy in 1947, and disrupted programs to control such fly-borne diseases as *Shigella* in Georgia, and trachoma in North Africa (Brown 1958). Subsequently, the housefly developed strains resistant to virtually every useful insecticide (e.g., DDT, BHC, chlordane, parathion, diazinon, coumaphos, trichlorfon, malathion, fenchlorphos, fenthion, and dimethoate in Denmark from 1947 to 1967) (Brown and Pal 1971), and failures of chemical control have become the rule rather than the exception.

It is evident that regular treatments with insecticides either as residual sprays or baits cannot long control as resourceful an insect as the housefly with its enormous biotic potential (each female lays an average of about 500 eggs and generations require only 20 days) and its rich gene pool. Pest-management practices are obviously necessary to control both the numbers of the fly and the diseases such as bacillary dysentery and trachoma that it transmits.

Ironically, sanitation, which is the core to pest management of this creature, has been thoroughly understood since the heyday of the horse-drawn vehicle. Such practices as burying manure, human excrement, and food wastes in sanitary landfills, or drying them so that their moisture content is less than 10% are very effective fly-control measures. It is said that their employment by an army of workers in mainland China has resulted in a virtually fly-free environment. The basic difficulties in applying these commonsense measures in underdeveloped countries are a general lack of sanitation in human excretion and heavy reliance on manure for fuel and building. The privy is the worst offender in providing a link between the human, the infective agent, and the housefly vector. Every effort should be made to flyproof the privy and to disinsectize it daily with lime or a similar agent.

In developed countries modern agricultural practices, which rely on synthetic nitrogenous fertilizers instead of utilizing manure, coupled with the practices of confining cattle in feed lots with as many as 100,000 animals or chickens in poultry pens with up to several hundred thousand birds, have made the sanitary disposal of animal excrement almost impossible, and commonly result in plagues of flies around livestock farms and in nearby residential areas.

Case History

Trachoma is a virus disease spread by the houseflies *Musca domestica* and *M. sorbens*. The flies are attracted to the copious secretion from the eyes of infected persons. Worldwide, trachoma is by far the most important cause of human blindness, and WHO has estimated that there are 80 million human victims. In areas of North Africa such as Morocco, there is nearly 100% infection of humans with the trachoma virus, which begins in the first months of life and more than 3.75 persons per 1000 become blind (Reinhards et al. 1968). The *Haemophilus* bacterium causes epidemic bacterial conjunctivitis, and this condition facilitates the transmission of trachoma. As a result of 13 years of study in Morocco with mass treatment campaigns involving as many as 4,500,000 people, Reinhards et al. (1968) developed a successful pest-management program involving a threefold attack on the housefly vector, the *Haemophilus* bacterium, and the trachoma virus. The use of fly sprays, chemoprophylaxis with systemic sulfonamides or intermittent antibiotic treatment with tetracycline eye ointment were not especially successful individually or in limited combination (Table 15.5). It was only when all these measures were applied together that a rapid rate of cure resulted.

Pest-Management Program: Trachoma control, like that of other vector borne diseases, can be successfully carried out only as a

Table 15.5 Effects of Fly Suppression (F), Systematic Sulfonamides, (S) and Tetracycline Ointment (T) on Arrest and Cure of Trachoma in Southern Morocco[a]

	AB_{50}[b]	
Treatment	Cicatrization	Cure
None	5.11	16.81
F	5.21	13.55
S	3.83	9.35
FS	3.98	11.90
T	3.43	7.10
FT	2.14	9.89
ST	1.63	6.33
FST	0.77	4.27

[a] Reinhards et al. 1968.

[b] Age (in years) at which 50% of cases of trachoma showed signs of arrest or cure.

widespread community effort involving public understanding and acceptance (Chapter 1) and adequate technological direction. The following components are involved in such a program (Reinhards et al. 1968):

1. Apply all practical measures for fly control, including sanitation, sanitary disposal of animal excrement, and so on.

2. Use insecticides for fly control sparingly and efficiently as space sprays and as sugar baits, alternating various chemicals to delay the onset of resistance.

3. Treat the human population with systemic sulfa drugs: sulfadiazine, sulfathiazole, sulfamerazine at 75 mg/kg body weight on the first day, and 50 mg/kg on the second, third, and fourth days.

4. Treat the eyes of the human population with 1% chlortetracycline ointment applied twice daily on three consecutive days and repeated at regular monthly intervals throughout the season of epidemic conjunctivitis.

C. Mites and Ticks

The Acarina, or mites and ticks, are involved in much misery for both humans and domestic animals. The Rocky Mountain wood tick or spotted fever tick, *Dermacentor andersoni* Stiles, attaches to wild and domestic animals, and its bite can cause inflammation, ulcer formation, and tick paralysis when the tick attaches to the base of the skull. This tick causes its most severe injury as the vector of the highly fatal Rocky Mountain spotted fever organism, *Rickettsia rickettsii*. In the eastern United States the vector of this disease is the American dog tick, *Dermacentor variabilis* (Say). The cattle ticks *Boophilus annulatus* and *B. microplus* are the vectors of the protozoan *Babesia bigemina,* which causes an important and often fatal disease of cattle, piroplasmosis. Various species of *Ornithodoros* ticks are vectors of several *Borrelia* spirochetes which cause relapsing fevers in man. The fowl tick, *Argas persicus* (Oken), is the vector of a spirochete, *Borrelia anserina,* which causes fowl spirochetosis in domestic poultry. The lesser members of the Acarina, the mites, are most notoriously represented by chiggers, Trombiculidae, whose bites cause an intense irritation totally out of proportion to their microscopic size. As discussed below, they are also vectors of an important rickettsial disease of man.

The control of these pests is very difficult, because of their inconspicuous nature, their complex metamorphosis, their attraction to rodent hosts, their enormous reproductive capacity (a single tick may produce 4000 to 7000 eggs), and the ability of many species to hibernate without food and water for months and sometimes years. Cattle ticks have been successfully controlled by rigorous dipping programs with insecticides, and suitable pasture rotations (Metcalf et al. 1962). However, resistance to suitable insecticides is widespread and has become a crucial problem.

Scrub typhus or tsutsugamushi fever is caused by *Rickettsia tsutsugamushi* (Hayashi) Ogata and is transmitted from rodent hosts (*Rattus* spp.) to man by the bites of chiggers, principally *Leptotrombidium deliense* and *L. akamushi*. The disease is endemic in Japan, Malaysia, Korea, Central Asia, Afghanistan, Nepal, Tibet, and southeastern Siberia. Its ecological relationships are complex, as described by Traub and Wisseman (1968a), and the incidence of the disease is favored by human activity in the conversion of jungle to roads, camps, and habitations; and through the abandonment of lawns and gardens—activities that affect the nature and density of rodent and mite populations and opportunities for their interactions with man.

Control measures that have been considered involve *area treatments* against the vector such as burning, bulldozing cover, the use of herbicides, and spraying and fogging the ground and low vegetative cover with persistent pesticides; *rodent control* by poison baits and systemic toxicants; *personal protection* by the use of chigger repellents such as benzyl benzoate on the body and on clothing; and *treatment of the disease* with broad-spectrum antibiotics. The latter is complicated by the anomalous nature of the symptoms of infection and consequent difficulties in diagnosis. Traub and Wisseman (1968b) found the most effective and practical control measure to be area spraying of ground and low vegetative cover with dieldrin at 2.5 lb per acre. This reduced the mite population on wild-caught rats by 97 to 98% for 6 months and by 91% for 2 years (Traub and Dowling 1961). Because of concern for the persistent pesticides, other compounds were evaluated, lindane being effective for 2 months, as were fenthion and propoxur; and malathion for about 2 weeks. Area burning had no long-term effectiveness on mite densities. It is apparent that scrub typhus can be uniquely controlled by the area application of dieldrin, as the incidence of the disease in human volunteers has been shown to be directly proportional to the density of the chigger population. However, the frequency and dosage of the pesticide treatment may be substantially reduced by supplementary rodent control and by areawide considerations of land use and of vegetative growth. The details of such pest-disease management programs need to be developed individually for each of the many geographic areas and ecological situations in which the disease occurs.

D. German Cockroach

This insect, *Blattella germanica* (L.), with 200-million-year antecedents, has become the number-one insect enemy of urban United States. The reasons for this ascendency are its omnivorous habits, allowing its survival on infinitesimal bits of carbohydrate, fat, and protein; its high biotic potential; its huge genetic base, permitting the rapid development of races resistant to every pesticide employed; and its secret and solitary habits, through which it avoids detection and destruction.

Case History

Blattella germanica females produce one to seven ootheca or capsules, each containing 25 to 30 eggs, during their lifetime, and there are two to three generations a year in the average heated building. Insects in a protected location in homes, markets, and warehouses are nearly free from the usual invertebrate and vertebrate natural enemies, except perhaps for the foot of man. Therefore the entrance of a single female carrying the egg capsule, which is dropped in a dark and secluded location shortly before the eggs are ready to hatch, is sufficient to rapidly infest any new habitation.

Cockroach control, largely involving *Blattella germanica,* comprises about 70% of the calls for the services of professional exterminators. The traditional reliance has been on sodium fluoride and pyrethrin dusting powders blown into cracks, crannies, crevices, and dark corners where *B. germanica* is to be found. With the development of chlordane in 1945, oil sprays and fogs of this material provided a high degree of residual effectiveness. However, resistance to chlordane rapidly spread through urban populations within several years. Subsequently, malathion, diazinon, dicapthon, ronnel, chlorpyrifos, and propoxur have been used successively by exterminators, with resistance developing to each in turn. More recently, there has been some use of 10% boric acid in sugar as a bait, and Kepone and mirex as fatty baits have had substantial vogue. Despite a ceaseless battle with insecticides, the German cockroach has become a disgusting plague in household kitchens, restaurants, groceries, and food warehouses. The problems are worst in old, dilapidated structures of the inner city which abound with cracks and crannies permitting shelter for the insect and allowing rapid reinfestation from adjacent structures. The constant irritation caused by these creeping, food-infesting insects is a major factor in exacerbating the tensions of ghetto dwellers. In addition, the German cockroach has been strongly implicated in the urban transmission of hepatitis virus, nematodes, and typhoid and dysentery.

Pest management of *Blattella germanica* is an urgent necessity in most city locations. The conventional insecticide approach seems woefully inadequate. Therefore the basic procedures in control of *B. germanica* and related cockroaches must rely on scrupulous sanitation, cockroachproofing, and prevention of reinfestation. A suggested program includes:

1. Prompt disposal of all loose foodstuffs, crumbs, and garbage by enclosure in tight, sanitary containers; enclosure of all cereals, bread, cookies, and so on, in tight containers
2. Construction of kitchens and food-handling facilities to avoid cracks and crevices and to provide smooth, uninterrupted surfaces wherever possible. Pipes should be sealed off wherever

they enter through floors and walls, and cracks should be filled with crack filler.

3. Care should be taken to inspect groceries and produce entering homes or food-handling establishments to be sure that gravid females are not unknowingly introduced.

4. When substantial cockroach populations are present, insecticide treatments are the only recourse. The most appropriate applications are poisoned baits, but spot treatments to cracks, crevices, and other sheltered spots may be necessary.

E. New Frontiers

Despite the immense progress made in the control of vector-borne diseases during the nearly 100 years since the discovery of the causative organisms of malaria (Laveran, Manson, and Ross) and Texas fever of cattle (Smith and Kilbourne), many challenges still face the medical entomologist and public health specialist. As we have seen, malaria control and eradication, although relatively secure in many subtropical and temperate zones, have experienced serious reverses in the tropical areas of the highest endemicity (Bruce-Chwatt 1970, 1971). Several lesser vector-borne diseases have retreated to relative insignificance, but control is precarious and dependent on subjugation of an ever-widening circle of insecticide-resistant vectors.

In addition, there are several major human vector-borne diseases such as mosquito-borne filariasis in Southeast Asia, tsetse fly–transmitted African trypanosomiasis (sleeping sickness), Triatominae-transmitted American trypanosomiasis (Chagas' disease), and *Simulium*-borne onchocerciasis in Africa where progress in total ecological understanding is meager and where the opportunities for control by pest and disease management are both imperative and challenging. In this section we briefly analyze two of these, *Simulium*-transmitted onchocerciasis and Triatominae-transmitted American trypanosomiasis.

Chagas' disease, or American trypanosomiasis, is found in the Americas in the area between 42° N latitude and 45° S latitude, where it infects about 30 million people, of whom 5 to 8 million suffer from permanent heart damage. Chagas' disease is caused by *Trypanosoma cruzi* and is transmitted between its animal hosts, wild and domestic, and to humans by 95 species of Triatominae (reduviid bugs), the most important of which are of the genera *Triatoma, Rhodnius,* and *Panstrongylus.* Human infection with the protozoan results from contact with the feces of the bugs, ingestion of contaminated food, drinking milk from infected animals, by blood transfusions, or congenitally. Chagas' disease is primarily a rural disease associated with poverty (Metcalf 1970).

Ecologically, Chagas' disease represents a series of complex interactions between (1) a zoonotic cycle maintained between wild mammals and wild Triatominae, (2) a zooanthropotic cycle (intradomicilliary) involving domestic mammals, humans, and wild or domestic Triatominae, and (3) an anthropotic cycle (periodomicilliary) between humans and domestic Triatominae, involving congenital infections and infection through blood transfusions. The disease moves between the various cycles by invasion of dwellings by wild vectors or hosts, infection of humans or domestic animals in the field, and use by humans of natural building materials or of food infested by wild vectors. There are localities in South America where 100% of the dwellings are infested by Triatominae, with sometimes as many as 2000 bugs per house, where more than 80% of the inhabitants are infected with *Trypanosoma cruzi,* and where 80% of the bugs are infected. Both sexes of the bug can transmit the disease for up to 20 days after biting an infected human or animal. Infected humans remain infective throughout their lifetime, and there is neither cure nor immunological protection (Metcalf 1970).

With these intricacies it seems remarkable that any sort of control measures are effective for Chagas' disease. However, residual spraying of dwellings with dieldrin at 1 g/m^2 or with benzene hexachloride (lindane) at 0.5 g/m^2 has reduced the number of infested houses by 90%, as well as the frequency of *Trypanosoma cruzi* infection in the vector. Three yearly treatments over all indoor areas, adjacent structures, cattle pens, and so on, are required. The second year only houses showing positive infection are treated. However, because of continual reinfestations, treatments must be virtually continuous. Dieldrin spray treatments are therefore too expensive and have resulted in severe chronic toxicity to spray crews exposed for long periods.

Onchocerciasis or "river blindness" is a parasitic disease of tropical Africa and the tropical Americas, caused by a filarian *Onchocerca volvulus.* The infection is transmitted among humans by the biting female blackflies of the genus *Simulium.* More than 20 million persons suffer from this disease and, in Upper Volta, for example, of 4.5 million inhabitants 400,000 have onchocerciasis and 1 in 10 suffers from severe eye lesions caused by migration of the microfilaria into the eyeball. In some West African populations, 50% of the inhabitants are said to be infected, 30% have impaired vision, and up to 20% are permanently blind (WHO

1971). Onchocerciasis has great economic importance not only because of human injury from infection and blindness, but also because of the enormous numbers of flies, their nearly intolerable biting, and the inherent fear of onchocerciasis; these factors have resulted in the abandonment of river valley areas of rich farmland of sub-Saharan West Africa (WHO 1973).

The relationship between *Onchocerca, Simulium,* and the human is complex. The female blackfly becomes infected when it bites a human and ingests microfilariae along with its blood meal. Although many of the microfilariae are digested, a few may succeed in penetrating through the fly's gut and entering the thoracic muscle where they pass through three larval stages over a period of about 1 week. Free-living larvae are formed, which pass from the fly's blood into the proboscis and infect a subsequent human host as the fly feeds. Infectivity is low, and a victim must receive numerous bites before pairs of adult filariae can survive and mate to produce a generalized infection. The adult filariae may live in the human body for as long as 15 years and reside in subcutaneous tissues where they form nodules which are of diagnostic importance. Young microfilariae, which are estimated to live about 2 years, reside in the skin where they cause itching and skin patches. They also invade the eye and may cause progressive loss of sight, culminating in blindness. Infection is cumulative, and severe symptoms are the result of chronic attacks by the blackfly over the human life-span.

Simulium blackflies belong to several species of which *S. damnosum* is the best known. The larvae breed in the flowing water of rivers, streams, and reservoirs, where they attach to silken threads spun over stones and rocks. The adults of the various species differ in flight range and biting habits, but *S. damnosum* is known to travel up to 100 km during the first few days after emergence. The adult fly populations are often very large, and they may emerge in enormous swarms which together with the irritating nature of their bite renders river areas almost uninhabitable.

Pest management of the *Simulium–Onchocerca* complex represents a highly complex problem and presents an enormous challenge. WHO has undertaken this task in the Volta basin of 700,000 km^2 in parts of seven countries—Mali, Upper Volta, Niger, Ivory Coast, Ghana, Togo, and Dahomey. The principal weapon is the application of blackfly larvicides to streams and rivers. DDT has been used in this way, but is a relatively inefficient larvicide and its use provides objectionable environmental pollution. For any sort of control program to succeed, it must be planned as a combined operation over a complete river basin which can be isolated successfully from its neighbors. The following elements are suggested for integration.

1. *Simulium* larviciding with highly effective biodegradable larvicides such as chlorpyrifos methyl (Methyl Dursban) or temephos (Abate) applied to small streams by drip can or over larger areas by aircraft spraying (Quellennec 1972). The necessity for maintaining an effective dosage in running water over long distances is a complicating factor.
2. Areawide treatment of infected humans with effective antifilarial drugs such as Suramin, which kills the adult filariae, and dimethylcarbamazine, which kills only the microfilariae
3. Encouragement of the widespread use of blackfly repellents such as diethyl toluamide (Off) and 2-ethylhexanediol (612) by farmers and fishermen exposed daily to attack by the biting flies (Chapter 8)
4. River channel improvement to clean and deepen waterways, with removal of stones, logs, and other obstructions that cause ripples attractive to blackfly larvae; construction of outlets of lakes, dams, and spillways to provide clear, unobstructed drops into deep pools
5. Development of traps to attract adult blackflies by light and/or chemicals
6. Surgical removal of nodules containing filariae is practical in selected instances
7. Biological control with insect pathogens such as viruses, *Coelomycidium* spp. microsporidians such as *Pleistophora* and *Thelohania,* and parasitic nematodes of the family Mermithidae, which are highly parasitic on populations of Canadian blackflies, may ultimately play an important role in substantially reducing *Simulium* populations (International Development Research Centre 1972).

REFERENCES

AID. 1973. U.S. Agency for International Development, Malaria Eradication Branch, Washington, D.C. Unpublished data.

Brown, A. W. A. 1958. Insecticide resistance in arthropods. Monograph Series no. 38. World Health Organization, Geneva, Switzerland. 240 pp.

Brown, A. W. A. 1971. Pest resistance to insecticides. Pages 457–552 *in* R. White-Stevens, ed., Pesticides in the environment. Vol. 1, pt. 2. Marcel Dekker, New York.

Brown, A. W. A., and R. Pal. 1971. Insecticide resistance in arthropods. Monograph Series no. 38. World Health Organization, Geneva, Switzerland. 491 pp.

Bruce-Chwatt, L. J. 1970. Global review of malaria control and eradication by attack on the vector. *Misc. Publ. Entomol. Soc. Amer.* **7**(1):7–23.

Bruce-Chwatt, L. J. 1971. Insecticides and the control of vector-borne diseases. *Bull. World Health Organ.* **44**:419–424.

Chapman, H. C., T. B. Clark, and J. J. Petersen. 1970. Protozoans, nematodes, and viruses of anophelines. *Misc. Publ. Entomol. Soc. Amer.* **7**(1):134–139.

Conway, G. R. 1970. Computer simulation as an aid to developing strategies for anopheline control *Misc. Publ. Entomol. Soc. Amer.* **7**(1):181–193.

Hamon, J., J. Mouchet, J. Brengues, and G. Chauvet 1970. Problems facing anopheline vector control: Vector ecology and behavior before, during, and after application of control measures. *Misc. Publ. Entomol. Soc. Amer.* **7**(1): 28–41.

Hayes, W. J., Jr. 1959. The toxicity of dieldrin to man—Report of a survey. *Bull. World Health Organ.* **20**:891–912.

Hess, A. O., and T. F. Hall. 1945. The relation of plants to malaria control on impounded waters with a suggested classification. *J. Nat. Malar. Soc.* **4**:20–46.

Hess, A. D., F. C. Harmston, and R. O. Hayes. 1970. *CRC Crit. Rev. Environ. Control.* Nov.

Huffaker, C. 1944. The temperature relations of the immature stages of the malaria mosquito, *Anopheles quadrimaculatus* Say, with a comparison of the developmental power of constant and variable temperatures in insect metabolism. *Ann. Entomol. Soc. Amer.* **37**:1–27.

International Development Research Centre. 1972. Preventing onchocerciasis through blackfly control. 6e, Ottawa, Canada.

MacDonald, G., C. B. Cuellar, and C. V. Foll. 1968. The dynamics of malaria. *Bull. World Health Organ.* **38**:743–755.

MacDonald, W. W., and C. P. Ramachandram. 1965. The influence of the gene f^m (filarial susceptibility *Brugia malayi*) on the susceptibility of *Aedes aegypti* to seven strains of *Brugia, Wuchereria,* and *Dirofilaria. Ann. Trop. Med. Parasitol.* **59**:64–87.

Metcalf, R. L. 1970. Role of pesticides in the integrated control of disease vectors. *Am. Zool.* **10**:583–593.

Metcalf, C. L., W. P. Flint, and R. L. Metcalf. 1962. Destructive and useful insects. 4th ed. McGraw-Hill, New York. 1087 pp.

NAS. 1973. Mosquito control: Some perspectives for developing countries. National Academy of Science, Washington, D.C.

Quellennec, G. 1972. Essais sur le terrain de nouvelles formulations d'insecticides OMS-708, resmethrin, et OMS-1155, contre les larves des simulies. *Bull. World Health Organ.* **46**:227–231.

Reinhards, J., A. Webber, B. Nizeties, and F. Maxwell-Lyons. 1968. Studies in the epidemiology and control of seasonal conjunctivitis and trachoma in Southern Morocco. *Bull. World Health Organ.* **39**:497–545.

Roberts, D. W. 1970. Coelomomyces, Entomophthora, Beauveria, and Metarrhizium as parasites of mosquitoes. *Misc. Publ. Entomol. Soc. Amer.* **7**(1): 140–159.

Southwood, T. R. E., G. Murdie, M. Yasuno, R. J. Tonn, and P. M. Reader. 1972. Studies on the life budget of *Aedes aegypti* in Wot Samphaya, Bangkok, Thailand. *Bull. World Health Organ.* **46**:211–226.

Traub, R., and M. A. C. Dowling 1961. The duration of efficacy of the insecticide dieldrin against the chigger vectors of scrub typhus in Malaya. *J. Econ. Entomol.* **54**:654–659.

Traub, R., and C. L. Wisseman. 1968a. Ecological considerations in scrub typhus. 1. Emerging concepts. *Bull. World Health Organ.* **39**:209–218.

Traub, R., and C. L. Wisseman. 1968b. Ecological considerations in scrub typhus. 3. Methods of area control. *Bull. World Health Organ.* **39**:231-237.

Trpis, M. 1973. Interaction between the predator *Toxorhynchites brevipalpis* and its prey *Aedes aegypti*. *Bull. World Health Organ.* **49**, 359–366.

TVA. 1947. Malaria control on impounded waters. U.S. Tennessee Valley Authority, Washington, D.C. 422 pp.

WHO. 1971. Vector control. *WHO Chronicle*. **25**(5).

WHO. 1972. Vector ecology. *World Health Organ. Tech. Rep. Ser.* 501.

WHO. 1973a. Safe use of pesticides. *World Health Organ. Tech. Rep. Ser.* 513.

WHO. 1973b. Manual on larval control operations in malaria programmes. World Health Organ. Geneva, Switzerland. 199 pp.

WHO. 1973c. Onchocerciasis and facts about river blindness. *World Health Mag.* October. pp 3–11.

EPILOGUE

16

PEST MANAGEMENT AND THE FUTURE

William H. Luckmann

In the beginning paragraphs of this book, we state that insect pest management is absolutely essential to the survival and future of modern insect control, and that the pest-management philosophy is relevant in all pest-control actions. Throughout the remaining chapters there is an evident dedication to insect pest management. We know what is needed; perhaps we will look back a decade or two hence and be astonished at our immaturity, approach, and design, but the sincerity, intent, and dedication is here.

To some, insect pest management is insecticide management; to others, it is remote-sensing or developing a resistant plant, managing a predator, or monitoring with a sex pheromone. The introductory chapters of this book aid in planning pest management, other chapters identify some of the tools and tactics that can be employed, and the final chapters illustrate examples of programs being developed or programs in action. Currently, single-factor programs (i.e., use of only one control agent) are the norm but, as the examples show, very sophisticated multifactor programs are being developed which provide economic and environmental planning alternatives, and not only predict potential for damage but also predict very accurately the chronology of plant and pest development and the factors that will daily suppress or enhance pest density and crop development.

There was an obvious awareness in the 1950s on the part of many leaders in insect control that sole reliance on insecticides was not adequate or wise, and the art and practice of modern insect pest management began to evolve at that time. The real push in insect pest-management research and funding began late in the 1960s. The terminology was quickly adopted and readily applied to almost any program and almost any course in pest control. Simply stated, pest management was acceptable to any and to all. And that quick embrace of pest management—not yet well-defined—poses the greatest peril for the future. For example, routine annual prophylactic soil treatment for corn rootworms applied because the field was planted in corn the previous year is not pest management. When such a procedure is called pest management, the grower is confused by the "new" program, since nothing really has changed. Any impetus in pest management is halted and, indeed, pest management is degraded. The greatest obstacle to the future success of pest management is the attitude of people—growers, industry personnel and, most important of all, applied research entomologists and extension personnel. It is not enough to just accept pest-management concepts and philosophy. They must be used. Insect pest management is people-oriented and, in order for it to be put into general practice, the control specialist must use and demonstrate pest-management technology and successfully teach it to others.

Many things will have a bearing on effective education and communication. One of these is the success of United States agriculture. In 1972, the average United States farmer produced food and fiber for himself and 51 others, with only 6% of the United States labor force engaged in farming, but this unique agricultural system and the people it serves now face a crisis. In 1973, there were about 3.9 billion people in the world to feed, an increase of more than 50% in 2 decades. In our effort to maximize yields, there may be an effort to maximize insect control with insecticides. No approach could be more wrong. We are entering an era of more intensive agriculture; soils and crops will be managed for production and not for pest control. But it seems reasonable to insist that some adjustments be made to favor pest management.

The economics of pest management and the economics of crop production will undoubtedly overshadow environmental concerns. Agricultural economists, with training and experience in plant protection and environmental quality, will be important members of pest-management teams. They will be a modifying influence on what is often an exaggerated view of the pest problem. Further, they can explore the cost and benefits of various pest-control actions such as scouting or crop pest insurance, and they can develop the guidelines within which the cost of pest management with its favorable consequences will be borne by all those who share its long-term benefits.

Pest-management specialists should be particularly realistic in regard to local economic needs in designing programs in developing countries. Where abundant manpower is available, its use can be an asset to pest management and an asset to the economics of the area. Except for pest outbreaks beyond reasonable control by hand labor, pest-management programs can be readily adopted in developing countries.

Catastrophic outbreaks of insects are rare, but they will occur at times in the future. Even the mass use of insecticides cannot prevent them, but only make the event less catastrophic for some. These kinds of things, if well explained, should not have an adverse effect on the acceptance of pest management. Indeed, pest-management researchers seeking genetic variability in crops and genetic vulnerability in pests can do much to prevent catastrophic outbreaks. However, mistakes in judgment and decision making in pest management will have an adverse effect. The desire on the part of a researcher, grower, state, or nation to develop a pest-management program quickly should not outweigh or precede the judgment and experience of professionals based on well-conducted research.

The value of the computer, the systems analyst, and the multidisciplinary team approach to pest management is obvious. Fortunately, some basic researchers outside agriculture are beginning to realize the rich intellectual content of agricultural systems, and the agricultural scientist is beginning to realize the importance of the basic scientist in pest management. Pest management can be the vehicle to stimulate and advance this cooperation.

The staff and the policies for developing pest-management teams and significant programs are already established in many institutions. First-echelon pest-management practices, upon which educational programs can be based and to which more sophisticated multipest programs can be added, are already in use in many areas. In general, we know the key pests and the problems they cause, and we know quite a bit about their life history. We know less about the behavior of pests and the biological processes that regulate them. Within the limits of what we know, we can begin to design programs. The zero, or near zero, approach of "no pest—no problem" is a starting point, providing we survey and monitor accurately. The tools of insect management listed at the end of Chapter 1 and referred to in practice and theory throughout this book can be put in many combinations. Guidelines for pest management for an insect(s) on a single crop may be illustrative and detailed to introduce the user to the pest(s) and methods of management. Integrating pest-management strategies may often begin a year in advance. Guidelines may contain portions or all of the following kinds of information:

1. Introductory statement on the philosophy of pest management and the economic, ecological, and environmental reasons for adopting the pest-management program
2. Identification of the pest problem
3. Illustration or photograph of the pest and a summary of life history and mortality factors (parasitoids, predators, diseases, weather)
4. Recommendation of cultural practices that minimize or avoid pest attack
5. A figure or table of data illustrating the relationship between pest density or pest damage and potential loss in yield or quality of the crop
6. A table of statistical data illustrating loss in yield versus monetary value of the crop
7. Illustration of the chronology of pest invasion and increase versus development of the crop
8. Detailed directions for surveying and monitoring for the pest, for accurately measuring pest density and/or damage, and for making a decision
9. Emergency control measures when pest density exceeds the economic threshold.

The future should be very rewarding for young men and women seeking careers in insect pest management. Some should pursue very specialized training, others should strive to be good field biologists with skills in observing and recording. As a pest-management specialist, you will frequently make decisions involving millions of dollars worth of crops and affecting the welfare of thousands of people.

Seventeen very fine people contributed to the writing and preparation of this book. They voluntarily gave of their time, because each of them is a dedicated person who has a sincere personal ambition to improve pest control through pest management. Hopefully, this book will aid in that purpose.

INDEX

Abate, 548
Acala cotton, 384-385
Acanthomyops claviger, 281
Acarina, 556
p-Acetoxyphenethyl methyl ketone, 284-285
Aculus schlechtendali, 473, 480, 493
Acyrthosiphon pisum, 23, 178
Adulticides, 547-550
Aedes aegypti, 22, 237, 535, 536, 537, 546, 547
  life table for, 536
Aedes nigromaculis, 237, 547, 553
Aedes spp., 293-294
Aerial spraying, 250
Aerosols, 237
African trypanosomiasis, 22, 529
Aggregation pheromones, 281-282, 288
Aggression, 281
Agroecosystems, 8, 11
Agrotis ipsilon, 12
Agrotis malefida, 414
Aircraft disinsectization, 550
Airplane application of insecticides, 418-419
Alabama argillacea, 290, 386, 404, 414
Alarm pheromones, 280
Alcidodes spp., 404
Aldicarb, 246, 259, 266, 289, 435
Aldrin, 5, 238, 240, 246, 248, 291, 419
Alfalfa, 22, 129, 134-135, 264, 267, 314, 446-466
  pollination of, 244
Alfalfa butterfly, 448
Alfalfa caterpillar, 23
Alfalfa leaf-cutting bee, 244
Alfalfa weevil, 23, 58, 257, 316, 449-454
Algorithm, 355-359, 369-371, 373
Alkali bee, 244
Allelochemics, 107, 113
Allomones, 113, 276
Alsophila pometaria, 195
Alternate hosts, 168
Amblyseius fallacis, 25, 492-501
Amblyseius hibisci, 169
American cockroach, 529
American dog tick, 556
American trypanosomiasis, 529, 559-560
Aminopterine, 303
Amlure, 284
Amphimallon majalis, 284
Anagrus armatus, 480
Anagrus epos, 169
Anal glands, 281
Analysis and modeling, 353-376, 461-466, 495-499, 519-526, 531-535
Anaphes flavipes, 268
Anasa tristis, 53
Anastrepha ludens, 303
Animal reservoir, 531, 556, 559
Annual crops, 160
Anobium, 291
Anopheles albimanus, 541, 552
Anopheles aquasalis, 541
Anopheles atroparvus, 541
Anopheles culicifacies, 541
Anopheles gambiae, 239
Anopheles pseudopunctipennis, 541
Anopheles quadrimaculatus, 537, 547, 552
Anopheles spp., 240, 245, 260, 293-294
Anopheles sundaicus, 541
Anthonomus grandis, 20, 52, 242,

258, 278, 280, 289, 290, 386, 389, 397-399, 404
Anthrenus piceus, 292
Anthrenus spp., 291
Antibiosis, 116, 119
Antibiotics, 555, 557
Anticarsia gemmatalis, 310
Antimalarial drugs, 530, 534
Antimetabolites, 119, 303
Ants, 281, 287
Anuraphis roseus, 473, 478-479
Aonidiella aurantii, 174
Apanteles, 23
Apanteles bucculatricis, 407
Apanteles glomeratus, 406
Apanteles laeviceps, 406
Apanteles marginiventris, 406-407
Apanteles militaris, 406
Aphanias dispar, 551
Aphelinus, 23
Aphelinus mali, 479
Aphelopus typhlocybae, 480
Aphidius, 23
Aphids, 338
Aphid vectors of plant diseases, 53-54
Aphiochaeta spp., 407
Aphis craccivora, 19
Aphis fabae, 128
Aphis gossypii, 386, 405
Aphis pomi, 109, 473, 479
Apholate, 303
Aphytis lingnanensis, 174
Aphytis melinus, 174
Aphytis mytilaspidis, 479
Apis mellifera, 260
Apple, 236, 242, 262, 264, 286-287, 332
pests of, 471-483
Apple aphid, 473, 479
Apple diseases, 484-485
Apple maggot, 472-475, 489
Apple orchard ecosystem, 472-481
Apple rust mite, 473, 480, 493
Apple scab, 471, 473-474
Applied ecology, 251
Archips argyrospilus, 255-256, 332
Archytas californiae, 406-407
Area burning, 557
Area-under-curve-technique, 368-369
Argas persicus, 556
Argentine ant, 58, 170
Argyrotaenia velutinana, 166, 242, 262, 277, 287, 472-475, 487-488
Army cutworm, 449
Armyworm, 13, 56, 197, 200-203
Arrestants, 113, 281-282
Artichoke plum moth, 20
Asparagus, 237
Asparagus beetle, 20
Assessing plant damage, 330-339
Aster yellows, 29
'Atlantic' alfalfa, 135
Aquatic plant types, 536
Atificial defoliation, 339
Attagenus megatoma, 278
Atta texana, 281
Attractants, 113, 260-261, 275-290
Austroasca terraereginae, 405
Autographa precationis, 110
Avocado, 169
Azinphosmethyl, 237, 246, 248, 257, 262, 263, 420
Aziridines, 303

Babesia bigemina, 556
Bacillus popilliae, 194-5, 205, 226
Bacillus sphaericus, 552
Bacillus thuringiensis, 23, 189, 191, 195, 197, 205-206, 215-218, 222-226, 253, 414, 552
Bacillus thuringiensis, production of, 215-216
Bacillus thuringiensis toxin, 206
Bacterial conjunctivitis, 554
Baculovirus, 199
Bagworm, 223
Bait traps, 321, 322, 328
Ballast and insect invaders, 58-59
Bark beetles, 287-288
Bathyplectes, 23
Bathyplectes curculionis, 257, 459-460
Bean aphid, 128
Bean leaf beetle, 52, 56, 114, 314
Bedbug, 529
Beet armyworm, 23, 403
Beet leafhopper, 53
Behavioral selectivity, 259-263
Belastomatidae, 551
Bemisia tabaci, 386
Beneficial insects, 403-413
Benefit/risk, 14, 236, 257, 269

'Benhur' wheat, 130
Benzil, 293
Benzyl benzoate, 292-293, 557
Berlese funnel, 315-316
BHC, 5, 239, 242, 419, 553
Billbug, 13
Bioconcentration, 245
Biodegradability, 252
Bioenvironmental controls, 236
Biological control, 139, 147-187, 243, 403-413, 459-460, 549-553
  advantages of, 158-159
  evaluation of, 178-181
  feasibility of, 159-162
  implementation of, 163-178
  principles in use, 182-183
  probability of, 500
  world wide, 148-150
Biological deserts, 244
Biomass, 62
Biosynthetic origins of plant components, 112
Biotypes for resistance, 129-131
Biting flies on cattle, 293-295
Blackberry, 169
Black carpet beetle, 278
Black cutworm, 12-13
Blackflies, 292-294, 560-562
Black scale, 177
Blattella germanica, 530, 557-559
Blissus leucopterus, 53, 291
Bombycol, 277
Bombyx mori, 109, 195, 276-277, 280, 283
Boophilus annulatus, 556
Boophilus microplus, 556
Bordeaux mixture, 290
Borrelia anserina, 556
Brevicomin, 279, 282
Brevicoryne brassicae, 129, 347
Broccoli, 237
Brown plant hopper, 120
Brown trout, 245
Brugia malayi, 537, 544
Brussels sprouts, 129
BT insecticide, 253
Buffalo, 445
'Buffalo' alfalfa, 134
Bumblebees, 244
Bumper sticker, 30
Busulfan, 303
Butoxypolypropylene glycol, 293-294
N-Butylacetanilide, 293-294
2,6-di-*tert*-Butyl-4-(α,α-dimethylbenzyl)-phenol, 253
t-Butyl-2-methyl-4-chlorocyclohexane-carboxylate, 284
2-(*p-tert*-Butylphenoxy) cyclohexyl 2-propynyl sulfite, 497

Cabbage, 16, 177, 237, 254, 332, 347
  food web of, 10
  life table for, 15
Cabbage aphid, 129, 347
Cabbage looper, 11, 65, 195-197, 254, 262, 277, 286, 290, 347
Cabbage maggot, 15, 254
Cabbageworms, 347
Cactoblastis cactorum, 171
Cacodylic acid, 288
Caddis flies, 245
Calcium arsenate, 418-419
California red scale, 174
'Caliverde' alfalfa, 128
Campoletis argentifrons, 406
Campoletis intermedius, 407
Caparidaceae, 51
Caproic acid, 284
Carabid beetles, 58-59
Carbaryl, 25, 237, 244, 246, 248, 420, 427, 544
Carbofuran, 246, 259, 266, 268
Carbon dioxide, 283
Carbophenothion, 246
Carmine spider mite, 238
Carnivores, 11, 44
Carpet beetles, 291-292
Cassava, 104
Catalpa sphinx, 418
Cattle, 301-302
Cattle ticks, 556
Cecidophyes gossypii, 405
Centaurea nigra, 41
Cephus cinctus, 137
Ceratitis capitata, 261, 284-285, 287
Ceratoma trifurcata, 52, 114
Ceratomia catalpae, 418
Cereal leaf beetle, 152, 167, 173, 175, 180, 258, 268, 314, 318-320, 363-367
Chagas' disease, 559-560
Chaoborus astictopus, 8

Chelonus texanus, 406-407
Chemical control, 32, 138, 481, 484
Chemical defenses of plants, 51
Chemical ecology, 288
Chemosterilants, 303
'Cherokee' alfalfa, 135
Chiggers, 293, 556-557
Chikungunga fever, 544
Chilo suppressalis, 117
Chinch bug, 13, 53, 56, 291
Chitin formation, 253
Chlorbenside, 6
Chlordane, 246, 291, 553, 558
Chlordimeform, 422
1-(4-chlorophenyl)-3-(2,6-diflurorbenzoyl)-urea, 253
Chlorpyrifos, 5, 246, 548
Chlorpyrifos methyl, 548, 562
Chlorquine, 530
Chlortetracycline, 554-555
Choristoneura fumiferana, 245, 278, 513
Christmas tree plantations, 509
Chromaphis juglandicola, 169
Chromosomal translocations, 299
Chrysolina brunsvicensis, 107
Chrysolina quadrigemina, 49
Chrysopa carnea, 413
Chrysopa oculata, 408
Chrysopa, spp., 155, 169, 409
Chrysops spp., 292
Cimex lectularius, 529
Cinnabar moth, 66
Circulifer tenellus, 53
Cirrospilus spp., 407
Citrus, 264
Citrus red mite, 5
Citrus red scale, 177
'Clark 63' soybeans, 118, 123
Clean cultivation, 25, 68
Clear Lake gnat, 8
Cleridae, 288
Closterocerus utahensis, 407
Clothes moths, 291-292
Clover bud weevil, 448
Clover head caterpillar, 448
Clover head weevil, 448
Clover leafhopper, 448
Clover leaf weevil, 448
Clover mite, 448
Clover root borer, 448
Clover root curculios, 448
Clover seed chalcid, 448
Clover seed weevil, 448
Clover stem borer, 448
Clubroot, 15
^{60}Cobalt, 301-302
Coccinella spp., 408
Coccinellidae, 23, 169
Cochliomyia hominivorax, 295, 301-303, 310
Coconut moth, 171
Coconut rhinoceros beetle, 284
Codlemone, 263
Codling moth, 19, 53, 166, 236, 245, 257, 262-263, 277, 286, 333, 471, 474-475, 485-487
'Cody' alfalfa, 134, 267
Coelomomyces spp., 552, 562
Coevolution, 107
Cohort life tables, 367
Coleomegilla maculata, 268
Coleoptera, 325
Coleotechnites milleri, 513, 514
Colias eurytheme, 23
Collembola, 327
Collops spp., 408
Colonization, 42-43, 44-52
Colorado potato beetle, 19, 109, 239, 314
Community energetics, 61
Community structure, 61
Community succession, 60-66
Competition, 75
Computer language, 362
Computer modeling, 355, 362, 430, 463, 499, 519, 532-534
Conditioned lethals, 298-299
Conotrachelus nenuphar, 254, 333, 472-475, 488
Contarinia sorghicola, 332
Continuous models, 360
Copidosoma truncatellum, 406
Copper acetoarsenite, 548
Corixidae, 551
Corn, 26-27, 120, 123, 132, 236, 238, 240, 244, 310, 331, 332, 338
Corn earworm, 13, 20, 123, 138, 259, 283, 310, 357
Corn leaf aphid, 13
Corn picker, 68
Corn root aphid, 13

Corn rootworms, 12-13, 238, 245, 259
Corn storage, 68
Corn tassel ratio, 27
Corydalus cornutus, 551
Cosmophila flava, 404
Cost/benefit, 14, 76-78, 238-239, 355, 542
Costs, of natural enemy release, 177
  of pest management, 82
Cotinis nitida, 284
Cotton, 6, 80-95, 168, 224, 236, 242, 264, 289, 310, 347
  agronomic practices of, 389-394
  ecosystem, 382-396
  insects, 379-436
    biological control of, 403-413
  key pests of, 397-403
  losses to, 380
  pests in Nicaragua, 386-387
  plant, 382-385
  production, 381
Cotton aphid, 418
Cotton boll weevil, 20, 52, 118, 236, 242, 258, 278, 280, 289, 290, 387, 389, 397-399, 422, 426
Cotton bollworm, 236, 242-243, 401, 409, 418, 420, 427
Cotton fleahopper, 301, 389, 419, 422
Cotton leaf perforator, 423
Cotton leafworm, 290
Cottonseed, 221
Cottony cushion scale, 58, 166, 242
Coumaphos, 553
Coumarin, 283
Cowpea aphid, 19
CPV, 204
Cranberry false blossom, 29
Creontiades femoralis, 386
Crickets, 287
Crioceris asparagi, 20
Crop islands, 52-60
Cropland soil, 248
Cropping practices, 160
Crop protection, patterns of, 6
Crop resistance, implementation of, 131-137
Crop rotation, 12-14, 31, 160
Cruciferae, 51
Cryptococcus fagi, 58
Cucurbita, 125
Cue-lure, 284-285
Culex fatigans, 300, 544, 547
Culex pipiens quinquefasciatus, 300, 544
Culex spp., 293-294
Culex tarsalis, 530, 552
Cultural practices, 31, 77, 139, 167, 392-394, 457-459
Curly top, 29
Current life tables, 367
Cutworms, 15, 287
Cyanogenetic glucosides, 104
Cynolebias bellottii, 551
Cytoplasmic plant resistance, 127
Cytoplasmic polyhedrosis viruses, 204

Dacus curbitae, 260, 284
Dacus dorsalis, 261, 282-283
Dahlbominus fuscipennis, 172
Dairy products, 455
Damages, probabilities of, 80
  by screwworm, 301
  to wood, 291
Danaus plexippus, 280
Dandelion, 47, 110
DDD, 8, 248
DDE, 239, 248
DDT, 5, 8, 195, 238, 239, 242, 245, 248, 249, 260, 263, 266, 290, 291, 292, 356, 419, 427, 481, 541, 542, 543, 547, 549, 550, 553
  banning of, 422, 427
  DDT'ase enzyme, 239
  in Lake Michigan, 17
  and predators, 24
  resistance, 239
Deaths from insecticides, 250
Decision-making index, 496
Decision theory, 83
Deet, 293
Defoliation, 371-373
Deleterious mutations, 299
Demeton, 6, 23, 247, 258, 266, 267
Dendroctonus brevicomis, 278, 282, 287-289, 520
Dendroctonus frontalis, 279, 280, 520
Dendroctonus ponderosae, 520
Dendroctonus spp., 279, 287
Density dependent factors, 157, 181
Density independent factors, 157
Dermacentor variabilis, 556
Dermestidae, 292

Descriptive models, 359-367
Deterrents, 113
Diabrotica longicornis, 12, 53, 346
Diabrotica spp., 238, 245, 259
Diabrotica undecimpunctata howardi, 290
Diabrotica virgifera, 12, 53, 75, 127, 240-241
Diacrisia virginica, 19
Diamondback, 52, 254, 283, 347
Diapause in cotton insects, 392-393
Diatraea saccharalis, 123
Diazinon, 237, 246, 248, 553
Dibutyl phthalate, 293
Dibutyl succinate, 293-294
Dicapthon, 558
Dichlorvos, 238, 547, 550
Dicofol, 6, 237, 246, 248, 253
Dieldrin, 5, 239, 246, 249, 291, 292, 419, 545, 551, 557
Dieldrin resistance, 239
N,N-Diethyl *m*-toluamide, 293
Differential equations, 361-362
2,4-Dihydroxy-7-methoxy-1,4-benzoxazine-3-one, 122
Dikrella cruentata, 169
DIMBOA, 26, 122
Dimethoate, 246, 553
Dimethylcarbamazine, 562
3,3-Dimethyl-Δ-α-cyclohexaneacetaldehyde, 278, 280
3,3-Dimethyl-Δ-β-cyclohexane ethanol, 278, 280
1,5-Dimethyl-6,8-dioxabicyclo-[3.2.1.]-octane, 279
Dimethyl phthalate, 290, 291, 293
4′-(Dimethyltriazeno)-acetanilide, 290
Dinitro-*o*-cresol, 237, 291
Diparopsis castanea, 404
Diparopsis watersi, 404
Diprion hercyniae, 58, 172
Dipropylpyridine-2,5-dicarboxylate, 293-294
Dirty field technique, 395-396
Discrete models, 360
Diseases in pest management, 189-227
Disparlure, 277, 285-289
Dispersal by wind, 57
Distance to nearest neighbor, 311
Disulfoton, 246, 257, 258
Diversity, 62, 68-69, 168
DMC, 6
8,10-Dodecadien-l-ol, 263, 277, 285-286
3,6,8-Dodecatrien-l-ol, 281
7-Dodecenyl acetate, 277, 286
8-Dodecenyl acetate, 277
Dolichoderine ants, 281
Domestication of plants, 104-105
Douglas fir, 291
Douglas fir tussock moth, 512, 513
Drainage, 545
Drought, 15
Dust and natural enemies, 170
Dutch elm disease, 22
D-Vac, 324-325
Dynamic economic evaluation, 85
Dynamic models, 359
Dysdercus spp., 405
Dytiscidae, 551

Earias insulana, 425
Earias spp., 404
Earwigs, 287
Eastern red cedar, 223
East Indian Teak, 291
Ecological homologues, 174
Ecological niche, 40-44, 174
Ecological plant resistance, 114
Ecological selectivity, 253-255
Ecology of pest management, 37-73
Economic injury level, 18-22, 163, 182, 331, 372-374, 530
Economics of pest management, 75-99
Economic threshold, 4, 18-23, 83, 87, 163, 213, 235, 236, 251, 267-269, 275, 331, 428-430, 530
Ecosystem concept, 8, 37-41, 64, 395-396
Ectoparasitoids, 151
Egg sampling, 315, 318-319
Egyptian alfalfa, 23
Empoasca, 387
Empoasca devastans, 405
Empoasca fabae, 11, 56, 57, 123, 239, 405, 449
Empoasca facialis, 125, 405
Empoasca solana, 238, 394
Encapsulation of parasitoids, 154
Encarsia formosa, 170
Encephalitides, 530, 544, 545, 547
Endoparasitoids, 151
Endosulfan, 237, 246, 260, 542

Endrin, 242, 246, 250, 347-348, 419, 422, 427
Energy flow in ecosystem, 44
Energy pathways, 63
Entomophthora, 212, 267
Environmental Protection Agency, 264, 511
Environmental quality, 8, 243-249, 264-266
Environmental Quality Improvement Act, 512
Enzymes in genetic fingerprinting, 175
EPA, 412
Ephemeroptera, 245
Epilachna varivestis, 20, 56, 117, 121
Epiplatys senegalensis, 551
Epizootics, 172, 211-212, 214
EPN, 5, 246, 420
7,8-Epoxy-2-methyloctadecane, 277
Equilibrium models, 46-47
Equilibrium position, 19
Equine encephalomyelitis, 237, 530
Eradication, of insects, 435
  of screwworm, 300-303
Eriosoma lanigerum, 105, 473, 479
Erythroneura elegantula, 169
Erythroneura spp., 479
Estigmene acrea, 414
Estigmene spp., 433
Ethion, 246
2-Ethyl-2-butyl-1,3-propanediol, 293-294
Ethylene thiourea, 303
2-Ethyl-1,3-hexanediol, 290, 292-293
Ethyl linoleate, 282
7-Ethyl-5-methyl-6,8-dioxabicyclo-[3.2.1.] -octane, 279
Ethyl oleate, 282
Ethyl palmitate, 282
1-(4′-ethylphenoxy)-6,7-epoxy-3,7-dimethyl-2-octene, 252
Ethyl stearate, 282
Eucelatoria armigera, 406-407
Eugenol, 284-285
Eulan CN, 292
Euphorocera claripennis, 407
European chafer, 284
European corn borer, 13, 26, 53, 56, 58, 65, 68, 115, 120, 137, 196, 254, 276, 290
European elm bark beetle, 22, 58
European pine sawfly, 214
European red mite, 6, 24, 240, 242, 262, 473, 478, 480, 490-501
European spruce sawfly, 58, 172, 214
Euryphagic insects, 106
Eurytopic insects, 106
Eutinobothrus spp., 404
Euxoa auxiliaris, 449
Evolutionary phase, 50
Excitants, 113
Excitorepellents, 290
Exenterus claripennis, 172
Exenterus vellicatus, 172
Exorista larvarum, 407
External costs, 94-96
Externalities, 17, 90
Extinction rates, 46
Extrinsic cycle, 537-539
Eye-spotted budmoth, 332

Fabric-eating insects, 291-292
Face fly, 294
Fallacy of composition, 96-97
Fall armyworm, 13, 132
Fall webworm, 20
Farmer, and economic reality, 78
  and pest-management decisions, 83-88
  strategies, 66-68
  and uncertainty, 79-83
β-Farnesine, 281
Federal Environmental Pesticide Control Act, 264, 511
Federal Insecticide, Fungicide, and Rodenticide Act, 511
Feeding deterrents, 290
Feeding stimulants, 113
Felted beech scale, 58
Feltia subterranea, 414
Fenchlorphos, 553
Fenitrothion, 548, 550
Fensulfothion, 259
Fenthion, 5, 548, 553, 557
Fertilization, 67, 116
Field checking, 432-433
Filariasis, 537, 544
Filmore Citrus Protective District, 177
Fish food organisms, 245
Flea beetles, 15, 51
Flight range, 281, 540
Flit MLO, 548
Floodwater mosquitoes, 5
5-Fluoroorotic acid, 303

5-Fluorouracil, 303
Flushing, 323
Fonofos, 259
Food and Drug Administration, 17, 248
Food lures, 282-283
Food web, 9, 44, 63, 247
Forage crops, 445-466
Forest, use constraints of, 512
Forest insects, 509-527
Forest nurseries, 509
Forest pests, economic impact of, 523, 525
Forest tent caterpillar, 214
Formica rufa, 281
FORTRAN, 362
Fowls, 556
Fowl spirochetosis, 556
Fowl tick, 556
Frego cotton, 388
Fruit tree leaf roller, 255-256, 332
Fumigants, 237
Fundulus heteroclitus, 551
Fungi, 206-208

Galleria mellonella, 196, 276
Gambusia affinis, 551
Gamma radiation, 301-302
Gardona, 246
Gas flamer, 459
Gasterosteus aculeatus, 551
Gastroenteric diseases, 553
Genetic Control, 32, 295-303
Genetic engineering, 299
Geocoris spp., 267, 408, 425
Geraniol, 285
German cockroach, 530, 557-559
Gilpinia, 199
Gilpinia hercyniae, 214
Gladiolus thrips, 56
Glossina spp., 22, 529, 539, 542
Glossina swynnertoni, 260
Glycine max, 11
Goats, 301
Gonia capitata, 406
Gossypium arboreum, 382
Gossypium barbadense, 117, 381, 382
Gossypium herbaceum, 382
Gossypium hirsutum, 381, 382
Gossyplure, 434
Gossypol, 388
Grain aphids, 57
Grandlure, 434-435
Granulars, 238, 245, 259
Granulosis capsule, 202
Granulosis viruses, 202-203
Grapes, 105, 353
Grape colaspis, 13
Grape leafhopper, 169
Grape phylloxera, 105
Grapholitha glycinivorella, 138
Grapholitha molesta, 277
Grasshoppers, 56, 109, 287, 448
Grazing pressure, 66
Greater wax moth, 196, 276
Greenbug, 139, 141, 257
Green cloverworm, 19, 209, 212, 344-346, 373
Greenhouse whitefly, 170
Green June beetle, 284
Green peach aphid, 29, 212, 238
Gregarious parasitoids, 151
Ground spraying, 458
GV, 202-203
Gypsy moth, 5, 20, 58, 245, 277, 285, 289, 310, 513
Gypsy moth, life table, 368

Haematobia irritans, 294
Haemophilus spp., 554
Hair pencils, 280
'Harosoy' soybeans, 117, 123
Hay, 22
Helicoverpa, 224
Heliothis armigera, 283, 404
Heliothis NPV, 224-225, 226
Heliothis spp., 224, 347, 388, 389, 410, 413, 421, 433
Heliothis virescens, 6, 168, 242, 357, 393, 401, 404
Heliothis zea, 6, 20, 80, 82, 123, 242, 246, 253, 259, 310, 357, 401, 404
Helopeltis schoutedeni, 405
Hemel, 303
Hemerobius, 408
Hemerocampa pseudotsugata, 512, 513
Hemisarcoptes malus, 479
Hemitarsonemus latus, 405
Heptachlor, 238, 240, 247, 419
Heptachlor epoxide, 248
Heptyl butyrate, 284
Herbicides, 67

Hessian fly, 53, 67, 105, 127, 130-131, 137
10,12-Hexadecadienol, 276, 277
7,11-Hexadecadienyl acetate, 277
Hexalure, 289, 434
Hexanoic acid, 281
Hexen-l-ol, 283
Hibiscus syriacus, 118
Hippodamia convergens, 53
Hippodamia spp., 408
Homeostasis, 61
Honeybee, 244, 246-247, 265, 267
Honeydew-feeding ants, 170
Horcias nobilellus, 405
Horn fly, 294
Host acceptance, 109
Host finding, 107
Host plant resistance, 11-12, 31, 103-141, 387-389, 460-461
Host preference, 540-541
Host recognition, 109
Host-selection process, 108
Host suitability, 109
Housefly, 283, 294, 303, 530, 553-555
Household pest control, 557-559
Human body louse, 238, 529, 542-543
Human fat, 248
Hydrogen cyanide, 237
10-Hydroxy-3,7-dimethyl deca-2,6-dienoic acid, 280
2-Hydroxyethyl octyl sulfide, 293-294
Hylemya brassicae, 254
Hyles lineata, 20
Hypera brunneipennis, 23
Hypera postica, 23, 58, 257, 449-454
Hypericin, 107
Hypericum, 107
Hyperparasitoids, 153
Hyphantria cunea, 20
Hyposoter annulipes, 406
Hyposoter exiguae, 406-407

Icerya purchasi, 58, 147, 166, 242
Ichneumonidae, 172
Idaho potatoes, 31
Identification of insect injury, 331-332
2-Imidazolidinone, 303
Imported cabbage worm, 52, 64, 254, 332, 347
Imported fire ant, 287
Infection of host, 539
Insect, affecting man and domestic animals, 529-562
  control, methods of, 31-32
  diseases, host range of, 190
    pathogenicity of, 194-198
    persistence of, 191-194
    transmission of, 190
  fragments, 21, 164
  frass, 330
  growth regulators, 5
  nests, 330
  numbers of, 446
  pathogens, 189-227
  plant interactions, 106-115
  resistance, in cotton, 387-388, 390-391
    to insecticides, 239-241
  scouts, 29
  viruses, 191, 198-204
Insecticides, 67, 236-239, 415-428, 454-457, 497, 546-550
  advantages of, 236-239
  broad spectrum, 227, 242, 243, 420
  cost of, 238
  environmental persistence of, 246-247, 265
  history of, 417-424
  improved application of, 255-259
  limitations of, 239-250
  mammalian toxicity of, 246-247, 265
  in pest management, 235-269
  production of, 5, 236
  registrations, 264
  resistance to, 5, 239-241, 420, 426-428, 531, 544, 545, 553
  selective use of, 236, 251-263, 415, 424-425
  tolerances, 248
Integrated control, 3, 424
Intersection line, 545
Ips spp., 279
Iridomyrmex humulis, 58, 170
Island biogeography, 45
2-Isopropenyl-1-methyl-cyclobutane ethanol, 278, 280
Isopropyl 11-methoxy-3,7,11-trimethyl-dodeca-2,4-dienoate, 252
Isoptera, 291
Isthmus of Panama, 303

Japanese beetle, 58, 194, 214, 284-285, 290

Juvenile hormone mimics, 252

Kairomones, 113, 276
K and r strategists, 172
Kepone, 558
Khapra beetle, 281
Klamath beetle, 49
Klamath weed, 107
Kolmogorov equation, 361

Lacewing flies, 267
Lactic acid, 283
Lady bird beetles, 53, 267
'Lahontan' alfalfa, 127, 128, 132, 137, 267
Lake George, 245
Lake Michigan, 249
  DDT in, 17
Lake Superior, 292, 294
Lake trout, 245, 249
Large milkweed bug, 49, 57
Larvicides, 546-548, 562
Laspeyresia pomonella, 19, 53, 245, 257, 262, 277, 286, 333, 471, 474-475, 485-487
Late planting, 67
LD_{50}, of insect pathogens, 194
  to rat, 249, 265
  values for insects, 427-428
Lead arsenate, 418
Leaf area, 431
Leafhoppers, 57, 169
Leave a pest residue, 23-25
Lepesia archippivora, 406
Lepidosaphes ulmi, 473, 479
Leptinotarsa decemlineata, 19, 50, 109, 239
Leptoconops spp., 292
Leptotrombidium akamushi, 556
Leptotrombidium deliense, 556
Leslie matrix, 362
Lesser knapweed, 41
Lethal gametes, 299
Lettuce, 237
Levuana iridescens, 171
LI-COR area meter, 334-336
Life tables, 15, 179, 254-256, 367-369, 515-517, 535-536
Light and suction trap, 312
Light traps, 245, 261-262, 321, 328-329, 433
Lima beans, 118
Limnanthaceae, 51
Lincoln index, 317, 539
Lindane, 5, 246, 291, 542, 544, 548, 550, 557
Lisiphlebus testaceipes, 139
Livestock, 446
Locusta migratoria, 237
Lodgepole needle miner, 513, 514
Lodgepole pine, 520, 524
Lodging of corn, 339
London purple, 417-418
Losses, probabilities of, 82
LT_{50} of insect pathogens, 196-7
Lyctus, 291
Lygus hesperus, 400, 402
Lygus lineolaris, 400
Lygus spp., 178, 389, 394-395, 422
Lygus vosseleri, 405

Macrosiphum euphorbiae, 132
Macrosteles fascifrons, 57
Malacosoma disstria, 195, 214
Malaise traps, 325
Malaria, 5, 238, 529, 531, 541, 544-547
  model of, 532-534
  parasite, life-cycle of, 538
Malathion, 237, 240, 248, 257, 420, 530, 547, 548, 550, 553, 557
Malpighian tubules, 205
Man, toxicity of insecticides to, 249-250
Management strategy, 83
Mandibular glands, 281
Manduca quinquemaculata, 261
Manduca sexta, 114, 261, 357
Mangrove islands, 43
Mating confusants, 289-290, 434
Matrices, 362
Mayetiola destructor, 53, 67, 105
6-MBOA, 122, 290
Measurement, of defoliation, 333-335
  of root damage, 335, 337
Mechanical control methods, 32
Mechanisms of plant resistance, 114-126
Medicago sativa, 22
Mediterranean fruit fly, 261, 284-285, 287
Megachile rotundata, 244
Melanicheneumon rubicundus, 406
Melanoplus bivittatus, 195
Melittia cucurbitae, 125

Melon fly, 260, 284
Mermithidae, 552
Metamorphosis, 253
Meteorus vulgaris, 406-407
Methoprene, 252
Methotrexate, 303
6-Methoxybenzoxazolinone, 122, 290
Methoxychlor, 23, 237, 247, 248, 252, 266
p-Methylacetophenone, 283
β-Methylanthraquinone, 291
Methyl bromide, 237
Methyl eugenol, 261, 282-284, 286
2-Methylheptanone, 281
Methyl parathion, 5, 6, 237, 247-249, 257, 347-348, 420, 428
4-Methylpyrrole-2-carboxylate, 281
Mevinphos, 237, 247, 266, 267
Mexican bean beetle, 20, 56, 117, 121
Mexican fruit fly, 303
Microbial insecticides, future of, 225-226
  registration of, 220-222
  safety of, 221
  standardization of, 217-218
Microfilariae, 560
Microplitis brassicae, 406
Microsporidians, 195, 208, 210, 224, 552, 562
Midge larvae, 245
Migrations, 53, 57
Migratory locusts, 237
Milk, 22
Mirex, 260, 287, 558
Mites, 13, 491-501, 556-557
Mitin FF, 292
Mitotic apparatus, 303
'Moapa' alfalfa, 127, 137, 267
Modeling, 181, 353-376, 429-432, 461-466, 495-499, 519-526, 532-535
Monarch butterfly, 280
Monocrotophos, 421
'Monon' wheat, 130
Monophagous insects, 106
Mosquito abatement districts, 530
Mosquito fish, 551-552
Mosquito oil, 548
Mosquitoes, 544-553
  larval habitats of, 535-537
Mothproofing, 29
Mountain pine beetle, 520, 524
Mountain whitefish, 245
'Mudgo' rice, 120
Multiparasitism, 153
Multiple matings, 298
Musca autumnalis, 294
Musca domestica, 5, 239, 283, 294, 303, 530, 553-555
Musca sorbens, 554
Musca vicina, 195
Mustard oils, 51-52
Mutations, 298
Myzus persicae, 29, 53-54, 212, 238

Nabis spp., 267, 408, 425
Naled, 247, 261, 266, 286
National Environmental Policy Act, 512
National Science Foundation, 462
Natural enemies, 32, 139, 147-183, 267-268, 403-413, 459-460, 492-499, 549-553
  conservation and enhancement of, 165
  destruction of, 244
  field collections, 173, 177
  identification of, 174
  importation and colonization, 170-173
  mass culture of, 176-178
  modeling of, 181
  multiple species introductions, 173-176
  release of, 176-177
Natural forests, 510
Naturalistic control, 156-159, 549-553
Negative binomial distribution, 341-342
Nematodes, 209-211, 552
Neoaplectana carpocapsae, 193
Neodiprion, 199
Neodiprion sertifer, 214
Neogregarines, 208-209
Neotran, 6
Nepidae, 551
Nepiera marginata, 407
Net benefits, 84, 87, 92
Nezara viridula, 405
Nicaragua, 386
Nicotine, 237
Nicotine sulfate, 419
Nigeria, 239
Nilaparvata lugens, 120
Nomia melanderi, 244
Nomuraea rileyi, 198, 209, 212
Nondegradable insecticides, 246
Noninclusion viruses, 204, 224

Nonpersistence of insecticides, 257
Nonpreference, 116
Nonspore-forming bacteria, 205
Nontarget species, 243-245
Northern cornrootworm, 12-13, 53, 65, 346
Norway rat, 299
Nosema algerae, 552
Nosema locustae, 195, 224
Nosema trichoplusiae, 195-6
Nothobranchius guentheri, 551
Notonectidae, 551
Notoxus calcaratus, 408
NPV, 199-202, 217, 221, 224, 253
Nuclear polyhedrosis viruses, 199-202, 217, 221, 224
Nutrient cycling, 61
Nymphalidae, 276

Occupational Health and Safety Act, 250
Octanoin, 285
Odonata, 551
Odor corridor, 280
Oligogenic plant resistance, 126
Oligophagous insects, 106
Olla spp., 408
Omite, 253, 497
Onchocerciasis, 294, 529, 560-562
Oncopeltus fasciatellus, 49
Oncopeltus fasciatus, 49, 57
O'nyong-nyong fever, 544
Opuntia aurantiaca, 171
Opuntia stricta, 171
Oriental fruit fly, 261, 282-283
Oriental fruit moth, 277
Orius, 425
Orius tristicolor, 408
Ornithodoros, 556
Oryctes rhinoceros, 284
Ostomidae, 288
Ostrinia nubilalis, 26, 53, 58, 115, 196, 254, 277, 290
Oulema melanopus, 167, 180, 258, 268, 363-367
Overwintering sites, 52
Ovex, 6, 253
Oviposition lures, 283
Oxycarenus spp., 405
Oxydemeton-methyl, 247, 258

Panchax panchax, 551
Panonychus citri, 5, 224
Panonychus ulmi, 6, 24-25, 240, 242, 262, 473, 478, 480, 490-501
Panstrongylus spp., 559
Parameter estimation, 342-343
Parasitoids, 58, 77, 151-154, 165-180, 267-268, 404-407, 459-460
Parathion, 6, 23, 250, 257, 420, 553
Paratrioza cockerelli, 290
Paris green, 417-418, 548
Pasture rotations, 556
Pastures, 66
Pea aphid, 23, 178, 448
Peach yellows, 29
Peanut butter, 287
Pears, 263
Pear psylla, 262
Pectinophora gossypiella, 242, 277, 389, 397, 400-401, 405
Pediculus humanus, 238, 529, 542-543
Pentachlorophenol, 291
Perennial crops, 160
Peridroma saucia, 414
Periplaneta americana, 529
Permissive factors, 110
Persistent pesticides, 17
Pest control strategies, 87
Pesticide residues, 245-249
Pesticides, interference with biological control, 166
Pest management, 4
  concepts of, 8-31
  decisions, 87
  districts, 29
  by parasitoids and predators, 147-187
  plant resistance in, 103-145
  and public policy, 89
  ratings of insecticides, 246-247, 265
  systems, 353-376, 461-466, 491-502, 520-527, 531-535
  tools of, 31-32
  of tree fruits, 471-502
Pest mosquitoes, 293
Phaseolus vulgaris, 120
Pheasant, 246-247, 265
Phenethyl propionate, 284-285
Phenetic plant resistance, 123-125
Phenological asynchrony, 114
Pheromones, 259-263, 276-282, 285-

290, 330, 433-435, 486-488
aggregation, 281-282, 288
alarm, 281
antennal receptors for, 280
emission rates, 281
orientation to, 280
trail marking, 281
Pheromone traps, 245, 262-263, 328, 330, 434, 486, 488
Phorate, 247, 258, 266
Phosphamidon, 247, 266
Photoperiod, 280
Photoplanimeters, 333-335
Phthirus pubis, 529
Phyllophaga spp., 12
Phyllotreta cruciferae, 51
Phyllotreta striolata, 51
Phylloxera, 353
Phylloxera vitifoliae, 105, 136
Physical control methods, 32
Phytoseiulus persimilis, 176
Phytoseiulus spp., 408
Pieris rapae, 52, 64, 195, 254, 347
Pima cotton, 381
Pink bollworm, 242, 277, 289, 397, 400-401, 409, 434
Pinosylvin monomethyl ether, 291
Pinus sylvestris, 291
Piperonyl butoxide, 237, 550
Piroplasmosis, 556
Pitfall traps, 328
Plague, 529
Plague flea, 529
Plant components, 110-113
Plant damage, assessing, 330-339
Plant growth and development, 430-431
Plant growth model, 465
Plant-herbivore relationships, 51
Plant resistance, 103-145
Plant varieties, 67
Planter-box treatment, 258
Plasmodium spp., 254, 530, 537-538, 544
Plathypena scabra, 19, 212, 373
Platyptilia carduidactyla, 20
Plecoptera, 245
Pleistophora, 562
Pleistophora schubergi, 210
Plictran, 497
Plum curculio, 254, 333, 472-475, 488
Plutella xylostella, 254, 283, 347
Podagrica spp., 404
Podosphaera leucotricha, 473-475
Poison baits, 287
Poison bran baits, 255
Poisson distribution, 340-341
Polistes spp., 169
Pollinators, 244
Polygenic plant resistance, 127
Polyhedra, 199
Polyhedrosis virus, 195
Polyphagous insects, 106
Popillia japonica, 58, 214, 284-285, 290
Population dynamics, 514
Population measurement, 310-330, 431-432, 539
Porthetria dispar, 5, 20, 58, 245, 277, 285, 289, 310, 513
Potato, 29-30, 132, 239, 264
Potato aphid, 132
Potato leafhopper, 11-12, 123, 239, 448-449, 473, 479
Potato leaf roll, 29
Potato psyllid, 290
Potato viruses, 20
Powder-post beetles, 291
Powdery mildew, 473-475
Prairies, 445
Praon, 23
Praon pollitans, 267
Predators, 32, 58, 63, 77, 150, 154-156, 169, 267-268, 404, 408-411, 492-501, 550-552
Predatory mites, 169
resistance of, 501
President's Science Advisory Committee, 25
Prickly pear cactus, 171
Private versus social values, 89
Producer's surplus, 91
Propiononitrile, 283
Propoxur, 240, 548, 550, 557-558
Propyl 1,4-benzodioxan-2-carboxylate, 284
Prospaltella perniciosi, 479
Prospaltella spp., 406
Protein hydrolyzate bait, 261, 286, 303
Protozoa, 208
Pseudaletia unipuncta, 56, 197, 200-203
Pseudatomoscelis seriatus, 389, 401
Pseudomonas aeruginosa, 195
Pseudoplusia includens, 212-213

Pseudotsuga menziesii, 291
Psorophora spp., 293
Psylla pyricola, 262
Pterocormus difficilis, 407
Ptychomyia remota, 171
Pubic louse, 529
Pulp wood, 509
Pyrethrins, 237, 290, 294, 314, 544, 548, 550

Quadrapidiotus perniciosus, 473, 478-479
Quarantine facilities, 172

R-20458, 252
Rainbow Trout, 246-247, 265
Ratoon cotton, 383
Rattus spp., 556
Recapture of marked insects, 317, 539
Recording dynamometer, 335, 337
Recurrent phenotypic selection, 134
Red-banded leafroller, 166, 242, 262, 277, 287, 472, 473-475, 487-488
Red clover, 448
Red mangrove, 42-43
Redwood, 291
Reesimermis nielseni, 552
Registered Pesticides, 5
Registration of pesticides, 264
Regulatory control methods, 32
Relapsing fevers, 556
Removal trapping, 317-321
Repellents, 113, 290-295, 557, 562
Reovirus, 199
Residual fumigants, 237
Residual spraying, 547-549
Residue persistence, 247
Resistant natural enemies, 162, 166
Resmethrin, 550
Reticulitermes spp., 281
Rhagoletis pomonella, 472-475, 489
Rhizophora mangle, 42
Rhodnius spp., 559
Rhopalosiphoninus staphyleae, 53-54
Rhopalosiphum incertus, 109
'Ribeiro' wheat, 130
Rice, 117
Rice stem borer, 117, 283
Rickettsia prowazekii, 542
Rickettsia rickettsii, 556
Rickettsia tsutsugamushi, 556
Rio Grande Valley, 242
River blindness, 560-561
Rocky Mountain wood tick, 556
Rodents, 15, 556, 560
Rodolia cardinalis, 147, 166, 242
'Rogers' barley, 141
Ronnel, 558
Rotary net, 313
Rubus leafhopper, 54
Rubus spp., 54, 169

Sacadodes pyralis, 386, 387, 404
Saccharophiles, 11
Safety in insecticide use, 250
Safflower, 394
Salmonella, 221
Sampling, air, 312-313
  for decision-making, 340
  habitat, 311
  and measuring, 309-348
  program, 339-346
  soil and litter, 314
  vegetation, 313-314
Sanitation, 553-555
San Jose scale, 473, 478-479
Saprobes, 11
Sarcophaga spp., 406
Schistocerca gregaria, 237
Schizaphis graminum, 139, 257
Schradan, 6
Scolothrips sexmaculatus, 408
Scolytidae, 287
Scolytus multistriatus, 22, 58
Screwworm, 295, 300-303, 310
Scrub typhus, 556-557
Sea Island cotton, 382
Secondary pests, 242-243, 403-404, 423-424
Seed corn beetles, 13
Seeds and planting, 67
Seed treatments, 238, 245, 258
Selection pressure, 61
Selective insecticides, 162, 252, 347, 424-425
Semiochemicals, 276
'Seneca' wheat, 130
Sensory sensillae, 280
Sequential sampling, 343-346, 515
Sequoia sempervirens, 291
Sex attractants, 434-435

Sex pheromones, 262, 263, 276-281, 285-286
Shaking and beating, 324
Shannon-Wiener index, 69
Shigella, 221, 553
Sialidae, 551
Silage corn, 338
Silkworm, 109, 276-277, 283
Simulium damnosum, 561
Simulium spp., 292, 294, 529, 539, 560-562
Sinea spp., 408
Sinigrin, 283
Sitona cylindricollis, 283
Six-spotted leafhopper, 448
Slugs, 13
Social economics, 88-98
Sodium aluminum fluosilicate, 292
Sod webworm, 210
Soft rot, 15
Soil barriers, 291
Soil core sampler, 315
Soil preparation, 67
Soil treatments, 291
Solanum tuberosum, 50, 132
Solenopsis saevissima richteri, 287
Solitary parasitoids, 151
Sooty mold fungus, 267
Sorghum, 332, 394
Sorghum midge, 332
Source reduction, 545
Southern armyworm, 277
Southern cornrootworm, 12-13
Southern garden leafhopper, 238
Southern pine beetle, 279, 520
Southwestern corn borer, 13
Soybeans, 11-12, 111, 114, 121-124, 224, 236, 264, 310, 331, 339, 371-373
Soybean looper, 118, 212
Soybean pod borer, 138
Spatial distribution, 340-341
Species diversity, 42
Sphenoptera gossypii, 404
Spilochalcis igneoides, 406
Spilonota ocellana, 332
Spirochetes, 556
Spissistilus festinus, 331
Spodoptera, 433
Spodoptera eridania, 277
Spodoptera exigua, 23, 403, 405, 414, 425
Spodoptera frugiperda, 132, 405, 414
Spodoptera littoralis, 387, 405, 425
Spodoptera ornithogalli, 405, 414
Spore-forming bacteria, 205-206
Spotted alfalfa aphid, 23, 137, 257-258, 266-267
Spotted cucumber beetle, 290
Spotted fever tick, 556
Spotted ladybird, 268
Spot treatments, 547, 548
Spruce budworm, 245, 278, 513
Squash bug, 53
Squash vine borer, 125
Stable fly, 294
Stages of plant succession, 61
Stalk destruction, 357
Stenophagic insects, 106
Stenotopic insects, 106
Sterile-male technique, 295-298, 300-303, 435, 487
Stethorus picipes, 408
Sticky traps, 326-327
Stinkbugs, 33s
Stomoxyx calcitrans, 292-294
Stone flies, 245
'Stowell's Evergreen' sweetcorn, 138
Strategists, *r* and *K*, 48-49, 60
Strip cutting, 395-396, 436
Strobane, 419
Stub cotton, 383
Sugar, 283
Sugar bait, 303
Sugar beets, 53, 238, 394
Sugar beet yellows, 53-54
Sugarcane, 123
Sugercane borer, 123
Sulfadiazine, 555
Sulfamerazine, 555
Sulfathizaole, 555
Sulphenone, 6
Sulfonamides, 554-555
Superparasitism, 153
Suppressants, 113
Suppression and regulatory tactics, 526
Suramin, 562
Surface Waters, 248
Survivorship curve, 516
Sweep net, 318-319, 323
Sweet clover weevil, 283
Syagrus spp., 404

Sylepta derogata, 405
Syrphid flies, 267
Syrphus spp., 408
System graph, 363-365
Systemics, 237
Systemic insecticides, 245, 257-259, 435
Systems dynamics, 375

Tabanus spp., 292
Tachinidae, 171
Taeniothrips simplex, 56
Tanguis cotton, 382
Tank mixes, 237
Taraxacum officinale, 47, 110
Tarnished plant bug, 448
Taxifolin, 291
Taylor's power law, 342
TDE, 419
Team alfalfa, 461
Tectona grandis, 291
Tedion, 6
Temephos, 5, 548, 562
Temperature and natural enemies, 170
Tepa, 303
Termites, 281, 291
Termite resistance in wood, 291
Terpenes, 282, 288
11-Tetradecenal, 278
9-Tetradecenyl acetate, 277
11-Tetradecenyl acetate, 277, 287
Tetradifon, 253
3,5-Tetradodecadienoic acid, 278
Tetraethyl pyrophosphate, 237, 247
Tetranychus mcdanieli, 176
Tetranychus spp., 25
Tetranychus telarius, 238, 405
Tetranychus urticae, 176, 262, 473, 480, 493
Tetrastichus julis, 152, 167, 173, 175, 268
Tetrasul, 253
Texas leaf-cutting ant, 281
TH6040, 253
Thelohania, 562
Therioaphis maculata, 23, 257-258, 266-267
Therion californicum, 406-407
Three-cornered alfalfa hopper, 331
Threshold response, 280
Thrips, 13
Thyridopteryx ephemeraeformis, 223
Ticks, 556
Timber treatments, 291
Timing, of harvest, 67
  spray applications, 262-263
  of treatments, 25
Tinea pellionella, 291
Tineidae, 291
Tineola bisselliella, 291
Tobacco, 114, 236, 261, 357
Tobacco budworm, 6, 168, 242-243, 357, 393, 401, 422, 426-428
Tobacco hornworm, 114, 261, 357
Tolerance, 116
  of pest damage, 17
Tomato, 224, 264
Toxaphene, 5, 237, 238, 243, 247, 248, 419, 427
Toxaphene-DDT, 420, 427
Toxorhynchites brevipalpis, 552
Trachoma, 553-555
Trachysphyrus tejonensis, 407
Trail-marking pheromones, 281
Traps, 261-263, 280, 286, 312-313, 325-330, 433-435
Trees, insect fauna of, 50-51
Trialeurodes vaporariorum, 170
Triatoma spp., 559
Triatominae, 529, 560
Trichlorfon, 242, 247, 266, 553
Trichogramma spp., 23, 177, 406, 411, 413, 485
Trichoplusia ni, 11, 64, 191, 195-7, 254, 262, 277, 286, 290, 347, 386, 387, 414
Trichoplusia spp., 433
Trichoptera, 245
Tricyclohexylhydroxytin, 497
Trimedlure, 284-285
Trioxys, 23
Trioxys pallidus, 169
Trioxys utilis, 267
Triphenyl tins, 303
Tripsacum dactyloides, 132
Trogoderma granarium, 281
Trombicula spp., 292-293
Tropaeolaceae, 51
Trophic level, 44
Trypanasoma cruzi, 559-560
Tsetse flies, 22, 260, 529, 539, 542
Tsutsugamushi fever, 556-557

Turkey' wheat, 130
Tween-80, 294
Two-spotted spider mite, 176, 262, 473, 480, 493
Typhlocyba pomaria, 473, 478-479
Typhlodromus occidentalis, 176
Typhus, 238, 529, 542-544
Tyria jacobaeae, 66

Ultralow-volume spraying, 530, 547
Ultraviolet light, 261-263, 328-329, 433
Undecanal, 276
U.S. Department of Agriculture, 461, 520

Vacuum harvesting of natural enemies, 177
Vacuum trapping, 324
Vector control, 22, 260, 292-295, 529-562
Vector-disease systems, 531-534
Vector dispersal, 540
Vector ecology, 530-542
Vector-host contact, 540-541
Vector population, 539
Vectorial capacity, 542
Vedalia beetle, 166
Velvet bean caterpillar, 310
Venturia inaequalis, 471, 473-474
Verbenol, 279, 282
Verbenone, 288
Verhulst-Pearl logistic equation, 361
Vertical pull technique, 335
Vespula spp., 284
Vibrio, 221
Viprion, 199
Viral hepatitis, 530, 537, 558
Viron/H, 253
Visual searches, 322
Visual traps, 328, 329
Voria ruralis, 406

Walnut aphid, 163, 169
Water level manipulations, 545-546, 562
Weak link in disease transmission, 548
Western corn rootworm, 12-13, 53, 75, 127, 240-241
Western grebes, 8
Western pine beetle, 279, 282, 287-289, 520
  life table of, 517
Wheat, 127, 130-131
Wheat stem sawfly, 137
White apple leafhopper, 473, 478-479
White grubs, 12-13
White-lined sphinx, 20
White potato leafhopper, 56
Wildlife, 245
'Will' barley, 141
'Winter majetin' apple, 105
Winthemia quadripustulata, 406
Wireworms, 258
Woolly apple aphid, 105, 473, 479
World Health Organization, 239, 240, 530, 533, 538, 543, 549, 554, 561
Wuchereria bancrofti, 537, 544

Xanthodes graellsii, 404
Xenopsylla cheopis, 529

Yellow fever, 22, 237, 544
Yellow jacket, 284
Yellowstone River, 245
Yellow woollybear, 19
Yield reduction, 335-339

Zectran, 247
Zelus spp., 408
Zootermopsis nevadensis, 281
Zero tolerance, 22